Engineering Materials

This series provides topical information on innovative, structural and functional materials and composites with applications in optical, electrical, mechanical, civil, aeronautical, medical, bio- and nano-engineering. The individual volumes are complete, comprehensive monographs covering the structure, properties, manufacturing process and applications of these materials. This multidisciplinary series is devoted to professionals, students and all those interested in the latest developments in the Materials Science field, that look for a carefully selected collection of high quality review articles on their respective field of expertise.

Indexed at Compendex (2021) and Scopus (2022)

Sabu Thomas · Poushali Das · Sayan Ganguly
Editors

Quantum Dots Based Nanocomposites

Design, Fabrication and Emerging Applications

Springer

Editors
Sabu Thomas
School of Chemical Sciences
Mahatma Gandhi University
Kottayam, Kerala, India

Poushali Das
School of Biomedical Engineering
McMaster University
Hamilton, ON, Canada

Sayan Ganguly
Department of Chemistry
University of Waterloo
Waterloo, ON, Canada

ISSN 1612-1317 ISSN 1868-1212 (electronic)
Engineering Materials
ISBN 978-3-031-54781-2 ISBN 978-3-031-54779-9 (eBook)
https://doi.org/10.1007/978-3-031-54779-9

This Springer imprint is published by the registered company Springer Nature Switzerland AG
The registered company address is: Gewerbestrasse 11, 6330 Cham, Switzerland

Paper in this product is recyclable.

Preface

Enter the captivating world of nanocomposites that leverage quantum dots! The book, named "Quantum Dots Based Nanocomposites: Design, Fabrication and Emerging Applications," provides a thorough investigation of the advanced topic that combines quantum research and nanotechnology. At the vanguard of scientific progress, the incorporation of quantum dots into nanocomposite materials holds the potential to profoundly transform multiple industries and unlock unparalleled opportunities.

The impetus for this work arises from the notable advancements observed in the synthesis, design and application of quantum dots in recent times. Quantum dots, due to their distinct electrical and optical properties that vary based on their size, have become essential components for developing innovative nanocomposites. These nanocomposites possess synergistic features that surpass the abilities of their individual components, resulting in improved performance and multifunctionality.

This volume explores the complexities of nanocomposites based on quantum dots. The book is organised in a manner that aims to offer a thorough comprehension of the principles of design, processes of manufacturing and the wide range of new applications in various sectors such as electronics, photonics, energy, medicine and others. Every chapter is meticulously designed to explore distinct facets, providing a combination of theoretical underpinnings and practical observations derived from recent research endeavours.

The initial segment of the book presents the fundamental notions of quantum dots, their distinctive quantum mechanical characteristics and the underlying rules governing their incorporation into nanocomposite structures. The following chapters examine several fabrication methods, investigating the intricacies of building these nanocomposites at the nanoscale to achieve the best possible performance. The book covers a wide range of techniques used in the production of nanocomposites based on quantum dots, including solution-based approaches and sophisticated nanofabrication processes.

The book's second portion centres on the varied applications that have arisen from the combination of quantum dots with different matrices. The range of possibilities, from sophisticated electrical gadgets to effective energy conversion systems and precise drug delivery in medicine, is extensive and continuously developing. The investigation of various applications demonstrates the adaptability and profound capacity for change of nanocomposites based on quantum dots.

As the editors of this book, we have had the honour of working together with prominent scholars and specialists in the subject, who have each provided their distinct viewpoints and specialised knowledge. The collaborative endeavour seeks to furnish an all-encompassing repository for researchers, scientists, engineers, and students who are enthusiastic to delve into the forefront of this swiftly advancing domain.

This book aims to be a beneficial reference for anybody interested in exploring the fascinating field of nanocomposites based on quantum dots. As we explore the complex aspects of design, fabrication and developing applications, we welcome you to join us in understanding the secrets and harnessing the immense possibilities that this sector offers for the future.

Enjoy your reading and exploration!

Kerala, India Sabu Thomas
Hamilton, Canada Poushali Das
Waterloo, Canada Sayan Ganguly

Contents

Introduction to Quantum Dots Based Nanostructures

Sayan Ganguly

Abstract This chapter provides an overview of the exciting new area of quantum dots composites, including information on their preparation, characteristics, and potential uses. Nanosized semiconductor particles are known as quantum dots exhibit size-dependent properties that make them a promising candidate for use in different composites. The chapter discusses several synthesis methods, emphasising the need of exact control over quantum dot size and composition. Quantum dot composites are discussed, along with their many potential uses in fields like optoelectronics, photovoltaic, biomedicine, and sensing. Improved optical, electrical, and mechanical qualities are only a few of the ways in which these composites could revolutionise several fields of technology. This section lays the groundwork for future study and development in this exciting topic.

Keywords Quantum dots composites · Semiconductor · Optoelectronics · Photovoltaics · Biomedicine · Sensing

1 Introduction

In 2004, Xu et al. purified SWCNTs to create luminous carbon nanomaterials [1]. Sun et al. later in 2006 developed fluorescent (FL) carbon NPs, dubbed carbon dots, in both a solvent-dispersed and a solid-extracted form [2]. Many studies have since been conducted to learn more about CDs from the perspectives of synthesis, characterisation, and application. Carbon dots (CDs) [3, 4], Carbon quantum dots (CQDs) [5, 6], Carbon nanodots (CNDs) [7], Graphene quantum dots (GQDs) [8, 9], and Polymer dots (PDs) [10] are all names that have been offered for these nanomaterials. Fluorescence is an important tool in the life sciences, where it is utilised to investigate cellular interactions and the dynamics between biomolecules. To label, detect, and visualise the bio-targets, several fluorophores are created for this method.

S. Ganguly (✉)
Department of Chemistry, University of Waterloo, 200 University Ave West, Waterloo, ON, Canada
e-mail: sayanganguly2206@gmail.com

S. Thomas et al. (eds.), *Quantum Dots Based Nanocomposites*, Engineering Materials,
https://doi.org/10.1007/978-3-031-54779-9_1

Small compounds, proteins, or quantum dots (QDs) are used as these fluorophores. Quantum dots (QDs) are semiconductor nanoparticles with diameters between 2 and 10 nm. The atoms that make them up are often found in either group II–VI or group III–V of the periodic table. QDs exhibit fluorescence in a different way, by quantizing their semiconductor energy level less than their nanometer sized radius, whereas tiny molecules' fluorescence contributes to delocalized electrons that can leap a band and stabilise the energy absorbed. There are primarily three reasons why QDs have drawn more great attention from biologists and chemists than tiny fluorescent molecules and protein fluorophores. First, the wavelength of the band-edge adsorption and fluorescence emission of QDs can be changed systematically by varying their size. Secondly, the photoluminescence spectra of QDs may be observed by a single excitation source over a broad wavelength range, from the visible to the near-infrared. Third, QDs can be used for dynamic imaging of living cells because to their long luminous life and great photostability. Qdots can be broken down into two major classes: elemental and compound. Nanostructured materials based on compound semiconductors are discussed, together with their optoelectronic and optical properties, and potential multimodal applications are highlighted.

There has been a rise in interest in studying composites made from CDs and other functional molecules/materials. The limitations of pure CDs can be greatly reduced and their usefulness expanded by using composites. CDs' many surface functional groups make it simple for them to build composites with other materials. Polymers [11], inorganic nanostructures [12], and biomaterials [13] have all been used successfully in the creation of CDs composites to date. Composites of polymers with fullerene, carbon nanofibers, carbon nanotubes, and graphene-based materials are also possible [14–16]. CDs may include polymers to make the composites with little drawbacks and maximal benefits, unlike other carbon nanostructures. Composites made from CDs have been created by incorporating the materials into polymer matrices, polymeric gel, and molecularly imprinted polymers [17]. CDs can interact strongly with polymers due to their tiny size and vast number of surface functional groups, allowing for homogenous dispersion within the polymers. These composites feature advantageous properties such as low toxicity, low photobleaching, high chemical stability, low absorption, good functionalization, and photobleaching resistance [18]. CD/polymer composites are more hydrophobic, durable, flexible, and strong than their individual components [19]. Some polymers with desirable properties, including as self-healing, increased mechanical performance, shape memory, anti-corrosion, stimuli-responsiveness, and electrochemical characteristics, may be made available through CD coating. Anticounterfeit, ultraviolet (UV) shields, optoelectronic devices, solar cells, supercapacitors, solid-state optical films, catalysts, and sensors are just some of the possible uses for CD/polymer composites [20].

The revolutionary field of quantum dot-based composites, on the cutting edge of nanotechnology, is introduced in Quantum Dot-Based Composites. Quantum dots' size-dependent optoelectronic qualities are highlighted in this chapter as we investigate their fundamental properties and fabrication procedures. Composites made of quantum dots are revealed to be transformative in a variety of fields. We dive into their significance in electronics, where they enhance transistor efficiency, improve LED and solar cell performance, and pave the path for high-efficiency devices. Their incorporation with photonics materials makes possible the development of technologies like tunable lasers, high-resolution screens, and cutting-edge imaging equipment. Quantum dot-based composites have several uses outside of technology, including in health (improvements in medication delivery, imaging, and diagnostics) and the environment (helping with clean energy production and environmental monitoring) and the energy sector (facilitating clean energy production). We recognise the difficulties, including toxicity and scalability, but also show glimpses of a bright future in which quantum dot composites may catalyse astonishing discoveries across industries. Case studies from the real world illustrate their current influence and provide readers a full picture of this revolutionary technology's potential to alter the field of materials science.

2 Synthesis Methods for QD/Polymer Composites

Both top-down and bottom-up synthesis approaches are viable options for the preparation of QDs. Laser ablation, electrochemical oxidation, chemical oxidation, arc discharge, and ultrasonic treatment are some of the physical and chemical processes that are used in top-down techniques [21]. These processes are used to destroy larger carbon compounds. The bottom-up procedures, on the other hand, are characterised by the transformation of tiny carbon molecules into QDs through the use of a chemical reaction, such as pyrolysis, plasma treatment, solvothermal therapy, hydrothermal treatment, microwave treatment, or the template route. Treatments such as hydrothermal, solvothermal, microwave, pyrolysis, and ultrasonic synthesis have been utilised in the process of preparing QDs prior to their incorporation with polymers. In the meantime, QD/polymer composites have been prepared using a variety of methods, such as stirring, sol–gel, drop casting, conventional solution casting, in situ chemical polymerization, reverse microemulsion polymerization, interfacial polymerization, cross-linking reaction, photopolymerization, polymer-assisted self-assembly, Schiff base reaction, bulk polymerization, electrospinning, thermal treatment, hydrothermal treatment, nonsolvent induced According to our review of the relevant published research, the hydrothermal treatment has been the one used most frequently for the manufacture of QDs for composites. The names of composites as well as the characteristics of the synthesis processes for QDs and QD/polymer composites are listed in Table 1. This section provides an overview of the synthesis processes described in the previous section for the production of QD/polymer composites.

Table 1 Different types of CDs based composites and their synthesis methods

Composite	Method	Temperature (°C)	References
PVA/CDs	Hydrothermal	260	[22]
MIPs/CDs	Hydrothermal	200	[23]
PVA/CDs	Hydrothermal	160	[24]
PPy/CDs	Hydrothermal	230	[25]
PANI/CDs	Hydrothermal	230	[25]
CDs/Silica	Hydrothermal	180	[26]
GQDs/polyimide	Hydrothermal	120	[27]
CDs/polyethylenimine/polyethersulfone	Hydrothermal	180	[28]
CDs/thermoplastic polyurethane	Hydrothermal	170	[29]
CDs/Fe_3O_4	Solvothermal	240	[30]
CDs/polyethyleneimine	Solvothermal	150	[31]

3 Properties of QDs

Nano-materials, including quantum dots, have been broken down into three different nanosystems and one bulk system. In other words, as n-dimensional materials with (3-n)-angles of quantum confinement, such as bulk mate-rials (n = 3), quantum wells (thin films; n = 2), quantum wires (nanotubes; n = 1), and quantum dots (nanoparticles; n = 0). Due to their distinct photophysical and electrical properties after photon excitation, quantum dots have found widespread application in a variety of devices and applications, including but not limited to light-emitting diodes (LEDs), solar cells, bioimaging, sensors, displays, quantum communications, and lasers [32]. The phenomenon of photoemission in quantum dots, which is influenced by the size of the dots, has been explained by considering the concepts of quantum confinement and quantum size effects. This explanation has been achieved by employing a square-well potential model. The initial phenomenon pertains to the condition of electrons that are contained within the exciton Bohr radius. Consequently, the freedom of an electron becomes significantly restricted within a quantum dot, resulting in a larger bandgap compared to a bulk system [33]. Quantum dot composites (e.g., core-shell organic-inorganic hybrids) and encapsulation can stabilise and improve photoluminescence quantum yields (s) due to the QDs' enhanced quantum confinement effects. For instance, in comparison to a core-type QD, the photo-emission of a core/shell-type QD is higher because the wide-bandgap inorganic shell serves to remove surface imperfections and dangling bonds. In addition, hetero-junctions can be formed between the core QDs and other inorganic materials with similar lattice constants and crystalline structures during the synthesis of core/shell-type QDs and encapsulated QDs, thereby providing further physical stabilisation.

Applications in electronics, photonics, and even medicine have resulted from these qualities. In this article, we will explain the significance of some of the most important physical features of quantum dots.

3.1 Size-Dependent Optical Properties

Semiconductor crystals on the nanoscale, such as cadmium selenide (CdSe), lead sulphide (PbS), or indium arsenide (InAs), are what we call quantum dots. The bandgap and, by extension, the optical properties of a quantum dot are strongly dependent on their size. Bigger bandgaps in smaller quantum dots lead to higher-energy emissions, while smaller bandgaps in bigger quantum dots lead to lower-energy emissions. This size-dependent behaviour is called the quantum confinement effect, and it allows scientists to tailor quantum dots for uses like LEDs and displays based on quantum mechanics.

3.2 Exceptional Photostability

Quantum dots have remarkable photostability, which allows them to maintain their light emission over extended periods of time without degrading significantly. This quality is especially useful in biological imaging and fluorescence microscopy, where quantum dots can be employed as highly efficient and stable probes. Quantum dots are superior to conventional organic dyes for researching biological processes due to their resistance to photobleaching.

3.3 High Quantum Yields

The efficiency with which quantum dots emit light, or their "quantum yield," is quite high. They may emit a considerable amount of absorbed photons as fluorescence, making them advantageous for applications such as biological labelling and LED technology. Surface states have a significant role in increasing radiative recombination processes, which contributes to quantum dots' high quantum yields as a result of their unusual electronic structure.

3.4 Tunable Emission Wavelength

The ability to adjust the quantum dot's emission wavelength is a major benefit. Scientists are able to precisely tune the colour of produced light by manipulating the quantum dots' size and composition. In multiplexed biological imaging, where distinct colours of quantum dots are utilised to track different cellular components, this characteristic is leveraged to create colourful and high-quality displays.

3.5 Broad Absorption Spectrum

Quantum dots can yield in light from a wide variety of wavelengths according to their large absorption spectrum. Because of this, quantum dots have found a particularly valuable application in solar cell technology, where they can be incorporated into photovoltaic devices to efficiently harvest sunlight. Quantum dots of varying sizes and compositions can be used to tailor the solar cell's absorption spectrum and, in turn, the cell's ability to convert light into usable energy.

3.6 Size-Dependent Electrical Properties

Quantum dots display size-dependent behaviour in both their optical and electrical properties. Quantum dots can function as quantum wells, wires, or dots, depending on their size and shape. The capacity to fine-tune the electrical behaviours of quantum dots is particularly useful in applications involving semiconductor devices like transistors and sensors.

3.7 Quantum Confinement Effects

The quantum confinement effect, coming from the reduced dimensionality of quantum dots, leads to discrete energy levels. Because of their distinct energy levels, quantum dots exhibit characteristically sharp absorption and emission peaks. This phenomenon stands in stark contrast to the continuous energy levels present in bulk materials. The novel optical and electrical features of quantum dots are based on their quantum confinement, which has implications for their employment in a wide range of technologies.

Quantum dots are a type of nanomaterial with many interesting physical properties. These include optical properties that vary with size, high quantum yields, a broad absorption spectrum, tunable emission wavelengths, and quantum confinement effects. Due to their unique characteristics, quantum dots have found use in a wide variety of technological contexts, including electronic displays, medical imaging, solar cells, and semiconductors. Their versatility as a platform for targeted engineering keeps nanotechnology at the forefront of scientific inquiry and development, pointing to promising new frontiers in the area.

4 Applications

4.1 Solar Energy Conversion

Using solar energy to generate electricity or hydrogen for fuel requires the application of cheap materials that can capture photon energy. There are currently two technologies that can harvest and convert solar energy: (i) PEC cells and (ii) photovoltaic solar cells. Heterostructures with an appropriate bandgap alignment and appropriate interfaces would be beneficial for high efficient charge separation in solar cells. QDs in semiconductor heterostructures can facilitate fast charge transfer. Heterostructures made out of QDs consist of two or more parts separated by an interface; these parts are either donors or acceptors. Donor materials are photoexcited to create electrons and holes, which are then transferred to the acceptor across the junction. If you're looking for a donor material, make sure it has a wide bandgap and a high absorption coefficient for the solar spectrum. In addition, the donor and acceptor conduction bands should be in the right place with respect to one another. The acceptor would serve as a conduit for charge carriers to travel from the donor to the electrode's conductive covering [34]. PECs, films containing QDs submerged in an electrolyte, and QD-sensitized solar cells are only a few of the QD-based solar cell topologies that have been proposed. The bandgap of QD-based solar cells can be adjusted, which is an advantage. The QD-based solar cells can overcome the Shockley-Queisser limit. Many semiconductor QDs have been synthesised for application in the sensitization of wide bandgap semiconductors. These materials include CdSe, CdS, PbS, InAs, etc. Because of their size, the QDs can only absorb light within a specific range of the sun's spectrum [35]. The creation of solar cells that utilise QDs and carbon nanostructures is only getting started. The next generation of solar cells may result from investigations investigating the use of these materials [36].

4.2 Photoelectrochemical Water Splitting

A polymer electrolyte membrane (PEC) cell is a type of solar cell that may be used to either directly produce electricity or electrolyze water into oxygen and hydrogen. Hydrogen is created by the PEC water splitting process, which uses solar energy to create chemical energy. It has been proposed that this cell could be used to convert solar energy into usable hydrogen fuel by shining light on a photoanode. PEC photoanodes typically consist of TiO2, ZnO, and WO3, which are all transition metal oxides. These oxides' photoanodes have poor performance because to low light absorption and rapid recombination. However, there have been some advancements in terms of light absorption and the decreased rate of recombination by employing semiconductor QDs. Semiconductor QDs have the disadvantages of being poisonous, expensive to produce, and unstable when used in photoanodes [37]. CQDs can be utilised in PEC photoanodes in place of semiconductor QDs, thereby reducing toxicity issues while increasing stability and production cost. The ability to use CQDs as a photosensitizer is enhanced by their broad spectrum absorption. In CQDs, the excited electron relaxes to the ground state via a gradual process of fluorescence or phosphorescence rather than thermal relaxation. As a result, the electron can be transported from the sensitizer to the contacting acceptor. Under irradiation from a 300 W Xe lamp with a L42 cutoff filter (using this filter causes just the visible and NIR portion of the spectrum to be transmitted), hydrogen generation is only detectable for the photoanode sensitised with CQDs. This demonstrates that the use of CQDs in the photoanode is necessary for hydrogen evolution to occur [38]. CQDs-sensitized TiO_2 was employed for water splitting. When exposed to visible light, the built photoanode showed up to five times the activity of TiO_2 without CQDs. Visible light irradiation of the CQDs causes a dramatic increase in their activity. In PEC, CQDs/TiO_2 photoanodes exhibit enhanced light absorption in proportion to the CQDs concentration [37]. In a PEC cell for water oxidation, N-doped CQDs have been employed for a hematite nanorods array on a Ti sheet. N-CDs@α-Fe_2O_3/Ti photoanode had a 3.4-fold increase in photocurrent density over pure α-Fe_2O_3/Ti. N-doped CQDs have an up-conversion fluorescent feature that allows them to absorb light at longer wavelengths and emit light at shorter wavelengths. Excitation of α-Fe_2O_3 and formation of electron-hole pairs can also occur in response to irradiation with a shorter wavelength [39].

4.3 Agriculture Sector

Pre-harvest (growth, production, and disease resistance) and post-harvest (food preservation and safety) are two areas where agriculture has benefited from QDs [40]. CDs play a part in the pre-harvest process because they act as nano-bio-fertilizers, increasing the efficiency with which plants convert sunlight into usable energy by activating additional photosynthetic genes like PsbP and PsiK [41]. QDs have been observed to potentiate growth regulators, boost cellulose and protein content, and improve germination rates [42]. Considering post-harvest, QDs show bacterial inhibitory effects. Fan et al. discovered that CDs made from kelp significantly reduced the number of Staphylococcus aureus and Escherichia coli on cucumbers during storage [43]. Mangoes were found to maintain their high quality even after being stored for an extended period of time when covered with QDs and chitosan, which was proven to prevent water loss and sugar acid conversion [44]. Therefore, it would be of great interest to design a programme for the widespread use of QDs in agricultural fertilisers and materials for coating agricultural goods, and to investigate their toxicity along the food chain (effects on soil, plants, animals, and humans). Plants have also benefited from QD-based genetic engineering. Schwartz and coworkers synthesised CDs from polyethyleneimine using the solvothermal technique. Tomatoes from the Nicotiana benthamiana 16C line were successfully dosed with CD/siRNA, resulting in gene silencing [45].

4.4 Anti-Counterfeit and Security

Forensic medicine makes use of CDs since it is a field that gives scientific proof in the fight against counterfeiting [4]. Orange-emitting CDs made from the chemical rhodamine B were used to quickly and efficiently visualise latent fingerprints after four months under ultraviolet (UV) light and in poor settings, such as curved and coloured surfaces, according to a prior study [46]. Moreover, CDs have been used in anti-counterfeit systems to identify counterfeit bills. Triple-mode photoluminescence when activated with 365 nm UV light was displayed by CD-based printable ink produced from luminescent m-phenylenediamine and poly(vinyl alcohol) to identify counterfeit banknotes [47]. CDs have been the subject of numerous drug/narcotic/explosive chemical detection tests [48]. When it comes to finding fakes, CDs are a need. So, it's important to equip counterfeit investigators with portable carbon-dot devices or sensors that can withstand high temperatures and moisture. After combining eco-friendly fluorescent CDs (made from turtle shell), researchers recently built colloid photonic crystal (CPC)/CD beads using P(St-MMA-AA) microspheres; the resulting fluorescent CPC/CD patterned films were used as a multi-signal anti-counterfeiting device [49].

5 Conclusions

As this chapter draws to a close, a thorough introduction to a fascinating and quickly developing area of materials science and nanotechnology has been presented. Due to their size-dependent features, quantum dots have enabled intriguing new possibilities in the development of high-performance composites for use in a wide variety of fields. In this chapter, we have looked at the basics of quantum dots and how they can be used to improve the properties of composites, such as their optical, electrical, and mechanical qualities. Colloidal, epitaxial, and organic–inorganic hybrid production methods for quantum dots were explored, and the need of having fine-grained control over their size and composition was emphasised. Additional uses of quantum dots composites, such as in optoelectronics, photovoltaics, biomedical imaging, and sensors, were explored in this chapter. We discovered that these nanocomposites have the potential to completely transform technology in various fields due to their better performance.

In conclusion, the investigation of quantum dots composites is not only interesting from a purely scientific viewpoint, but also has tremendous potential in a variety of real-world settings. We can look forward to significant advancements that will define the future of materials science and technology as researchers continue to investigate and create in this area. This chapter has set the foundation for further investigation and experimentation in the field of quantum dots composites, opening the door for new discoveries and advancements.

References

1. Xu, X., Ray, R., Gu, Y., Ploehn, H.J., Gearheart, L., Raker, K., Scrivens, W.A.: Electrophoretic analysis and purification of fluorescent single-walled carbon nanotube fragments. J. Am. Chem. Soc. **126**(40), 12736–12737 (2004)
2. Sun, Y.-P., Zhou, B., Lin, Y., Wang, W., Fernando, K.S., Pathak, P., Meziani, M.J., Harruff, B.A., Wang, X., Wang, H.: Quantum-sized carbon dots for bright and colorful photoluminescence. J. Am. Chem. Soc. **128**(24), 7756–7757 (2006)
3. Das, P., Ganguly, S., Ahmed, S.R., Sherazee, M., Margel, S., Gedanken, A., Srinivasan, S., Rajabzadeh, A.R.: Carbon dot biopolymer-based flexible functional films for antioxidant and food monitoring applications. ACS Appl. Polym. Mater. **4**(12), 9323–9340 (2022)
4. Das, P., Ganguly, S., Mondal, S., Ghorai, U.K., Maity, P.P., Choudhary, S., Gangopadhyay, S., Dhara, S., Banerjee, S., Das, N.C.: Dual doped biocompatible multicolor luminescent carbon dots for bio labeling, UV-active marker and fluorescent polymer composite. Luminescence **33**(6), 1136–1145 (2018)
5. Das, P., Bose, M., Das, A.K., Banerjee, S., Das, N.C.: One-step synthesis of fluorescent carbon dots for bio-labeling assay. In: Macromolecular Symposia, vol. 1, p. 1800077. Wiley Online Library (2018)
6. Molaei, M.J.: Principles, mechanisms, and application of carbon quantum dots in sensors: a review. Anal. Meth. **12**(10), 1266–1287 (2020)
7. Mocci, F., de Villiers, E.L., Olla, C., Cappai, A., Casula, M.F., Melis, C., Stagi, L., Laaksonen, A., Carbonaro, C.M.: Carbon nanodots from an in silico perspective. Chem. Rev. **122**(16), 13709–13799 (2022)
8. Das, P., Ganguly, S., Banerjee, S., Das, N.C.: Graphene based emergent nanolights: a short review on the synthesis, properties and application. Res. Chem. Intermed. **45**, 3823–3853 (2019)
9. Ganguly, S., Das, P., Banerjee, S., Das, N.C.: Advancement in science and technology of carbon dot-polymer hybrid composites: a review. Funct. Compos. Struct. **1**(2), 022001 (2019)
10. Xia, C., Zhu, S., Feng, T., Yang, M., Yang, B.: Evolution and synthesis of carbon dots: from carbon dots to carbonized polymer dots. Adv. Sci. **6**(23), 1901316 (2019)
11. Feng, Z., Adolfsson, K.H., Xu, Y., Fang, H., Hakkarainen, M., Wu, M.: Carbon dot/polymer nanocomposites: From green synthesis to energy, environmental and biomedical applications. Sustain. Mater. Technol. **29**, e00304 (2021)
12. Paquin, F., Rivnay, J., Salleo, A., Stingelin, N., Silva, C.: Multi-phase semicrystalline microstructures drive exciton dissociation in neat plastic semiconductors. arXiv:13108002 (2013)
13. Bui, T.T., Park, S.-Y.: A carbon dot–hemoglobin complex-based biosensor for cholesterol detection. Green Chem. **18**(15), 4245–4253 (2016)
14. Pinto, A.M., Magalhães, F.D.: Graphene-Polymer Composites, vol. 13. MDPI (2021)
15. Ganguly, S., Mondal, S., Das, P., Bhawal, P., Das, T.K., Ghosh, S., Remanan, S., Das, N.C.: An insight into the physico-mechanical signatures of silylated graphene oxide in poly (ethylene methyl acrylate) copolymeric thermoplastic matrix. Macromol. Res. **27**, 268–281 (2019)
16. Ganguly, S., Ghosh, S., Das, P., Das, T.K., Ghosh, S.K., Das, N.C.: Poly (N-vinylpyrrolidone)-stabilized colloidal graphene-reinforced poly (ethylene-co-methyl acrylate) to mitigate electromagnetic radiation pollution. Polym. Bull. **77**, 2923–2943 (2020)
17. Zhou, Y., Sharma, S.K., Peng, Z., Leblanc, R.M.: Polymers in carbon dots: a review. Polymers **9**(2), 67 (2017)
18. De, B.: Carbon dots and their polymeric nanocomposites. In: Nanomaterials and Polymer Nanocomposites, pp. 217–260. Elsevier (2019)
19. Konwar, A., Gogoi, N., Majumdar, G., Chowdhury, D.: Green chitosan–carbon dots nanocomposite hydrogel film with superior properties. Carbohyd. Polym. **115**, 238–245 (2015)
20. Bera, D., Qian, L., Tseng, T.-K., Holloway, P.H.: Quantum dots and their multimodal applications: a review. Materials **3**(4), 2260–2345 (2010)

21. Wang, R., Lu, K.-Q., Tang, Z.-R., Xu, Y.-J.: Recent progress in carbon quantum dots: synthesis, properties and applications in photocatalysis. J. Mater. Chem. A **5**(8), 3717–3734 (2017)
22. Abdullah Issa, M., Z. Abidin Z,: Sustainable development of enhanced luminescence polymer-carbon dots composite film for rapid Cd^{2+} removal from wastewater. Molecules **25**(15), 3541 (2020)
23. Wang, M., Gao, M., Deng, L., Kang, X., Yang, L., Quan, T., Xia, Z., Gao, D.: Composite material based on carbon dots and molecularly imprinted polymers: a facile probe for fluorescent detection of 4-nitrophenol. NANO **15**(08), 2050105 (2020)
24. Carvalho, J., Santos, L.R., Germino, J.C., Terezo, A.J., Moreto, J.A., Quites, F.J., Freitas, R.G.: Hydrothermal synthesis to water-stable luminescent carbon dots from acerola fruit for photoluminescent composites preparation and its application as sensors. Mater. Res. **22** (2019)
25. Devadas, B., Imae, T.: Effect of carbon dots on conducting polymers for energy storage applications. ACS Sustain. Chem. Eng. **6**(1), 127–134 (2018)
26. Kazemifard, N., Ensafi, A.A., Rezaei, B.: Green synthesized carbon dots embedded in silica molecularly imprinted polymers, characterization and application as a rapid and selective fluorimetric sensor for determination of thiabendazole in juices. Food Chem. **310**, 125812 (2020)
27. Li, S., Li, C., Su, B., Hu, M.Z., Gao, X., Gao, C.: Amino-functionalized graphene quantum dots (aGQDs)-embedded thin film nanocomposites for solvent resistant nanofiltration (SRNF) membranes based on covalence interactions. J. Membr. Sci. **588**, 117212 (2019)
28. Yang, W.-J., Shao, D.-D., Zhou, Z., Xia, Q.-C., Chen, J., Cao, X.-L., Zheng, T., Sun, S.-P.: Carbon quantum dots (CQDs) nanofiltration membranes towards efficient biogas slurry valorization. Chem. Eng. J. **385**, 123993 (2020)
29. Bai, J., Ren, W., Wang, Y., Li, X., Zhang, C., Li, Z., Xie, Z.: High-performance thermoplastic polyurethane elastomer/carbon dots bulk nanocomposites with strong luminescence. High Perform. Polym. **32**(7), 857–867 (2020)
30. Wang, M., Fu, Q., Zhang, K., Wan, Y., Wang, L., Gao, M., Xia, Z., Gao, D.: A magnetic and carbon dot based molecularly imprinted composite for fluorometric detection of 2, 4, 6-trinitrophenol. Microchim. Acta **186**, 1–11 (2019)
31. Bhattacharya, S., Phatake, R.S., Nabha Barnea, S., Zerby, N., Zhu, J.-J., Shikler, R., Lemcoff, N.G., Jelinek, R.: Fluorescent self-healing carbon dot/polymer gels. ACS Nano **13**(2), 1433–1442 (2019)
32. Yue, Z., Lisdat, F., Parak, W.J., Hickey, S.G., Tu, L., Sabir, N., Dorfs, D., Bigall, N.C.: Quantum-dot-based photoelectrochemical sensors for chemical and biological detection. ACS Appl. Mater. Interfaces **5**(8), 2800–2814 (2013)
33. Bryant, G.W.: Excitons in quantum boxes: correlation effects and quantum confinement. Phys. Rev. B **37**(15), 8763 (1988)
34. Selinsky, R.S., Ding, Q., Faber, M.S., Wright, J.C., Jin, S.: Quantum dot nanoscale heterostructures for solar energy conversion. Chem. Soc. Rev. **42**(7), 2963–2985 (2013)
35. Rühle, S., Shalom, M., Zaban, A.: Quantum-dot-sensitized solar cells. ChemPhysChem **11**(11), 2290–2304 (2010)
36. Kamat, P.V.: Quantum dot solar cells. Semiconductor nanocrystals as light harvesters. J. Phys. Chem. C **112**(48), 18737–18753 (2008)
37. Xie, S., Su, H., Wei, W., Li, M., Tong, Y., Mao, Z.: Remarkable photoelectrochemical performance of carbon dots sensitized TiO_2 under visible light irradiation. J. Mater. Chem. A **2**(39), 16365–16368 (2014)
38. Zhang, X., Wang, F., Huang, H., Li, H., Han, X., Liu, Y., Kang, Z.: Carbon quantum dot sensitized TiO_2 nanotube arrays for photoelectrochemical hydrogen generation under visible light. Nanoscale **5**(6), 2274–2278 (2013)
39. Huang, X., Yang, L., Hao, S., Zheng, B., Yan, L., Qu, F., Asiri, A.M., Sun, X.: N-Doped carbon dots: a metal-free co-catalyst on hematite nanorod arrays toward efficient photoelectrochemical water oxidation. Inorgan. Chem. Front. **4**(3), 537–540 (2017)
40. Agnol, L.D., Neves, R.M., Maraschin, M., Moura, S., Ornaghi, H.L., Jr., Dias, F.T.G., Bianchi, O.: Green synthesis of Spirulina-based carbon dots for stimulating agricultural plant growth. Sustain. Mater. Technol. **30**, e00347 (2021)

41. Wang, Y., Xie, Z., Wang, X., Peng, X., Zheng, J.: Fluorescent carbon-dots enhance light harvesting and photosynthesis by overexpressing PsbP and PsiK genes. J. Nanobiotechnol. **19**(1), 260 (2021)
42. Wang, H., Zhang, M., Song, Y., Li, H., Huang, H., Shao, M., Liu, Y., Kang, Z.: Carbon dots promote the growth and photosynthesis of mung bean sprouts. Carbon **136**, 94–102 (2018)
43. Fan, K., Zhang, M., Fan, D., Jiang, F.: Effect of carbon dots with chitosan coating on microorganisms and storage quality of modified-atmosphere-packaged fresh-cut cucumber. J. Sci. Food Agric. **99**(13), 6032–6041 (2019)
44. Xiao, D., Pu, H., Tian, H., Yang, D., Yang, Y., Li, H.: Application of carbon dots-chitosan coating in preservation of mango. Food Ferment. Ind. **45**(22), 130–135 (2019)
45. Schwartz, S.H., Hendrix, B., Hoffer, P., Sanders, R.A., Zheng, W.: Carbon dots for efficient small interfering RNA delivery and gene silencing in plants. Plant Physiol. **184**(2), 647–657 (2020)
46. Yin, J.-Y., Liu, H.-J., Jiang, S., Chen, Y., Yao, Y.: Hyperbranched polymer functionalized carbon dots with multistimuli-responsive property. ACS Macro Lett. **2**(11), 1033–1037 (2013)
47. Jiang, K., Zhang, L., Lu, J., Xu, C., Cai, C., Lin, H.: Triple-mode emission of carbon dots: applications for advanced anti-counterfeiting. Angew. Chem. Int. Ed. **55**(25), 7231–7235 (2016)
48. Verhagen, A., Kelarakis, A.: Carbon dots for forensic applications: a critical review. Nanomaterials **10**(8), 1535 (2020)
49. Guo, J., Li, H., Ling, L., Li, G., Cheng, R., Lu, X., Xie, A.-Q., Li, Q., Wang, C.-F., Chen, S.: Green synthesis of carbon dots toward anti-counterfeiting. ACS Sustain. Chem. Eng. **8**(3), 1566–1572 (2019)

Discovery and History of Quantum Dots

Konatham Teja Kumar Reddy, B. Madhavi Latha, Jeetendra Kumar Gupta, A. Chaitanya, P. Srinivasa Babu, Vamseekrishna. G., Yeluguri Manikanta, and Alapati Sahithi

Abstract Quantum dots are a type of semiconducting material that possesses distinctive electronic and optical characteristics at the nanoscale level. As a result, they have found extensive utility in a diverse array of scientific and technological domains. The present chapter provides a historical account of quantum dots, beginning with their initial discovery in the 1980s and extending to contemporary times. The chapter provides an overview of the different techniques employed in the synthesis and fabrication of quantum dots, which include colloidal synthesis, vapor-phase synthesis, and epitaxial growth. The text delineates the various categories of quantum dots, including core-shell and alloyed quantum dots, and elucidates how their characteristics can be adjusted to cater to particular use cases. Moreover, the chapter explores the diverse implementations of quantum dots in domains such as

K. T. K. Reddy
Department of Pharmacy, University College of Technology, Osmania University, Amberpet, Hyderabad, Telangana 500007, India

B. Madhavi Latha
Department of Pharmaceutical Chemistry, School of Pharmacy, Nalla Narasimha Reddy Education Society's Group of Institution, Chowdariguda, Ghatkesar, Ranga Reddy 500088, India

J. K. Gupta
Institute of Pharmaceutical Research, GLA University Mathura, Bharthia, Uttar Pradesh, India

A. Chaitanya · A. Sahithi (✉)
Department of Pharmaceutical Analysis, School of Pharmacy, Nalla Narasimha Reddy Education Society's Group of Institution, Chowdariguda, Ghatkesar, Ranga Reddy 500088, India
e-mail: sahithi.a@pharma.nnrg.edu.in

P. Srinivasa Babu
Department of Pharmaceutics, Vignan Pharmacy College, Vadlamudi, Guntur, Andhra Pradesh 522213, India

Vamseekrishna. G.
Department of Pharmaceutical Analysis, NRI College of Pharmacy, Pothavarappadu (V), Agiripalli (M), Krishna (D.T), Andhra Pradesh 521212, India

Y. Manikanta
School of Pharmaceutical Sciences, Velsinstitute of Science Technology and Advanced Studies (VISTAS), Pallavaram, Tamilnadu 600117, India

S. Thomas et al. (eds.), *Quantum Dots Based Nanocomposites*, Engineering Materials,
https://doi.org/10.1007/978-3-031-54779-9_2

photovoltaic technology, illumination, medical visualisation, and quantum information processing. The text also addresses the potential hazards linked to quantum dots, such as apprehensions regarding toxicity and ecological consequences.

1 Introduction

1.1 What Are Quantum Dots?

Quantum dots are incredibly small nanocrystals with sizes in the region of a few nanometers that are formed of semiconducting materials. They can be compared to synthetic atoms because of the distinct electrical and optical characteristics they display that are not present in their bulk counterparts [1]. Due to their tiny size, which causes the material's electrical characteristics to become quantized, quantum dots have peculiar properties [2]. Particularly, the quantization of the electronic energy levels occurs when a quantum dot is created and its electrons are constrained within a constrained area of space [3]. Due to this, discrete energy levels are created, and these levels are governed by the quantum dot's size, composition, and shape [4]. These energy levels are known as the "quantum confinement effect," and they give birth to a variety of intriguing features that may be tweaked and controlled by altering the quantum dot's size and make-up [5]. Optoelectronics is one of the most promising fields in which quantum dots can be used. In example, photovoltaic cells and highly efficient light-emitting diodes (LEDs) can be produced using quantum dots [6–8]. The colour of the light emitted by the LED may be precisely modified due to the controllable quantum dot size [9, 10]. Quantum dots can also be employed as down-converters to boost solar cells' efficiency by expanding the quantity of light that can be absorbed. Biomedicine is a promising area for using quantum dots. Quantum dots are fluorescent probes that can be utilised for imaging and disease diagnostics [11–13]. Due to their tiny size, quantum dots can enter cells and tissues and produce high-resolution photographs of biological structures. Quantum dots can also be functionalized with particular targeting molecules to bind to certain cells or tissues with a specific preference, enabling tailored drug administration [14]. Quantum dots are an exciting area of study that could revolutionise a variety of industries, including optoelectronics, healthcare, and energy [15–22]. Due to their special qualities, such as the quantum confinement effect, quantum dots are highly adaptable materials that can be tweaked and controlled to fit various needs [23]. They will therefore probably continue to play a significant role in many fields of science and technology going forward.

The distinctive characteristics and potential uses of quantum dots render them significant in both technological and research domains [24]. Nanoparticles of semiconductors can be accurately designed to exhibit distinct electronic and optical characteristics, rendering them advantageous in various domains such as optoelectronics, biomedicine, and energy [14]. Quantum dots possess the potential to yield light-emitting diodes (LEDs) and photovoltaic cells with exceptional efficiency in the field of optoelectronics. Quantum dots possess the potential to enhance colour accuracy and diminish power consumption in displays and lighting applications due to their capability to emit light at highly specific wavelengths [25]. Furthermore, quantum dots possess the potential to serve as down-converters, thereby enhancing the efficacy of photovoltaic cells through the amplification of light absorption [26]. Quantum dots have been identified as potential fluorescent probes for imaging and diagnosis of diverse diseases in the field of biomedicine [27]. The diminutive dimensions of quantum dots facilitate their infiltration into tissues and cells, thereby affording high-fidelity visual representations of biological structures [28–30]. Furthermore, quantum dots possess the ability to be functionalized with particular targeting molecules, which can facilitate the selective binding to specific cells or tissues. This feature allows for the possibility of targeted drug delivery [16, 20, 31]. Quantum dots possess the potential to enhance the efficacy of solar cells by augmenting the extent of light absorption in the field of energy. Researchers have the ability to produce materials that can absorb light across a wide range of wavelengths by adjusting the size and composition of quantum dots [32]. Quantum dots hold significant importance in the realm of fundamental research in physics and materials science, in addition to their specific applications [33]. Quantum dots can be employed by researchers to investigate the conduct of electrons within a restricted area, thereby facilitating an enhanced comprehension of quantum mechanics and the emergence of novel materials and technologies [34]. Quantum dots possess distinctive characteristics and hold significant potential for utilisation in diverse domains, rendering them crucial in both technological and research contexts [35]. Nanoparticles possess the capability to bring about a significant transformation in domains such as optoelectronics, biomedicine, and energy. It is highly probable that they will persist in playing a crucial role in various fields of science and technology in the forthcoming years.

1.2 Early History of Quantum Dots

The inception of quantum dots can be traced back to the early 1980s, when the Russian physicist Alexei Ekimov and the American physicist Louis E. Brus conducted groundbreaking research in this field. In the year 1982, Ekimov made a discovery regarding small semiconductor crystals that were referred to as "zero-dimensional" entities. These crystals displayed quantum confinement effects that were not observable in bulk semiconductors [36]. In 1983, Brus made independent discoveries that yielded similar effects and subsequently introduced the term "quantum dots" to refer to these minute nanocrystals. The research conducted by Brus was centred on the optical characteristics of said materials, demonstrating that they discharged light at distinct wavelengths which were contingent on their dimensions. During the decades of 1980s and 1990s, scholars conducted further investigations on the characteristics of quantum dots and devised novel approaches for their synthesis and manipulation. In 1994, a group of researchers from Bell Labs conducted a demonstration that showcased the potential of quantum dots in the development of light-emitting diodes (LEDs) with superior efficiency. This breakthrough discovery paved the way for novel applications in the field of optoelectronics. Subsequent to that time, scholars have achieved noteworthy progress in the production and analysis of quantum dots, resulting in novel implementations in the domains of biomedicine, energy, and other related areas [2]. Quantum dots represent a burgeoning field of study that is currently experiencing rapid growth, owing to their numerous potential applications and substantial commercial appeal [37–39].

Quantum mechanics underpins quantum confinement effects in quantum dots. Knowing electron behaviour in quantum dots requires knowing the wave-particle duality of matter and energy quantization in restricted systems. Electrons travel freely in a three-dimensional lattice structure in bulk semiconductor materials. Electrons can occupy any energy level in these continuous materials. However, nanoscale semiconductor materials restrict electron movement in all three dimensions. Electron energy levels are quantized by confinement, depending on the quantum dot's size and shape [40]. Quantized energy levels in quantum dots provide them unique features not found in bulk semiconductors. Quantum dots produce light at specified wavelengths depending on their size and shape, hence their emission spectra are crisp and narrow. Quantum dots are potential for LEDs and solar cells due to their emission spectrum tenability [41, 42]. The effective mass approximation and k-p approach are used to calculate quantum dot electron energy levels and wave functions. These models take into account quantum confinement, semiconductor crystal structure, and composition. Quantum physics predicts that electron energy levels become quantized in restricted systems, which underlies quantum confinement effects in quantum dots [17, 43, 44]. Quantization of energy levels in quantum dots produces unique electrical and optical features that have many uses in science and technology [45, 46].

The investigation of size-dependent properties of nanocrystals was initiated in the 1980s through early experimental studies. The phenomenon of the shift in absorption and emission spectra towards higher energies was observed by researchers in certain semiconductor nanocrystals, as the size of the nanocrystals decreased. The observed blue-shift in the spectra has been ascribed to the quantum confinement effect, which emerges from the confinement of electrons and holes within a limited volume of the nanocrystal [47]. The phenomenon in question leads to the quantization of energy levels of both electrons and holes, thereby causing alterations in the optical and electronic characteristics of the nanocrystals that are dependent on their size [48]. The initial investigations initiated a novel area of inquiry regarding the characteristics and prospective uses of nanocrystals. Apart from the blue-shifted absorption and emission spectra, the scholars also noted alterations in various characteristics of nanocrystals, including their magnetic behaviour and conductivity, which were dependent on their size. As particle size decreases in certain materials, such as gold and silver, the melting point exhibits a corresponding decrease [49]. The observed phenomenon, commonly referred to as "melting point depression", can be attributed to the heightened surface area-to-volume ratio of the particles. This characteristic renders the particles more vulnerable to surface melting and thermal fluctuations [50]. Size-dependent properties of nanocrystals have been investigated, including their mechanical characteristics, such as hardness and ductility, as well as their catalytic activity in chemical reactions. The impact of the morphology and constitution of nanocrystals on their characteristics, as well as their interplay with other substances and biological systems, has been examined by scholars. In general, the initial experimental investigations concerning the size-dependent characteristics of nanocrystals have initiated a novel area of inquiry into the attributes and prospective uses of materials at the nanoscale. Currently, the investigation of nanocrystals and other nanomaterials is a swiftly expanding field of study with numerous potential applications in domains such as electronics, energy, biomedicine, and other related fields.

1.3 First Synthesis of Quantum Dots

Louis E. Brus and his team accomplished the initial prosperous amalgamation of quantum dots in 1984. The researchers employed solution-phase methodologies to produce colloidal quantum dots composed of semiconductors, specifically CdSe, CdTe, and CdS [51]. The process of synthesis entailed the introduction of precursors, namely cadmium and selenium, into a heated solution of a coordinating solvent, such as tri-n-octylphosphine oxide (TOPO) and trioctylphosphine (TOP). Subsequent to the reaction, a coordinating ligand, namely hexadecylamine (HDA), was introduced to serve as a capping agent with the purpose of regulating the dimensions and morphology of the quantum dots [52]. The quantum dots obtained exhibited a high degree of uniformity in size, superior quantum efficiency, and a robust luminescent property, rendering them a desirable candidate for a diverse array of applications. The triumph of this particular method of synthesis has paved the way for further exploration into the synthesis and characteristics of colloidal quantum dots. Currently, solution-phase synthesis is a frequently employed technique for the production of quantum dots, with numerous modifications and advancements implemented over time to enhance regulation of the quantum dots' size, shape, and characteristics. Brief explanation of the synthesis technique.

2 Key Contributions from Early Researchers

The initial investigators of quantum dots made significant advancements that facilitated their current utilisation in various domains. The identification of the quantum confinement effect stands out as a significant contribution. During the initial years of the 1980s, Alexei Ekimov and Louis E. Brus conducted research and made observations that revealed the absorption and emission spectra of specific semiconductor nanocrystals underwent a shift towards higher energies with a decrease in the size of the nanocrystals. The observed blue-shifted spectra can be explained by the quantum confinement effect, which is a result of the confinement of electrons and holes within a limited volume of the nanocrystal. The phenomenon under consideration leads to the quantization of energy levels of both electrons and holes thereby causing alterations in the optical and electronic characteristics of the nanocrystals that are dependent on their size. The development of synthetic methods for quantum dots was deemed a significant contribution. The synthesis of colloidal quantum dots through solution-phase techniques was pioneered by Louis E. Brus and R. Murray. The techniques employed in this study entailed the introduction of precursor substances into a coordinating solvent, succeeded by the incorporation of a capping agent to regulate the dimensions and morphology of the quantum dots. The quantum dots that were produced exhibited a limited range of sizes, a high quantum yield, and robust luminescence, rendering them appealing for utilisation in a diverse array of applications. In the initial stages of research, quantum dots were subjected to diverse methodologies to determine their properties [53]. The optical properties of quantum dots were investigated using absorption and emission spectroscopy, while their structural properties were examined through X-ray diffraction and transmission electron microscopy. The researchers additionally examined the impacts of dimensions, morphology, and constitution on the characteristics of quantum dots. Early researchers have demonstrated the potential applications of quantum dots in various fields. The distinct optical and electronic characteristics of quantum dots were demonstrated by Paul Alivisatos and Moungi Bawendi, which could be utilised for various purposes including optoelectronics, sensing, and biolabeling [54]. The research conducted by the authors established the fundamental principles for the emergence of quantum dots as a distinct area of study, and facilitated their integration into a diverse array of contemporary applications. In brief the finds can be uttered like these:

- The quantum confinement effect was initially observed by early researchers. This phenomenon causes the energy levels of electrons and holes in a nanocrystal to become quantized. Consequently, the optical and electronic properties of the nanocrystal undergo size-dependent changes [55].
- The researchers additionally devised synthetic procedures for quantum dots through solution-phase methodologies, yielding nanocrystals characterised by a limited size range, elevated quantum efficiency, and robust luminescence.
- Quantum dots were characterised by early researchers through a range of techniques such as absorption and emission spectroscopy, X-ray diffraction, and transmission electron microscopy.
- The researchers conducted an investigation into the impact of quantum dot size, shape, and composition on their respective properties.
- The initial researchers exhibited the potential uses of quantum dots in domains such as optoelectronics, sensing, and biolabeling, thereby establishing the groundwork for their current utilisation in a diverse array of applications.

3 Advancements in Quantum Dot Synthesis

Various techniques have been devised for the production of quantum dots, each possessing unique merits and drawbacks. Several frequently employed techniques include:

1. The colloidal synthesis method is a commonly employed technique for the production of quantum dots in a liquid medium. The present technique involves the dissolution of precursors in a suitable solvent, subsequent to which a reducing agent and stabilising agent are introduced. Quantum dots are generated via a process of nucleation and growth.
2. The sol-gel synthesis technique entails the chemical reactions of hydrolysis and condensation of metal alkoxides in a solution, resulting in the formation of a substance with a gel-like consistency. Subsequently, the gel is subjected to a drying process and subsequently exposed to heat in order to generate the quantum dots.
3. The process of chemical vapour deposition entails the application of a slender layer of a precursor substance onto a substrate, succeeded by the reaction of the precursor with a reactant in the gas phase, resulting in the formation of quantum dots.
4. The electrochemical synthesis technique entails the deposition of metal ions onto a substrate while a reducing agent is present, resulting in the formation of quantum dots.
5. The process of plasma synthesis entails the utilisation of a plasma discharge to produce exceedingly reactive species that can interact with precursor gases, resulting in the formation of quantum dots.
6. The process of laser ablation entails utilising a laser beam to ablate a solid target, thereby producing plasma that can interact with a gas-phase reactant to create quantum dots.

Each of the aforementioned techniques possesses unique benefits and drawbacks with respect to regulating the dimensions, morphology, and chemical makeup of the quantum dots, as well as their potential for expansion and consistency in results. The selection of the appropriate technique is contingent upon the particular usage scenario and the intended characteristics of the quantum dots.

4 Discussion of the Role of Surface Chemistry in Quantum Dot Synthesis

Quantum dots are semiconductor materials at the nanoscale level that possess distinctive electronic and optical characteristics, rendering them exceedingly appealing for utilisation in various domains, including but not limited to biological imaging, optoelectronics, and energy. The process of synthesising quantum dots is intricate and demands meticulous regulation of the nanocrystals' size, shape, and composition. The significance of surface chemistry in quantum dot synthesis cannot be overstated, as it has a profound impact on the growth kinetics, stability, and properties of the nanocrystals [56]. The manipulation of the surface chemistry of quantum dots can be achieved through the incorporation of ligands or surface coatings during the process of synthesis. The ligands exhibit an interaction with the surface of the quantum dots, thereby influencing their size, shape, and properties. Ligands serve multiple pivotal functions in the process of quantum dot fabrication [57]. Primarily, they function as capping agents that impede the agglomeration of quantum dots, which is a pivotal element in regulating the dimensions of the nanocrystals. The presence of ligands induces steric hindrance among the quantum dots, thereby constraining their capacity to approach one another and coalesce into aggregates. The growth kinetics of nanocrystals can be influenced by ligands, which have the ability to modulate the surface energy of quantum dots. The quantum dot's surface energy plays a crucial role in determining its propensity to accept atoms or molecules during the growth process, thereby influencing its size and morphology [58]. The modulation of surface energy through the use of ligands can effectively decelerate the rate of nanocrystal growth, ultimately resulting in the production of smaller and more homogenous particles. In addition, it should be noted that ligands have the potential to modify the surface charge of quantum dots, thereby affecting their stability and interactions with chemical or biological surroundings [4, 59–61]. The determination of nanoparticle stability in a solution is reliant on the crucial factor of surface charge. The manipulation of surface charge by ligands can establish a protective enclosure around quantum dots, thereby impeding their aggregation or interaction with other molecules present in the solution. The selection of ligands employed in the process of synthesis plays a crucial role in determining the surface chemistry of quantum dots [62]. This, in turn, has a significant impact on their properties and potential applications. Hydrophilic ligands, such as carboxylic acids or amines, can generate a water-soluble surface, which facilitates the utilisation of quantum dots in biological imaging and sensing applications. In contrast, ligands that exhibit hydrophobic properties have the ability to augment the stability of quantum dots in solvents that lack polarity, thereby conferring utility in optoelectronic domains.

Over the years, there have been notable developments in the production of quantum dots. Presented below is a concise summary of some of the principal progressions.

The initial methods of synthesising quantum dots involved colloidal synthesis and were accomplished in the early 1990s. Subsequently, additional methodologies, including sol-gel synthesis and chemical vapour deposition, were also established. The technique of size-tunable synthesis was developed by researchers in the late 1990s, which enabled accurate manipulation of the size of quantum dots. The aforementioned outcome was attained via alterations in the reaction parameters, including but not limited to adjustments in temperature, duration, and concentration of the precursor substances [63]. The development of high-quality synthesis methods in the early 2000s resulted in significant progress in the production of quantum dots with superior optical and electronic properties, as reported by researchers. The aforementioned outcome was attained through the utilisation of materials with high levels of purity and the optimisation of synthesis conditions to reduce the occurrence of defects [64]. The development of quantum dots has garnered increasing attention in recent times, with a focus on utilising novel materials like perovskites and metal-organic frameworks. The aforementioned materials possess distinctive benefits, including elevated quantum yields and adjustable bandgaps [65]. The escalating demand for quantum dots across diverse applications has prompted a concentration on the advancement of large-scale synthesis techniques. Recent developments in this field encompass the implementation of continuous flow synthesis and microwave-assisted synthesis techniques.

5 Properties of Quantum Dots

The optical properties of quantum dots are dependent on their size, shape, and composition, which give rise to their distinctive characteristics. The quantization of energy levels arises due to the confinement of electrons and holes within a small volume in quantum dots, resulting in size-dependent properties. As the dimensions of the quantum dot decrease, the energy levels exhibit greater discreteness, leading to an increase in the bandgap and a corresponding shift in the absorption and emission spectra towards higher energies [66]. The phenomenon being referred to is commonly recognised as the quantum confinement effect. The surface area-to-volume ratio of a quantum dot is influenced by its size, which in turn affects its reactivity and stability, as well as its bandgap.

The properties of quantum dots can be significantly influenced by their shape [67–70]. Quantum dots that are anisotropic in nature, such as nanorods or nanowires, demonstrate absorption and emission spectra that are dependent on polarisation. This is attributed to the alignment of the dipoles along the longitudinal axis of the particle [71]. The electronic structure and optical properties of a quantum dot can be altered by its shape, resulting in modifications to its energy levels. Quantum dots possessing faceted geometries may manifest distinct surface terminations, thereby influencing their surface chemistry and stability.

The properties of a quantum dot can be significantly influenced by its composition, particularly its chemical makeup. The determination of the bandgap of a quantum dot is contingent upon the disparity in energy levels between the conduction and valence bands, a factor that is subject to the quantum dot's size and composition [72]. Cadmium-based quantum dots are known to demonstrate a greater bandgap in comparison to lead-based quantum dots of equivalent size. The chemical composition of a quantum dot can have a significant impact on its surface chemistry, thereby influencing its stability and reactivity.

The optical characteristics of quantum dots have garnered significant attention owing to their potential utility in diverse domains [73]. Quantum dots possess a narrow emission spectra and high quantum yields, rendering them suitable for deployment as fluorescent probes. Furthermore, the tunability of quantum dots' emission spectra can be achieved through alterations in their size, shape, and composition, rendering them remarkably versatile [74]. Quantum dots possess the capability to be deliberately designed to emit at particular wavelengths within the visible or near-infrared spectrum, thereby rendering them advantageous for employment in imaging and sensing applications.

To summarise, the characteristics of quantum dots are significantly influenced by their dimensions, morphology, and chemical makeup. Comprehending the aforementioned characteristics holds paramount significance in customising quantum dots for particular uses in domains such as optoelectronics, biomedicine, and energy transformation. Furthermore, current investigations within the discipline are concentrated on enhancing the stability and quantum yield of quantum dots, while also delving into novel applications for these distinctive nanomaterials.

6 Overview of the Application of Quantum Dots in Various Fields

Semiconductor nanoparticles known as quantum dots possess distinctive electronic and optical characteristics that render them highly advantageous instruments across a range of disciplines. Nanocrystals are commonly composed of materials like cadmium selenide, cadmium telluride, and indium arsenide, and exhibit a size distribution ranging from a few to several hundred nanometers. The field of optoelectronics has witnessed a significant utilisation of quantum dots. Quantum dots possess distinctive optical characteristics owing to their high surface-to-volume ratio, which is attributed to their diminutive size. Particularly, they possess the ability to produce light at exact wavelengths, rendering them suitable for implementation in superior quality screens, light-emitting diodes, and photovoltaic cells. The capacity of emitting light in various hues, coupled with their elevated luminosity, renders them advantageous in the context of illuminative purposes. Quantum dots have the potential to serve as a labelling tool for biological molecules and cells in the field of biological imaging, thereby facilitating their visualisation and monitoring. Compared to conventional organic dyes, they present various benefits, including enhanced luminosity, improved resistance to photodegradation, and narrower emission spectra. Moreover, the diminutive dimensions of quantum dots render them capable of infiltrating tissues to a significant extent, thereby rendering them well-suited for deployment in the realm of in vivo imaging. The utilisation of quantum dots in the domain of energy production is currently under investigation. Quantum dots have the ability to absorb a wider spectrum of light wavelengths in comparison to conventional solar cells, thereby enhancing their capacity to transform solar energy into electrical energy. Furthermore, quantum dots exhibit promising prospects for utilisation in energy storage systems, including batteries and supercapacitors. Quantum dots are currently under investigation in the realm of information technology due to their potential to serve as qubits, which are fundamental units in the construction of quantum computers.

The utilisation of quantum dots is applicable in the advancement of memory and data storage devices that possess high-speed and high-capacity capabilities. Quantum dots exhibit potential for utilisation in sensing and detection applications. These devices have the capability to function as sensors for detecting alterations in environmental factors such as temperature, pressure, and others. Furthermore, they have the capability to detect and distinguish diverse categories of molecules such as biomolecules, contaminants, and explosives. Quantum dots exhibit distinctive electronic and optical characteristics that render them highly favourable for diverse applications spanning multiple domains. Anticipated progress in research endeavours is poised to yield noteworthy breakthroughs in the utilisation and creation of these nanoparticles across domains encompassing energy, healthcare, and information technology.

7 Quantum Dot Technology

The distinct characteristics exhibited by quantum dots have resulted in their utilisation across a diverse array of technological domains (Fig. 1). The principal implementations of quantum dots are as follows.

Quantum dots have found extensive application in the realm of display technology, particularly in LED backlit displays. The utilisation of these elements results in the

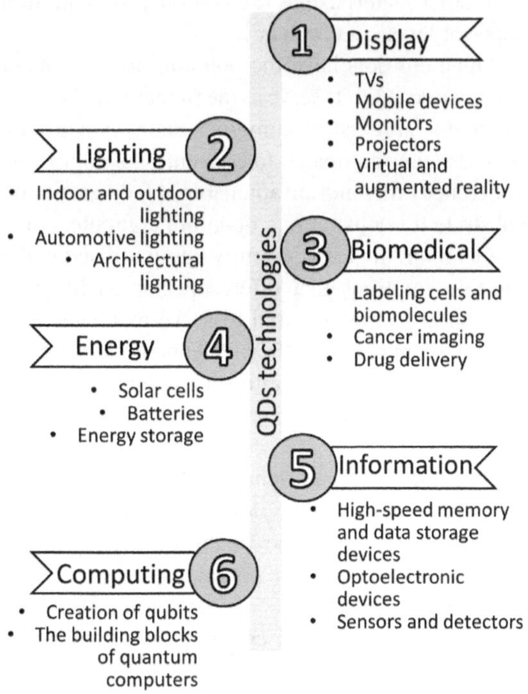

Fig. 1 Various types of applications of QDs nanoparticles

generation of hues that are more pronounced, accompanied by an elevated degree of luminosity and differentiation [75]. Conventional LED displays typically employ white LEDs in conjunction with colour filters to generate diverse hues; however, this approach may lead to energy dissipation and a reduced range of colours. In contrast, displays utilising quantum dots employ blue light emitting diodes (LEDs) in conjunction with quantum dots that generate red and green light upon being stimulated by the blue light source. The outcome of this is an expanded range of colours and a display that consumes less energy.

Quantum dots exhibit distinctive optical characteristics that render them valuable for medical imaging purposes. Fluorescent tags can be utilised for the purpose of identifying and tracking particular cells or molecules within the human body. Quantum dots possess the ability to be tailored to emit light at precise wavelengths, rendering them highly suitable for imaging purposes that necessitate superior resolution and sensitivity [76]. Furthermore, they are currently under development as contrast agents to facilitate more accurate imaging of neoplasms and other anomalous tissues.

The utilisation of quantum dots is being explored as a means of enhancing the efficacy of solar cells through the capture of a wider range of light wavelengths, which can subsequently be converted into electrical energy [77]. Solar cells conventionally comprise crystalline silicon and possess a limited capacity to absorb light wavelengths. The incorporation of quantum dots into the solar cell enables the capture of a wider range of light wavelengths, thereby facilitating the conversion of light energy into electrical energy [49]. The utilisation of quantum dots in conjunction with other materials has the potential to yield multi-junction solar cells that exhibit superior levels of efficiency.

Quantum dots have the potential to serve as qubits within quantum computing frameworks. Qubits serve as the fundamental units of quantum computers and possess the ability to exist in numerous states concurrently, thereby facilitating expedited processing as compared to conventional computers. Quantum dots exhibit extended coherence times and are amenable to facile manipulation, rendering them a propitious substrate for constructing quantum computers on a macroscopic scale.

Quantum dots are currently being utilised in the development of energy-efficient lighting solutions that offer superior quality [16, 49, 69]. Light-emitting diodes (LEDs) have the potential to serve as a substitute for conventional fluorescent or incandescent bulbs, as they are capable of emitting light with a broader spectrum of colours and a greater luminosity. Quantum dot lighting possesses the added benefit of enhanced energy efficiency, thereby resulting in considerable cost reductions in the long run.

Quantum dots are currently under development for various biotechnology purposes such as drug delivery, biomolecule detection, and disease diagnosis. Nanoparticles possess the capability to selectively target particular cells or molecules, thereby rendering them a potent instrument in the realm of nanomedicine. Quantum dots possess the capability of being tailored to selectively bind with particular proteins or nucleic acids, thereby enabling their application in the detection of biomolecules or diagnosis of diseases.

In general, quantum dots exhibit a broad spectrum of potential applications across diverse domains, and continued scientific inquiry and innovation is anticipated to yield further compelling utilities in the times ahead. Quantum dots possess distinctive characteristics that render them highly adaptable and multifaceted, thereby positioning them as a technology with considerable potential for diverse applications.

The field of optoelectronics has shown interest in utilising quantum dots due to their distinctive optical characteristics. Quantum dots exhibit a quantized energy spectrum owing to their diminutive size, resulting in the emission of light at distinct wavelengths. The aforementioned characteristic renders them highly suitable for utilisation in light-emitting diodes and laser devices, wherein meticulous regulation of the emanated wavelengths holds paramount significance. Quantum dots exhibit a broad absorption spectrum, enabling them to capture light across a diverse range of wavelengths. The aforementioned characteristic holds practical significance in down-conversion scenarios, wherein photons with high energy levels are transformed into photons with lower energy levels that can be assimilated by conventional semiconductor materials, such as silicon.

Furthermore, quantum dot light-emitting diodes (QLEDs) possess the capability to offer superior visual displays featuring an extensive range of colours. Quantum dots have the capability to emit light across a diverse range of colours, encompassing those within the visible spectrum. Quantum dot light-emitting diodes (LEDs) exhibit high efficiency and are capable of generating vivid and intense hues, rendering them a compelling substitute for conventional LEDs in lighting and display contexts.

Quantum dots have been identified as a potential material for advanced solar cells owing to their distinctive characteristics. Quantum dots possess a significant benefit in their ability to effectively absorb light across a wide spectrum of wavelengths. This implies that the light-capturing capacity of conventional semiconductor materials, such as silicon, is restricted to a limited range of wavelengths, whereas the aforementioned materials can capture a broader spectrum of light. Furthermore, quantum dots possess the capability to function as down-converting layers. This feature facilitates the enhancement of the light-capturing capacity of conventional silicon solar cells by transforming high-energy photons into lower-energy photons that can be assimilated by silicon.

Quantum dots possess the capability to enhance the efficacy of solar cells by enabling them to apprehend light in the infrared segment of the spectrum, which is beyond the reach of silicon-based solar cells. The reason for this is that quantum dots can be customised to possess a band gap that can be adjusted to the targeted wavelength spectrum. The integration of quantum dots into photovoltaic cells is being investigated by scholars as a means of enhancing their efficacy and diminishing the expense of solar power.

The utilisation of quantum dots as qubits, which are the fundamental units of quantum computers, is a viable option in the field of quantum computing. The utilisation of the spin of an electron in a quantum dot as a qubit and the exploitation of the quantum properties of these systems for the purpose of executing quantum computations can be achieved. The scalability of quantum dots and their compatibility with current semiconductor manufacturing processes make them a promising technology. This implies that they possess the capability of being manufactured in significant quantities and incorporated into pre-existing electronic apparatus.

Nevertheless, there are still obstacles to be overcome with regards to preserving the consistency of the qubits for extended durations. The reason for the susceptibility of qubits to decoherence is attributed to the high sensitivity of quantum dots to their surrounding environment, whereby any form of noise or fluctuations can result in the loss of coherence. Scholars are currently engaged in the development of methodologies aimed at regulating the ambient conditions surrounding quantum dots, with the objective of mitigating the deleterious effects of noise on their operational efficacy.

The distinctive optical characteristics of quantum dots render them a desirable option for application in the fields of biotechnology and medical imaging. Fluorescent probes have the potential to serve as tracking agents for cellular or molecular movements within biological systems. Quantum dots exhibit exceptional stability and emit bright and persistent light, rendering them a highly suitable option for various imaging applications. Moreover, quantum dots have the potential to serve as contrast agents in medical imaging, offering superior resolution and heightened sensitivity in comparison to conventional imaging techniques.

Quantum dots possess potential for utilisation in drug delivery and therapeutic applications. Due to their diminutive dimensions, nanoparticles possess the ability to infiltrate cells and tissues. Additionally, their surface can be modified with targeting molecules, thereby enabling the transportation of therapeutic agents or drugs to precise cells or tissues.

Quantum dots exhibit potential applications in diverse fields including catalysis, sensing, and data storage performance in various chemical reactions. The electronic and optical properties of nanoparticles, which are dependent on their size, have been found to have a significant impact on their catalytic activity. As a result, they have been employed as catalysts in a variety of chemical reactions.

8 Commercialization of Quantum Dots

The inception of quantum dots' commercialization dates back to the latter part of the 1990s. Subsequently, there has been a notable surge in their manufacturing and utilisation. Quantum dots are extensively utilised in display technology as a primary application, facilitating the creation of energy-efficient and high-quality displays. Quantum dots have been employed in medical imaging due to their ability to selectively bind to particular tissues and produce more precise images compared to conventional imaging methods. Quantum dots have recently gained traction in the realm of quantum computing as a potential application, serving as fundamental units or qubits for quantum computers. The potential application of quantum dots in solar cells is currently under investigation, with the aim of enhancing the efficiency of the cells and decreasing the expenses associated with solar energy production. The commercialization of quantum dots has encountered various obstacles, such as the exorbitant expenses associated with their manufacturing and apprehensions regarding their plausible toxicity. Notwithstanding the challenges, scholars and enterprises are endeavouring to surmount them by devising novel production techniques and investigating measures to enhance the safety of quantum dots for human utilisation.

The process of commercialising quantum dots has been in progress for a number of years, and it is anticipated that the quantum dots market will experience substantial growth in the near future. Several prominent corporations are engaged in the process of commercialising quantum dots.

1. Nanosys is a prominent enterprise that specialises in the advancement and monetization of quantum dots. The corporation provides a variety of display technology products, such as quantum dot films, LED backlights, and colour conversion layers.
2. QD Vision is a prominent enterprise that specialises in the advancement and dissemination of quantum dots. The corporation provides quantum dot merchandise that caters to display technology, lighting, and biomedical applications.
3. Samsung is a significant participant in the market for quantum dot displays. The QLED (Quantum Dot LED) technology employed by the company is utilised in its premium-grade televisions and monitors.
4. LG is a significant participant in the quantum dot display industry. The NanoCell TVs manufactured by the company utilise quantum dot technology to augment colour precision and luminosity.
5. Quantum Materials Corp is a corporation that specialises in the research, development, and manufacturing of quantum dots for a wide range of applications, including but not limited to display, lighting, solar cells, and biomedical imaging.
6. Crystalplex Corporation is a specialised enterprise that focuses on the advancement of quantum dots for the purpose of biomedical imaging and diagnostics.
7. NN-Labs is a corporation that specialises in the production of top-notch quantum dots for various purposes such as LED illumination, display technology, and biomedical imaging.
8. Ocean NanoTech is a commercial enterprise that specialises in the manufacture and distribution of quantum dots, which are utilised in various fields such as imaging, diagnostics, and sensing.
9. Quantum Solutions is a corporation that specialises in the production of top-tier quantum dots utilised in various applications, including but not limited to solar cells, LED lighting, and display technology.

Other leading companies are also in market which is depicted in Table 1.

Table 1 Different companies those are working in quantum dots and their focus on the market

Company	Industry	Focus
Max Planck Institute	Academic/research	Quantum computing, sensing
Nanosys	Materials science	Quantum dot films for display technology
QD Vision	Materials science	Quantum dots for display technology and lighting
Quantum Materials Corp	Materials science	Quantum dots for display technology, solar cells, and biomedical imaging
NN-Labs	Materials science	Quantum dots for led lighting, display technology, and biomedical imaging
Ocean NanoTech	Materials science	Quantum dots for biomedical imaging, diagnostics, and sensing
Crystalplex Corporation	Biotechnology	Quantum dots for biomedical imaging and diagnostics

9 Challenges and Opportunities in the Commercialization of Quantum Dots

Quantum dots exhibit several characteristics that render them a highly favourable material for a diverse array of applications, such as optoelectronics, biotechnology, and energy harvesting. Notwithstanding, there exist a number of obstacles and prospects that necessitate resolution in order to effectively bring quantum dots to market.

Challenges:

The commercialization potential of quantum dots is limited due to the relatively high production cost of producing high-quality ones. The elevated expense can be attributed to the intricate process of synthesis, utilisation of costly initial components, and the requirement for specialised equipment. The presence of heavy metals, such as cadmium, in certain quantum dots has raised concerns regarding their potential toxicity during production and use. The development of non-toxic and biocompatible quantum dots is imperative for their effective utilisation in biomedical applications. The long-term stability of quantum dots is limited due to their susceptibility to degradation, which ultimately impacts their performance. The cause of this instability can be attributed to various factors, including exposure to light, oxygen, and temperature fluctuations. This presents a notable obstacle that must be overcome in order to facilitate the commercialization of these products.

Opportunities:

Quantum dots exhibit distinctive optical characteristics that render them advantageous for various optoelectronic applications, including but not limited to display technology, lighting, and solar cells. Quantum dot displays exhibit superior colour gamut, colour accuracy, and energy efficiency in comparison to conventional LCD displays. The field of biotechnology has shown great potential in utilising quantum dots as effective instruments for biomedical imaging and drug delivery. Quantum dots possess high luminosity and stability, and their emission characteristics can be adjusted to precise wavelengths, rendering them valuable for visualising particular biological structures and functions. Quantum dots possess the capability to absorb light across a broad spectrum of wavelengths, rendering them a viable option for solar energy harvesting. These can be utilised to augment the efficacy of photovoltaic cells, and to fabricate novel substances for energy retention.

To summarise, the commercialization of quantum dots encounters various obstacles such as high production expenses, apprehensions regarding toxicity, and durability. Quantum dots have emerged as a promising material for a diverse array of applications, including optoelectronics, biotechnology, and energy harvesting, presenting significant opportunities in these fields. The realisation of the complete potential of quantum dots in the future will necessitate the resolution of challenges and the development of novel applications.

10 Future of Quantum Dots

The particular optical, electrical, and physical characteristics of quantum dots render them a propitious domain of investigation. Quantum dots are utilised in optoelectronics for various purposes, including but not limited to display technology, lighting, and solar cells. Ongoing research endeavours are focused on enhancing the performance and efficiency of quantum dot displays, exploring novel lighting technologies, and optimising the efficacy of solar cells. Quantum dots exhibit promising potential in the realm of biomedical applications, specifically in the areas of drug delivery and imaging. Scholars are currently exploring methods to enhance the biocompatibility of quantum dots and mitigate their toxicity. Additionally, they are devising novel imaging techniques and targeting strategies for drug administration. Quantum dots exhibit promise in the realm of quantum computing owing to their capacity to confine and manoeuvre individual electrons. Scholars are currently investigating potential applications of quantum dots in the development of qubits, which serve as the fundamental building blocks of quantum information, with the aim of enhancing the efficacy of quantum computing platforms. Quantum dots have potential applications in energy storage, including but not limited to batteries and capacitors. Ongoing research endeavours are focused on the development of novel materials and devices for the purpose of energy storage, utilising quantum dots. Quantum dots exhibit potential applications in the realm of environmental monitoring and remediation, including the detection of pollutants and the augmentation of water purification processes. Ongoing research endeavours are focused on the development of novel sensing and treatment technologies utilising quantum dots. In recent decades, scholars have achieved notable advancements in comprehending the conduct of quantum dots and devising novel techniques for their amalgamation and integration into apparatus. The following discourse outlines potential advancements and future trajectories in the realm of quantum dot research and its practical applications. Quantum dot sensors have been employed for the detection of minute variations in temperature, pressure, and magnetic fields. Currently, scholars are investigating the potential of quantum dots as biosensors for the identification of biomolecules, including proteins and DNA. The potential applications of these biosensors encompass medical diagnostics and drug discovery. The utilisation of quantum dots in solar cells has been shown to increase their efficacy by augmenting light absorption and subsequent conversion into electrical energy. Scientists are currently engaged in enhancing the reliability and expandability of solar cells that are based on quantum dots, with the aim of rendering them feasible for commercial purposes. The utilisation of quantum dots as qubits in quantum computing is a promising avenue for exploration. Quantum dots are being investigated by researchers as a potential foundation for constructing quantum computers that exhibit greater stability and scalability in comparison to alternative qubit configurations. Quantum dot-based light-emitting diodes (LEDs) have been developed as a promising solution for efficient and adjustable lighting and display applications. Scholars are currently engaged in enhancing the effectiveness and chromatic accuracy of light-emitting diodes that utilise quantum dots, with the aim of rendering them comparable to conventional LEDs. Quantum dots have the potential to serve as memory devices in electronic systems. Quantum dots are being investigated by scholars as a potential non-volatile memory storage alternative that can be downsized to dimensions smaller than those of existing memory technologies. To attain these advancements and fully exploit the capabilities of quantum dots, it is imperative for researchers to enhance their comprehension of the fundamental physics of quantum dots and devise novel techniques for their synthesis and integration into devices. Furthermore, the progression of fabrication and manufacturing methodologies will play a crucial role in the expansion of quantum dot-centered devices for commercial purposes.

11 Conclusion

The potential application of quantum dots in solid-state lighting, specifically in LED lights, is a noteworthy impact. The utilisation of quantum dots as colour conversion agents enables the generation of white light that exhibits superior energy efficiency and colour rendering characteristics in comparison to conventional lighting sources.

Quantum dots are currently being investigated in the realm of biology and medicine due to their potential applications in targeted drug delivery and gene therapy. Biological molecules can be applied as a coating to enable targeted delivery of drugs or genes to affected cells or tissues, thereby minimising the risk of unintended impacts. Quantum dots possess the capability to transform the domain of quantum cryptography by facilitating the creation and transmission of secure quantum keys for data encryption. Similar to other nascent technologies, quantum dots present certain potential hazards, including apprehensions regarding toxicity and ecological consequences. Scholars are currently investigating methods to tackle these issues and guarantee the secure and ethical advancement of quantum dot technology. To conclude, quantum dots represent an intriguing and auspicious realm of investigation that harbours the possibility of transforming diverse domains of science and technology. Due to their distinctive characteristics and adaptability, they are deemed indispensable instruments for scientists and professionals who are engaged in the advancement of novel applications and technologies.

References

1. Bacon, M., Bradley, S.J., Nann, T.: Graphene quantum dots. Part. Part. Syst. Charact. **31**(4), 415–428 (2014)
2. Ganguly, S., et al.: Advancement in science and technology of carbon dot-polymer hybrid composites: a review. Funct. Compos. Struct. **1**(2), 022001 (2019)
3. Mansur, H.S.: Quantum dots and nanocomposites. Wiley Interdisc. Rev. Nanomed. Nanobiotechnol. **2**(2), 113–129 (2010)
4. Das, P., et al.: Heteroatom doped photoluminescent carbon dots for sensitive detection of acetone in human fluids. Sens. Actuators, B Chem. **266**, 583–593 (2018)
5. Semonin, O.E., Luther, J.M., Beard, M.C.: Quantum dots for next-generation photovoltaics. Mater. Today **15**(11), 508–515 (2012)
6. Bozyigit, D., Yarema, O., Wood, V.: Origins of low quantum efficiencies in quantum dot LEDs. Adv. Func. Mater. **23**(24), 3024–3029 (2013)
7. Latief, U., et al.: Luminescent Manganese/Europium doped ZnS quantum dots: tunable emission and their application as fluorescent sensor. J. Alloy. Compd. **910**, 164889 (2022)
8. Mukherjee, C., et al.: Layered T full adder using quantum-dot cellular automata. In: 2015 IEEE International Conference on Electronics, Computing and Communication Technologies (CONECCT). IEEE (2015)
9. Piloto, A.M.L., et al.: Cellulose-based hydrogel on quantum dots with molecularly imprinted polymers for the detection of CA19-9 protein cancer biomarker. Microchim. Acta **189**(4), 134 (2022)
10. Ahmed, S.R., et al.: Target specific aptamer-induced self-assembly of fluorescent graphene quantum dots on palladium nanoparticles for sensitive detection of tetracycline in raw milk. Food Chem. **346**, 128893 (2021)
11. Sol-Fernández, D., et al.: Digging on the application of novel rod-like particles of polydopamine-CdTe quantum dots for photohermal cancer therapy (2021)
12. Reimann, S.M., Manninen, M.: Electronic structure of quantum dots. Rev. Mod. Phys. **74**(4), 1283 (2002)
13. Kouwenhoven, L.P., Austing, D., Tarucha, S.: Few-electron quantum dots. Rep. Prog. Phys. **64**(6), 701 (2001)
14. Bailey, R.E., Smith, A.M., Nie, S.: Quantum dots in biology and medicine. Phys. E **25**(1), 1–12 (2004)
15. Bonilla, J.C., et al.: Applications of quantum dots in food science and biology. Trends Food Sci. Technol. **53**, 75–89 (2016)
16. Das, P., et al.: Carbon dot biopolymer-based flexible functional films for antioxidant and food monitoring applications. ACS Appl. Polym. Mater. **4**(12), 9323–9340 (2022)
17. Das, P., et al.: Tailor made magnetic nanolights: fabrication to cancer theranostics applications. Nanosc. Adv. **3**(24), 6762–6796 (2021)
18. Das, P., et al.: Carbon-dots-initiated photopolymerization: an in situ synthetic approach for MXene/poly (norepinephrine)/copper hybrid and its application for mitigating water pollution. ACS Appl. Mater. Interfaces **13**(26), 31038–31050 (2021)
19. Rajabzadeh, A.R., et al.: Magnetic Quantum Dots for Bioimaging. CRC Press (2023)

20. Das, P., et al.: Waste-derived sustainable fluorescent nanocarbon-coated breathable functional fabric for antioxidant and antimicrobial applications. ACS Appl. Mater. Interfaces (2023)
21. Das, P., et al.: MXene/0D nanocomposite architectures: design, properties and emerging applications. Mater. Today Nano **24**, 100428 (2023)
22. Das, P., et al.: Naturally derived carbon dots in situ confined self-healing and breathable hydrogel monolith for anomalous diffusion-driven phytomedicine release. ACS Appl. Bio Mater. **5**(12), 5617–5633 (2022)
23. Das, P., et al.: One-step synthesis of fluorescent carbon dots for bio-labeling assay. In: Macromolecular Symposia. Wiley Online Library (2018)
24. Jamieson, T., et al.: Biological applications of quantum dots. Biomaterials **28**(31), 4717–4732 (2007)
25. Jelinek, R.: Carbon Quantum Dots, pp. 29–46. Carbon quantum dots. Springer International Publishing, Cham (2017)
26. Lent, C.S., Tougaw, P.D.: A device architecture for computing with quantum dots. Proc. IEEE **85**(4), 541–557 (1997)
27. Van Dam, J.A., et al.: Supercurrent reversal in quantum dots. Nature **442**(7103), 667–670 (2006)
28. Parak, W.J., Pellegrino, T., Plank, C.: Labelling of cells with quantum dots. Nanotechnology **16**(2), R9 (2005)
29. Shiohara, A., et al.: On the cyto-toxicity caused by quantum dots. Microbiol. Immunol. **48**(9), 669–675 (2004)
30. Güçlü, A.D., et al.: Graphene Quantum Dots. Springer (2014)
31. Das, P., et al.: Carbon dots for heavy-metal sensing, pH-sensitive cargo delivery, and antibacterial applications. ACS Appl. Nano Mater. **3**(12), 11777–11790 (2020)
32. Tian, P., et al.: Graphene quantum dots from chemistry to applications. Mater. Today Chem. **10**, 221–258 (2018)
33. Smith, A.M., Nie, S.: Next-generation quantum dots. Nat. Biotechnol. **27**(8), 732–733 (2009)
34. Bimberg, D., Pohl, U.W.: Quantum dots: promises and accomplishments. Mater. Today **14**(9), 388–397 (2011)
35. Das, P., et al.: Dual doped biocompatible multicolor luminescent carbon dots for bio labeling, UV-active marker and fluorescent polymer composite. Luminescence **33**(6), 1136–1145 (2018)
36. Bawendi, M.G., Steigerwald, M.L., Brus, L.E.: The quantum mechanics of larger semiconductor clusters ("quantum dots"). Annu. Rev. Phys. Chem. **41**(1), 477–496 (1990)
37. Ganguly, S., et al.: Microwave-synthesized polysaccharide-derived carbon dots as therapeutic cargoes and toughening agents for elastomeric gels. ACS Appl. Mater. Interfaces **12**(46), 51940–51951 (2020)
38. Ganguly, S., et al.: Microwave assisted green synthesis of Zwitterionic photolumenescent N-doped carbon dots: an efficient 'on-off' chemosensor for tracer Cr (+ 6) considering the inner filter effect and nano drug-delivery vector. Colloids Surf. A **579**, 123604 (2019)
39. Saravanan, A., et al.: Green synthesis of multifunctional carbon dots with antibacterial activities. Nanomaterials **11**(2), 369 (2021)
40. Dai, X., et al.: Solution-processed, high-performance light-emitting diodes based on quantum dots. Nature **515**(7525), 96–99 (2014)
41. Ganguly, S., et al.: Strongly blue-luminescent N-doped carbogenic dots as a tracer metal sensing probe in aqueous medium and its potential activity towards in situ Ag-nanoparticle synthesis. Sens. Actuators, B Chem. **252**, 735–746 (2017)
42. Liu, Z., et al.: Micro-light-emitting diodes with quantum dots in display technology. Light Sci. Appl. **9**(1), 83 (2020)
43. Sun, Q., et al.: Bright, multicoloured light-emitting diodes based on quantum dots. Nat. Photon. **1**(12), 717–722 (2007)
44. Das, P., et al.: Graphene based emergent nanolights: a short review on the synthesis, properties and application. Res. Chem. Intermed. **45**, 3823–3853 (2019)
45. Ahmed, S.R., et al.: Graphdiyne quantum dots for H2O2 and dopamine detection. ACS Appl. Nano Mater. **6**(10), 8434–8443 (2023)

46. Maity, P.P., et al.: Capra cartilage-derived peptide delivery via carbon nano-dots for cartilage regeneration. Front. Bioeng. Biotechnol. **11** (2023)
47. Zhao, G., et al.: Orbital mixing between colloidal quantum dots and surface-bound molecules. J. Phys. Chem. Lett. **13**(51), 11892–11898 (2022)
48. Xu, Q., et al.: Recent progress of quantum dots for energy storage applications. Carbon Neutrality **1**(1), 13 (2022)
49. Abdellatif, A.A., et al.: Biomedical applications of quantum dots: overview, challenges, and clinical potential. Int. J. Nanomed. 1951–1970 (2022)
50. Rainò, G., et al.: Ultra-narrow room-temperature emission from single CsPbBr 3 perovskite quantum dots. Nat. Commun. **13**(1), 2587 (2022)
51. Efros, A.L., Brus, L.E.: Nanocrystal quantum dots: from discovery to modern development. ACS Nano **15**(4), 6192–6210 (2021)
52. Raja, A., et al.: Energy transfer from quantum dots to graphene and MoS2: the role of absorption and screening in two-dimensional materials. Nano Lett. **16**(4), 2328–2333 (2016)
53. Shirasaki, Y., et al.: Emergence of colloidal quantum-dot light-emitting technologies. Nat. Photon. **7**(1), 13–23 (2013)
54. Somers, R.C., Bawendi, M.G., Nocera, D.G.: CdSe nanocrystal based chem-/bio-sensors. Chem. Soc. Rev. **36**(4), 579–591 (2007)
55. Ramalingam, G., et al.: Quantum Confinement. IntechOpen (2020)
56. Hines, D.A., Kamat, P.V.: Recent advances in quantum dot surface chemistry. ACS Appl. Mater. Interfaces **6**(5), 3041–3057 (2014)
57. Kilina, S.V., Tamukong, P.K., Kilin, D.S.: Surface chemistry of semiconducting quantum dots: theoretical perspectives. Acc. Chem. Res. **49**(10), 2127–2135 (2016)
58. Bilan, R., et al.: Quantum dot surface chemistry and functionalization for cell targeting and imaging. Bioconjug. Chem. **26**(4), 609–624 (2015)
59. Ganguly, S., et al.: Sonochemical green reduction to prepare Ag nanoparticles decorated graphene sheets for catalytic performance and antibacterial application. Ultrason. Sonochem. **39**, 577–588 (2017)
60. Das, P., et al.: Surface quaternized nanosensor as a one-arrow-two-hawks approach for fluorescence turn "on–off–on" bifunctional sensing and antibacterial activity. New J. Chem. **43**(16), 6205–6219 (2019)
61. Ganguly, S., Margel, S.: Bioimaging probes based on magneto-fluorescent nanoparticles. Pharmaceutics **15**(2), 686 (2023)
62. Devi, S., et al.: Quantum dots: an emerging approach for cancer therapy. Front. Mater. **8**, 798440 (2022)
63. Bukowski, T.J., Simmons, J.H.: Quantum dot research: current state and future prospects. Crit. Rev. Solid State Mater. Sci. **27**(3–4), 119–142 (2002)
64. Coe-Sullivan, S.: Quantum dot developments. Nat. Photonics **3**(6), 315–316 (2009)
65. Winnik, F.M., Maysinger, D.: Quantum dot cytotoxicity and ways to reduce it. Acc. Chem. Res. **46**(3), 672–680 (2013)
66. Lim, S.Y., Shen, W., Gao, Z.: Carbon quantum dots and their applications. Chem. Soc. Rev. **44**(1), 362–381 (2015)
67. Ganguly, S., Margel, S.: Fluorescent quantum dots-based hydrogels: Synthesis, Fabrication and multimodal biosensing. Talanta Open 100243 (2023)
68. Ganguly, S.: Reinforcement mechanisms of quantum dot-polymer composites. In: Quantum Dots and Polymer Nanocomposites, pp. 151–169. CRC Press (2022)
69. Ganguly, S.: 13 Challenges prospects of and magnetic future quantum dots. Magnetic Quantum Dots for Bioimaging 269 (2023)
70. Das, P., et al.: Converting waste Allium sativum peel to nitrogen and sulphur co-doped photoluminescence carbon dots for solar conversion, cell labeling, and photobleaching diligences: a path from discarded waste to value-added products. J. Photochem. Photobiol. B. **197**, 111545 (2019)
71. Jang, E., Jang, H.: Quantum dot light-emitting diodes. Chem. Rev. **123**(8), 4663–4692 (2023)

72. Liu, S.-F., et al.: 3D nanoprinting of semiconductor quantum dots by photoexcitation-induced chemical bonding. Science **377**(6610), 1112–1116 (2022)
73. Wu, X.-G., et al.: Industry outlook of perovskite quantum dots for display applications. Nat. Nanotechnol. **17**(8), 813–816 (2022)
74. Karagianni, A., et al.: Application of carbon-based quantum dots in photodynamic therapy. Carbon **203**, 273–310 (2023)
75. Manikandan, V., Lee, N.Y.: Green synthesis of carbon quantum dots and their environmental applications. Environ. Res. **212**, 113283 (2022)
76. Kaur, A., et al.: Nanocomposites of carbon quantum dots and graphene quantum dots: environmental applications as sensors. Chemosensors **10**(9), 367 (2022)
77. Deng, S., et al.: Rational design of $ZnMn_2O_4$ quantum dots in a carbon framework for durable aqueous zinc-ion batteries. Angew. Chem. **134**(12), e202115877 (2022)

Synthesis Strategies of Doped QDs

Konatham Teja Kumar Reddy, A. Venkata Badarinath, Jeetendra Kumar Gupta, Yeluguri Manikanta, Nagam Santhi Priya, G. Vamseekrishna, Jimmy Alex, and Alapati Sahithi

Abstract The implementation of synthesis strategies for doped QDs (QDs) is of paramount importance in the customization of their electronic and optical characteristics, thereby facilitating their utilisation in a diverse array of fields such as optoelectronics, biomedicine, energy, sensing, and others. This abstract presents a succinct summary of the synthesis methodologies utilised in the fabrication of doped QDs (QDs). The aforementioned tactics encompass codoping, post-synthetic doping, ion exchange, alloying, and core-shell architectures. The process of codoping entails the concurrent introduction of multiple dopants during the synthesis of QDs, thereby enabling accurate regulation of their concentration and amalgamation. The process

K. T. K. Reddy
Department of Pharmacy, University College of Technology, Osmania University, Amberpet, Hyderabad, Telangana 500007, India

A. Venkata Badarinath
Department of Pharmaceutics, Santhiram College of Pharmacy, Kurnool, Nandyala, Andhra Pradesh NH-18, India

J. K. Gupta
Institute of Pharmaceutical Research, GLA University Mathura, Bharthia, Uttar Pradesh, India

Y. Manikanta
School of Pharmaceutical Sciences, Vels Institute of Science Technology and Advanced Studies (VISTAS), Pallavaram, Tamilnadu 600117, India

N. S. Priya
Department of Pharmaceutics, Vignan Pharmacy College, Vadlamudi, Guntur, Andhra Pradesh 522213, India

G. Vamseekrishna
Department of Pharmaceutical Analysis, NRI College of Pharmacy, Pothavarappadu (V), Agiripalli (M), Krishna (D.T), Andhra Pradesh 521212, India

J. Alex
Department of Pharma Chemistry, Sri Shanmugha College of Pharmacy, Sankari, Salem, Tamilnadu 637304, India

A. Sahithi (✉)
Department of Pharmaceutical Analysis, School of Pharmacy, Nalla Narasimha Reddy Education Society's Group of Institution, Chowdariguda, Ghatkesar, Ranga Reddy 500088, India
e-mail: sahithi.a@pharma.nnrg.edu.in

S. Thomas et al. (eds.), *Quantum Dots Based Nanocomposites*, Engineering Materials,
https://doi.org/10.1007/978-3-031-54779-9_3

of post-synthetic doping involves the introduction of dopants onto pre-synthesized QDs (QDs) through surface modification or ligand exchange procedures. The process of ion exchange involves the substitution of dopant ions for the original cations or anions present in the lattice of QDs. The process of alloying involves the combination of distinct semiconductor materials, with the inclusion of dopants during the alloying procedure. Core-shell architectures facilitate the encapsulation of doped QDs (QDs) within a safeguarding shell, thereby providing the ability to regulate dopant diffusion and associated characteristics. The utilisation of synthesis strategies offers a degree of adaptability in customising the characteristics of doped QDs to suit particular applications. The objective of future investigations in this domain is to devise novel doping methodologies and refine current techniques with the aim of augmenting the efficacy and broadening the scope of applications of doped QDs.

Keywords Doped QDs · Codoping · Encapsulation · Biomedicine · Optoelectronics

1 Introduction

QDs are semiconductor particles at the nanoscale that display distinctive optical and electronic characteristics as a result of quantum confinement phenomena [1, 2]. In recent times, there has been a notable surge in interest towards them owing to their prospective utilisation in diverse domains, including optoelectronics, biomedicine, and renewable energy [3–6]. The extent to which their properties can be adjusted is restricted by their dimensions and constituents. The utilisation of impurities to dope QDs has emerged as a promising strategy to overcome their limitations and improve their performance. The process of doping QDs entails the incorporation of dopants or impurities into the crystal lattice of the semiconductor material [1]. The alteration of the electronic structure and bandgap of QDs through this process can result in modifications of their optical and electronic characteristics. Doping has the potential to enhance the stability, biocompatibility, and solubility of QDs, thereby expanding their applicability across various domains [7]. Numerous methodologies have been devised to fabricate doped QDs with regulated composition and characteristics. The aforementioned strategies can be categorised into two distinct groups: (1) doping after synthesis and (2) co-doping during the synthesis process. The process of post-synthesis doping pertains to the incorporation of dopants into QDs that have already undergone synthesis. The aforementioned methodology facilitates the alteration of characteristics of extant QDs while preserving their dimensions and morphology. Ion implantation is a frequently utilised technique for post-synthesis doping [8]. The process of introducing dopants into QDs involves ion bombardment. The ion implantation methodology is characterised by a high degree of precision and enables the incorporation of a diverse array of dopants. Nevertheless, it has the potential to impair the crystal structure of the QDs and diminish their efficacy in emitting light.

Molecular doping represents an alternative strategy for post-synthesis doping [9]. The present technique involves the introduction of dopants into the QDs through either surface adsorption or lattice diffusion. The utilisation of this particular methodology is comparatively less deleterious to the crystal structure of QDs in contrast to ion implantation, while also having the capability to incorporate a more extensive spectrum of dopants. Nevertheless, the process of molecular doping may result in the formation of quantum dot aggregates, thereby diminishing their stability. The process of co-doping in the synthesis of QDs refers to the concurrent integration of dopants into the synthesis procedure. The aforementioned methodology facilitates meticulous regulation of the QDs' composition and characteristics, while concurrently mitigating the incidence of flaws and impairment to the crystal lattice. The hot-injection method is a frequently employed technique for co-doping. The present methodology involves the injection of a blend of precursor substances and dopants into a heated solution to produce QDs with regulated composition and characteristics. The hot-injection technique is a remarkably effective approach that enables the production of doped QDs of exceptional quality while providing accurate regulation over their characteristics [10]. The one-pot synthesis method represents an alternative approach to co-doping. The present methodology involves the introduction of a blend of precursor substances, dopants, and ligands into a solitary reaction container to effectuate the production of doped QDs in a unified step [11]. The aforementioned methodology is characterised by its straightforwardness and effectiveness, as well as its capacity to generate substantial volumes of doped QDs. The intricate reaction environment may pose challenges in regulating the characteristics of the QDs produced. Various dopants have been employed to alter the characteristics of QDs, encompassing metals, metal oxides, rare-earth elements, and organic molecules. The incorporation of dopants has the potential to modify the optical and electronic characteristics of QDs, including their quantum yield, emission wavelength, and bandgap [12–14]. The selection of a dopant is contingent upon the intended characteristics of the QDs and their particular usage. To sum up, the introduction of impurities into QDs through doping has surfaced as a propitious method for altering their characteristics and amplifying their efficacy. Various techniques have been devised to fabricate doped QDs with regulated composition and characteristics, such as doping after synthesis and co-doping during synthesis [15–20].

2 Different Doping Strategies of QDs

Doping is an effective tool for modifying the characteristics of QDs. Doping of impurities into QDs can be broadly categorised as either extrinsic or intrinsic doping.

2.1 Extrinsic Doping Strategies

The traditional technique of doping, which involves the introduction of impurity atoms, proved to be ineffective in several initial instances due to the tendency of the impurities to be expelled from the small crystalline cores. Additionally, the thermal ionization of impurities, which generates free carriers, was impeded by strong confinement. A potential avenue for addressing these limitations involves the extrinsic introduction of dopant ions. The study conducted by Guyot-Sionnest et al. [21, 22] explored into the examination of colloidal QDs composed of different materials, namely CdSe, CdS, and ZnO. The results of the study demonstrated that these materials can undergo n-type reduction through the use of Na, K, biphenyl radical anions, or electrochemical doping, wherein electrons occupy the quantum confined states of the conduction band. The present study showcases the impact of sodium biphenyl on the infrared and visible absorption spectra of CdSe nanocrystals. The results revealed a noticeable reduction in the intensity and broadening of the first and second excitonic peaks, indicating bleaching. The manifestation of the 1Se–1Pe infrared absorption peak at 0.3 eV provides evidence for the n-type nature of the nanocrystals. Additional prospects encompass the investigation of the electrical conductivity of n-type QDs films, attributed to their increased inter-QD electron transfer, which may yield photovoltaic or electronic implementations [21, 23]. Another technique for extrinsic doping involves the utilization of a Bronsted base, such as hydrazine, to modify the surface of the quantum dot (QD). This process replaces the bulky oleic acid groups, resulting in a reduction of intermolecular spacing. Additionally, the dangling bonds on the surface are passivated by the lone pair of electrons on the base, which is comparable to primary amines. The improvement of conductivity in QD solids is facilitated by this process, thereby enhancing the efficacy of QDs in diverse applications such as field effect transistors. The conductance of PbSe nanocrystal solids was observed to escalate by ten orders of magnitude subsequent to hydrazine treatment. The technique has been demonstrated to have a broad range of applications on various materials; however, its widespread applicability is constrained by the instability of the ligands involved [24]. In recent times, hydrazine molecules have been substituted with more stable and environmentally friendly molecules, thereby showcasing the extended practicality of this approach [25].

2.2 Intrinsic Doping

Extrinsic doping is not a very feasible choice for imparting functionalities such as magnetic, optical, or optoelectronic functions, despite the modest results that have been achieved with this technique. As a result, it is essential to find procedures that are capable of producing effective intrinsic doping of dopant ions [26]. While doping has been widely conducted and has yielded high-quality QDs (QDs) in certain instances,

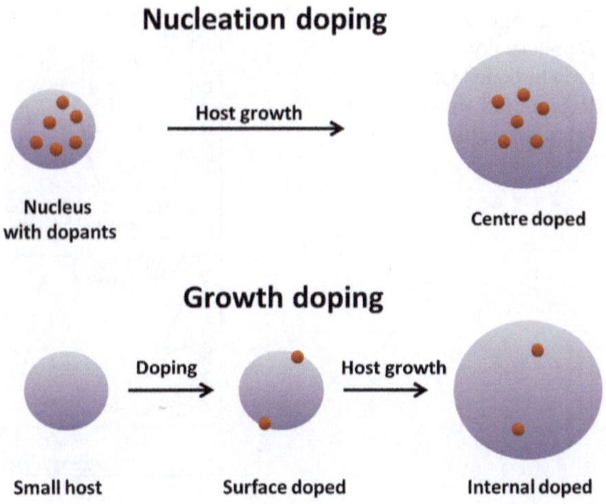

Fig. 1 Graphical representation of the proposed radial positions of the dopant in nucleation and growth doping

Pradhan et al. were the first to propose mechanistically significant techniques for doping Mn and Cu in ZnSe-based systems [26]. The authors proposed techniques to dissociate the dopant and host precursors to account for their distinct reactivities. These techniques involve nucleation or growth doping to achieve efficient doping of the QDs, as illustrated in Fig. 1.

In the context of nucleation doping, the initial step involves the formation of a small dopant compound. In the context of nucleation doping, a small dopant compound was initially generated. Subsequently, the aforementioned core nuclei were coated with a host shell, leading to the formation of either graded alloys or sharp interfaces, dependent upon the prevailing conditions. In the context of growth doping, the authors presented evidence of the emergence of small host QDs, succeeded by the adsorption of the dopant ion on the surface at a reduced temperature, and the introduction of the dopant metal precursor. In order to encapsulate the dopant ions, isocrystalline or heterocrystalline shells were formed, which demonstrates the extensive applicability of this method [3, 27]. As shown in Fig. 2, effective doping and decoupling from nucleation and growth are demonstrated by the rise in the photoluminescence (PL) intensity of the dopant emission and the reduction in the PL intensity of the host (top, left). This prototype demonstration of the proposed mechanism was carried out for Cu or Mn-doped ZnSe QDs. The PL undergoes a redshift as a consequence of the regrowth of ZnSe on the surface of the doped QDs (top, right). Because it took 100 min to finish, the procedure described above would have been tough to control if it hadn't been for the decoupling technique. On the other hand, nucleation doping was accomplished for Mn-doped ZnSe by utilizing a variety of dopant precursors. Due to the fact that Mn^{2+} is a harder Lewis acid than Zn^{2+}, it is anticipated that the Mn^{2+} precursor will exhibit lower levels of reactivity when combined with the same ligand. When manganese stearate was utilised as the nucleation precursor, zinc acetate was preferred over zinc stearate for the production

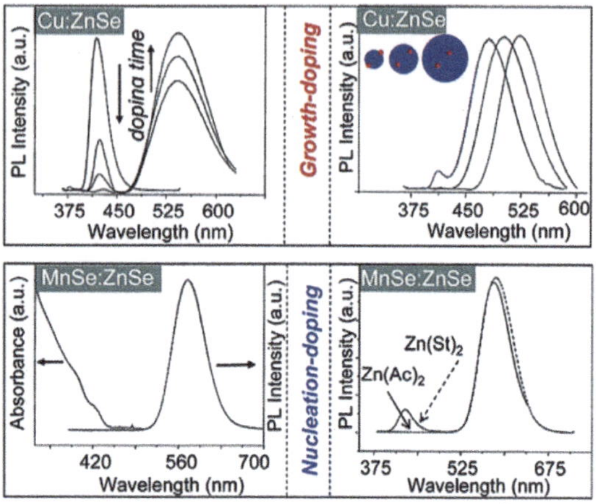

Fig. 2 Spectroscopic data for nucleation- and growth-doped QDs. Reproduced with permission from Ref. [26]

of MnSe: ZnSe nanocrystals. Zinc stearate necessitated a higher reaction temperature, leading in homogenous nucleation of ZnSe and a band gap PL. The main disadvantage of this approach is high-temperature annealing during shell formation, which causes dopants to escape and eventually leach out of the surface, leading to self-purification. This was specifically demonstrated in the instance of Cu-doped ZnSe QDs. The overcoating process carried out at a temperature of 210 °C effectively confined the dopants within the host lattice, thereby preventing any diffusion of dopants from the nucleus to the surface. Nevertheless, it was observed that upon annealing at 220 °C or above, the expulsion of dopants from the lattice occurred for all sizes. Furthermore, a notable reduction of the dopant photoluminescence was observed during annealing in the temperature range of 20–80 °C, as the overcoating temperature was raised. The process of overcoating ZnSe at elevated temperatures leads to a significant diffusion of Cu dopants toward the surface. Therefore, despite the prolonged endeavours of diverse groups on different systems to acquire internally doped QDs, the attainment of internally doped QDs has remained elusive.

The process of cation exchange reactions is extensively utilised in the doping of QDs, wherein the cation of the QDs is replaced with a dopant cation [28–30]. The extent of cation exchange can be tuned by varying the solution concentration of the incoming dopant cations and the type of ligands bound to them. These processes, however, are not trivial in semiconductor QDs, as evidenced by the large energies involved in doping. Furthermore, clustering or directional doping based on reactivity is inherent in this doping strategy. While the directed nature has been employed constructively to generate various hetero-structure topologies, clustering of dopant ions does not allow for homogenous impurity doping [31–33]. Another technique called diffusion doping, originally developed by Saha and colleagues, employs the self-purification mechanism instead of attempting to address it as an obstacle, a commonly perceived notion [34]. Using the SILAR (successive ionic layer adsorption and reaction) approach, a tiny magnetic core was over-coated with a thick shell

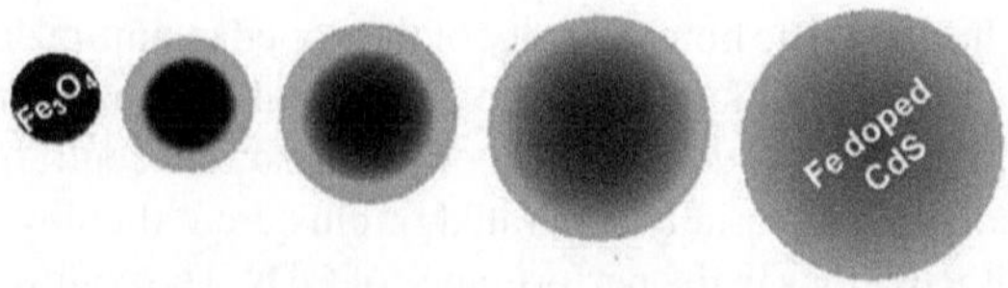

Fig. 3 Dopant diffusion into the CdS matrix to produce Fe-doped CdS QDs. Reproduced with permission from Ref. [34]

of the host semiconductor. High-temperature annealing of this core-shell arrangement causes the core to diffuse at the interface and into the semiconductor shell, eventually allowing it to diffuse out of the system. As illustrated in Fig. 3, the model system examined by Saha et al. comprised a Fe3O4 core overcoated with a CdS shell. The strained interface with a lattice mismatch of around 4% and high-temperature annealing causes core diffusion into the shell with a decrease in core size, eventually leading to uniformly doped QDs. Furthermore, the technique's generality and universality were shown for various dopant ions, such as M^{2+} (Fe^{2+}, Ni^{2+}, Co^{2+}, Mn^{2+}). The efficiency of inside-out diffusion doping is determined by the core's capacity to enter the CdS lattice, followed by the metal ion's ability to diffuse within the lattice and the ease of cation exchange [35].

3 Factors Influencing Doping Efficiency

The distinctive electronic and optical characteristics of QDs (QDs) have generated considerable interest in contemporary times. Nanoscale semiconductor crystals exhibit significant potential across a diverse array of applications, encompassing electronics, optoelectronics, and biomedicine. The deliberate incorporation of impurities into QDs, commonly known as doping, is a pivotal technique in customising their characteristics and augmenting their efficacy. This discourse aims to investigate the variables that impact the efficacy of doping in QDs.

4 Role of Doping Precursor Concentration

The degree of dopant incorporation into the quantum dot lattice during synthesis is directly influenced by the concentration of the doping precursor. Elevated concentrations of doping precursors may augment the likelihood of the integration of dopant atoms into the quantum dot (QD) framework. On the other hand, reduced concentrations of precursors may yield a restricted quantity of dopants, thereby causing inadequate levels of doping. Attaining the targeted dopant concentration requires precise management of the precursor concentration throughout the synthesis procedure. The distribution and homogeneity of dopants within a quantum dot ensemble are influenced by the concentration of the doping precursor. An increased concentration of precursors frequently leads to a homogeneous dispersion of dopants within the QDs. The homogeneity of the doped quantum dot (QD) population is a favourable attribute that facilitates the attainment of consistent and foreseeable characteristics. In contrast, reduced levels of precursor concentrations may give rise to dopants that are either spatially separated or clustered, thereby causing non-uniform doping and fluctuations in the performance of QDs. The control of size and composition in doped QDs can be influenced by the concentration of doping precursors. The growth kinetics and nucleation process of QDs in various synthesis methods, including colloidal synthesis, are influenced by the concentration of the precursor. Elevated concentrations of precursors typically facilitate accelerated rates of growth and increased sizes of QDs. Furthermore, it should be noted that the dopant material has the potential to impact the nucleation and growth patterns of the QDs. As a result, it becomes imperative to fine-tune the precursor concentration in order to attain the desired size and composition of the QDs. The optical and electronic characteristics of doped QDs are influenced by the concentration of the doping precursor. The concentration of dopant atoms within the QD lattice can exert an influence on optical properties, including emission wavelength and quantum yield. Elevated levels of precursor concentrations have the potential to augment the transfer of energy and the dynamics of charge carriers, ultimately resulting in alterations to the optical characteristics. The electronic characteristics of doped QDs (QDs), including energy levels and charge transport, are significantly influenced by the concentration of dopants, which ultimately affects the overall functionality of these QDs. The concentration of the doping precursor has an impact on the stability of QDs that have been doped, as well as their impurity effects. Elevated levels of dopant concentrations have the potential to introduce impurities and defects, thereby impacting the structural integrity and stability of QDs (QDs). Elevated precursor concentrations have the potential to cause non-uniform or inadequate doping, thereby leading to heightened impurity scattering or carrier recombination. These consequences can negatively impact the efficiency and stability of QDs.

Nair et al. have exhibited that the synthesis of metal- and nonmetal-codoped graphene QDs (GQDs) with improved optical properties can be achieved through the co-reaction of lignosulfonate (Fig. 4), which contains heteroatoms such as sulphur (S) and nitrogen (N) in its structure, and a metal precursor [36]. The GQDs derived from lignosulfonate synthesis exhibited notable blue fluorescence, displaying a substantial quantum yield of 23%. This phenomenon is ascribed to the presence of S and N doping, which was verified through the utilisation of X-ray photoelectron spectroscopy and Fourier transform infrared spectroscopy analyses. The lignosulfonate-derived GQDs were subjected to in situ doping to facilitate their modification. Additionally, a metal atom dopant was introduced to further engineer the GQDs, resulting in an improved quantum yield of 31%. This is noteworthy as it represents the highest quantum yield achieved for any GQDs derived from lignin.

A technique has been reported for the production of diluted magnetic semiconductor QDs (DMS-QDs) consisting of colloidal ZnO. This method involves the alkaline-activated hydrolysis and condensation of zinc acetate solutions in dimethyl sulfoxide (DMSO) [37]. The critical nuclei exhibit quantitative exclusion of dopants, while the nanocrystals' subsequent growth incorporates them in an almost isotropic

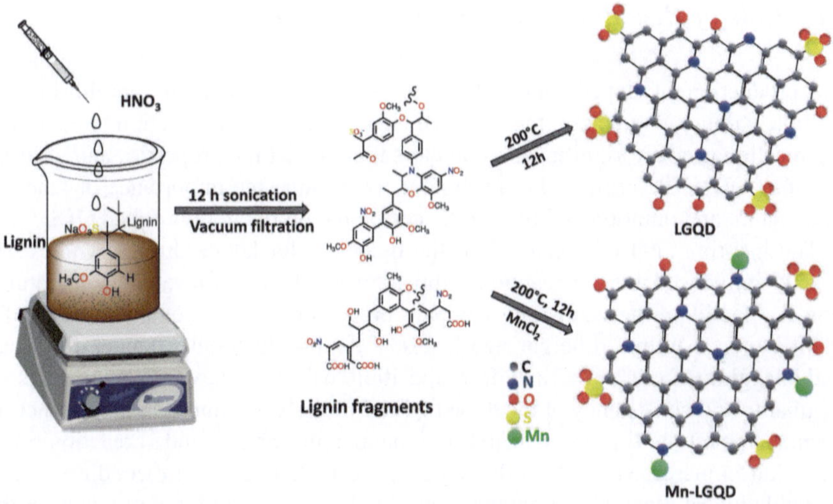

Fig. 4 A diagrammatic representation of the hydrothermal process for the synthesis of LGQDs and Mn-LGQDs. Reproduced with the permission from Ref. [36] © 2020 American chemical society

manner. The reduction in nanocrystal diameters that occurs as a result of doping can be elucidated by the Gibbs-Thompson relationship, which describes the correlation between lattice strain and crystal solubility. The process of synthesising graphene QDs (GQDs) involved the utilisation of femtosecond laser ablation in liquid, with laser induced graphene serving as the carbon source. The synthesis of Nitrogen-doped graphene QDs (N-GQDs) was achieved through the introduction of ammonia water into the graphene suspension [38]. The structure of GQDs/N-GQDs is composed of a graphitic core that exhibits oxygen and nitrogen functionalities, with a particle size that is smaller than 10 nm. This has been demonstrated through the use of various analytical techniques such as x-ray photoelectron spectroscopy, Fourier infrared spectrometer spectroscopy, and transmission electron microscopy. Lead halide perovskite QDs (PQDs) doped with transition metal cations are a promising group of materials for advanced photonic devices in the future. This is because of their fascinating properties induced by dopants and enhanced stabilities against photo-, moisture-, and thermal factors. The multifaceted nature of metal cation-doped PQDs in colloidal synthesis presents a challenging task when utilising conventional batch reactors. Bateni et al., employed a modular microfluidic platform that was equipped with a mobile spectral monitoring probe to investigate the kinetics and elucidate the chemical transformation mechanism of the rapid cation doping of cesium lead chloride QDs [39]. The authors additionally presented a design framework that enables accurate adjustment of the optical characteristics of QDs (QDs) through a post-synthetic doping mechanism in a flow-based system. The present study has developed a modular microfluidic platform that, in conjunction with facile doping chemistry, can facilitate the uninterrupted production of metal cation-doped PQDs possessing optoelectronic properties tailored to specific applications.

4.1 Role of Reaction Temperature and Time

The production of doped QDs (QDs) necessitates a meticulous equilibrium of multiple parameters to attain the intended characteristics. The reaction temperature and reaction time are significant parameters that impact the properties and efficacy of doped QDs. The present discourse aims to examine the importance of reaction temperature and duration and their influence on the production of doped QDs.

The incorporation of dopants into the quantum dot lattice during synthesis is directly influenced by the reaction temperature and time. Elevated temperatures typically facilitate the diffusion of dopant precursors into the quantum dot (QD) architecture, resulting in heightened levels of doping. Excessive temperatures can lead to agglomeration or thermal decomposition, which can have adverse effects on the quality and consistency of the doped QDs. Similarly, the duration of the reaction permits ample time for the inclusion of dopant atoms, and extended reaction times can result in increased levels of doping given suitable temperature conditions. The manipulation of reaction temperature and duration exerts a notable impact on the regulation of size and composition of doped QDs. Elevated temperatures frequently facilitate accelerated nucleation and growth kinetics, leading to the formation of comparatively larger QDs. In contrast, reduced temperatures could hinder the developmental mechanism and result in the formation of diminutive QDs. The overall size distribution and size homogeneity of the doped QDs are influenced by the reaction time in a similar manner. The manipulation of temperature and reaction time allows for precise regulation of the dimensions and constitution of QDs, which is imperative in customising their optical and electronic characteristics. The crystal structure and stability of doped QDs are also influenced by the reaction temperature and duration. The kinetics of nucleation and growth processes, as well as the crystal structure of QDs, is impacted by temperature. Elevated temperatures have the potential to facilitate the process of crystal growth and enhance the degree of crystallinity, thereby resulting in heightened stability. Nevertheless, elevated temperatures have the potential to cause imperfections or separation of phases. The growth dynamics are influenced by the reaction time, which provides adequate time for the development of a clearly defined crystal structure. Achieving stable and structurally well-defined doped QDs (QDs) requires the optimisation of temperature and reaction time.

The manipulation of reaction temperature and time is a crucial aspect in customising the optical and electronic characteristics of doped QDs. The manipulation of reaction conditions enables regulation of the quantum yield, energy levels, and emission wavelength of the QDs. Elevated temperatures have the potential to facilitate energy transfer mechanisms and the dynamics of charge carriers, leading to alterations in optical characteristics. Furthermore, the duration of the reaction time has an impact on the creation of imperfections and obstructions within the QDs, potentially altering their electronic characteristics, including the movement of charge carriers and the rate of recombination.

The kinetics of the synthesis process and the controllability of doped QDs are influenced by the reaction temperature and time. Incomplete doping or undesired side reactions may occur due to rapid temperature changes or shorter reaction times. Conversely, extended reaction times or reduced temperatures may facilitate the occurrence of secondary reactions or undesired contaminants. Attaining the ideal reaction parameters in terms of temperature and duration facilitates superior regulation of the synthesis procedure, culminating in the production of doped QDs of exceptional quality.

4.2 Role of QD Size and Shape

The optical, electronic, and chemical properties of QDs (QDs) are significantly influenced by their size and shape. The size and shape parameters are critical factors in determining the doping efficiency and resulting characteristics of synthesised doped QDs. The present discourse aims to investigate the significance of quantum dot dimensions and configurations in the production of doped QDs.

The doping efficiency during synthesis is influenced by the size of QDs. In general, QDs that is smaller in size exhibit a greater surface-to-volume ratio, thereby augmenting the interaction between the dopant precursor and the surface of the quantum dot. The enhanced surface interaction promotes the integration of dopant atoms into the lattice of the quantum dot, leading to an enhancement in the efficiency of doping. In contrast, QDs of greater size exhibit a diminished surface-to-volume ratio, thereby limiting the surface area accessible for dopant integration and potentially resulting in decreased doping efficacy.

The doping process is also influenced by the shape anisotropy of QDs, which may take on various forms such as spherical, rod-shaped, or plate-like. The surface energies and reactivity of crystallographic facets on various quantum dot shapes can differ. The surface characteristics have the potential to impact the adsorption and diffusion of dopant precursors in the course of synthesis. The phenomenon of preferential doping on specific facets due to shape anisotropy can result in a non-uniform distribution of doping within the ensemble of QDs. Comprehending the doping behaviour that is contingent on the shape is of utmost importance in regulating the characteristics of doped QDs.

The electronic and optical properties of QDs are directly impacted by their size, thereby exerting an influence on the doping behaviour. As the size of the quantum dot diminishes, the quantum confinement effect becomes increasingly conspicuous, resulting in distinct energy levels and adjustable optical characteristics. The size-dependency of QDs significantly impacts their energy levels and bandgap, thereby influencing the arrangement of dopant energy levels in the QD band structure. The properties of QDs that are dependent on their size are of utmost importance in the optimisation of doping techniques and the attainment of desired functionalities. The carrier dynamics, which encompass charge carrier generation, recombination, and transport, are notably impacted by the dimensions of QDs. The discrete energy levels and reduced non-radiative recombination pathways of smaller QDs typically result in faster carrier relaxation and radiative recombination rates. The utilisation of doping has the potential to modify the dynamics of carriers through the alteration of charge transfer processes and the modification of energy states. Comprehending the carrier dynamics that are dependent on the size is of utmost importance in order to enhance doping techniques and achieve the highest possible efficiency of doped QDs. The stability and surface passivation requirements of QDs are also influenced by their size. QDs with larger dimensions exhibit a greater density of surface defects and trap states, owing to their increased surface area (Table 1). The introduction of impurities or dopant-induced defects due to doping can have an impact on the surface passivation requirements. QDs exhibit a greater surface-to-volume ratio, which necessitates the implementation of more efficient surface passivation techniques to counteract surface-related recombination and improve their stability. The management of size-dependent stability and surface passivation is a crucial factor in attaining sustained efficacy and robustness of doped QDs.

5 Strategies to Optimize Doping Efficiency

Various approaches can be utilised to enhance doping efficacy in QDs (QDs) for optimal performance. The present strategies are centred on augmenting the integration of dopants into the quantum dot lattice and attaining the targeted levels of doping. The following are several fundamental methods for enhancing doping efficiency:

- The selection of a suitable dopant material is a critical factor in achieving optimal doping efficiency. Efficient energy level alignment and charge transfer can be achieved by selecting a dopant material that possesses electronic properties that are compatible with those of the QD matrix. Furthermore, it is imperative that the dopant material demonstrates favourable solubility within the quantum dot (QD) matrix and maintains stability throughout the doping procedure. Thorough evaluation of the dopant material characteristics is crucial for the effective integration of said material into the quantum dot lattice.
- In the field of engineering, it is imperative to optimise the doping precursor in order to enhance the efficiency of doping. The process entails the careful selection of dopant precursors with high reactivity and favourable thermodynamics for the doping reaction. Advanced engineering methodologies, such as the manipulation or creation of ligands, can be utilised to amplify the responsiveness and durability of the precursor, thereby facilitating its interaction with the quantum dot surface and promoting effective doping.
- The optimisation of different process parameters can enhance the efficiency of doping. The aforementioned parameters encompass temperature, duration of reaction, concentration of precursor, and atmosphere of reaction. Optimising these parameters according to the particular doping scheme and target doping concentrations can improve the integration of dopants into the quantum dot crystal structure. Modulating the temperature and reaction duration can exert an impact on the

Table 1 Sum up the major strategies and their examples

Strategies	Description	Examples/considerations	References
Selection of suitable dopant material	Choose dopant materials with compatible electronic properties to achieve desired energy level alignment and charge transfer	ZnSe QDs doped with Mn^{2+}, CdSe QDs doped with Cu^{+} ions	[40]
Precursor engineering	Modify dopant precursors to enhance reactivity, stability, and solubility in the QD matrix, improving dopant incorporation efficiency	Ligand design, precursor synthesis optimization, chelating agents for stabilization	[41, 42]
Control over doping process parameters	Optimize temperature, reaction time, precursor concentration, and atmosphere to enhance diffusion kinetics and achieve desired doping levels	Higher temperatures for increased diffusion, longer reaction times for improved incorporation	[1, 43, 44]
Surface engineering and passivation	Employ surface ligands, shell layers, or surface treatments to passivate QD surface, reducing surface defects and improving doping efficiency	Ligand exchange, shell growth, surface passivation techniques	[45, 46]
Core-shell doping	Utilize core-shell QDs to selectively dope the core or shell regions, enabling precise control over doping levels and charge transfer processes	CdSe/CdS core-shell QDs doped with transition metal ions	[47]
Hybrid doping approaches	Combine multiple doping techniques, such as ion implantation, chemical doping, or cation exchange, to enhance doping efficiency synergistically	Sequential doping processes, simultaneous doping with different dopants	[48]
Characterization and feedback	Employ various characterization techniques (e.g., spectroscopy, microscopy, electrical measurements) to assess doping efficiency and optimize	Photoluminescence spectroscopy, TEM/SEM imaging, current-voltage characteristics	[49, 50]

diffusion and kinetics of dopant incorporation, whereas fine-tuning the precursor concentration guarantees the attainment of the intended doping levels.

- The efficiency of doping can be hindered by non-radiative recombination or impurity scattering caused by surface defects and trap states. Therefore, surface engineering and passivation are crucial in addressing this issue. The implementation of proficient surface engineering and passivation methodologies can alleviate these concerns and amplify the efficacy of doping. The implementation of surface ligands, shell layers, or surface treatments can serve as a means of passivating the surface of QDs (QDs), thereby reducing surface defects and enhancing carrier confinement. The implementation of surface engineering techniques enhances the general stability, optical characteristics, and doping efficacy of QDs.

- Core-shell doping is a technique that can be utilised to improve doping efficiency in QDs (QDs). Specifically, the efficiency can be enhanced by selectively doping either the core or shell regions of the QDs. The utilisation of a core-shell structure enables a regulated milieu for the inclusion of dopants, thereby facilitating meticulous modulation of the doping levels. Through the strategic introduction of dopants into either the core or shell, it is feasible to enhance charge transfer mechanisms, mitigate the emergence of dopant-induced flaws, and enhance the efficacy of doping.
- Hybrid doping approaches refer to the combination of various doping techniques with the aim of optimising their effectiveness. Multiple dopants can be introduced or the doping process can be enhanced through the use of sequential or simultaneous doping techniques. The simultaneous utilisation of ion implantation, chemical doping, or cation exchange methods can result in a synergistic enhancement of doping efficiency, as each technique can leverage its unique advantages.
- The utilisation of characterization techniques, such as spectroscopy, microscopy, and electrical measurements, can offer significant insights into the effectiveness of doping. The assessment of various doping strategies and optimisation of the doping process can be achieved by monitoring the optical and electronic properties of doped QDs (QDs). The iterative enhancement of doping efficiency is facilitated by the approach that is driven by feedback.

6 Applications of Doped QDs

6.1 Bioimaging

QDs that have been doped have been utilised in diverse areas such as bioimaging and biosensing. The following are several significant implementations:

The utilisation of doped QDs as imaging agents in biological systems is highly advantageous due to their distinctive optical properties in the field of bioimaging. Due to their diminutive size, exceptional photostability, and customizable emission spectra, they are well-suited for a diverse array of imaging methodologies, including fluorescence microscopy and in vivo imaging. Through the conjugation of doped QDs with targeting molecules, scientists are able to selectively label and monitor different cellular components, biomolecules, or particular disease markers within living organisms. The utilisation of doped QDs in biosensing applications based on Fluorescence Resonance Energy Transfer (FRET) is feasible. Förster resonance energy transfer (FRET) is a phenomenon of energy transfer that transpires non-radiatively between a donor fluorophore and an acceptor fluorophore when they are in close proximity. Through the utilisation of Förster resonance energy transfer (FRET), scientists can effectively track and analyse targeted biochemical reactions or interactions by combining a doped quantum dot with an acceptor fluorophore.

This technique allows for the detection of biomolecular events with exceptional sensitivity and spatial resolution. Proximity-based assays employ doped QDs to measure the spatial proximity of target molecules. The detection and quantification of target molecules can be achieved by conjugating target-specific molecules, such as antibodies or aptamers, to doped QDs. This approach relies on the proximity-induced changes in the optical properties of the QDs. This approach is useful for various applications, including protein-protein interaction studies and the detection of biomarkers in clinical samples. The application of doped QDs in single-molecule tracking provides exceptional photostability, facilitating the extended and precise visualisation of singular molecules. This property is particularly useful in single-molecule tracking experiments, where the movement and behavior of individual biomolecules (such as proteins or nucleic acids) can be monitored in real-time. Through the utilisation of doped QDs, scientists can acquire significant knowledge regarding the dynamics, interactions, and cellular processes of these molecules. The integration of doped QDs with other imaging modalities, such as magnetic resonance imaging (MRI) or computed tomography (CT), can result in the development of multimodal imaging agents. These hybrid agents can provide complementary information and enable enhanced imaging capabilities, such as combining the high-resolution imaging of doped QDs with the anatomical information from MRI or CT, leading to improved diagnosis and monitoring of diseases. In general, doped QDs have demonstrated their versatility and significance in the fields of bioimaging and biosensing, providing improved sensitivity, photostability, and multiplexing abilities for various biological investigations. A facile hydrothermal technique was employed to synthesise nitrogen-sulfur doped graphene QDs (NS-GQDs). The stability of fluorescence signals exhibited by NS-GQDs remains high even in the presence of diverse metal ions [51]. Aspartic acid and cysteine were employed as carbon precursors and heteroatomic sources of nitrogen and sulphur, respectively. The NS-GQDs that were generated exhibited a quantum yield of up to $19.3 \pm 1.7\%$ and a peak emission of 480 nm when excited by 400 nm. The results depicted in Fig. 5a demonstrate a noteworthy decrease in the fluorescence of Pdots within cells upon the introduction of 200 μM Cu^{2+}. This finding suggests that Cu^{2+} has a quenching effect on Pdots in the context of bioimaging. On the other hand, it was observed that the fluorescence imaging ability of cells stained with NS-GQDs remained unaffected upon treatment with 200 μM Cu^{2+}, as depicted in Fig. 5b. The findings indicate that the NS-GQDs that were created exhibit exceptional resistance to metal ions in the context of bioimaging.

6.2 *Optoelectronics*

The unique properties and potential applications of doped QDs (QDs) have garnered a lot of attention in optoelectronics. The wide applicability of QDs and their ongoing works have been tabulated in Table 2. Researchers have found that by doping or

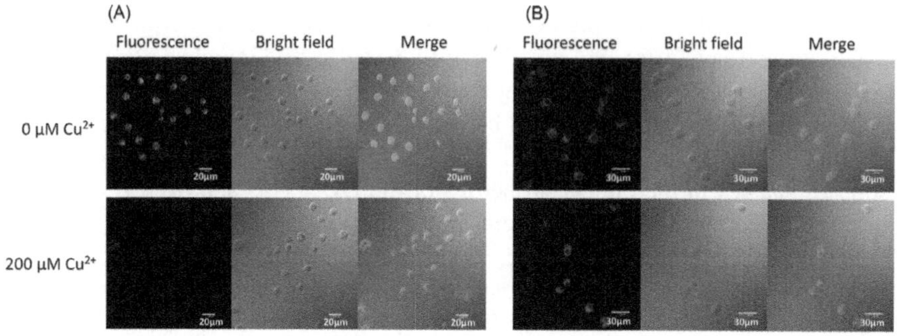

Fig. 5 LSCM images of RAW 264.7 cells incubated with **a** 10.0 μg/mL Pdots, **b** 1.0 mg/mL NS-GQDs with 200 μM Cu^{2+}. Reproduced with the permission from Ref. [51] © 2019 American chemical society

introducing impurities into the QDs, they can alter their electronic and optical properties, expanding their potential for use in optoelectronic devices. Some examples of where doped QDs have found use in optoelectronics are as follows:

Doped QDs can function as emitters in light-emitting diodes (LEDs). Researchers have found that they can tune the QDs' emission wavelength across the visible and near-infrared spectrum by adjusting their size and composition. QDs are promising candidates for next-generation display technologies because their efficiency, stability, and colour purity can be further improved by doping them with specific impurities.

Doped QDs may improve the efficiency of solar cells, which is important in the field of photovoltaics. It is possible to increase the solar cell's absorption range by incorporating doped QDs into the active layer. Dopants can also help with charge separation and transport, which boosts the solar cell's overall efficiency.

Doped QDs are an active material that can be used in photodetectors. Scientists can design QDs to be exceptionally sensitive to certain colours of light by adjusting the bandgap and the amount of doping. This paves the way for the creation of photodetectors with enhanced responsivity, rapid response, and minimal noise.

When it comes to lasers, doped QDs can be used as gain media. In order to achieve population inversion and stimulated emission, the dopants introduce energy levels within the QD bandgap. Doped QD lasers are useful for telecommunications, spectroscopy, and medical diagnostics because they can operate at a wide range of wavelengths, from the near-infrared to the visible and even the mid-infrared.

When it comes to quantum computing and information processing, doped QDs can serve as qubits. Implementing quantum logic operations requires manipulating electron spins and charge states in the QDs, which is made possible through the controlled introduction of dopants. Doped QDs have the potential to contribute to the growth of quantum technologies by enabling scalable and coherent qubit systems.

Table 2 Various applications of QDs device making and fabrications

Application	Description	Benefits	References
Light-emitting diodes	Doped QDs serve as emitters in LEDs, enabling tunable and efficient displays with enhanced color purity	Wide color gamut, high efficiency, long operational lifetime, solution processability	[52]
Photovoltaics	Doped QDs enhance the absorption range and charge transport in solar cells, improving overall efficiency	Broadband light absorption, enhanced charge collection, compatibility with flexible substrates	[53–55]
Photodetectors	Doped QDs enable the development of highly sensitive and fast photodetectors with low noise characteristics	High responsivity, wide spectral range, low dark current, ultrafast response time	[56]
Lasers	Doped QDs act as gain media, facilitating population inversion and stimulated emission for lasers in various wavelengths	Wavelength tunability, low threshold current, high quantum efficiency, compact size	[57]
Quantum computing	Doped QDs can be used as qubits for quantum computing and information processing, leveraging controlled electron spins and charge states	Scalability, long coherence times, high-fidelity operations, interfacing with solid-state platforms	[58]
Biological imaging	Doped QDs offer excellent optical properties for bioimaging, including high brightness, photostability, and tunable emission wavelengths	Bright and stable fluorescence, multiplexed imaging capabilities, targeted molecular imaging, high signal-to-noise ratio	[13]
Sensing applications	Doped QDs are utilized for sensing applications, such as chemical and biological sensing, due to their high sensitivity and selectivity	High sensitivity and selectivity, label-free detection, real-time monitoring, miniaturization potential	[59]
Display technologies	Doped QDs enable the development of vibrant and energy-efficient displays with wide color gamut and high contrast ratios	Vivid and accurate colors, energy efficiency, high contrast ratio, flexible and transparent displays	[60, 61]

(continued)

Table 2 (continued)

Application	Description	Benefits	References
Optical data storage	Doped QDs have potential applications in optical data storage due to their stable and controllable optical properties	High storage density, long-term data retention, fast data writing and reading, compatibility with existing technology	[62]
Energy-efficient lighting	Doped QDs can be used in solid-state lighting to achieve energy efficiency and long lifetimes, such as in quantum dot-based down-converters for LEDs	High color quality, energy efficiency, long operational lifetime, warm white lighting	[63, 64]

6.3 Photovoltaics

The utilisation of doped QDs is currently being extensively investigated for their potential application in photovoltaics [65–68], with a particular focus on enhancing light absorption and overall conversion efficiency. The bandgap of QDs can be precisely adjusted to enhance the absorption of a wider range of wavelengths in the solar spectrum by introducing various materials or impurities. Furthermore, it has been observed that QDs that have been doped possess the capability to produce numerous excitons from a solitary high-energy photon. This occurrence is commonly referred to as multiple exciton generation (MEG) and has the potential to augment the effectiveness of solar cells. Moreover, the introduction of precise impurities can result in the formation of intermediate bands within the bandgap of QDs, which facilitates the absorption of photons possessing energies lower than the material's bandgap. The utilisation of intermediate band solar cells presents a potentially auspicious pathway towards achieving enhanced efficiency in the conversion of solar energy. In addition, doped QDs have the potential to serve as sensitizers in the context of QDs sensitised solar cells [69–71], wherein they are capable of absorbing photons and subsequently transferring the resulting electrons to a photoactive material. This methodology enables the utilisation of economical and solution-deposited substances while upholding elevated conversion efficacy. Ongoing investigation and innovation in the application of doped QDs exhibit significant promise in propelling the domain of photovoltaics forward.

The synthesis of N, S-doped carbon QDs (N, S-CQDs) was reported through a straightforward hydrothermal treatment approach utilising ascorbic acid and ammonium persulfate as reagents [72]. The enhancements in performance and the removal of the light-soaking phenomenon in ZnO:N, S-CQDs cells are ascribed to the passivation of surface defects in ZnO by N, S-CQDs (Fig. 6). This conclusion is supported by fluorescence spectroscopy and scanning Kelvin probe microscopy. The power conversion efficiency of cells with N, S-CQDs-modified ZnO ETL was observed to be 9.31%, indicating a significant improvement compared to the reference ZnO cells.

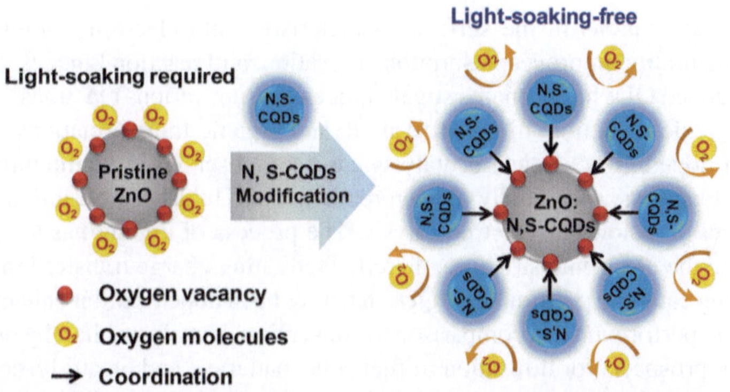

Fig. 6 The mechanism for the passivation of surface defects on ZnO nanoparticles through the utilisation of N, S-functionalized carbon QDs (CQDs). Reproduced with the permission from Ref. [72] © 2019 American chemical society

Si QDs (QDs) were synthesised in SiO2 using a co-sputtering technique, resulting in multiple layers with a narrow size distribution. The present study investigates the structural, electrical, and optical characteristics of multilayer films composed of Si QDs (QDs) and SiO2. The films were prepared by introducing varying concentrations of boron (B) during the sputtering process [73]. It was observed that the intensity of photoluminescence (PL) decreased as the concentration of boron increased. Yang et al. conducted a study in which they produced a range of nitrogen-doped carbon QDs (N-CQDs) through hydrothermal synthesis at varying temperatures, utilising Eichhornia crassipes (ECs) as the source material [74]. The researchers conducted an experimental fabrication of N-CQDs sensitised solar cells, resulting in lower power conversion efficiency when compared to state-of-the-art devices that are based on silicon, polymer, dye, CdS, CdSe, or perovskite.

6.4 Catalysis

Doped QDs have exhibited considerable potential in the domain of catalysis, wherein they can augment the efficacy and specificity of diverse chemical reactions. Several potential uses of doped QDs in catalysis have been identified. The utilisation of doped QDs as photocatalysts is a promising approach for facilitating chemical reactions through the conversion of light energy. The process of doping allows for accurate manipulation of the electronic configuration, thereby facilitating the production and transmission of charge carriers. The aforementioned ability amplifies the effectiveness of photocatalytic processes, such as the division of water for the purpose of producing hydrogen or the breakdown of pollutants. QDs, upon doping with particular atoms or functional groups, exhibit remarkable catalytic activity and selectivity for diverse reactions in the realm of heterogeneous catalysis. The process of doping induces modifications in the surface characteristics and electronic configuration, thereby facilitating improved adsorption, activation, and reaction kinetics. The utilisation of doped QDs has been investigated for catalysing various reactions, including carbon dioxide reduction, nitrogen fixation, and organic transformations. The utilisation of doped QDs as electrocatalysts has been observed in facilitating electrochemical reactions, specifically oxygen reduction and hydrogen evolution [75–77]. This process is known as electrocatalysis. The process of doping has the potential to enhance the electronic structure, thereby facilitating charge transfer kinetics and augmenting catalytic performance. QDs that have been doped exhibit enhanced electrocatalytic performance in comparison to conventional catalysts, thereby presenting promising prospects for utilisation in fuel cells, batteries, and electrolyzers. Doped QDs have the ability to imitate the catalytic activity of enzymes, which is commonly referred to as artificial enzyme mimetics or nanozymes. The catalytic properties of enzymes, such as oxidase, peroxidase, or catalase-like activities, can be emulated by doping QDs with particular ions or functional groups [78, 79]. Nanozymes present several benefits, including enhanced stability, customizable properties, and economical feasibility in contrast to their natural enzyme counterparts. These attributes create opportunities for their implementation in various domains, such as biosensing, environmental restoration, and medical research.

Prekodravac et al. presented a rapid and efficient method for synthesising nitrogen-doped carbon QDs in an aqueous medium without the use of metal catalysts. This was achieved through the utilisation of a microwave reactor [80]. A synthesis was conducted on a glucose water solution utilising a microwave reactor under low temperature and applied microwave power for a duration of only 1 min. The N-doped carbon QDs that were synthesised exhibit noteworthy photocatalytic efficacy in eliminating hazardous organic dye (Rose Bengal) when subjected to visible light radiation. A significant degradation of approximately 93% of the dye is attained within a brief period of 30 min of radiation. The synthesis of fluorescent nitrogen doped carbon dots (N-CDs) was discussed in a report, using a straightforward hydrothermal method [81]. The N-CDs exhibit remarkable catalytic efficacy in the sodium borohydride-mediated reduction of methylene blue. Elsewhere, a report was made on the green synthesis of fluorescent nitrogen-doped carbon dots (N-CDs) utilising *Actinidia deliciosa* (A. deliciosa) fruit extract as a carbon precursor and aqueous ammonia as a nitrogen dopant [82]. Chandrasekaran et al. have documented a straightforward and environmentally-friendly hydrothermal method for synthesising nitrogen-doped carbon dots (N-CDs) utilising *C.grandis* and aq.NH3 as carbon and nitrogen sources, respectively [83]. The catalytic activity of NaBH4 in the reduction of methyl orange was affected by the synthesised N-CDs. The rate constant of the reduction process of organic dye, specifically methyl orange, by NaBH4 in the presence of the synthesised green catalyst was also ascertained.

7 Prospects for Future Developments and Applications

The potential for advancements and utilisation of doped QDs (QDs) in the future is exceedingly auspicious. The following are prospective domains of progress.

Doped QDs (QDs) can be customised to demonstrate distinct optical and electronic characteristics, rendering them highly suitable for advanced optoelectronic devices. Enhancements in synthesis methodologies have the potential to facilitate the creation of lasers, LEDs, and photodetectors based on QDs, which exhibit superior efficiency, colour purity, and stability.

Doped QDs (QDs) exhibit significant promise in sensing applications owing to their exceptional sensitivity and selectivity. The optical properties of materials can be adjusted to enable the detection of particular molecules or ions by introducing precise dopants. Prospective advancements could result in the production of exceedingly responsive and easily transportable sensors based on QDs (QDs) for the purposes of monitoring the environment, diagnosing medical conditions, and ensuring food safety.

Doped QDs (QDs) have the potential to significantly contribute to the advancement of renewable energy technologies through energy harvesting and storage. By enhancing their electronic characteristics, they can be employed in high-performance photovoltaic cells for the conversion of solar energy into electrical energy. Moreover, the utilisation of doped QDs (QDs) can be investigated for the purpose of energy storage in various applications, including supercapacitors and electrochemical reactions as catalysts.

Doped QDs (QDs) have garnered significant interest as contrast agents in bioimaging and as targeted drug delivery systems for biomedical applications. The distinctive optical characteristics of nanoparticles, in conjunction with surface alterations, facilitate the accurate visualisation of biological structures and the regulated dispensation of therapeutic agents. Subsequent progressions may potentially result in the creation of diagnostic and therapeutic instruments that are safer and more effective in the field of medicine.

Doped QDs (QDs) exhibit potential for utilisation in quantum computing and quantum information processing, as they possess the capability to confine and manipulate individual charges and spins. Advancements in dopant engineering and control have the potential to facilitate the achievement of quantum computing architectures that are both scalable and resilient.

Doped QDs (QDs) have potential for utilisation in environmental applications, specifically in photocatalysis for the purpose of water purification or pollutant degradation. Moreover, these can be incorporated into lighting systems that are designed to be energy-efficient, thereby mitigating energy usage and environmental repercussions. Prospective progressions may facilitate the creation of more environmentally friendly technologies that entail diminished resource consumption and enhanced efficacy.

In general, the potential applications and advancements of doped QDs are extensive and offer significant promise in diverse scientific and technological domains.

Further investigation, refinement of manufacturing techniques, and examination of innovative additives and structures will facilitate the realisation of their complete potential and broaden their pragmatic implementations.

8 Present Market of Doped QDs

The market for doped QDs (QDs) is still in its early stages but is projected to experience substantial growth in the coming years. The unique characteristics and potential applications of doped QDs (QDs) are the primary drivers behind the increasing demand for them across diverse industrial sectors. The market for QDs (Fig. 7), including doped QDs, is expected to grow significantly in the coming years. Several market research reports indicate that the QDs market is anticipated to attain a significant valuation in the billions by the year 2026, exhibiting a compound annual growth rate (CAGR) exceeding 20%. The growth in question is being propelled by several factors, including but not limited to the escalating demand for displays that are energy-efficient, the progress made in healthcare imaging technologies, and the increasing acceptance of sensors that are based on QDs. While it's challenging to provide specific market figures, here are some key insights into the current market of doped QDs:

Doped QDs (QDs) are increasingly being utilised in the optoelectronics sector, specifically in display technologies like quantum dot LED (QLED) displays. QLED displays provide several benefits, including superior colour precision, a broad range of colours, and low power consumption. Doped QDs are progressively gaining popularity in the realm of consumer electronics, encompassing devices such as televisions, smartphones, and monitors.

Doped QDs (QDs) are employed as fluorescent markers in biomedical imaging owing to their exceptional optical characteristics, including high luminosity and

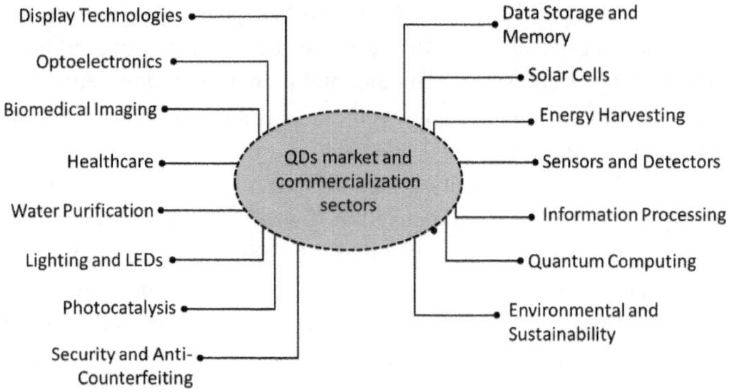

Fig. 7 Different types of application areas for QDs based products

resistance to photobleaching. They offer enhanced imaging capabilities for applications in biological research, diagnostics, and drug delivery. Ongoing research and commercialization efforts have identified the biomedical imaging sector as a significant market for doped QDs.

Energy and Solar Cells: Doped QDs hold promise in the field of energy conversion and storage. The tunable bandgap and high absorption efficiency of these materials have rendered them a subject of exploration for their potential application as constituents of next-generation solar cells. Doped QDs offer the potential for increased solar cell efficiency and reduced production costs, driving their adoption in the renewable energy sector.

Sensors and Detectors: Doped QDs are being developed for various sensing applications. They exhibit high sensitivity, selectivity, and stability, making them suitable for environmental monitoring, biosensing, and chemical detection. Doped QD-based sensors and detectors have the potential to provide rapid, accurate, and cost-effective solutions for diverse industries.

Research and Development: The market for doped QDs is also supported by ongoing research and development activities. Researchers are exploring new doping strategies, optimizing synthesis techniques, and investigating novel applications. This continuous innovation fuels the growth of the doped QD market and opens up possibilities for emerging applications.

It's important to note that the market for doped QDs is dynamic and evolving, with ongoing technological advancements and increasing commercialization efforts. The integration of doped QDs into various industrial applications is anticipated to result in an increase in market size and revenue.

9 Conclusion

To summarise, the synthesis strategies utilised for doped QDs comprise a variety of methods with the objective of integrating dopants into the QD lattice to attain specific electronic and optical characteristics. The aforementioned tactics encompass codoping, post-synthetic doping, ion exchange, alloying, and core-shell architectures. The process of codoping entails the concurrent introduction of multiple dopants during the synthesis of QDs, thereby enabling accurate regulation of both dopant concentration and combination. The process of post-synthetic doping pertains to the introduction of dopants onto pre-synthesized QDs through surface modification or ligand exchange procedures. The process of ion exchange entails the substitution of the initial cations or anions present in the quantum dot lattice with dopant ions. The process of alloying involves the integration of distinct semiconductor materials, along with the inclusion of dopants during the alloying procedure. The utilisation of core-shell structures allows for the confinement of doped QDs within a safeguarding shell, thereby facilitating the regulation of dopant diffusion and associated characteristics.

The implementation of synthesis strategies provides a malleable and adaptable approach to customise the characteristics of doped QDs to meet particular utilisation requirements. The optimisation of doped QDs for various applications such as optoelectronics, photovoltaics, sensing, and bioimaging can be achieved by manipulating dopant concentration, distribution, and combination, as per the research findings. In summary, the implementation of synthesis strategies involving doped QDs has resulted in notable progressions within the realm of nanotechnology and nanomaterials. The deliberate incorporation of impurities into QDs has presented novel opportunities for the manipulation of their optical and electronic characteristics, resulting in improved functionalities and innovative applications. Further investigation in this domain exhibits potential for enhancing these amalgamation approaches, broadening the scope of dopants and substrates, and unleashing the complete potential of doped QDs in various domains.

References

1. Makkar, M., Viswanatha, R.: Frontier challenges in doping quantum dots: synthesis and characterization. RSC Adv. **8**(39), 22103–22112 (2018)
2. Mocatta, D., Cohen, G., Schattner, J., Millo, O., Rabani, E., Banin, U.: Heavily doped semiconductor nanocrystal quantum dots. Science **332**(6025), 77–81 (2011)
3. Grandhi, G.K., Tomar, R., Viswanatha, R.: Study of surface and bulk electronic structure of II–VI semiconductor nanocrystals using Cu as a nanosensor. ACS Nano **6**(11), 9751–9763 (2012)
4. Hanif, K.M., Meulenberg, R.W., Strouse, G.F.: Magnetic ordering in doped Cd1-x Co x Se diluted magnetic quantum dots. J. Am. Chem. Soc. **124**(38), 11495–11502 (2002)
5. Grandhi, G.K., Swathi, K., Narayan, K., Viswanatha, R.: Cu doping in ligand free CdS nanocrystals: conductivity and electronic structure study. The J. Phys. Chem. Lett. **5**(13), 2382–2389 (2014)
6. Shanker, G.S., Tandon, B., Shibata, T., Chattopadhyay, S., Nag, A.: Doping controls plasmonics, electrical conductivity, and carrier-mediated magnetic coupling in Fe and Sn codoped In_2O_3 nanocrystals: local structure is the key. Chem. Mater. **27**(3), 892–900 (2015)
7. Baker, S.N., Baker, G.A.: Luminescent carbon nanodots: emergent nanolights. Angew. Chem. Int. Ed. **49**(38), 6726–6744 (2010)
8. Larson, L.A., Williams, J.M., Current, M.I.: Ion implantation for semiconductor doping and materials modification. Rev. Accel. Sci. Technol. **4**(01), 11–40 (2011)
9. Salzmann, I., Heimel, G., Oehzelt, M., Winkler, S., Koch, N.: Molecular electrical doping of organic semiconductors: fundamental mechanisms and emerging dopant design rules. Acc. Chem. Res. **49**(3), 370–378 (2016)
10. Sinatra, L., Pan, J., Bakr, O.M.: Methods of synthesizing monodisperse colloidal quantum dots. Mater. Matters **12**, 3–7 (2017)
11. Salaheldin, A.M., Walter, J., Herre, P., Levchuk, I., Jabbari, Y., Kolle, J.M., Brabec, C.J., Peukert, W., Segets, D.: Automated synthesis of quantum dot nanocrystals by hot injection: mixing induced self-focusing. Chem. Eng. J. **320**, 232–243 (2017)
12. Xu, Q., Kuang, T., Liu, Y., Cai, L., Peng, X., Sreeprasad, T.S., Zhao, P., Yu, Z., Li, N.: Heteroatom-doped carbon dots: synthesis, characterization, properties, photoluminescence mechanism and biological applications. J. Mater. Chem. B **4**(45), 7204–7219 (2016)
13. Das, P., Ganguly, S., Margel, S., Gedanken, A.: Tailor made magnetic nanolights: fabrication to cancer theranostics applications. Nanoscale Adv. **3**(24), 6762–6796 (2021)
14. Das, P., Ganguly, S., Margel, S., Gedanken, A.: Immobilization of heteroatom-doped carbon dots onto nonpolar plastics for antifogging, antioxidant, and food monitoring applications. Langmuir **37**(11), 3508–3520 (2021)

15. Das, P., Ganguly, S., Saravanan, A., Margel, S., Gedanken, A., Srinivasan, S., Rajabzadeh, A.R.: Naturally derived carbon dots in situ confined self-healing and breathable hydrogel monolith for anomalous diffusion-driven phytomedicine release. ACS Appl. Bio Mater. (2022)
16. Das, P., Ganguly, S., Ahmed, S.R., Sherazee, M., Margel, S., Gedanken, A., Srinivasan, S., Rajabzadeh, A.R.: Carbon dot biopolymer-based flexible functional films for antioxidant and food monitoring applications. ACS Appl. Polym. Mater. **4**(12), 9323–9340 (2022)
17. Das, P., Ganguly, S., Saha, A., Noked, M., Margel, S., Gedanken, A.: Carbon-dots-initiated photopolymerization: an in situ synthetic approach for MXene/poly (norepinephrine)/copper hybrid and its application for mitigating water pollution. ACS Appl. Mater. Interfaces **13**(26), 31038–31050 (2021)
18. Das, P., Maruthapandi, M., Saravanan, A., Natan, M., Jacobi, G., Banin, E., Gedanken, A.: Carbon dots for heavy-metal sensing, pH-sensitive cargo delivery, and antibacterial applications. ACS Appl. Nano Mater. **3**(12), 11777–11790 (2020)
19. Ganguly, S., Das, P., Itzhaki, E., Hadad, E., Gedanken, A., Margel, S.: Microwave-synthesized polysaccharide-derived carbon dots as therapeutic cargoes and toughening agents for elastomeric gels. ACS Appl. Mater. Interfaces **12**(46), 51940–51951 (2020)
20. Saravanan, A., Maruthapandi, M., Das, P., Ganguly, S., Margel, S., Luong, J.H., Gedanken, A.: Applications of N-doped carbon dots as antimicrobial agents, antibiotic carriers, and selective fluorescent probes for nitro explosives. ACS Appl. Bio Mater. **3**(11), 8023–8031 (2020)
21. Shim, M., Guyot-Sionnest, P.: N-type colloidal semiconductor nanocrystals. Nature **407**(6807), 981–983 (2000)
22. Shim, M., Wang, C., Guyot-Sionnest, P.: Charge-tunable optical properties in colloidal semiconductor nanocrystals. J. Phys. Chem. B **105**(12), 2369–2373 (2001)
23. Yu, D., Wang, C., Guyot-Sionnest, P.: N-Type conducting CdSe nanocrystal solids. Science **300**(5623), 1277–1280 (2003)
24. Talapin, D.V., Murray, C.B.: PbSe nanocrystal solids for n-and p-channel thin film field-effect transistors. Science **310**(5745), 86–89 (2005)
25. Kovalenko, M.V., Bodnarchuk, M.I., Zaumseil, J., Lee, J.-S., Talapin, D.V.: Expanding the chemical versatility of colloidal nanocrystals capped with molecular metal chalcogenide ligands. J. Am. Chem. Soc. **132**(29), 10085–10092 (2010)
26. Pradhan, N., Goorskey, D., Thessing, J., Peng, X.: An alternative of CdSe nanocrystal emitters: pure and tunable impurity emissions in ZnSe nanocrystals. J. Am. Chem. Soc. **127**(50), 17586–17587 (2005)
27. Grandhi, G.K., Viswanatha, R.: Demystifying complex quantum dot heterostructures using photogenerated charge carriers. J. Phys. Chem. Lett. **8**(9), 2043–2048 (2017)
28. Barrows, C.J., Chakraborty, P., Kornowske, L.M., Gamelin, D.R.: Tuning equilibrium compositions in colloidal Cd1–x Mn x Se nanocrystals using diffusion doping and cation exchange. ACS Nano **10**(1), 910–918 (2016)
29. Son, D.H., Hughes, S.M., Yin, Y., Paul Alivisatos, A.: Cation exchange reactions in ionic nanocrystals. Science **306**(5698), 1009–1012 (2004)
30. Beberwyck, B.J., Surendranath, Y., Alivisatos, A.P.: Cation exchange: a versatile tool for nanomaterials synthesis. J. Phys. Chem. C **117**(39), 19759–19770 (2013)
31. Robinson, R.D., Sadtler, B., Demchenko, D.O., Erdonmez, C.K., Wang, L.-W., Alivisatos, A.P.: Spontaneous superlattice formation in nanorods through partial cation exchange. Science **317**(5836), 355–358 (2007)
32. Sadtler, B., Demchenko, D.O., Zheng, H., Hughes, S.M., Merkle, M.G., Dahmen, U., Wang, L.-W., Alivisatos, A.P.: Selective facet reactivity during cation exchange in cadmium sulfide nanorods. J. Am. Chem. Soc. **131**(14), 5285–5293 (2009)
33. Demchenko, D.O., Robinson, R.D., Sadtler, B., Erdonmez, C.K., Alivisatos, A.P., Wang, L.-W.: Formation mechanism and properties of CdS-Ag2S nanorod superlattices. ACS Nano **2**(4), 627–636 (2008)
34. Saha, A., Shetty, A., Pavan, A., Chattopadhyay, S., Shibata, T., Viswanatha, R.: Uniform doping in quantum-dots-based dilute magnetic semiconductor. J. Phys. Chem. Lett. **7**(13), 2420–2428 (2016)

35. Saha, A., Makkar, M., Shetty, A., Gahlot, K., Pavan, A., Viswanatha, R.: Diffusion doping in quantum dots: bond strength and diffusivity. Nanoscale **9**(8), 2806–2813 (2017)
36. Nair, A.N., Chava, V.S.N., Bose, S., Zheng, T., Pilla, S., Sreenivasan, S.T.: In situ doping-enabled metal and nonmetal codoping in graphene quantum dots: synthesis and application for contaminant sensing. ACS Sustain. Chem. Eng. **8**(44), 16565–16576 (2020). https://doi.org/10.1021/acssuschemeng.0c05789
37. Schwartz, D.A., Norberg, N.S., Nguyen, Q.P., Parker, J.M., Gamelin, D.R.: Magnetic quantum dots: synthesis, spectroscopy, and magnetism of Co^{2+}- and Ni^{2+}-doped ZnO nanocrystals. J. Am. Chem. Soc. **125**(43), 13205–13218 (2003). https://doi.org/10.1021/ja036811v
38. Shen, L., Zhou, S., Huang, F., Zhou, H., Zhang, H., Wang, S., Zhou, S.: Nitrogen-doped graphene quantum dots synthesized by femtosecond laser ablation in liquid from laser induced graphene. Nanotechnology **33**(11), 115602 (2021)
39. Bateni, F., Epps, R.W., Abdel-latif, K., Dargis, R., Han, S., Volk, A.A., Ramezani, M., Cai, T., Chen, O., Abolhasani, M.: Ultrafast cation doping of perovskite quantum dots in flow. Matter **4**(7), 2429–2447 (2021)
40. Wang, H., Gao, J., Zhang, M., Liu, P., Guo, Y., Li, H., Zhao, G., Hu, S., Cheng, Z., Zang, J.: ZnSe/ZnS core-shell quantum dots doped with Mn^{2+} ions for magnetic state-manipulated light sources. ACS Appl. Nano Mater. **5**(6), 8448–8456 (2022)
41. Shaik, S.A., Sengupta, S., Varma, R.S., Gawande, M.B., Goswami, A.: Syntheses of N-doped carbon quantum dots (NCQDs) from bioderived precursors: a timely update. ACS Sustain. Chem. Eng. **9**(1), 3–49 (2020)
42. Soheyli, E., Behrouzi, S., Sharifirad, Z., Sahraei, R.: Multi-colored type-I Ag-doped ZnCdS/ZnS core/shell quantum dots with intense emission. Ceram. Int. **45**(9), 11501–11507 (2019)
43. Schwartz, D.A., Norberg, N.S., Nguyen, Q.P., Parker, J.M., Gamelin, D.R.: Magnetic quantum dots: synthesis, spectroscopy, and magnetism of Co^{2+}-and Ni^{2+}-doped ZnO nanocrystals. J. Am. Chem. Soc. **125**(43), 13205–13218 (2003)
44. Das, P., Ganguly, S., Agarwal, T., Maity, P., Ghosh, S., Choudhary, S., Gangopadhyay, S., Maiti, T.K., Dhara, S., Banerjee, S.: Heteroatom doped blue luminescent carbon dots as a nano-probe for targeted cell labeling and anticancer drug delivery vehicle. Mater. Chem. Phys. **237**, 121860 (2019)
45. Das, P., Ganguly, S., Bose, M., Mondal, S., Choudhary, S., Gangopadhyay, S., Das, A.K., Banerjee, S., Das, N.C.: Zinc and nitrogen ornamented bluish white luminescent carbon dots for engrossing bacteriostatic activity and Fenton based bio-sensor. Mater. Sci. Eng. C **88**, 115–129 (2018)
46. Tavakoli, M.M.: Surface engineering of Pbs colloidal quantum dots using atomic passivation for photovoltaic applications. Proc. Eng. **139**, 117–122 (2016)
47. Radovanovic, P.V., Gamelin, D.R.: Electronic absorption spectroscopy of cobalt ions in diluted magnetic semiconductor quantum dots: demonstration of an isocrystalline core/shell synthetic method. J. Am. Chem. Soc. **123**(49), 12207–12214 (2001)
48. Tejwan, N., Saini, A.K., Sharma, A., Singh, T.A., Kumar, N., Das, J.: Metal-doped and hybrid carbon dots: a comprehensive review on their synthesis and biomedical applications. J. Control. Release **330**, 132–150 (2021)
49. Jing, H.H., Bardakci, F., Akgöl, S., Kusat, K., Adnan, M., Alam, M.J., Gupta, R., Sahreen, S., Chen, Y., Gopinath, S.C.: Green carbon dots: synthesis, characterization, properties and biomedical applications. J. Funct. Biomater. **14**(1), 27 (2023)
50. Stroyuk, O., Raievska, O., Langner, S., Kupfer, C., Barabash, A., Solonenko, D., Azhniuk, Y., Hauch, J., Osvet, A., Batentschuk, M.: High-throughput robotic synthesis and photoluminescence characterization of aqueous multinary copper–silver indium chalcogenide quantum dots. Part. Part. Syst. Charact. **38**(10), 2100169 (2021)
51. Schroer, Z.S., Wu, Y., Xing, Y., Wu, X., Liu, X., Wang, X., Pino, O.G., Zhou, C., Combs, C., Pu, Q., Wu, M., Zhao, J.X., Chen, J.: Nitrogen–sulfur-doped graphene quantum dots with metal ion-resistance for bioimaging. ACS Appl. Nano Mater. **2**(11), 6858–6865 (2019). https://doi.org/10.1021/acsanm.9b01309

52. Bozyigit, D., Wood, V.: Challenges and solutions for high-efficiency quantum dot-based LEDs. MRS Bull. **38**, 731–736 (2013)
53. Blokhin, S., Sakharov, A., Nadtochy, A., Pauysov, A., Maximov, M., Ledentsov, N., Kovsh, A., Mikhrin, S., Lantratov, V., Mintairov, S.: AlGaAs/GaAs photovoltaic cells with an array of InGaAs QDs. Semiconductors **43**, 514–518 (2009)
54. Nozik, A.J., Beard, M.C., Luther, J.M., Law, M., Ellingson, R.J., Johnson, J.C.: Semiconductor quantum dots and quantum dot arrays and applications of multiple exciton generation to third-generation photovoltaic solar cells. Chem. Rev. **110**(11), 6873–6890 (2010)
55. Sanehira, E.M., Marshall, A.R., Christians, J.A., Harvey, S.P., Ciesielski, P.N., Wheeler, L.M., Schulz, P., Lin, L.Y., Beard, M.C., Luther, J.M.: Enhanced mobility CsPbI3 quantum dot arrays for record-efficiency, high-voltage photovoltaic cells. Sci. Adv. **3**(10), eaao4204 (2017)
56. Oertel, D.C., Bawendi, M.G., Arango, A.C., Bulović, V.: Photodetectors based on treated CdSe quantum-dot films. Appl. Phys. Lett. **87**(21), 213505 (2005)
57. Ledentsov, N.N., Grundmann, M., Heinrichsdorff, F., Bimberg, D., Ustinov, V., Zhukov, A., Maximov, M., Alferov, Z.I., Lott, J.: Quantum-dot heterostructure lasers. IEEE J. Sel. Top. Quantum Electron. **6**(3), 439–451 (2000)
58. De Rinaldis, S., D'Amico, I., Biolatti, E., Rinaldi, R., Cingolani, R., Rossi, F.: Intrinsic exciton-exciton coupling in GaN-based quantum dots: application to solid-state quantum computing. Phys. Rev. B **65**(8), 081309 (2002)
59. Molaei, M.J.: Principles, mechanisms, and application of carbon quantum dots in sensors: a review. Anal. Meth. **12**(10), 1266–1287 (2020)
60. Frecker, T., Bailey, D., Arzeta-Ferrer, X., McBride, J., Rosenthal, S.J.: Quantum dots and their application in lighting, displays, and biology. ECS J. Solid State Sci. Technol. **5**(1), R3019 (2015)
61. Bourzac, K.: Quantum dots go on display. Nature **493**(7432), 283 (2013)
62. Bimberg, D., Pohl, U.W.: Quantum dots: promises and accomplishments. Mater. Today **14**(9), 388–397 (2011)
63. Pode, R.: Organic light emitting diode devices: an energy efficient solid state lighting for applications. Renew. Sustain. Energy Rev. **133**, 110043 (2020)
64. Shirasaki, Y., Supran, G.J., Bawendi, M.G., Bulović, V.: Emergence of colloidal quantum-dot light-emitting technologies. Nat. Photon. **7**(1), 13–23 (2013)
65. Liu, F., Ding, C., Zhang, Y., Kamisaka, T., Zhao, Q., Luther, J.M., Toyoda, T., Hayase, S., Minemoto, T., Yoshino, K.: GeI2 additive for high optoelectronic quality CsPbI3 quantum dots and their application in photovoltaic devices. Chem. Mater. **31**(3), 798–807 (2019)
66. Liu, B., Wang, D., Zhang, Y., Fan, H., Lin, Y., Jiang, T., Xie, T.: Photoelectrical properties of Ag_2S quantum dot-modified TiO_2 nanorod arrays and their application for photovoltaic devices. Dalton Trans. **42**(6), 2232–2237 (2013)
67. Jabeen, U., Adhikari, T., Shah, S.M., Pathak, D., Kumar, V., Nunzi, J.-M., Aamir, M., Mushtaq, A.: Synthesis, characterization and photovoltaic applications of noble metal—doped ZnS quantum dots. Chin. J. Phys. **58**, 348–362 (2019)
68. Veerathangam, K., Pandian, M.S., Ramasamy, P.: Photovoltaic performance of Ag-doped CdS quantum dots for solar cell application. Mater. Res. Bull. **94**, 371–377 (2017)
69. Rühle, S., Shalom, M., Zaban, A.: Quantum-dot-sensitized solar cells. ChemPhysChem **11**(11), 2290–2304 (2010)
70. Mora-Sero, I., Giménez, S., Fabregat-Santiago, F., Gómez, R., Shen, Q., Toyoda, T., Bisquert, J.: Recombination in quantum dot sensitized solar cells. Acc. Chem. Res. **42**(11), 1848–1857 (2009)
71. Tian, J., Cao, G.: Semiconductor quantum dot-sensitized solar cells. Nano Rev. **4**(1), 22578 (2013)
72. Wang, Y., Yan, L., Ji, G., Wang, C., Gu, H., Luo, Q., Chen, Q., Chen, L., Yang, Y., Ma, C.-Q., Liu, X.: Synthesis of N, S-doped carbon quantum dots for use in organic solar cells as the ZnO modifier to eliminate the light-soaking effect. ACS Appl. Mater. Interfaces **11**(2), 2243–2253 (2019). https://doi.org/10.1021/acsami.8b17128

73. Hao, X., Cho, E.-C., Flynn, C., Shen, Y., Park, S.-C., Conibeer, G., Green, M.A.: Synthesis and characterization of boron-doped Si quantum dots for all-Si quantum dot tandem solar cells. Sol. Energy Mater. Sol. Cells **93**(2), 273–279 (2009)
74. Yang, Q., Duan, J., Yang, W., Li, X., Mo, J., Yang, P., Tang, Q.: Nitrogen-doped carbon quantum dots from biomass via simple one-pot method and exploration of their application. Appl. Surf. Sci. **434**, 1079–1085 (2018)
75. Tian, L., Li, Z., Wang, P., Zhai, X., Wang, X., Li, T.: Carbon quantum dots for advanced electrocatalysis. J. Energy Chem. **55**, 279–294 (2021)
76. Liu, Z.Q., Cheng, H., Li, N., Ma, T.Y., Su, Y.Z.: $ZnCo_2O_4$ quantum dots anchored on nitrogen-doped carbon nanotubes as reversible oxygen reduction/evolution electrocatalysts. Adv. Mater. **28**(19), 3777–3784 (2016)
77. Fei, H., Ye, R., Ye, G., Gong, Y., Peng, Z., Fan, X., Samuel, E.L., Ajayan, P.M., Tour, J.M.: Boron-and nitrogen-doped graphene quantum dots/graphene hybrid nanoplatelets as efficient electrocatalysts for oxygen reduction. ACS Nano **8**(10), 10837–10843 (2014)
78. Lopez-Cantu, D.O., González-González, R.B., Melchor-Martínez, E.M., Martínez, S.A.H., Araújo, R.G., Parra-Arroyo, L., Sosa-Hernández, J.E., Parra-Saldívar, R., Iqbal, H.M.: Enzyme-mimicking capacities of carbon-dots nanozymes: properties, catalytic mechanism, and applications—a review. Int. J. Biol. Macromol. **194**, 676–687 (2022)
79. Nirala, N.R., Abraham, S., Kumar, V., Bansal, A., Srivastava, A., Saxena, P.S.: Colorimetric detection of cholesterol based on highly efficient peroxidase mimetic activity of graphene quantum dots. Sens. Actuators B Chem. **218**, 42–50 (2015)
80. Prekodravac, J., Vasiljević, B., Marković, Z., Jovanović, D., Kleut, D., Špitalský, Z., Mičušik, M., Danko, M., Bajuk-Bogdanović, D., Todorović–Marković B,: Green and facile microwave assisted synthesis of (metal-free) N-doped carbon quantum dots for catalytic applications. Ceram. Int. **45**(14), 17006–17013 (2019)
81. Arul, V., Edison, T.N.J.I., Lee, Y.R., Sethuraman, M.G.: Biological and catalytic applications of green synthesized fluorescent N-doped carbon dots using Hylocereus undatus. J. Photochem. Photobiol. B. **168**, 142–148 (2017)
82. Arul, V., Sethuraman, M.G.: Facile green synthesis of fluorescent N-doped carbon dots from Actinidia deliciosa and their catalytic activity and cytotoxicity applications. Opt. Mater. **78**, 181–190 (2018). https://doi.org/10.1016/j.optmat.2018.02.029
83. Chandrasekaran, P., Arul, V., Sethuraman, M.G.: Ecofriendly synthesis of fluorescent nitrogen-doped carbon dots from coccinia grandis and its efficient catalytic application in the reduction of methyl orange. J. Fluoresc. **30**, 103–112 (2020)

Quantum Dots Based Nanocomposites

Kumaraswamy Gandla, Mahesh Namballa, P. S. Dhivya, K. Praveen Kumar, P. Kumar Nallasivan, Ravindran Muthukumarasamy, and Swathy Govindaswamy

Abstract This chapter explores the topic of quantum dots-based nanocomposites, providing a synopsis of their fabrication, characterisation, and varied practical applications. The unique qualities of quantum dots as building blocks are discussed, along with their relevance in fields such as photovoltaics, optoelectronics, biomedicine, and environmental monitoring. This chapter focuses on the cutting-edge techniques used to create them and highlights the importance of adapting their size, shape, and composition to a variety of uses. Characterization methods are presented in detail for understanding charge transfer, energy transfer, and interfacial interactions in nanocomposites. Examples of their use in solar cells, LEDs, targeted medication delivery, bio imaging, and environmental clean-up demonstrate their potential for far-reaching influence. Future directions for realising the full potential of quantum dots-based nanocomposites in advancing nanotechnology towards a sustainable and

K. Gandla (✉)
Department of Pharmaceutical Analysis, Chaitanya (Deemed to be University), Kishanpura, Hanamkonda, Telangana 506001, India
e-mail: drkumaraswamygandla@gmail.com

M. Namballa
Department of Pharmacy, School of Health Sciences, The Assam Kaziranga University, Jorhat, Assam, India

P. S. Dhivya
Department of Biomedical Science, Bharathidasan University, Tiruchirappalli 620024, India

K. P. Kumar
Department of Pharmaceutical Chemistry, School of Pharmaceutical Sciences, Govt. of NCT of Delhi, Delhi Pharmaceutical Sciences and Research University (DPSRU), Mehrauli, Pushp Vihar, New Delhi 110017, India

P. K. Nallasivan
Department of Pharmaceutical Chemistry, Karpagam Academy of Higher Education, Eachanari, Coimbatore, Tamilnadu 641021, India

R. Muthukumarasamy
Faculty of Pharmacy and Health Sciences, University Kuala Lumpur Royal College of Medicine, 30450 Jalan Greetown, IpohPerak, Malaysia

S. Govindaswamy
Department of Pharmacy Practice, KMCH College of Pharmacy, Coimbatore 641014, India

S. Thomas et al. (eds.), *Quantum Dots Based Nanocomposites*, Engineering Materials,
https://doi.org/10.1007/978-3-031-54779-9_4

efficient future are also discussed in this chapter, along with obstacles relating to safety, scalability, and technological hurdles.

Keywords Quantum dots · Bio imaging · Synthesis · Electrochemical · Characterization

1 Introduction

Quantum dots refer to semiconductor particles at the nanoscale level that exhibit distinct characteristics as a result of confinement effects in the quantum realm [1]. The optical behaviour of quantum dots is known to be dependent on their size, which is considered to be a remarkable property. The confinement of electrons and holes within the dot gives rise to the "quantum size effect" phenomenon [2]. The energy levels of charge carriers become quantized and discrete as the quantum dot size decreases. As a result, the bandgap, which governs the amount of energy needed for an electron to move from the valence band to the conduction band, exhibits an upward trend as the size of the dot decreases [3]. The tunability of the band gap in quantum dots results in a distinctive characteristic, namely the capacity to produce light emissions of varying hues contingent upon their dimensions. Through the manipulation of synthesis parameters, researchers possess the ability to precisely engineer quantum dots, resulting in the emission of light spanning the visible spectrum, as well as the ultraviolet and infrared regions [4–6]. The high photoluminescence quantum yield is a notable characteristic of quantum dots. Quantum dots exhibit high efficiency in emitting light when stimulated by photons, rendering them a desirable choice for utilisation in optoelectronic devices like light-emitting diodes (LEDs) and displays [7]. The narrow spectral linewidths exhibited by these entities render them appropriate for deployment in biological imaging and sensing applications, wherein the attainment of accurate and luminous fluorescence signals is of paramount importance [8]. Quantum dots exhibit remarkable photostability, indicating their ability to retain their optical characteristics for extended durations without notable deterioration. The maintenance of stability is of paramount significance in imaging and sensing applications, given the indispensable requirement of precise and uniform measurements [9].

The noteworthy characteristics of quantum dots include their electrical properties. Quantum dots are characterised by discrete energy levels owing to their diminutive size, which is commonly known as "Coulomb blockade" [10]. The aforementioned characteristic of the property facilitates the manipulation of individual electrons or holes confined within the dot, thereby rendering them suitable for utilisation in quantum computing and single-electron transistors [11]. Moreover, quantum dots manifest distinct electronic interactions commonly referred to as "quantum confinement effects". The aforementioned impacts lead to alterations in the electronic band configuration and the emergence of distinct energy states. Quantum dots' electronic characteristics can be manipulated through impurity doping, which can introduce supplementary energy levels and alter their conductivity [12]. The properties of quantum dots are also influenced by their size and shape [13]. The bandgap and consequent emission colour of a quantum dot are determined by its size. In addition, quantum dots exhibit diverse morphologies, including but not limited to spherical, rod-shaped, and tetrapodal, which can significantly impact their electronic and optical properties. The diverse range of shapes exhibited by quantum dots enables the precise modulation of their characteristics, thereby facilitating the development of tailored materials utilising quantum dots for targeted functionalities [14].

Quantum dots possess distinctive characteristics and adjustability, providing numerous benefits and serving a wide range of purposes in diverse domains. Nanoparticles composed of semiconducting materials have surfaced as a ground-breaking technology, offering a multitude of advantages spanning from cutting-edge imaging and sensing capabilities to energy acquisition and quantum computation [15]. Quantum dots possess a significant advantage in terms of their optical properties, which can be adjusted based on their size [16]. Through the manipulation of quantum dot dimensions and constituents, researchers are able to meticulously design the emission wavelength, spanning a wide range encompassing the visible, infrared, and ultraviolet regions. The capacity to adjust the properties of the material allows for the production of visual displays with vivid hues, imaging systems with high resolution, and light-emitting diodes (LEDs) that are energy-efficient. Quantum dot LEDs have exhibited enhanced colour fidelity, energy efficiency, and extended operational longevity in contrast to conventional display technologies [17]. Quantum dots demonstrate a notable benefit in terms of their elevated photoluminescence quantum yield [18]. Quantum dots exhibit high efficiency in light emission upon excitation by photons, rendering them suitable for utilisation in biological imaging and sensing applications. The precise and sensitive detection of biomarkers is facilitated by the bright and narrow-band fluorescence signals, which have contributed to advancements in diagnostics, drug discovery, and personalised medicine [19, 20]. Quantum dots have been found to be useful in solid-state lighting due to their high photoluminescence quantum yield, which results in improved energy efficiency and decreased power consumption. Quantum dots exhibit remarkable photostability, indicating their ability to sustain their optical characteristics for extended durations without substantial deterioration [21]. The maintenance of stability is of utmost importance in various applications, including imaging and sensing, wherein precision and uniformity of measurements hold significant value. The utilisation of imaging technologies based on quantum dots provides a means of visualising biological structures in real-time with high resolution [14]. Furthermore, the electronic properties of quantum dots can be altered by doping them with impurities, which presents an opportunity to create transistors and sensors with exceptional efficiency. The distinctive characteristics and adaptability of quantum dots have resulted in a diverse array of practical uses. Quantum dot-based photovoltaic cells present various benefits in the realm of solar energy harvesting, including but not limited to the ability to adjust bandgaps, the capacity for multiple exciton generation, and heightened absorption of sunlight. The aforementioned characteristics facilitate the development of solar cells that are economically viable and highly efficient, capable of converting a wider spectrum of solar radiation into electrical energy [13].

Moreover, quantum dots have exhibited potential in the field of catalysis, facilitating chemical reactions with enhanced efficiency and selectivity [22]. Nanoparticles possess the potential to be customised in terms of their dimensions and chemical makeup to augment their catalytic efficacy, rendering them highly advantageous in domains such as fuel synthesis, elimination of pollutants, and retention of hydrogen [3]. The extensive benefits and potential uses of quantum dots are continually expanding as scholars discover novel prospects. Quantum dots possess a range of advantageous properties, including tunable optical characteristics, high photoluminescence quantum yield, exceptional photostability, and controllable electrical features [5, 8, 23–31]. These attributes render quantum dots a highly valuable resource in various fields, such as optoelectronics, imaging, sensing, energy harvesting, and quantum computing. With ongoing advancements and research, quantum dots are poised to revolutionize various industries and contribute to the development of innovative technologies with significant societal impact [32–34].

The objective of this chapter is to furnish a thorough and all-encompassing exposition on nanocomposites that are based on quantum dots. The emphasis will be on their synthesis, properties, and wide-ranging applications. The distinctive optical and electronic characteristics of quantum dots, which are dependent on their size, in conjunction with the adaptable nature of nanocomposite materials, present promising prospects for progress in diverse domains such as optoelectronics, sensing, catalysis, and biomedical applications. The introductory section of this chapter will provide an overview of the essential principles and characteristics of quantum dots. Emphasis will be placed on their optical behaviour, which is dependent on their size, as well as their remarkable photoluminescence quantum yield and photostability. The forthcoming chapter will explore the various methods utilised for the synthesis of quantum dots, encompassing colloidal synthesis, epitaxial growth, and chemical vapour deposition. The chapter will analyse the benefits, difficulties, and recent advancements associated with each technique. The focus of this study will be on the techniques employed for regulating the dimensions, morphology, and constituent elements of quantum dots, thereby facilitating meticulous adjustment of their optical and electronic characteristics. The final section of the chapter will provide a succinct overview of the present obstacles and forthcoming prospects in the realm of nanocomposites that utilise quantum dots. The objective of this chapter is to furnish the readers with a thorough comprehension of the techniques employed in the synthesis of nanocomposites based on quantum dots, their characteristics, and their potential applications. This is intended to encourage additional research and advancements in this swiftly expanding domain.

2 Chemical Synthesis Techniques

Synthesising quantum dot polymer composites at the colloidal level is a flexible strategy for creating new materials with interesting characteristics and functions. Quantum dots (QDs) are synthesised by incorporating them into a polymer matrix via a colloidal technique, which allows for nanoscale QD dispersion. The resultant composite materials are promising for use in domains including optoelectronics, sensors, and energy devices due to their improved optical, electrical, and mechanical capabilities. QDs synthesis is the initial stage of colloidal synthesis. Nanocrystal semiconductors such as cadmium selenide (CdSe), lead sulphide (PbS), and indium phosphide (InP) are what make up quantum dots. Several chemical techniques, including hot-injection, sol-gel, and microwave-assisted synthesis, are available for synthesising these nanocrystals. The desired qualities and the specific QD material dictate the synthesis process chosen. QDs' optical and electrical characteristics may be fine-tuned by manipulating their size, shape, polarity, and surface chemistry during production. The QDs' surfaces are often changed with organic ligands after synthesis. These ligands are essential for improving the QDs' compatibility and stability in the polymer matrix. The QDs are more dispersed in the polymer matrix because they are surrounded by a protective layer. To improve the polymer's compatibility, ligands are often exchanged for their native counterparts using a ligand exchange method. The quantum dots with their functionalized surfaces are then suspended in an appropriate solvent. Uniform and steady dispersion needs precise regulation of process variables such temperature, solvent, and mixing methods. Sonication, magnetic stirring, or high-shear mixing are only some of the methods used to effectively disperse QDs and avoid their clumping together. The polymer matrix can be made independently of the QD colloidal solution. The composite material's intended use and desirable characteristics will dictate the polymer of choice. Polymers including poly(methyl methacrylate), polystyrene, and polyvinyl alcohol (PVA) are frequently employed. Solution casting, melt blending, and in-situ polymerization are all viable options for synthesising the polymer matrix. The composite material is formed by combining the QD colloidal solution with the polymer matrix by mixing procedures. Getting the appropriate characteristics requires evenly dispersing QDs throughout the polymer matrix. Various mixing methods are used to disperse QDs evenly throughout the polymer matrix, including stirring, ultrasonication, and high-shear mixing. The resultant material is malleable and may be shaped into films, fibres, or whatever else is needed. Films can be formed using a variety of methods, including spin coating, dip coating, and solvent evaporation. Processing parameters may be adjusted to customise the composite's qualities to a variety of uses by influencing its thickness, shape, and orientation. In order to improve the interfacial interactions between the quantum dots and the polymer matrix, post-treatment techniques may be used. The connection between QDs and the polymer matrix, for instance, can be increased by thermal annealing, leading to enhanced mechanical and electrical characteristics. The composite material's overall stability and durability can be increased by the use of crosslinking procedures. Quantum dot synthesis, surface functionalization, colloidal dispersion, polymer matrix preparation, composite creation, and post-treatment are the main processes in the colloidal synthesis of quantum dot polymer composites. To get the required characteristics and functions of the composite material, optimisation of synthesis parameters and processing conditions is essential. The resultant polymer-quantum-dot composites show great promise for use in several areas, including electronics, photonics, sensing, and energy conversion.

3 Sol-Gel Synthesis

The sol-gel method for synthesising quantum dot-polymer composites is a highly innovative and adaptable approach for producing nanocomposites that exhibit exceptional optical and material characteristics. The composites under consideration integrate the distinct attributes of quantum dots (QDs) and polymers, thereby presenting a diverse array of potential applications in various domains including optoelectronics, sensing, photonics, and biomedical engineering. This article aims to examine the fundamental principles underlying sol-gel synthesis and its utilisation in the fabrication of quantum dot-polymer composites.

The sol-gel synthesis technique is a solution-based approach that facilitates the precise development of nanoscale materials by means of the creation of colloidal solutions or sols. The aforementioned procedure entails the hydrolysis and condensation of metal alkoxides or metal chlorides, resulting in the formation of a cohesive network or gel. The sol-gel method offers various significant benefits in the context of QD-polymer composites. These advantages encompass the capacity to finely adjust the dimensions and structure of quantum dots, as well as their distribution throughout the polymer matrix. Consequently, this technique facilitates the augmentation of properties and functionalities in the resulting composites.

The initial stage in the sol-gel synthesis process of QD-polymer composites entails the preparation of the quantum dots. Quantum dots are semiconductor crystals at the nanoscale that demonstrate optoelectronic properties that vary with their size. Quantum dots possess the advantageous capability of being able to finely regulate their emission spectra over a broad spectrum of wavelengths by means of precise control over their size and composition. This characteristic renders them highly suitable for a diverse array of optoelectronic applications.

Frequently employed materials for quantum dots encompass cadmium selenide (CdSe), cadmium sulphide (CdS), lead sulphide (PbS), and indium phosphide (InP). The synthesis of quantum dots typically involves a distinct reaction, frequently conducted in the presence of surfactants, with the aim of impeding aggregation and regulating the size of the particles. After the synthesis process, the quantum dots are subsequently integrated into the sol-gel matrix. In the subsequent stage, the polymer precursor, typically consisting of a metal alkoxide or a silane-based compound, is combined with the quantum dot solution in order to generate a uniform blend. The aforementioned mixture is subsequently exposed to a sol-gel process, encompassing hydrolytic and condensation reactions. During the process of hydrolysis, the metal alkoxide undergoes a chemical reaction with water, resulting in the formation of metal hydroxide species. Following this, the metal hydroxides undergo condensation, resulting in the formation of a three-dimensional network that incorporates the quantum dots within the developing sol-gel matrix.

The sol-gel process provides precise manipulation of the dispersion of quantum dots within the polymer matrix, thereby guaranteeing homogeneous distribution and inhibiting the formation of agglomerates. The attainment of uniformity plays a critical role in the realisation of the desired characteristics and the improvement of the overall functionality of the composite material. Following the conclusion of the sol-gel process, it is customary to subject the resultant QD-polymer composite to annealing or curing procedures. These treatments serve the purpose of eliminating any residual solvent and enhancing the mechanical characteristics of the material. The annealing process additionally facilitates the promotion of cross-linking within the polymer matrix, leading to heightened levels of stability and durability. The sol-gel synthesis technique for producing QD-polymer composites presents numerous advantages in comparison to alternative conventional methods. One of the key benefits lies in the capacity to manipulate the dimensions and morphology of quantum dots, thereby exerting a direct influence on their optical characteristics. Moreover, the sol-gel process facilitates the integration of a diverse array of polymers, thereby facilitating the customization of the material's mechanical and chemical characteristics to meet specific application demands. The optical properties of the QD-polymer composites are distinct due to the quantum confinement effect of the quantum dots that are embedded within them. The aforementioned phenomenon results in the ability to adjust emission wavelengths and improve light absorption, rendering these composites exceedingly appealing for utilisation in various applications, including light-emitting diodes (LEDs), solar cells, and photodetectors. In addition, QD-polymer composites have exhibited promising capabilities in the field of biomedical applications, specifically in areas such as bioimaging and drug delivery. The adjustable emission properties of quantum dots can be employed for precise imaging of biological tissues and cells, while the polymer matrix offers a biocompatible medium for the encapsulation of drugs and regulated release. In summary, the utilisation of sol-gel synthesis for the production of QD-polymer composites exhibits significant potential in the development of versatile nanomaterials possessing customised characteristics. The precise manipulation of quantum dot dimensions and distribution within the polymer matrix presents novel opportunities for the advancement of optoelectronic devices, sensors, and biomedical tools. As further investigation and enhancement of this synthesis method are undertaken by researchers, it is anticipated that there will be a proliferation of compelling and influential applications of quantum dot-polymer composites in forthcoming times.

A report discussed the synthesis of Cu-In-S quantum dots (QDs) that exhibit a yellow emission and are encapsulated with double ZnS shells. These QDs demonstrate a remarkably high quantum yield of 92%. Additionally, the report covers the preparation of a composite plate consisting of QDs embedded in polymethylmethacrylate, which is free-standing. Furthermore, the application of this QD plate in the fabrication of white-light-emitting diodes (WLEDs) based on quantum dots is explored [35]. The synthesis of a quantum dot-silica monolith (QD-SM) (Fig. 1) material with exceptional luminescence and photostability was achieved through a two-step process involving the initial surface exchange of quantum dots (QDs) followed by base-catalyzed sol-gel condensation of silica [36]. The utilisation of a propylamine catalyst played a crucial role in preserving the initial luminescent properties of the quantum dots (QDs) within the sol-gel condensation process.

Kim presented findings on the production of a film composed of quantum dots (QDs) encapsulated in siloxane (referred to as QD-silox film). This film demonstrated

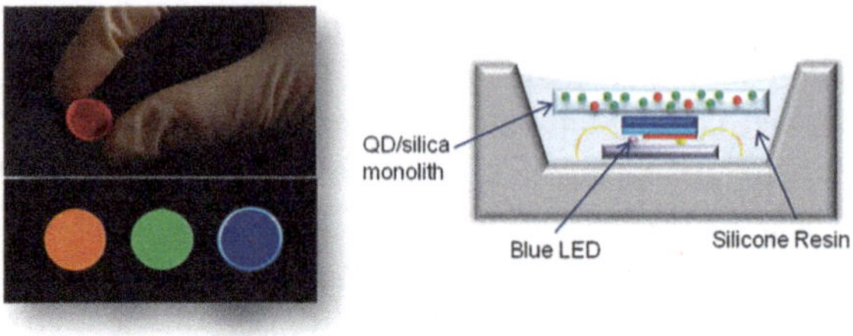

Fig. 1 Fabricated QDs-polymer composites and their graphical illustration of the fabrication. Reproduced with the permission from Ref. [36]© 2013 American chemical society

remarkable stability in terms of emission intensity, maintaining its performance for a duration exceeding one month, even under conditions of elevated temperature and humidity. The QD-silox films undergo solidification through a free radical addition reaction involving oligosiloxane resin and ligand molecules on the QDs [37]. The QD-oligosiloxane resin demonstrates optical clarity even after a storage period of 40 days, which is in stark contrast to alternative resins containing QDs that undergo turbidity and eventually precipitate. This study introduces a novel and straightforward technique for the synthesis of layer-by-layer (LbL) self-assembled films that integrate quantum dots (QDs) and exhibit strong photoluminescence (PL) across a wide spectral range, ranging from blue to red. The utilisation of sol-gel-derived glass layers, which are produced through the hydrolysis of 3-aminopropyltrimethoxysilane (APS) or 3-mercaptopropyltrimethoxysilane (MPS), has been employed as a means of connecting QD layers [38]. Characterization techniques such as absorption, photoluminescence spectroscopy, transmission electron microscopy, and atomic force microscopy were utilised to investigate the properties of the prepared films. The results of these analyses indicated that the QDs in the films exhibited a nearly close-packed coverage and were not found to be aggregated. The photoluminescence (PL) efficiencies of the QDs, specifically CdTe or ZnSe, both stabilised with thioglycolic acid, when dispersed in the films, were approximately 50% lower compared to the PL efficiencies of the initial colloidal solutions. However, the PL efficiencies of the dispersed QDs reached 24% before accounting for a refractive index correction. Phenyl-vinyl-oligosiloxane base resins, which incorporate quantum dots (QDs), were synthesised through an in situ condensation reaction involving sol-gel precursors and QDs. This reaction was carried out both with and without the addition of 3-(mercaptopropyl)trimethoxysilane [39]. Subsequently, the base resins underwent a curing process with the crosslinker resin via hydrosilylation in order to produce phenyl hybrimer (QDPH) resins that incorporate QDs. Our findings indicate that the QDPH resin incorporating a thiol group demonstrated superior transmittance and quantum efficiency compared to the resin without the thiol group. This can be attributed to the more uniform dispersion of quantum dots within the resin. The authors present a straightforward and efficient method for synthesising composites of QDs and polymers. In this study, we incorporated fluorescent semiconducting CdSe/ZnS QDs into various commercially accessible polymers through a straightforward single-step process [40]. It was observed that the introduction of an excessive amount of free octadecylamine (ODA) into the solution containing QDs and silica led to the formation of a gel. The duration of gelation was observed to be contingent upon the concentration of free ODA. In a similar vein, it was observed that the emission of QD-polymer composites was contingent upon the concentration of free ODA. In a separate study, researchers have successfully synthesised films composed of silica/siloxane embedded with quantum dots (QDs). This was achieved through the process of silica growth and siloxane encapsulation, utilising a straightforward sol-gel reaction [41]. The quantum dots (QDs) were immobilised within an amorphous silica matrix and chemically dispersed within an oligo-siloxane resin (Fig. 2), thereby effectively addressing the issue of aggregation during extended periods of storage. Furthermore, the absence of a solvent in the composite eliminated the requirement for a solidification process. The dual improved S-QD/siloxane film exhibited enhanced long-term performance in challenging conditions such as high temperature (120 °C), high humidity (85% relative humidity), exposure to ethanol, and application of a current bias (100 mA), without the need for an additional encapsulation layer. This performance surpasses that of the previously reported OA-QD/siloxane film.

4 Physical Synthesis Techniques

The utilisation of physical synthesis techniques in the production of QDs-polymer composites presents a distinctive and efficient strategy for the development of nanocomposites with customised properties suitable for a diverse array of applications. The aforementioned methodologies encompass the direct physical amalgamation or integration of quantum dots into polymer matrices, thereby facilitating meticulous regulation of the composite's configuration, dimensions, and quantum dot dispersion. This article aims to examine several prominent physical synthesis techniques employed for the production of quantum dot-polymer composites, and to elucidate their importance across diverse domains.

Solution mixing is a commonly employed and straightforward technique for the preparation of composites consisting of QDs and polymers. The present methodology involves the combination of pre-synthesized quantum dots and a polymer solution, resulting in the formation of a uniform mixture. Subsequently, the amalgamation is applied onto a substrate through either casting or spin-coating techniques, thereby resulting in the formation of the composite film. The utilisation of solution mixing facilitates the seamless integration of diverse quantum dots and polymers, thereby facilitating the customization of composite characteristics to cater to specific applications, including but not limited to optoelectronics, sensors, and bioimaging.

The process of in situ polymerization entails the concurrent formation of polymer chains and quantum dots within a shared reaction medium. The utilisation of this methodology facilitates the seamless incorporation of quantum dots into the polymer matrix during the polymerization procedure, leading to a significant enhancement in compatibility and enhanced dispersion of QDs within the composite material.

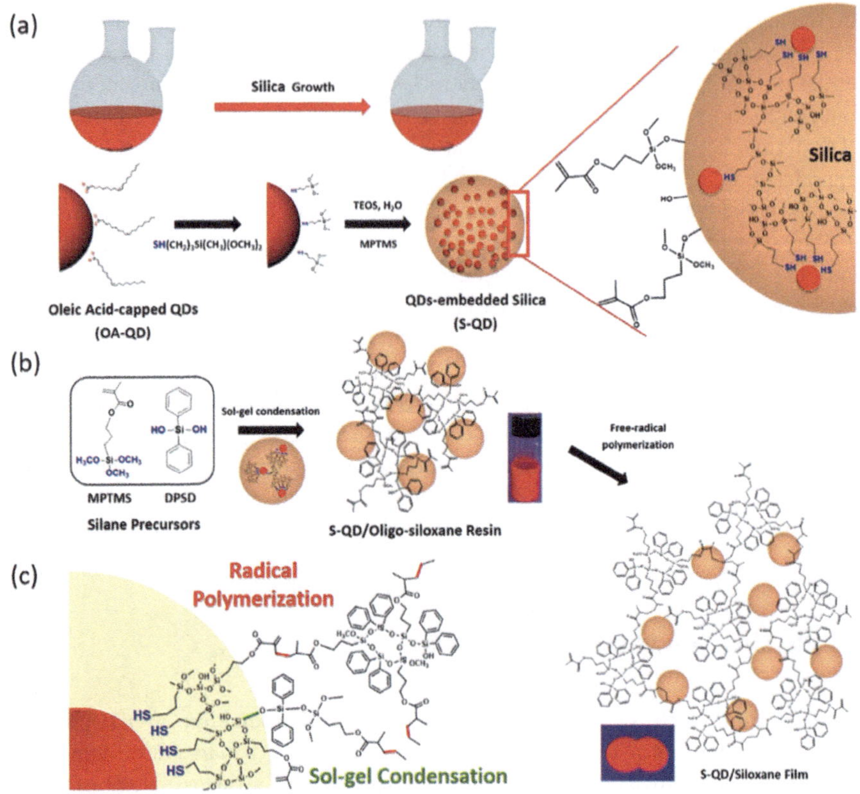

Fig. 2 Design concept of an S-QD/siloxane film. **a** Synthesis process and chemical structure of S-QD. **b** Sol–gel condensation reaction between MPTMS, DPSD, and S-QD. **c** Fabrication of an S-QD/siloxane film by UV-initiated free-radical polymerization between methacrylate groups of S-QD and oligo-siloxane resin. S-QD is chemically bonded to oligo-siloxane resin by radical polymerization and sol–gel condensation. Inset photographs are S-QD/oligo-siloxane resin and S-QD/siloxane film under 365 nm UV light. Reproduced with the permission from Ref. [41]© 2019 American chemical society

The utilisation of in situ polymerization proves to be highly advantageous in cases where monomers possess functional groups capable of interacting with the surface of quantum dots, thereby augmenting the overall stability and properties of the composite material.

Electrospinning is a highly adaptable and scalable methodology employed for the production of nanofibers or non-woven mats through the utilisation of polymer solutions. The production of QD-polymer nanofibers or mats can be achieved by introducing quantum dots into the polymer solution prior to the electrospinning process. The aforementioned technique presents a notable surface area-to-volume ratio and distinctive fibre morphology, rendering it appealing for implementation in various fields such as tissue engineering, sensors, and flexible electronics.

The technique of layer-by-layer (LBL) assembly is a method of controlled deposition that entails the sequential adsorption of species with opposite charges, such as polyelectrolytes and quantum dots, onto a substrate. The construction of multilayered quantum dot-polymer films is achieved by employing alternating deposition techniques, which enable meticulous manipulation of the quantum dot layer count and polymer thickness. The utilisation of LBL assembly presents the opportunity to manipulate optical properties and create customised multilayered structures with specific functionalities.

Emulsion polymerization is a method employed for the production of polymer particles characterised by well-regulated dimensions and minimal size variations. The formation of QD-polymer nanocomposite particles can be achieved by introducing quantum dots into the monomer droplets during the process of emulsion polymerization. The aforementioned particles have the potential to undergo additional processing in order to produce composite films or coatings that exhibit improved optical and mechanical characteristics.

The process of melt mixing entails the direct dispersion of quantum dots into a molten polymer through the utilisation of mechanical stirring or twin-screw extruders. This methodology is specifically well-suited for thermoplastic polymers and is frequently employed within the plastics industry. Melt mixing presents an economically viable and scalable approach for the fabrication of quantum dot-polymer composites, rendering it well-suited for the mass production of various items including LEDs, displays, and wearable electronics.

The physical synthesis techniques employed in the production of QDs-polymer composites offer numerous advantages when compared to alternative methods, such as chemical synthesis or in situ QD growth. Initially, the utilisation of pre-synthesized quantum dots in physical synthesis facilitates meticulous regulation of their dimensions, morphology, and chemical makeup. The implementation of this control mechanism plays a crucial role in the adjustment of the optical and electronic characteristics of the resultant composite material. Furthermore, physical synthesis methods possess inherent advantages such as simplicity, scalability, and the absence of intricate reaction conditions, rendering them appealing for industrial utilisation. The utilisation of physical synthesis techniques in the development of QDs-polymer composites presents a flexible and efficient approach for the fabrication of nanocomposite materials with customised characteristics. Various methods, such as solution mixing, in situ polymerization, electrospinning, and melt mixing, offer distinct advantages that are well-suited for specific applications in the fields of optoelectronics, photonics, sensors, and biomedical engineering. As scholars persist in investigating and enhancing these methodologies, the utilisation of QDs-polymer composites exhibits considerable potential in propelling numerous technologies and fostering innovation across a range of industries.

5 Vapor Deposition

Vapour deposition is a highly effective and extensively employed physical synthesis technique utilised for the production of QDs-polymer composites, offering meticulous regulation over the deposition process. The procedure entails the deposition of vaporised quantum dot materials onto a substrate, succeeded by the integration of the deposited quantum dots into a polymer matrix. The utilisation of vapour deposition presents a multitude of benefits in the fabrication process of quantum dot-polymer composites. This technique allows for improved regulation of quantum dot dimensions, concentration, and spatial arrangement within the polymer matrix. Consequently, this level of control facilitates the customization of material characteristics to suit diverse applications. The principles underlying vapour deposition are based on the process of vaporising quantum dot precursors. This vaporisation can be accomplished using different techniques, including thermal evaporation, sputter deposition, chemical vapour deposition (CVD), or electron beam deposition. Thermal evaporation involves the application of heat to the quantum dot precursors contained within a crucible, resulting in their evaporation and subsequent formation of a vapour plume. Subsequently, the quantum dots in a vaporised state are guided towards the substrate, where they undergo condensation and subsequently adhere as a thin film. The process of sputter deposition entails subjecting a target comprising quantum dot material to the impact of high-energy ions, resulting in the expulsion of atoms and their subsequent deposition onto the substrate. CVD, in contrast, employs reactive gases to produce quantum dot precursors in the vapour phase. These precursors subsequently undergo reactions with the substrate or other gases, resulting in the direct formation of quantum dots on the surface of the substrate. Electron beam deposition is a technique that employs an electron beam to vaporise the quantum dot material, thereby enabling meticulous regulation of the evaporation rate and facilitating the deposition of intricate nanostructures. The precise control over the growth of quantum dots is a significant advantage provided by vapour deposition. The manipulation of deposition conditions, including temperature, pressure, and precursor concentration, by researchers enables the attainment of a specific quantum dot size. This process leads to the uniform distribution of quantum dot sizes and the controlled density of quantum dots within the polymer matrix. The ability to finely adjust the optical, electronic, and mechanical characteristics of the composite material is crucial, rendering it exceedingly appealing for a wide range of applications in the fields of optoelectronics, photonics, and sensors. Furthermore, the utilisation of vapour deposition techniques results in enhanced dispersion and uniformity of quantum dots within the polymer matrix, surpassing certain solution-based methodologies. The direct deposition of quantum dots onto the substrate surface mitigates the likelihood of clustering and agglomeration, resulting in heightened stability and enhanced material performance. Achieving a uniform dispersion of quantum dots throughout the polymer matrix is of utmost importance in order to enhance the optical and electronic characteristics of the composite. Consequently, vapour deposition has emerged as a favoured technique for fabricating QD-polymer composites with superior quality. In addition, vapour deposition demonstrates compatibility with a diverse array of polymer materials, encompassing thermoplastics, thermosets, and elastomers. The inherent flexibility of composites enables researchers to customise their properties to suit particular applications, including but not limited to flexible electronics, light-emitting devices, and nanoscale sensors. The capacity to fabricate composite materials using a diverse range of polymers facilitates the advancement of multifunctional materials that are well-suited for a wide array of industry requirements. It can be stated that vapour deposition is an extensively adaptable and manageable physical synthesis method employed for the production of quantum dot-polymer composites. Vapour deposition facilitates the customization of material properties for diverse applications in optoelectronics, photonics, sensors, and related fields by providing precise manipulation of quantum dot size, density, and distribution within the polymer matrix. This level of control over nanocomposite properties is crucial for achieving optimal performance and functionality. The ongoing investigation and enhancement of this methodology possess significant potential for propelling nanotechnology forward and fostering innovation in diverse sectors. New heterogeneous CVD polymerization techniques are based on conventional wet-chemical chain- and step-growth mechanisms. Because of their similarities to inorganic CVD techniques, hybrid device fabrication is simplified. CVD polymers unite the fields of microfabrication with the building of complex systems of chemical, biological, and nanoparticles [42]. Covalent grafting can create robust interfaces, allowing for high-resolution (60 nm) patterning on flexible substrates. CVD polymerization is suitable for thermally sensitive substrates like paper, textiles, and plastics due to its use of low-energy input to drive selective chemistry, minimal vacuum, and room-temperature substrates. The potential for lowering environmental, health, and safety implications associated with solvents makes CVD approaches attractive for the production of insoluble and infusible films made from materials including fluoropolymers, electrically conductive polymers, and controllably crosslinked networks. High biocompatibility and low cellular toxicity composite material consisting of polymer coated quantum dots (QDs) and carbon nanotubes was synthesised, as reported by Balsak et al. X-ray photoelectron spectroscopy, laser scanning confocal microscopy, transmission electron microscopy, and scanning electron microscopy were used to characterise the MWCNT-QD-(-poly(glycidyl methacrylate)) (pGMA) hybrids that were synthesised [43].

6 Electrochemical Synthesis

Due to its widespread use in fields as varied as optoelectronics, sensing, bioimaging, and energy storage, the production of quantum dots (QDs) and their integration into polymer matrices has attracted a lot of attention in recent years. Electrochemical synthesis is a versatile and ecologically friendly process for creating QDs-polymer composites with perfect control over the size, shape, and composition of quantum dots, making it a standout among the numerous synthesis methods. This article delves into the benefits and possible uses of the electrochemical synthesis process for QDs-polymer composites. The production of quantum dots in a polymer matrix can be electrochemically synthesised by driving redox processes with an electric current. An electrolytic cell with a working electrode, a counter electrode, and an electrolyte solution is commonly used for the procedure. In most cases, indium tin oxide (ITO) is used for the working electrode, while platinum or graphite is used for the counter electrode. The QDs' building blocks—metal ions and ligands—are already present in the electrolyte solution. There are two primary methods for the electrochemical synthesis of QDs-polymer composites: in situ synthesis and ex situ synthesis. The polymer matrix is directly involved in the electrochemical reaction during in situ synthesis, resulting to the creation of QDs and polymer at the same time. Benefits of this method include increased homogeneity, precise regulation of QD sizing and distribution, and robust QD–polymer matrix interactions. In order to synthesise anything where it will ultimately be employed, an electrochemical reaction is performed with a monomer that has functional groups capable of interacting with the metal ions. The QDs within the expanding polymer chain are formed when metal ions are produced during the redox process and coordinate with the monomers. Separate production of the quantum dots is followed by their incorporation into a pre-formed polymer matrix using electrochemical methods in ex situ electrochemical synthesis. In this method, the desired quantum dot qualities can be synthesised in advance of their incorporation into the polymer, providing greater control over the final product. Applying a potential to the polymer matrix during the ex situ synthesis helps the QDs migrate from the solution to the polymer surface, where they can be incorporated. There are a number of benefits to using electrochemical synthesis to create QDs-polymer composites instead of more traditional chemical approaches. To begin, the procedure is straightforward, economical, and extensible, thus it may be used for mass manufacture of QDs-polymer composites. Second, the size, shape, and composition of quantum dots may be precisely controlled through electrochemical production, providing greater scope for tuning their properties. The ability to modify the QDs' properties is essential for enhancing their functionality in targeted settings. Thirdly, the environmental impact of electrochemical synthesis is low because it may be carried out in benign settings without the use of harmful reagents. Several different fields have benefited from the electrochemical synthesis of QDs-polymer composites. These composites are used to create high-efficiency LEDs, PV cells, and lasers in the field of optoelectronics. Quantum dots (QDs) are attractive candidates for next-generation electronic devices due to improvements in charge transport characteristics and stability achieved through their incorporation into polymer matrices. In addition, QDs-polymer composites have shown promising results as sensing materials thanks to their sensitivity to a wide range of analytes and enhanced selectivity made possible by polymer functionalization. QDs-polymer composites have great brightness, photostability, and biocompatibility, which paves the way for accurate imaging and tailored drug administration in bioimaging and biomedical applications. These composites have also been investigated for potential use in energy storage devices like supercapacitors and rechargeable batteries due to their high surface area and electrochemical characteristics. QDs-polymer composites with customised characteristics can be easily and effectively synthesised using electrochemical synthesis. Electrochemical synthesis of quantum dots is appealing because of its environmental friendliness and the fact that its size, shape, and composition can be controlled. It is anticipated that the expanding use of QDs-polymer composites synthesised using electrochemical methods will further drive innovation in the disciplines of optoelectronics, sensing, bioimaging, and energy storage as this area of study develops.

By simultaneously producing the corresponding precursors using cadmium sacrificial anode electrochemical methods, a unique, flexible, and clean colloidal processing strategy was devised for synthesising CdX (X = Te, Se, or S) quantum dots (QDs) [44]. The 3-mercaptopropionic acid (MPA)-stabilized CdX QDs were synthesised in a room-temperature, one-pot electrochemical technique utilising an aqueous medium in around 16 min. Very stable colloidal quantum dots with uniform narrow-size distributions and an average nanoparticle diameter of around 3.0 nm were created, as evidenced by UV–vis, PL, FTIR, TEM-EDS, XRD, DLS, XPS, and Zeta potential characterizations of the CdX quantum dots conjugates. Poly(vinyl alcohol) (PVA) was used as the matrix polymer and coal-derived graphene quantum dots (GQDs) were used to create luminescent polymer composite materials, which were cast from aqueous solutions (Fig. 3) [45]. Fluorescent characteristics are transferred to the polymer matrix from the coal-derived GQDs, and solid state fluorescence is observed in the resulting composite films. The PVA/GQD nanocomposites' optical, thermal, and fluorescent characteristics have all been investigated. At GQD concentrations between 1 and 5 wt%, the composite films display high optical transparency (78–91%) and good dispersion of the nanoparticles. CdSe quantum dots (QDs) were placed into a polymer matrix to create a film with electrochromic, electrofluorescent, and light-induced colouring properties. The film's excellent electrochromic features include low redox potential, strong optical contrast, and fast response time; these benefits are the result of a combination of those presented by CdSe QDs and those presented by conjugated electrochromic polymers [46]. Since the composite film may change between a non-fluorescent and a fluorescent state, with PL intensity greater than twice that of a pure polymer film, higher performance levels are attained. MoS2 is a promising luminophore, and its insertion into PS nanostructures may alter their luminescence properties [47]. To improve the electroluminescence property of PS nanostructures, the current study adopts electrochemical anodization of nanostructured porous silicon, followed by the integration of MoS2 quantum dots (QDs) by spin coating.

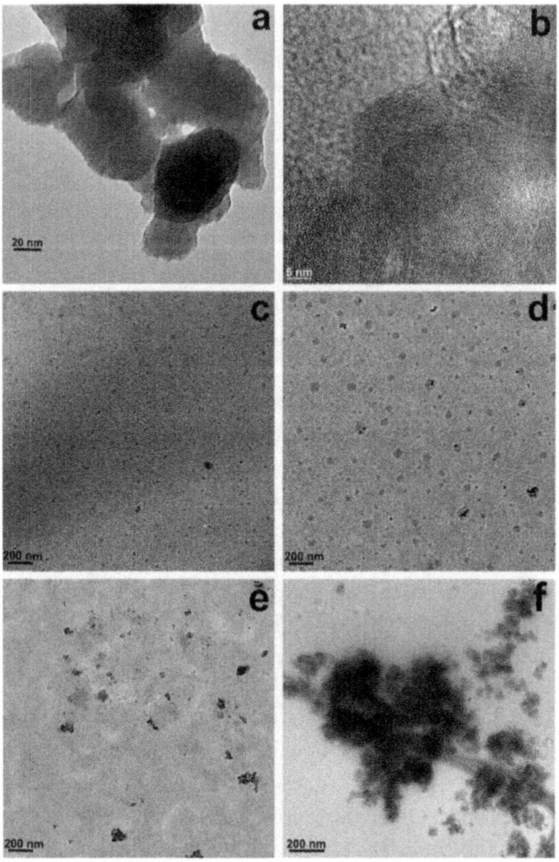

Fig. 3 TEM images of the **a**, **b** GQDs and PVA/GQD composites containing **c** 1 wt %, **d** 3 wt %, **e** 5 wt %, and **f** 10 wt % GQDs. Reproduced with the permission from Ref. [45]© 2015 American chemical society

7 Characterization Techniques for Quantum Dots-Polymer Composites

Nanometer-sized semiconductor quantum dots (QDs) have attracted significant interest for a variety of uses due to their unusual electrical and optical properties. These QDs, when mixed with polymers, create QDs-polymer composites that are superior to both of their constituent parts. The potential of these composites cannot be realised without first adequately characterising them [19]. This article delves into some of the most important characterisation methods used to investigate QDs-polymer composites and learn more about their properties and behaviour.

Spectroscopic methods are indispensable for studying QDs-polymer composites [30, 48, 49]. The optical characteristics of these composites are often analysed using

UV-Vis spectroscopy. It helps scientists evaluate the quantum dots' bandgap, size distribution, and photoluminescence quantum yield by providing data about their absorption and emission spectra [6, 33]. The effect of the polymer on QD luminescence can be deduced by the use of fluorescence spectroscopy, which also aids in analysing the emission behaviour of QDs within the polymer matrix. Spectroscopic methods provide essential information for deciphering the composite's optical characteristics and energy transfer pathways.

Another excellent characterization tool is transmission electron microscopy (TEM), which may be utilised to see the nanoscale shape and distribution of quantum dots within a polymer matrix [50]. By using TEM, scientists may get clear visuals of the QDs and evaluate how they are dispersed throughout the polymer matrix. This method is useful for learning about the size, shape, and distribution of QDs, as well as their propensity to aggregate or cluster [51]. Understanding the charge transfer mechanisms and the overall composite structure is made much easier with the assistance of TEM's evaluation of the interface between the QDs and the polymer.

Quantum dot and polymer matrix crystal structures can be investigated with the use of X-ray Diffraction (XRD) [52]. Quantum dot (QD) and polymer (poly) crystalline phases, lattice parameters, and crystallite size can all be determined from the XRD patterns. The effect of the polymer on the crystal structure can be seen in the diffraction pattern changes, which can be used to improve the composite material [27].

Quantum dot–polymer chemical interactions are typically studied with Fourier–Transform Infrared Spectroscopy (FTIR). Understanding the intermolecular interactions, bonding, and potential alterations to the polymer's structure due to the presence of QDs is made possible with the help of FTIR spectra, which provide information about distinctive functional groups and bond vibrations.

Quantum dots in a polymer matrix can be characterised with great precision using Dynamic Light Scattering (DLS). Insight into particle size distribution, aggregation, and surface charge can be gleaned from DLS measurements of the hydrodynamic size and stability of QDs. This information is crucial for determining the longevity of the QDs in the polymer matrix and for finding the most effective method of dispersing them.

QDs-polymer composites' heat stability and breakdown behaviour are investigated using thermogravimetric analysis (TGA). Researchers can evaluate the impact of the polymer on the thermal characteristics of QDs by analysing the weight loss as a function of temperature, thereby identifying the thermal degradation temperature of the composite. This knowledge is crucial for the development and implementation of these composites in high-temperature settings.

To fully grasp the viscoelastic behaviour of QDs-polymer nanocomposites, rheological investigations are necessary. The ability to optimise processing conditions and forecast the performance of the composite in real-world applications is greatly aided by the information provided by rheological approaches, which shed light on the composite's flow and deformation behaviour. For QDs-polymer composites used in electronics, it is crucial to characterise their electrical and optoelectronic properties. Current-voltage (I-V) measurements, impedance spectroscopy, and charge carrier mobility measurements are all useful for elucidating the composite's electrical conductivity and charge transport behaviour. When used to devices like solar cells and transistors, these numbers reveal essential information for improving the composites' performance. The ability to accurately characterise QDs-polymer composites is essential for gaining insight into their characteristics and behaviour. Optimising these composites for different uses requires data from a wide range of measures, including spectroscopy, microscopy, thermal, rheological, and electrical. The continued development of these characterisation methods will aid in the progress of QDs-polymer composites, thereby releasing their potential for application in forward-thinking technologies and research.

8 Conclusion

This book chapter has shed light on the synthesis, characterisation, and wide variety of applications of quantum dots-based nanocomposites; a fascinating topic indeed. Quantum dots, with their novel features, have proven to be an invaluable tool in the field of nanotechnology, allowing for significant progress in many other areas. Controlled synthesis of quantum dots, allowing for customization of their size, shape, and composition, has been a recurrent theme throughout this chapter, demonstrating how this factor greatly influences their performance and potential applications. Solar cells have improved light absorption and charge transport efficiency thanks to the inclusion of quantum dots into nanocomposites, leading to remarkable advancements in photovoltaics.

The incorporation of quantum dots in optoelectronic devices has also ushered in a new era of energy-efficient and visually stunning display and lighting systems. Nanocomposites based on quantum dots have shown promising results in a number of biomedical applications, including targeted drug delivery, bioimaging, and theranostics. This chapter has also highlighted the significance of nanocomposites based on quantum dots in environmental applications, highlighting their potential as novel tools for pollution monitoring and environmentally friendly cleanup procedures. One example of their participation in developing eco-friendly solutions is the use of solar energy for environmental remediation. While this chapter has shown the tremendous promise of quantum dot-based nanocomposites, it has also brought to light some difficulties that need to be addressed. There is a pressing need for more research and development in the areas of safety, scalability, and the optimisation of fabrication procedures.

Looking forward, the promise of nanocomposites based on quantum dots is undeniable. Researchers, scientists, and engineers interested in exploring the possibilities of this cutting-edge technology will find this chapter to be an invaluable resource. We can unlock even more useful applications and pave the path for quantum dots-based nanocomposites to become vital instruments in designing a sustainable, efficient, and linked world by tackling problems and expanding on the knowledge offered here.

The ultimate goal of the development of nanocomposites based on quantum dots is to advance nanotechnology into uncharted territory and bring about a better, more promising future.

References

1. García de Arquer, F.P., et al.: Semiconductor quantum dots: technological progress and future challenges. Science **373**(6555), eaaz8541 (2021)
2. Wagner, A.M., et al.: Quantum dots in biomedical applications. Acta Biomater. **94**, 44–63 (2019)
3. Lim, S.Y., Shen, W., Gao, Z.: Carbon quantum dots and their applications. Chem. Soc. Rev. **44**(1), 362–381 (2015)
4. Michler, P.: Quantum Dots for Quantum Information Technologies, vol. 237. Springer (2017)
5. Ganguly, S., et al.: Microwave-synthesized polysaccharide-derived carbon dots as therapeutic cargoes and toughening agents for elastomeric gels. ACS Appl. Mater. Interfaces **12**(46), 51940–51951 (2020)
6. Saravanan, A., et al.: Applications of N-doped carbon dots as antimicrobial agents, antibiotic carriers, and selective fluorescent probes for nitro explosives. ACS Appl. Bio Mater. **3**(11), 8023–8031 (2020)
7. Duan, L., et al.: Quantum dots for photovoltaics: a tale of two materials. Adv. Energy Mater. **11**(20), 2100354 (2021)
8. Das, P., et al.: Carbon dots for heavy-metal sensing, pH-sensitive cargo delivery, and antibacterial applications. ACS Appl. Nano Mater. **3**(12), 11777–11790 (2020)
9. Lv, W., et al.: Improving the stability of metal halide perovskite quantum dots by encapsulation. Adv. Mater. **31**(28), 1900682 (2019)
10. Brotons-Gisbert, M., et al.: Coulomb blockade in an atomically thin quantum dot coupled to a tunable Fermi reservoir. Nat. Nanotechnol. **14**(5), 442–446 (2019)
11. Liu, Y., et al.: Advances in carbon dots: from the perspective of traditional quantum dots. Mater. Chem. Front. **4**(6), 1586–1613 (2020)
12. Zhu, S., et al.: The photoluminescence mechanism in carbon dots (graphene quantum dots, carbon nanodots, and polymer dots): current state and future perspective. Nano Res. **8**, 355–381 (2015)
13. Devi, P., Saini, S., Kim, K.-H.: The advanced role of carbon quantum dots in nanomedical applications. Biosens. Bioelectron. **141**, 111158 (2019)
14. Benítez-Martínez, S., Valcárcel, M.: Graphene quantum dots in analytical science. TrAC, Trends Anal. Chem. **72**, 93–113 (2015)
15. Albash, T., Lidar, D.A.: Adiabatic quantum computation. Rev. Mod. Phys. **90**(1), 015002 (2018)
16. National Academies of Sciences, E. and Medicine. Quantum computing: Progress and prospects (2019)
17. Molaei, M.J.: Principles, mechanisms, and application of carbon quantum dots in sensors: a review. Anal. Meth. **12**(10), 1266–1287 (2020)
18. Das, P., et al.: Heteroatom doped blue luminescent carbon dots as a nano-probe for targeted cell labeling and anticancer drug delivery vehicle. Mater. Chem. Phys. **237**, 121860 (2019)
19. Ganguly, S., et al.: Advancement in science and technology of carbon dot-polymer hybrid composites: a review. Funct. Compos. Struct. **1**(2), 022001 (2019)
20. Ganguly, S., et al.: Microwave assisted green synthesis of Zwitterionic photolumenescent N-doped carbon dots: An efficient 'on-off' chemosensor for tracer Cr (+ 6) considering the inner filter effect and nano drug-delivery vector. Colloids Surf. A **579**, 123604 (2019)
21. Maxwell, T., et al.: Quantum dots. In: Nanoparticles for Biomedical Applications, pp. 243–265. Elsevier (2020)

22. Yuan, Y., et al.: Quantum dot photocatalysts for organic transformations. J. Phys. Chem. Lett. **12**(30), 7180–7193 (2021)
23. Sun, H., et al.: Deciphering a nanocarbon-based artificial peroxidase: chemical identification of the catalytically active and substrate-binding sites on graphene quantum dots. Angew. Chem. Int. Ed. **54**(24), 7176–7180 (2015)
24. Das, P., et al.: Waste-derived sustainable fluorescent nanocarbon-coated breathable functional fabric for antioxidant and antimicrobial applications. ACS Appl. Mater. Interfaces (2023)
25. Rajabzadeh, A.R., et al.: Magnetic Quantum Dots for Bioimaging. CRC Press (2023)
26. Das, P., et al.: Naturally derived carbon dots in situ confined self-healing and breathable hydrogel monolith for anomalous diffusion-driven phytomedicine release. ACS Appl. Bio Mater. **5**(12), 5617–5633 (2022)
27. Das, P., et al.: Carbon dot biopolymer-based flexible functional films for antioxidant and food monitoring applications. ACS Appl. Polym. Mater. **4**(12), 9323–9340 (2022)
28. Das, P., et al.: Tailor made magnetic nanolights: fabrication to cancer theranostics applications. Nanoscale Adv. **3**(24), 6762–6796 (2021)
29. Das, P., et al.: Carbon-dots-initiated photopolymerization: an in situ synthetic approach for MXene/poly (norepinephrine)/copper hybrid and its application for mitigating water pollution. ACS Appl. Mater. Interfaces **13**(26), 31038–31050 (2021)
30. Das, P., et al.: Immobilization of heteroatom-doped carbon dots onto nonpolar plastics for antifogging, antioxidant, and food monitoring applications. Langmuir **37**(11), 3508–3520 (2021)
31. Das, P., et al.: Biocompatible carbon dots derived from κ-carrageenan and phenyl boronic acid for dual modality sensing platform of sugar and its anti-diabetic drug release behavior. Int. J. Biol. Macromol. **132**, 316–329 (2019)
32. Gao, J., et al.: Advances, challenges and promises of carbon dots. Inorgan. Chem. Front. **4**(12), 1963–1986 (2017)
33. Das, P., et al.: Graphene based emergent nanolights: a short review on the synthesis, properties and application. Res. Chem. Intermed. **45**, 3823–3853 (2019)
34. Das, P., et al.: Dual doped biocompatible multicolor luminescent carbon dots for bio labeling, UV-active marker and fluorescent polymer composite. Luminescence **33**(6), 1136–1145 (2018)
35. Jang, E.-P., et al.: Preparation of a photo-degradation-resistant quantum dot–polymer composite plate for use in the fabrication of a high-stability white-light-emitting diode. Nanotechnology **24**(4), 045607 (2013)
36. Jun, S., Lee, J., Jang, E.: Highly luminescent and photostable quantum dot–silica monolith and its application to light-emitting diodes. ACS Nano **7**(2), 1472–1477 (2013)
37. Kim, H.Y., et al.: Quantum dot/siloxane composite film exceptionally stable against oxidation under heat and moisture. J. Am. Chem. Soc. **138**(50), 16478–16485 (2016)
38. Yang, P., Li, C., Murase, N.: Highly photoluminescent multilayer QD—glass films prepared by LbL self-assembly. Langmuir **21**(19), 8913–8917 (2005)
39. Yoon, C., et al.: Highly luminescent and stable white light-emitting diodes created by direct incorporation of Cd-free quantum dots in silicone resins using the thiol group. J. Mater. Chem. C **3**(26), 6908–6915 (2015)
40. Weaver, J., et al.: Synthesis and characterization of quantum dot–polymer composites. J. Mater. Chem. **19**(20), 3198–3206 (2009)
41. Kim, Y.H., et al.: Two-step-enhanced stability of quantum dots via silica and siloxane encapsulation for the long-term operation of light-emitting diodes. ACS Appl. Mater. Interfaces **11**(25), 22801–22808 (2019)
42. Alf, M.E., et al.: Chemical vapor deposition of conformal, functional, and responsive polymer films. Adv. Mater. **22**(18), 1993–2027 (2010)
43. Baslak, C., et al.: Biocompatible multi-walled carbon nanotube–CdTe quantum dot–polymer hybrids for medical applications. J. Lumin. **160**, 9–15 (2015)
44. Freitas, D.V., et al.: Toward greener electrochemical synthesis of composition-tunable luminescent CdX-based (X= Te, Se, S) quantum dots for bioimaging cancer cells. Sens. Actuators, B Chem. **250**, 233–243 (2017)

45. Kovalchuk, A., et al.: Luminescent polymer composite films containing coal-derived graphene quantum dots. ACS Appl. Mater. Interfaces **7**(47), 26063–26068 (2015)
46. Liu, J., et al.: Trifunctional CdSe quantum dots-polymer composite film with electrochromic, electrofluorescent and light-induced coloration effects. Sol. Energy Mater. Sol. Cells **177**, 82–88 (2018)
47. Shrivastava, M., et al.: Electrochemical synthesis of MoS2 quantum dots embedded nanostructured porous silicon with enhanced electroluminescence property. Opt. Mater. **73**, 763–771 (2017)
48. Ganguly, S.: 10 Quantum Dots–Rubber Composites. Quantum Dots and Polymer Nanocomposites: Synthesis, Chemistry, and Applications, p. 189 (2022)
49. Ganguly, S., Margel, S.: Fluorescent Quantum Dots-Based Hydrogels: Synthesis, Fabrication and Multimodal Biosensing. Talanta Open, p. 100243 (2023)
50. Koc, M.A., et al.: Characterizing photon reabsorption in quantum dot-polymer composites for use as displacement sensors. ACS Nano **11**(2), 2075–2084 (2017)
51. Ge, Y., et al.: Beta-lead oxide quantum dot (β-PbO QD)/polystyrene (PS) composite films and their applications in ultrafast photonics. Nanoscale **11**(14), 6828–6837 (2019)
52. Ganguly, S.: 13 challenges prospects of and magnetic future quantum dots. Magnetic Quantum Dots for Bioimaging, p. 269 (2023)

Properties of Quantum Dots Based Nanocomposites

Ashish Kumar Bhui, Sanvedna Shukla, Satish Sen, and Amar Dey

Abstract One of the first nanotechnological materials to be combined with nanocomposites as quantum dots (QDs), whose potential for use in a variety of commercial pharmaceutical and medical goods has been greatly anticipated. Within a wide range of swift developing field of material science and technology, these are some of the crucial scientific developments of the current era. Due to their exceptional physicochemical features, quantum dot-based nanocomposites have received a lot of attention recently. Furthermore, at this extent, not only the quantum confinement but also the surface effects become crucial, making it possible to manipulate the quantum dot's surface or diameter so that its various characteristics can be easily explored. Quantum confinement has an impact on the dot's photon emission and absorption. These features have been used in a variety of analytical tests, including electrochemical sensing platforms, because of their exceptional properties paired with their simplicity in synthesis and combination with nanocomposite materials.

Keywords Quantum dots · Nano composites · Quantum confinement · Electrochemical sensing platform

Abbreviations

QDs	Quantum dots
DNA	Deoxyribonucleic acid

A. K. Bhui (✉)
Govt. Shahid Kaushal Yadav College, Gunderdehi, Balod, Chhattisgarh, India
e-mail: bhuiashishkumar@gmail.com

S. Shukla
Govt. Digvijay College, Rajnandgaon, Chhattisgarh, India

S. Sen
Govt. VYT College, Durg, Chhattisgarh, India

A. Dey
Bhilai Institute of Technology, Durg, Chhattisgarh, India

S. Thomas et al. (eds.), *Quantum Dots Based Nanocomposites*, Engineering Materials,
https://doi.org/10.1007/978-3-031-54779-9_5

GQDs	Graphene Quantum dots
CT	Charge transfer
TDDFT	Time dependent density functional theory
TPP GQD	Tetraphenylporphyrin Graphne Quantum dots
DSSC	Dye-sensitized solar cell
ESP	Surface's electrostatic potential
NiTPP	Ni(II)tetraphenylporphyrin
E/G	Edge type binding
P/G	Graphene basal plane
NGs	Nanographenes
PCM	Polarizable continuum model
HBC	Hexa-peri-hexabenzocoronene
GNR	Graphene nanoribbon
PBE	Perdew-Burke-Ernzerhof exchange-correlation energy functional theory
SCC-DFTB	Self-consistent charge density-functional–based tight-binding theory
DOSs	Densities of states
CQDs	Carbon quantum dots
PL	Photoluminescence
LOD	Limit of detection
ECL	Electrochemiluminescence
L-CQDs	L-Tryptophan capped quantum dots
CuNCs-CNQDs	Copper nanoclusters@nitrogen-doped carbon dots
ZnO QDs	Zinc oxide quantum dots
VOCs	Volatile organic compounds

1 Introduction

Recent advances in nanomaterials have shown substantial relevance in various areas including optics, informatics, environmental health, biomedicine, electronics energy, and pharmaceuticals [1, 2]. Due to their potential to overcome cell gaps and enter cells, nanoparticles of a size equivalent to cell organelles or lipid biomolecules, deoxyribonucleic acid (DNA), or proteins have garnered researchers' attention. Nanoparticle application has been widely extended in biosensing, tumor cell imaging, and cancer treatment after being coupled with biomolecular affinity ligands prime among them are gold, silver, or iron oxide nanoparticles [3–6]. The semiconductor nanocrystal quantum dot (QD) is extensively probed [7–9]. The Review chapter exclusively deals with various quantum dots in terms of preparation and properties used in the modern context.

2 Quantum Dots and Nanocomposites

Through quantum confinement effect semiconductor nanoparticles as quantum dots (QDs) encompass unique optical and electrical properties. For several sectors, including nanotechnology, electronics, and photonics, they have sparked a great deal of interest. Nanocomposites, which are substances made of a mixture of many types of nanoparticles distributed in a matrix material, can include quantum dots. Quantum dot-containing nanocomposites have several benefits and possible uses. Apart from having a size in the ultra range, wavelength at broad excitation level, superior photostability and optimised size-tuneable-fluorescence, they have been studied as a substitute or supplement to organic dyes [10]. Newly developed QDs possessing properties of near-infrared emission can monitor animals in vivo without radioactive reagents [11]. Pristine QD's nanotoxicity and restricted functionalities limit its biological potential. QD's biocompatibility and roles beyond fluorescent labeling may be improved by combining it with nanomaterials and proteins. QD might be biocompatible and have other uses by combining it with nanomaterials and proteins. Heavy metal leaching can be prevented by coating QD with biocompatible materials without disturbing its optical properties. Nanocomposites that incorporate quantum dots (QDs) are materials that merge the distinctive characteristics of quantum dots with those of matrix material. The term "quantum dots" refers to nanocrystals made of semiconducting materials that exhibit unique opto-electronic features based on size. are The phenomenon of quantum confinement is often considered responsible for all these traits which in turn are based on emission features that are size adjustable, better photoluminescent quantum yield and broad absorption spectra. The incorporation of quantum dots into a matrix material, such as polymers, ceramics, or metals, results in the formation of nanocomposites. These nanocomposites can display unique or improved characteristics when compared to the individual components. Nanocomposites that utilize quantum dots possess several noteworthy characteristics and practical uses. Quantum dots are distinguished by their optical features, which include their high luminescence efficiency and controllable color-changing light emission. These emission features can be harnessed in various applications of optoelectronics, display technologies, and light-emitting devices when integrated into a matrix. The electrical behavior of quantum dots is distinct due to quantum confinement effect. Nanocomposites can be utilized in various applications including energy storage, electronics and sensing due to their ability to display improved electrical conductivity or customized electronic properties when integrated into a matrix. Using quantum dot-containing nanocomposites might be one way to boost light absorption and charge separation in solar cells. The use of quantum dots with a tunable bandgap is advantageous in capturing a broader range of solar radiation. The matrix material serves to provide structural support and facilitate electron transport pathways. The use of quantum dot-based nanocomposites as catalysts is effective due to considerable surface area, soaring surface reactivity and capacity to regulate charge transfer processes. These are used in various fields including photocatalysis, water splitting, and chemical synthesis. Nanoparticles possess distinctive optical characteristics and surface chemistry that render them highly desirable for variety of functions such as fluorescent labeling, molecular sensing, and targeted therapy in biological systems. Thus nanocomposites incorporating quantum dots (QDs) signify a category of hybrid materials that amalgamate the distinctive characteristics of both QDs and a matrix material. Quantum dots, sometimes referred to as semiconductor nanocrystals, display varying optical and electrical characteristics based on their size. These properties are frequently attributed to the phenomenon of quantum confinement. The aforementioned properties demonstrated by the subject under discussion include emission that can be adjusted according to size, a photoluminescence quantum yield that is notably high, and absorption spectra that are broad as shown in Fig. 1. Integration of quantum dots into matrix material, like polymers, ceramics, or metals, results in the formation of nanocomposites that may display superior or unique characteristics in comparison to the individual constituents. Nanocomposites that incorporate quantum dots exhibit several noteworthy characteristics and have a variety of potential uses. Quantum dots exhibit remarkable optical properties, characterized by their high luminescence efficiency and tunable emission of light across a range of colors, which is attributed to their size-dependent behavior. The integration of these emission characteristics within a matrix can be harnessed for implementation in optoelectronics, display methodologies, and devices that emit light. The integration of nanocomposites into a matrix can result in improved electrical conductivity or customized electronic characteristics. The utilization of nanocomposites containing quantum dots has been proposed as a means of improvising light absorption features along with charge separation in photovoltaic systems. The utilization of quantum dots possessing a modifiable bandgap has the potential to capture a broader spectrum of solar radiation. The matrix material serves the dual purpose of providing structural reinforcement and facilitating electron transport pathways. Biomedical research on quantum dot nanocomposites has been conducted extensively, especially in the domains of drug delivery, biosensing, and biomedical imaging. These substances are utilized in various fields including chemical production, water electrolysis and photocatalysis. Quantum dot nanocomposites has been materialized proficiently in recent times wide areas of biomedicine, particularly in biomedical imaging, drug delivery, and biosensing. Owing to their distinctive optical characteristics and surface chemistry, they are highly enviable for application in varied biological system tasks such as targeted therapy, molecular sensing, and fluorescent labeling. It is noteworthy that the distinct characteristics and utilisations of nanocomposites that rely on quantum dots are subject to variability contingent upon the matrix material implemented, dimensions and constitution of the quantum dots and the methods of fabrication used. Continued investigation in this area endeavours to further examine and utilize the capabilities of these nanocomposites for a diverse array of technological implementations.

Some researched quantum dots based nanocomposites with properties and applications are listed as follows.

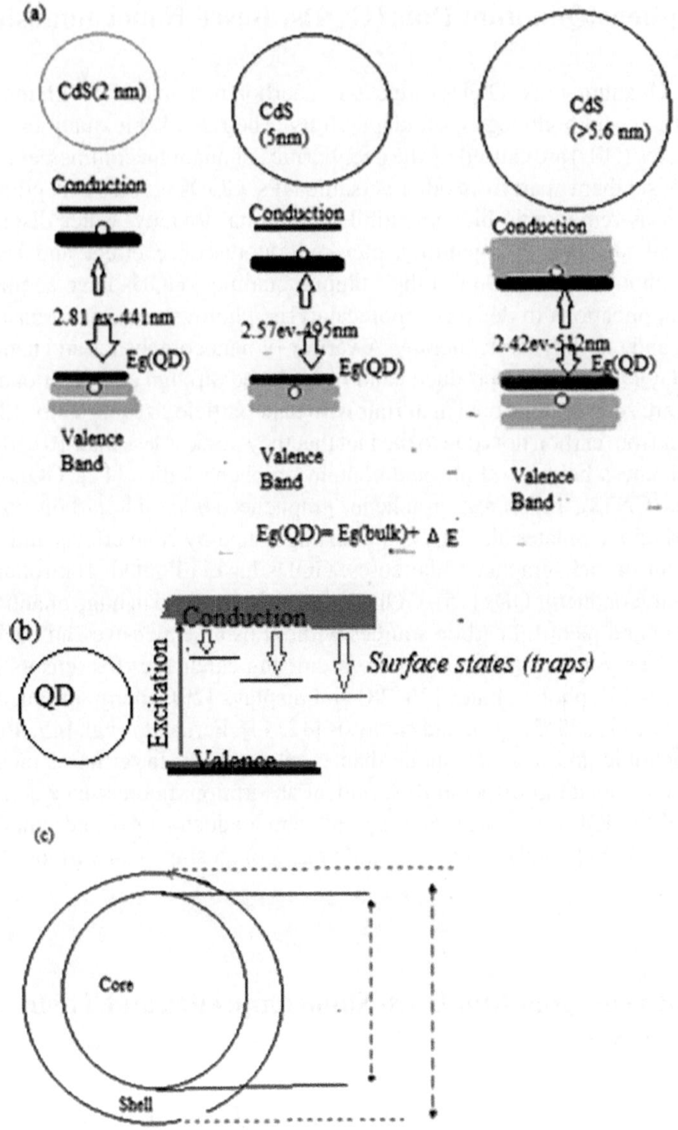

Fig. 1 A diagrammatic representation of different solid band structures **a** quantum confinement effect due to variation in size of quantum dots **b** electronic energy states of QDs bandgaps along with surface trap sites **c** The electronic view showing heterojunction formed by two semiconductor [12]

3 Graphene Quantum Dots (GQDs) Based Nanocomposites

Graphene Quantum dots (GQDs) are class of carbon nanomaterial with many diversified features and high degree of adaptablility. The remarkable qualities of carbon nanoparticles (NPs) are caused by the peculiarities of quantum confinement and edge effects that set them apart from other versatile NPs. GQDs possess exceptional properties such as remarkable biocompatibility, minimal toxicity, water dispersibility, high optical absorbance capability, elevated fluorescence effect and substantial degree of photo stability. Due to their unique qualities, GQDs have acquired their potential applications in delivery, photocatalysis, photovoltaics, nanosensor manufacturing, and photodynamic therapy. A variety of nanocomposites and nanohybrids based on GQDs have been produced and/or analysed through computational means.

GQDs are zero-dimensional materials with nanoparticle sizes < 10 nm [13]. GQDs are distinct from carbon dots due to the fact that they possess less than 10 carbon atom layers, thickness below 100 nm and contain graphene lattice [14]. GQDs, carbon nanotubes (CNTs), fullerenes, graphene, graphene oxide, and carbon dots (CDs) are all carbon nanomaterials. GQDs are distinguished by edge effects and quantum confinement in their graphene-like cores. GQDs have a limited, controllable band gap like semiconductor QDs [15]. GQDs may be synthesised in huge quantities from inexpensive and plentiful carbon sources without using expensive and ecologically damaging heavy metals [16]. GQDs are promising candidates for sensors [17–22], imaging [23–25], photovoltaics [26–28] and displays [29], energy production [30–34], medicine [13, 25, 35–42], and catalysis [43, 44], Fernando et al. [45, 46]. GQDs are hydrophobic and less poisonous than graphene [47]. In terms of biocompatibility, aqueous medium dispersibility, optical absorption, fluorescence activity, and photostability GQDs do better than typical semiconductor QDs and other carbon-based NPs [47–51]. GQDs create versatile nanocomposites with various chemicals [37, 51, 52].

4 Porphyrin-Quantum Dots-Nanocomposites and Their Derivatives

Porphyrins are promising dye-sensitized solar cell (DSSC) sensitizer owing to their easily mutable chemical structures apart from other features like ample absorption capacity in the near UV–vis range and fair degree of photostability [53]. Compared to other sensitizers Porphyrins absorption is weaker in solar light energy range of 400–600 nm [18, 54–56]. Thus these are good solar energy harvesting sensitizer which absorbs UV to infrared light. Thus, adjusting porphyrin optical absorption improves the efficiency of dye-sensitized solar cells (DSSC). This can be done by different ligands exchange to porphyrins with or by producing nanocomposites/nanohybrides/nanoconjugates. Due to their greater light absorption, dye sensitizer-quantum dot nanocomposites harvested solar energy more efficiently than their separate components (e.g. [49]). Thus, the amount of solar light absorption can be increased by nanocomposites containing GQDs and porphyrins or their derivatives. Relative enhancement in absorbance was studied through Computation method for tetraphenylporphyrin (TPP)-GQD nanocomposites as solar cell sensitizers to improve DSSC efficiency by observing optical absorbance of diverse GQD/porphyrin ratios 1:1,1:2,1,3, 1:4 and TPP:3GQDs ratio [57]. Based on Li et al.'s extensive explanation of edge-functionalized GQDs, the cis-coronene $C_{54}H_{18}$ was used as a model. The model involved One TPP phenyl binded by amide linkage with GQD, and the rest involved TPP functionalizion at their para-positions. Dispersion adjustments were needed to understand TPP-GQD interactions, The Gaussian 09 B3LYP/Lanl2dz technique optimised the gas phase structures of GQD, TPP, GQD-TPPs, and TPP-GQDs. Harmonic frequencies were computed to test optimised structural stability. Table 1 depicts the step wise synthesis of GQD.

Table 1 Typical process used to produce QDs for bioapplications [68]

Steps of process	Process
(i) Coreshell QDs (CdSe–ZnS	CdSe (Core) with ZnS (Shell) (*Size 2–10 nm*) → capping with capping ligand TOPO (Triooctylphosphine Oxide)
(ii) Organic colloidal stabilization [ligand tri-*n*-octylphosphine oxide (TOPO)];	Capped core → Incepton of PEG (polyethynene glycol) [HO-(CH_2-CH_2-)nCH_2-OH] polymer chain and TOPO polymer chain to surface of core
(iii) Hydrophilic polymer attachment (PEG	[((CdSe $_{core}$)ZnS($_{Shell}$))TOPO($_{capping\ ligand}$)]PEG$_{polymer}$ 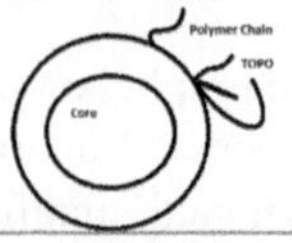
(iv) Bioconjugated with affinity Ligands as targeting molecule (immunoglobulin-G)	Bioconjugated with affinity Ligands as targeting molecule (immunoglobulin-G)
(v) Formation of integrated hybrid biocompatible nanocomposite (micellar structure)	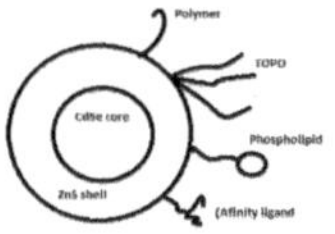 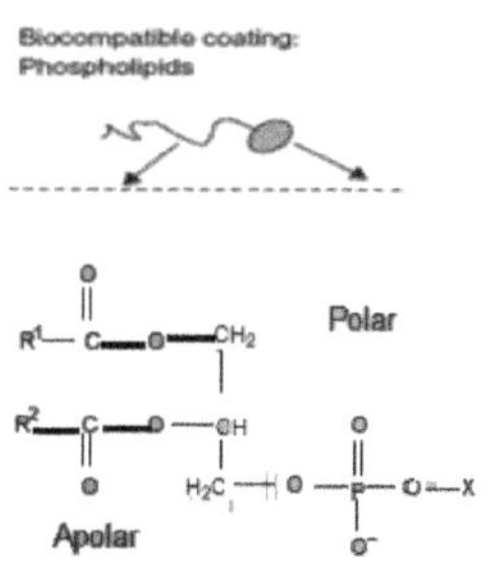

Optical absorption characteristics of optimised GQD, TPP, and nanocomposites geometries were studied through Octopus coded computation of TDDFT (Time dependent density functional theory) having for higher GQD-TPPs and TPP-GQD nanocomposites using the 'local-density approximation' and 'Hartwigsen-Goedecker-Hutter pseudopotential' [58]. All structures absorbed infrared. TDDFT demonstrated that adding TPPs or GQDs increased optical absorption. QD-rich nanocomposites have better optical absorption. GQD-3TPPs had a somewhat greater absorption than GQD-4TPPs. GQD-TPPs nanocomposites had greater optical absorptions than those having equal QD number. The lowest unoccupied molecular orbital energies (LUMO) of all nanocomposites exceeded the minimum energy of conduction band of TiO_2, meeting the DSSC criteria. 'Spatial charge separation and electron transport parameters' demonstrated sluggish 'electron-hole recombination' for GQD: TPP conjugates maintaining the ratios 1:2, 1:3 and 1:4 respectively. Despite the same electron transfer driving force, these complexes GQD-TPP, TPP-2GQDs, and TPP-3GQDs were estimated to be rapid. Due of its near UV optical absorption, GQD-3TPPs was the most promising nanocomposites for DSSCs. Karachevtsev et al. [59] performed DFT and MD analyses of meso-5,10,15,20-tetraphenyl porphyrin flattening owing to graphene absorption in the same year. Graphene noncovalent functionalization with organic molecules moieties with extended conjugated systems (porphyrins and their derivatives) can tune its electronic properties and expand its nanotechnology applications [60, 61]. Several research groups computationally examined naphthalene, pyrene, benzene, and other flat-conjugated aromatic hydrocarbon adsorption [43, 62, 63]. These compounds were shown to attach through planar stack conformations to graphene, leading to rise in interaction energies with contact area. Noncovalently bonded molecules and graphene did not transfer charge (CT).

Noncovalently bonded species weakly altered graphene's electrical structure [43, 59, 62, 63]. As predicted, dispersion forces dominated intermolecular interactions in these stacks. Electrostatic interactions contributed significantly less. Graphene functionalization with electron-withdrawing groups and electron-donating groups increased the interaction energies of nanocomposites. The adsorbed molecule's asymmetry also affected BEs. Porphyrins were widely studied in graphene noncovalent functionalization [24, 64–66]. However, basic structural and interaction energy problems remained unresolved [59]. Thus, understanding these systems' interconnections was vital. Further study on effect of porphyrins and meso-5, 10, 15, 20-TPP on graphene was done to determine how adsorption modifies TPP structure [59]. Gaussian 09 involving M05-2X/6-31G(d) was used in to study graphene-P complexes apart from graphene-TPP and their components. BSSE adjustments were also computed [59]. Similar observations was obtained conclusively stating that M05-2X estimated interaction energy without BSSE adjustment matched MP2 interaction energies better. The $C_{216}H_{36}$ fragment, big enough to adsorb the TPP molecule, modelled graphene. MD contained in NAMD package equipped with Charmm27 force field was used to model the production of graphene-P nanohybrids and graphene-TPP. The graphene surface modelling was done using a 120×124 Å graphene sheet rectangular in shape having $C_{5682}H_{212}$ composition. The nanohybrids were placed in water box 148 × 145 × 33 Å size containing 20,000 water molecules. MD simulations frozen graphene atom coordinates. After 500 stages of energy minimization, the simulation ran for 30 ns (1 fs step). Periodic boundary conditions with 303 K temperature and 1 atm pressure were used in simulation. Electrostatic Long-range interactions were computed using Particle Mesh Ewald. PBEs and TPP with graphene have similar DFT energies of −21.5 kcal/mol and −22.6 kcal/mol. Interactions of Porphyrin core-stacking with graphene offered high contribution to its interaction energy. Because the TPP phenyls are outside of the core plane, steric hindrances led the TPP core to approach the graphene surface less than the original porphyrin. The binding energy decrease caused by this impact was countered by the higher van der Waals interactions of the phenyls with graphene, giving TPP or porphyrin the same interaction energies with graphene. The dihedral angle between the core plane and the phenyl planes decreases from 64 to 50° as a result of TPP interactions with the graphene surface, which twist and deform the phenyls and porphyrin core. Through the creation of a small negative potential zone between the porphyrin molecule and the surface, TPP adsorption altered the electrostatic potential (ESP) of the graphene surface. Simulations of MD confirmed DFT findings that TPP and P were stacked on graphene in aqueous conditions. Porphyrins' in-plane motions on graphene result in low-energy barriers and excellent surface mobility. By combining 2D nanomaterials like graphene with aromatic chemicals like porphyrins, novel nanocomposites were created by taking advantage of the conformational behaviour of non-planar molecules on graphene. Graphene and Ni(II)-tetraphenylporphyrin (NiTPP) were covalently bonded by Hou et al. [67] to form a graphene-porphyrin nanohybrid.

They investigated the graphene reactivity of a novel ß-NiTPP-N2+ (the diazonium group connected to the ß-pyrrolic position of the porphyrin core) and meso-NiTPP-N2+ (the diazonium in 4-position of a peripheral phenyl). Due to steric restrictions, the second species could not attach to graphene's surface, only functionalizing GQD edge dangling bonds. The structural characteristics of the graphene-NiTPP hybrid were discovered by structural optimisation at the PM3 level for two model species: Covalent bonding between ß-NiTPP and the graphene moiety edge allows for a center-to-edge joining. A strain-free structure was formed by the edge type binding (E/G-NiTPP), in which the ß-NiTPP unit was oriented perpendicular to the graphene moiety plane, whereas deformations were caused by the covalent binding to the graphene basal plane (P/G).

A geometrically selective bottom-up synthesis of curved nanographenes (NGs) was described by Fernández-García et al. [45]. Different reaction conditions activated seven-membered ring formation, giving the NG negative curvature and creating two topologically unique curved NGs. Electrochemical, photophysical, Raman, X-ray crystallographic, and computational techniques were used to study structure-property connections for these two novel NGs.

These NGs that are curved and have a fused corannulene unit might represent GQD nanocomposites or models. To determine the type of absorptions seen in the

UV-vis spectra, TDDFT calculations were performed using the B3LYP-D3/def2-SVP-optimized grid for both NGs (the polarizable continuum model (PCM) model accounting for CHCl3 effects). The vertical excitations of these compounds were at calc = 479 and 471 nm, or 2.59 and 2.63 eV. These results were excellent with the experimental optical gap values (from excited state measurements): 2.61 and 2.59 eV, and moderately good with the experimental data, exp = 456 (2.72 eV) and 441 nm (2.81 eV). Electron transport from the HOMO (dominated by the polyaromatic hydrocarbon (PAH) domain) to the LUMO (a *-MO fully delocalized over the structure) was a feature of both vertical transitions. The -* transition in the 450 nm absorption was indicative of the prolonged -conjugation of the novel compounds. According to Umeyama et al. [56], these NGs' optical gap values are comparable to those of the reference compound, hexa-tert-butyl-hexa-peri-hexabenzocoronene. The study of graphene nanoribbons (NGs) with sp^2-carbon networks of varying sizes, notably hexa-peri-hexabenzocoronene (HBC) and its rectangularly-extended variant, a short graphene nanoribbon (GNR), covalently functionalized by photoactive Zn-porphyrins, has been recently described, both theoretically and experimentally. 42 sp^2-hybridized carbon HBCs are the smallest NG unit. Because of their distinct zones of -conjugation, graphenes, graphene oxides, chemically transformed graphenes, and GQDs are structurally heterogeneous at the molecular level.

Due to their various regions of conjugation, edge structures, and defects, graphenes, graphene oxides, chemically converted graphenes, and GQDs are structurally heterogeneous at the molecular level (Allen et al. 2010) [69–72].Well-defined extended HBC derivatives provide excellent GQD models, and they may be used to investigate the structure-property correlations of the sp2-carbon species mentioned above [56, 72, 73]. Spectroscopic analysis revealed that the photodynamic properties of porphyrin-linked NGs are strongly influenced by the size of the NG unit [56]. Exclusive energy transfer (EnT) from the first singlet excited state (S1) of the NG to the porphyrin unit was brought about by photoexcitation of the porphyrin–HBC system. The first and second singlet excited states (S1 and S2) of the porphyrin unit displayed opposing selective EnT from the porphyrin–GNR system to the NG. It was demonstrated that the corresponding ultrafast EnTs from the carotenoids S2 and S1 states to chlorophylls in naturally existing photosynthetic light-harvesting systems were replicated by the ultrafast efficient EnTs from the porphyrin S2 and S1 states to GNR. These characteristics make the nanocomposites suitable for solar energy conversion devices and optoelectronics. B3LYP/6-31G* spin-restricted calculations were carried out. Every structure was optimised without any constraints related to symmetry. DFT calculations revealed that the HBC moiety in ZnP–HBC complexes had a larger band gap than ZnP. The ZnP units dominated the ZnP–HBC HOMOs and LUMOs. In contrast to the ZnP–HBC systems, the HOMO and LUMO of the ZnP–GNR were located on the GNR moiety.

Theoretical research on porphyrin-sensitized GQD solar cells' feasibility are also noteworthy [74]. Functionalizing GQDs with porphyrins altered their electrical energy levels, therefore the authors investigated. GQD-porphyrin nanocomposites' electrical structure was investigated through the use of self-consistent charge density-functional–based tight-binding (SCC-DFTB). Slater-type orbitals and the Perdew-Burke-Ernzerhof (PBE) exchange-correlation energy functional were employed as basis sets. Here's how GQDs were modelled. A huge graphene sheet was created first. (ii) A tiny graphene QD was selected and sliced. (iii) Maintaining symmetry created bigger QDs. GQDs with symmetry up to 10 nm were constructed this manner. GQD-porphyrin nanocomposites were modelled by attaching porphyrin molecules to GQD through amide linkage. Functional groups at phenyl parapositions were employed in metal-free TPP molecules. Hexagonal and trigonal GQDs containing three TPP molecules were examined. The DFTB+code's SCC-DFTB technique calculated these GQD-TPP nanocomposites' total energy. To avoid erroneous interactions with nanocomposite periodic replicas, huge supercells with extensive vacuum regions in all directions were employed. In the Slater-Kirkwood model, dispersion interactions were added. Calculations of absorption spectra using the system containing a single TPP molecule. Nanocomposites can have type-I or type-II band energy alignment, depending on the size of GQDs. Smaller GQD nanocomposites exhibited type-II band energy alignment, where the GQD contributed to the LUMO and the TPP dominated the composite system HOMO. Systems with higher GQDs, however, aligned the type-I band energy. Systems for aligning type-II bands showed evidence of charge separation. Charge carrier recombination was modest in these systems. Smaller GQD nanocomposites were thought to offer greater photovoltaic properties. Porphyrin-GQD nanocomposites with bigger GQDs and electron-donating groups $eOCH_3$ and eNH_2 were also studied. Type-II alignment was identified in the nanocomposites. An extensive analysis of densities of states (DOSs) for nanocomposite materials revealed that the size of GQDs increased the energy gap between the LUMOs of porphyrin and GQD (a measure of electron injection efficiency from photoexcited porphyrin to GQD). Consequently, porphyrin-to-GQD electron transmission is enhanced by GQD size.

The intriguing physico-chemical properties of carbon-based quantum dots and nanocomposites, which are produced via edge effects and quantum confinement, have drawn attention to their potential as sensors. Top-down and bottom-up methods for CQDs, GQDs, and their nanocomposites include laser ablation, arc-discharge, chemical oxidation, ultrasonication, oxidative cleavage, microwave synthesis, thermal decomposition, solvothermal or hydrothermal method, stepwise organic synthesis, carbonisation from small molecules or polymers, and impregnation. Carbon-based compounds are extensively utilised for environmental cleaning because of their large specific surface area, adaptable structure, low toxicity, and superior stability. Semiconductor nanoparticles (NPs) with quantized energy states that are smaller than 10 nm are known as quantum dots (QDs) [75, 76]. More atoms are present on QD surfaces than in bulk materials because of 3-dimensional truncation. According to Reshma et al. [77], Hu et al. [78], QDs have good luminescent and electrical characteristics, such as narrow emission, broad and continuous absorption spectrum, and light stability. QDs based on semiconductors are durable inorganic fluorescence probes. In terms of biocompatibility, chemical inertness, and low cytotoxicity, CQDs and GQDs outperform semiconductor QDs such as CdSe dots, ZnO dots, and others

[79, 80], Alaghmandfard et al. [81]. They show simple synthesis, bioconjugation, and wavelength-dependent luminescence emission [82, 83]. Band gaps that are not zero improve optical characteristics. On their surfaces, functional groups produce edge effects. Wide photoluminescence emission, significant excitation wavelength dependency, and broad absorption bands are characteristics of carbon-based QDs [84]. The majority of researchers [85–87], (Zhang et al. 2019) have employed them in electrocatalysis. Light-emitting LEDs [80], cell labelling [88], cancer imaging and therapy [89], and cell imaging [90].

5 A Comparison Between Graphene and Carbon Quantum Dots (CQDS/GQDS)

With exceptional conductivity, chemical stability, environmental friendliness, high photostability, broadband optical absorption, low toxicity, photobleaching resistance, large surface area, and simplicity of modification, carbon nanoparticles (CQDs) smaller than 10 nm are a remarkable material. The most notable feature of CQDs is their photoluminescence [91]. After prolonged stimulation, CQDs' fluorescence emission intensity remains consistent due to their excellent fluorescence stability. Because of their special qualities, CQDs are perfect for sensing applications [92]. Because of their many carboxyl groups, CQDs can incorporate chemically reactive groups for functionalization or connect with other polymeric, organic, biological, inorganic, or natural components for surface modification. According to Ying Lim et al. [93], surface functionalization promotes solubility in both aqueous and non-aqueous environments, whereas surface passivation improves fluorescence. Because it creates a new energy state inside the band gap, high oxygen levels cause a change in the width of CQDs [94].

GQDs are small graphene particles with lateral dimensions of less than 100 nm, forming two-dimensional nanocrystals. Oxygen, hydrogen, and numerous layers under 10 nm are also present. The band gap energy of GQDs may be adjusted between 0 and 6 eV. Modifying the lateral dimension or surface chemical properties can tune the band gap because of conjugated edge effects and the quantum confinement effect [95]. Because of their superior optoelectronic qualities, affordability of production, and biocompatibility, GQDs can take the position of quantum dots based on metal chalcogenides. Graphene has better thermal and electrical conductivity than semi-conductor quantum dots because of its bonding both above and below the atomic plane [96–98]. Among the numerous great qualities of GQDs are tunable photoluminescence (PL), extraordinary spin property, exceptional UV-blocking capabilities, and high photo stability [99]. Redshifting from 522 to 575 nm, the pH of the GQDs solution rises from 1.0 to 11.0 [100]. This results from the deprotonation of functional groups that include oxygen, such as carboxyl, hydroxyl, and epoxy. Because red florescence has a peak at 626 nm, it happens at pH levels higher than 11.0. Being pH and temperature sensitive, and even a minor alteration can cause a fast fluorescence shift. Biomedical and bioanalytical applications of CQDs and GQDs are growing due to their photoluminescence and ability to interact with biomolecules. CQDs and GQDs' unique optical features might help create sensitive bioanalysis systems for point-of-care diagnostics. Several forms of carbon dots for medicinal use may improve their use. In point-of-care viral testing, CQDs-based biomacromolecule detection has shown great performance and feasibility [101]. Li et al. [102] detected SARS-CoV-2 using Gd^{3+} doped CQDs and specific antibodies. GQDs, gold nanoclusters coupled with antibodies, were used by Yuan et al. [103] to detect infections.

On comparing both CQDs and GQDs it is observed that CQDs and GQDs differ structurally. Amorphous CQDs are sp^3 hybridised, while crystalline GQDs are sp^2. CQDs glow due to surface imperfections. Quantum confinement fluoresces 2–20 nm GQDs [104]. GQDs differ from CDs in that their graphene sheets are less than 100 nm and 10 layers thick. Quasi-spherical carbon nanoparticles are CDs [105]. GQDs are perfect spherical sp^2 carbon nanosheets, whereas CQDs feature a carbogenic core that mimics graphene nanolayers and amorphous carbon. Carbogenic spherical structures can have a crystalline core with both sp^2 and sp^3 carbons or a disordered structural core with mostly sp^3. Conjugated QD domains, which mimic large aromatic systems, cause bandgap transitions. In the following two portions, high-density electrons in sp^2 hybridised orbitals absorb light and generate electronic transitions. Inorganic semiconductor QDs' excellent optical and electrical properties are employed to make sensors. Carbon-based QDs offer good selectivity and sensitivity for detecting a wide range of analytes. Carbon-based QDs are non-toxic and soluble in aqueous solvents, in contrast to other QDs. Due to their exceptional optical properties and surface flaws, CQD and GQD-based nanocomposites (NCs) are perfect for sensors that detect and destroy dangerous pollutants. Since CQDs and GQDs have a limit of detection (LOD) of nanomolar, picomolar, or even femtomolar, they may be used in precision sensor systems [106]. There are sensors that are photochemical, chemiluminescent, and electrochemical. Atomic absorption spectroscopy (AAS) and inductively coupled plasma mass spectrometry (ICP-MS) are two expensive and rigid detection methods for metal ions and chemical species. Photoluminescence-based sensors are better for these kinds of applications because of their low cost, simplicity, high sensitivity, and selectivity. Low LODs are one benefit of CQD and GQD fluorescence sensors [107]. If the LOD is low, the sensor is extremely sensitive. CQDs and GQDs photoluminescence sensors have strong selectivity, which enables them to identify certain ions or chemical compounds in a mixture of other substances. CQDs-based sensors have identified Fe^{3+} [90], Hg^{2+} [108], Pb^{2+} [109], and organophosphorus pesticides [27].

Because of its low equipment requirements, high sensitivity, broad linear range, and lack of background scattering light, chemiluminescence sensors are becoming more and more common in sensing applications [110]. Fe^{2+}, Cu^{2+}, H_2O_2, and 4-nitrophenol have all been detected using chemiluminescence sensors [111–114].

Because of its easier setup, higher sensitivity, and label-free nature, electrochemiluminescence (ECL), often referred to as electrogenerated chemiluminescence, has found increased application in analytical chemistry. In order to illustrate the ECL emission feature required to create ECL sensors, researchers demonstrated a variety of semi-conductor nano-crystals, such as QDs, CQDs, and GQDs. Reaction location and timing may be regulated when light-emitting samples are produced in situ, close to the electrode surface, because the ECL has nearly no background [19]. Ag/CQDs have property mutability. Hg^{2+} ions transform into Hg and break down into small fragments that combine with Ag NPs to create an AgHg amalgam layer on the surface, which reduces absorbance. Ag/CQDs' CQDs expedite the electron transfer from Ag NPs to Hg^{2+} ions. Ag/CQDs' CQDs expedite the electron transfer from Ag NPs to Hg^{2+} ions. L-tryptophan capped CQDs (L-CQDs) in aqueous solution were employed by Wan et al. [1] to detect Hg^{2+} ions (LOD of 11 nm) in 1 min. An electronic structural alteration was induced in L-CQDs via interactions between the amino, carbonyl, and Hg^{2+} moiety. Excitons recombine as a result of the charge transfer process, which lowers fluorescence emission. To identify Pb^{2+} ions in porphyria, copper nanoclusters@nitrogen-doped carbon dots (CuNCs-CNQDs) were employed. The self-calibration signal was provided by CNQDs by preserving fluorescence in Pb^{2+}. CuNCs-CNQDs nanohybrids, with a fluorescence intensity ratio difference between fluorophores-based LOD of 0.0031 mgL-1, were employed as ratiometric probes to detect Pb^{2+} at concentrations ranging from 0.010 to 2.5 mgL-1. The sensor detected Hg^{2+} in seafood with 86.45–114.47% recoveries and 0.20–1.92% RSDs. It was also accurate, dependable, and selective. The detection of heavy metals [115], glucose [116], pesticides [117–122], DNA [23], and nitro-explosives [123].

6 Epoxy-Graphene Quantum Dots Nanocomposites

In order to create graphene quantum dots, multiple weight percentages of graphene were introduced to epoxy resin along with a curing agent. This demonstrated the reinforcing effects of graphene quantum dots on different characteristics of epoxy resins. The resulting nanocomposites were characterised using FTIR, photoluminescence, UV-visible absorption, FESEM, TGA, tensile testing, and DMTA analysis. These techniques demonstrated that the incorporation of graphene quantum dots into the epoxy resin improved the nanocomposites' mechanical characteristics, viscoelastic properties, and thermal stability. Table 2 illustrates the epoxy graphene quantum dot fabrication process.

The excitation and emission peaks of GQDs and Ep/GQDs nanocomposites were detected at 350 nm and 450 nm, respectively, according to their FTIR Photoluminescence (PL) spectra. The photoluminescence curve indicates that when the electrons in the functional groups C=O and C=C on the edges of GQDs are excited, they may produce strong visible radiation [124]. According to research on the Ep/GQDs nanocomposites' fluorescence characteristics, excitation happened about 355 nm, and the emission peak happened at about 460 nm. The data indicate a strong correlation between the quantity of GQDs incorporated into the polymer matrix and variations in emission intensity. This is because there is a direct correlation between

Table 2 Synthesis of epoxy graphene quantum dots through series of different steps of potential chemical reactions [125]

Steps of synthesis of epoxygraphene QDs	Analyte	Process	Resultant product
(i)	Citric acid 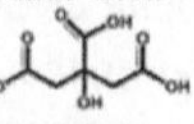	Microwave irradiation 800 W, 7 min	Graphene quantum dots (GQDs)
(ii)	GQDs	Rotary operation for 1 h, Oven 12 h, $60^{0}C$ at 7 pH 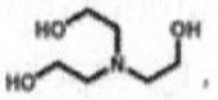	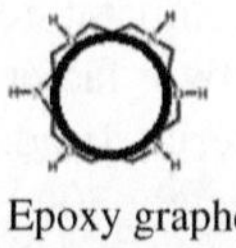 Epoxy graphene quantum dot
(iii)	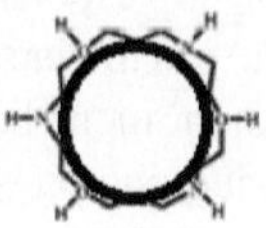	Chemical interaction with polymer matrix 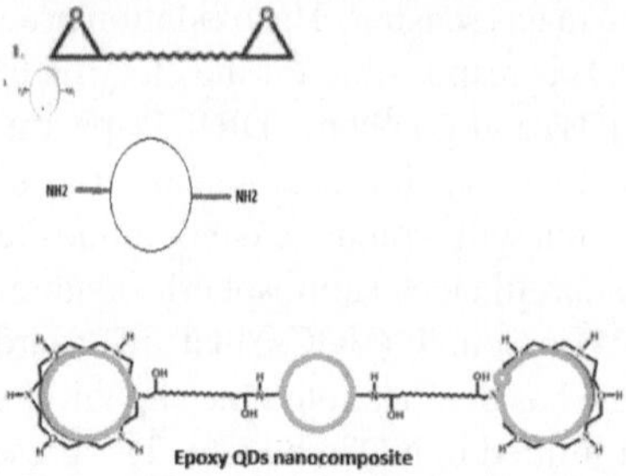	

the volume of GQDs in the polymer matrix and their increased fluorescence. The mechanical property data indicates that when the filler content increases, the tensile strength of the nanocomposites increases due to the addition of GQDs nanoparticles in the polymer matrix. Comparing Ep/GQDs 5 wt% to the other samples, the results show that it has a higher tensile strength. Numerous reasons might account for the reported rise in nanocomposites' tensile strength. These include the strong chemical interactions that take place between the GQDs and epoxy matrix, as well as the uniform dispersion of nanoparticles in the polymer matrix. Van der Waals forces, hydrogen bonds, and covalent bonds are a few examples of these interactions that work to fortify the nanocomposites and enhance their mechanical qualities. In particular, covalent interactions arise between the curing agent and epoxide groups, while hydrogen bonds form between the epoxy and carbonyl, carboxyl, and hydroxyl groups on the margins of the GQDs. DMTA analysis was used to analyse the viscoelastic characteristics. The results showed that the presence of GQD nanoparticles in the epoxy structure led to strong chemical interactions between the polymer matrix and the nanoparticles. The nanoparticles also filled up gaps and previously empty regions. It was uncovered that the storage modulus has improved.

7 Zinc Oxide Quantum Dots/Graphene Nanocomposites

Using a straightforward solution-based method, graphene nanocomposites embellished with unique zinc oxide quantum dots (ZnO QDs) were created. Nucleation and development on the surface of a graphene template create zinc oxide quantum dots (QDs), with a size of around 5 nm. By varying the reaction duration and precursor concentration, the density of their dispersion may be easily regulated. At room temperature, the formaldehyde sensing properties of pure graphene are enhanced by a factor of 4 when incorporated with ZnO QDs to form a nanocomposite material. In addition, the nanocomposite-based sensors exhibit rapid response (approximately 30 s) and recovery (approximately 40 s) characteristics, as well as exceptional selectivity and stability at ambient temperature. The synergistic interaction between ZnO quantum dots and graphene is responsible for the reported increase in gas sensing. The oxidation process of the gas under study on the surface of ZnO QDs is responsible for the electron transfer phenomena between ZnO quantum dots (QDs) and graphene. DRIFT spectra results demonstrates the reliability of the proposed gas sensing mechanism. The utilisation of ZnO quantum dots (QDs) in conjunction with graphene nanocomposites as sensors presents promising prospects for the surveillance of atmospheric contamination, particularly with respect to volatile organic compounds (VOCs) that are hazardous and toxic. As previously documented, graphene has demonstrated the capability to detect various types of gases, including but not limited to NO2 and CO. The present study reports on the gas sensing properties of a nanocomposite consisting of ZnO quantum dots and graphene at room temperature. The targeted gas molecule for detection was formaldehyde (HCHO).

8 Chitosan Based Carboxymethyl Nanocomposites with Chemically Bonded Quantum Dots

Using carboxymethyl chitosan (CMCS), a biocompatible nanocomposite consisting of magnetic nanoparticles (MNPs) and fluorescent quantum dots (QDs), magnetic-fluorescent nanoparticles (MFNPs) have been produced. Covalent bonds are an efficient way to connect magnetic nanoparticles (MNPs) and fluorescent quantum dots (QDs) to the carboxymethyl chitosan (CMCS) surface. The composite exhibits high colloidal stability in physiological conditions and keeps its advantageous magnetic and fluorescence features. The CMCS composite was used to build nanocarriers that had minimal cytotoxicity, prolonged drug release, and pH-dependent drug release characteristics. The reported CMCS composites hold great potential for effective cancer targeting, medication delivery, and in situ cellular imaging. These MFNPs effectively conjugate QDs and MNPs to the surface of CMCS through covalent bonding. The composite exhibits high colloidal stability in physiological conditions and keeps its advantageous magnetic and fluorescence features. Folate (FA) was added to the nanocomposites to provide a delivery system with a targeting function.

FA is a particular targeting ligand. Doxorubicin (DOX), a chemotherapy medication, was loaded to provide the therapeutic effect through electrostatic and hydrophobic interactions. The overall DOX release profile indicates pH sensitivity. Both confocal laser scanning microscopy examination and flow cytometry analysis indicated that these nanocomposites were taken up by cancer cells via FA receptor-mediated endocytosis pathway. Thus CMCS-based nanocomposites created in this study show enormous promise for efficient drug administration, cancer targeting, and in situ cellular imaging.

Nanocomposites with several functional features are created by combining nanomaterials with various physicochemical properties. For drug delivery and cell monitoring, a lot of work has recently been put into designing and creating nanocomposites using fluorescent QDs and magnetic nanoparticles (MNPs) [126, 127]. Opportunities for controlled drug administration and release, multimodal bioimaging, selective cell targeting, and diagnostic and therapeutic capabilities are presented by the resultant composites [128]. Such multi-functional composites are frequently created via.

Effectively, surface-modified QDs, MNPs, and folateFA have been combined with CMCS matrix to form nanocomposites for targeted drug delivery and cellular imaging. The presence of -COOH and -NH2 groups in CMCS allows for the formation of amido linkage (-CONH-) with QDs, MNPs, or foalte. In contrast to physical interactions like electrostatic force, strong chemical bonds between the constituents ensure the stability of the generated nanocomposites under physiological settings. The resulting composite possesses both the advantageous magnetic properties of MNPs and the strong fluorescence of QDs. Some possible advantages of in vitro research include low cytotoxicity, delayed and pH-dependent drug release patterns, high cellular uptake efficiency, and receptor-specific endocytosis.

9 CdS and CdSe Quantum Dots in Poly (3-Hexylthiophene) Based Nanocomposites

The efficiency of optical stimulation to form electron hole pairs and the accompanying radiative decay may be controlled by varying the size, shape, and distribution of nanoparticles because of the quantum size effect [129]. The effectiveness of optical excitation is controlled by quantum dot-polymer interactions as a result of the impacts of quantum size. Nanocrystal and polymer mixtures have shown promising breakthroughs in solar technology. When the nanocrystal has a notable bandgap, both the conjugated polymer and the nanocrystal contribute to the absorption in composites. Excellent hole-transporting polymers, such poly(3-hexylthiophene) (P3HT), can be combined with electron-transporting nanocrystals to take advantage of inorganic semiconductors' high transport capabilities without sacrificing organic materials' ability to be processed in a solution. An appreciable portion of solar energy is absorbed by composite devices that incorporate inorganic nanocrystals because they have a greater affinity for electrons than conjugated polymers. Conducting poly(3-hexylthiophene) (P3HT) matrix has been used to create CdS and CdSe nanoparticles. The spectroscopic results demonstrate that the electrical structure of the polymer is significantly influenced by the characteristics of the inserted semiconducting nanoparticles. SEM and TEM images show the ordered shape of the CdS and CdSe nanoparticles in the presence of the polymer matrix. P3HT shows a prominent absorption band at 451 nm in the superimposed UV-vis spectra of (A) P3HT, (B) P3HT/CdS, and (C) P3HT/CdSe nanocomposites, which may be caused by the excitation of electrons in the p-conjugated system. Similarly, CdS and CdSe nanocomposites show higher intensity absorption bands at 448 and 446 nm, which correspond to band gaps of 2.76 and 2.78 eV, respectively, suggesting that the p-p* transitions of conjugated systems are more prominent.

There is a modest blue shift of 3 and 5 nm, respectively, following the introduction of CdS and CdSe nanoparticles, indicating that the band gap in the composites is larger than that of pure P3HT. Short-range optical absorption curves offer theories on the development and synthesis of nanoparticles. The structure and apparent morphology of the nanocomposites differ significantly from those of the polymer, as shown by SEM pictures of the polymer matrix. Similar nanoparticles with consistent interparticle separation are seen in the CdS/P3HT nanocomposite, but pores with distinct edges and varying distribution are seen in the polymer. Interestingly, the CdSe/P3HTsample exhibits somewhat larger particles with variable inter-particle spacing. These characteristics clearly demonstrate the trapping of nanoparticles in the polymer matrix and the apparent shape supports nanocrystal-polymer composites of greater sizes.

TEM scans reveal a few more peaks, which are related to the P3HT polymer's increased crystallinity as a result of the nanoclusters' existence. To sum up, P3HT serves as an effective matrix for encapsulating CdS and CdSe nanoparticles in terms of physical, thermal, and electrical properties. This is particularly useful for constructing various devices intended for optical and optoelectronic uses. Morphological investigations using SEM and TEM show the development of nanocomposites with ordered Q-dots with an average particle size in the range of 3–5 nm. UV tests showed that the conjugation present in the polymer matrix creates considerable overlapping between electron-hole pair wave functions, despite TGA measurements demonstrating the nanocomposites' greater heat stability. A comparison of the cyclic voltammograms for pure polymer, CdS/P3HT, and CdSe/P3HT nanocomposites in the presence and absence of light demonstrates the capacity of the quantum dots immobilised in polymer to trap the intermediate radical cation at a given potential.

10 Poly(Methyl Methacrylate)/ZnO Nanocomposites Based on KH570 Surface Modified ZnO Quantum Dots

Transparent luminous ZnO embedded PMMA polymer has been generated by grafting poly(methyl methacrylate) (PMMA) onto the surface of ZnO QDs after treatment with g-methacryloxypropyl trimethoxy silane (KH570) [130]. The finished ZnO/PMMA nanocomposites with KH570-modified nanoparticles show better dispersibility while retaining the ZnO nanoparticles' enhanced luminosity. The PMMA/ZnO nanocomposite films that are generated have great UV shielding efficacy, enhanced thermal stability, and good transparency.

The characteristic absorption peaks of –CH_2– stretching vibration, CO stretching vibration, –SiOH stretching vibration, Zn–O–Si, and Si–O–Si symmetrical stretching vibration at 2953.92, 1700.43, 1184.56, 943.47, and 817.71 cm^{-1} are obtained at 3394.35 cm^{-1} due to hydroxyl absorption. These findings demonstrate that the introduction of active groups onto the surface of the nanoparticles led to the hydrolysis-condensation formation of a thin layer of organic silica nanonetwork. The FT-IR spectra of ZnO nanoparticles modified by KH570 following PMMA grafting are displayed in Fig. 1c. The C O stretching vibration and CH3 stretching vibration from PMMA are shown by the lines at 1735 and 1454 cm^{-1}. Stretching vibration bands of PMMA C-O-C are located at 1148, 1197, 1246, and 1283 cm^{-1}. It demonstrates how PMMA was grafted onto ZnO nanoparticles by hydrolysis–condensation, forming a thin layer of organic silica nanonetwork.

Transmission electron microscopy (TEM) provides evidence for the concept that KH570-surfaced modification and polymer play a major role in stabilising and spreading the nanoparticles because of the covalent link that occurs between them.

11 Nanocomposites Based on an Oxadiazole–Carbazole Copolymer (POC) and InP/ZnS Quantum Dots

Derivatives of carbazoles are used to modify the emission characteristics of polymers in the CIE diagram, therefore prolonging the life of devices. These derivatives produce a strong blue emission because they have an electron transporter unit (oxadiazole) and a hole transporter unit (carbazole). Studies [131], Peng and Zhang (1999), Bugatti et al. (2006), Acierno et al. (2002, 2003, 2004), Concilio et al. (2010) have all reported on this. The POC polymer has intercalated aliphatic lengths of active oxadiazole/cabazole units, which improve solubility in a range of organic solvents and guarantee favourable film shape. The two different kinds of red emitters QDs, InP/ZnS(570) and InP/ZnS(627), which are referred to as POC-InP/ZnS(570) and InP/ZnS(627), were combined to create the nanocomposites. The pure polymer, pure InP/ZnS(570) quantum dots (QDs), and nanocomposites with QDs at 25, 50, and 75 weight percent (wt%) were all subjected to Fourier-transform infrared (FTIR) spectra. The carbazole/oxadiazole rings' aromatic C-H signals may be detected at

2928, 2853, and 1758, 1609, and 1488 cm^{-1}, in that order. Oleylamine-capped InP/ZnS quantum dots (QDs) have a few unique peaks in their infrared spectra, and these peaks are positioned similarly to those of the POC QDs that have been investigated earlier. These peaks are specifically seen at 1,461 and 1,563 cm^{-1}, which represent the oleylamine's N–H bending, and at 2,846 and 2,922 cm^{-1}, which represent the C–H stretching. The assessment of POC's valence (HOMO) and conduction (LUMO) energy levels verified the viability of an efficient electron/energy transfer between POC and QDs. There seems to be a substantial preference for the flow of electrons and energy between the two materials. The POC-InP/ZnS(570) Nanocomposite's spectra show that peaks linked to the emission of POC and QD coexist. Concurrently, we find that the polymer peak in every nanocomposit has a little blue shift when compared to the pure polymer. Furthermore, when the concentration of QD rises, the polymer peak's intensity decreases. The concentration of QDs in the nanocomposites positively correlates with the intensity of the QD emission peak, which is detected at a wavelength of 655 nm. The facts stated above suggest that the bulk material's polymer chain organisation is altered by quantum dots (QDs). Moreover, the polymer and the nanoparticles have an efficient energy transfer mechanism. The study displays the POC-InP/ZnS(627) nanocomposites' photoluminescence spectra, which have been normalised according to the quantity of photons absorbed at the excitation wavelength. At 400 nm, the nanocomposites were energised. A main peak that is stimulated at 400 nm is visible in the emission spectrum. Emissions have also been detected at four different wavelengths. Changes in polymer organisation brought about by an increase in QD number might be the cause of the observed increase in polymer fluorescence lifetime. This ultimately causes the polymer's relaxation decay pathway to emerge in a unique way. Quantum dots (QDs) are much smaller than the extended polymer chains, which makes it easier for them to be incorporated into the chains. This event also suggests that a high efficiency of energy transmission to the QDs can be achieved.

12 CuInS2 Quantum Dots@Graphene Oxide Nanocomposites

The fabrication of the Ky2-CuInS2 quantum dots (QDs)@graphene oxide (GO) nanocomposites involved the conjugation of amino terminal kanamycin-binding Ky2 aptamer with carboxy groups present on the surface of CuInS2 QDs, which were modified using mercaptopropionic acid. N-hydroxysuccinimide and 1-(3-dimethylaminopropyl)-3-ethylcarbodiimide hydrochloride were present when this was accomplished. The nucleobases and graphene oxide (GO), the Ky2-CuInS2 quantum dots were easily attached to the GO surface by pi-pi stacking contact. The "turn-off" and "turn-on" fluorescence characteristics of Ky2-CuInS2 QDs are used by the sensor. For kanamycin, under ideal circumstances, a linear response was found in the range of 0.3–45 nmol L^{-1} (0.174–26.1 mg L^{-1}), with a low detection limit of

Table 3 Tabular illustration for fabrication of Ky2-CuInS2 QDs (Liu et al. 2016)

Various steps for fabrication of Ky2-CuInS2 QDs	Process	Product
Step 1—CuInS2 QDs 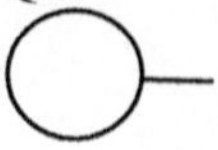 COOH	Conjugation of carboxy groups present on the surface of CuInS2 with mercaptopropionic acid, with amino terminal kanamycin-binding Ky2 aptamer 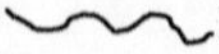 NH_2 in the presence of 1-(3-dimethylaminopropyl)-3-ethylcarbodiimide hydrochloride and N-hydroxysuccinimide	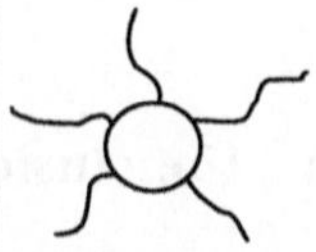 Ky2 CuInS2 QDs
Ky2 CuInS2 QDs	Affixed on graphene oxide (GO) surface with/without kanamycin	Ky2 CuInS2 QUANTUM DOTS@graphene oxide nanocomposites

0.12 nmol L^{-1} (0.070 mg L^{-1}), exceeding the MRL set by the EU. Table 3 shows the tabular representation used in the Ky2-CuInS2 QD production process.

13 Polyurethane-Based Bulk Nanocomposites from 1-Thioglycerol-Stabilized CdTe Quantum Dots

N_2H_4-enhanced growth promotes the production of CdTe quantum dots (QDs) in an aqueous solution. Thioglycerol (TG) stabilised the QDs, and their photoluminescence enhanced. The synthesis procedure led to a significant decrease in the time scale of the size evolution of the quantum dots. The process made it easier for QDs to develop at room temperature and lower temperatures, which prevented TG from decomposing and sulphur from becoming trapped inside the QDs. The QDs' hydroxyl coverage from the TG ligand made them miscible with PU prepolymers in dimethyl sulfoxide. Bulk nanocomposites were formed when hydroxyl coverage enabled the bonding of polyurethane (PU) to the surface of quantum dots (QDs) by the interaction of hydroxyl (-OH) and isocyanate (-NCO) groups resulting in the formation of bulk nanocomposites of CdTe QD-PU. These luminous nanocomposites are made of prepolymers and CdTe QDs stabilised with thioglycerol (TG). An important part was the optimisation of the TG-stabilized QD synthesis process, which was accomplished in an aqueous solution via a N2H4-promoted method. This method produced a notable improvement in the PL intensity as well as a significant decrease in the preparation time. Because of the hydroxyl coverage from TG, QDs and PU prepolymers in DMSO were miscible without the need for ligand exchange. This made it possible for the reaction between -OH and -NCO to initiate polymerization. More investigation revealed that the polyurethane (PU)-based nanocomposites included an equal distribution of quantum dots (QDs). The QDs' size-dependent photoluminescence (PL) characteristics were also preserved. The ease and feasibility of synthesising TG-stabilized QDs and polymerizing them currently suggests that luminescent nanocomposites could have practical applications in the display and optical fields.

14 Conclusion

The features and uses of the newly developed nanocomposites based on quantum dots as sensors have been covered in length in this review chapter. However, research into nanocomposites based on QDs is still in its early phases. Even though there has been a lot of study on synthesis approaches over the past ten years, it is still challenging to discover a scalable, effective, yet straightforward manufacturing process for their high-quality synthesis. And even other features require attention.

References

1. Wan, X., Li, S., Zhuang, L., Tang, J.: L-Tryptophan-capped carbon quantum dots for the sensitive and selective fluorescence detection of mercury ion in aqueous solution. J. Nanopart. Res. **18**, 202 (2016)
2. Qiao, Y., Li, C.M.: Nanostructured catalysts in fuel cells. J. Mater. Chem. **21**, 4027–4036 (2011)
3. Hao, R., Xing, R.J., Xu, Z.C., Hou, Y.L., Gao, S., Sun, S.H.: Synthesis, functionalization, and biomedical applications of multifunctional magnetic nanoparticles. Adv. Mater. **22**, 2729–2742 (2010)
4. Haun, J.B., Yoon, T.J., Lee, H., Weissleder, R.: Magnetic nanoparticle biosensors. Wiley Interdiscip. Rev. Nanomed. Nanobiotechnol. **2**, 291–304 (2010)
5. Chen, W., Li, C.M., Yu, L., Lu, Z.S., Zhou, Q.: In situ AFM study of electrochemical synthesis of polypyrrole/Au nanocomposite. Electrochem. Commun. **10**, 1340–1343 (2008)
6. Mi, C.C., Zhang, J.P., Gao, H.Y., Wu, X.L., Wang, M., Wu, Y.F., Di, Y.Q., Xu, Z.R., Mao, C.B., Xu, S.K.: Multifunctional nanocomposites of superparamagnetic (Fe_3O_4) and NIR-responsive rare earth-doped up-conversion fluorescent (NaYF4: Yb, Er) nanoparticles and their applications in biolabeling and fluorescent imaging of cancer cells. Nanoscale **2**, 1141–1148 (2010)
7. Lu, Z.S., Zhu, Z.H., Zheng, X.T., Qiao, Y., Guo, J., Li, C.M.: Biocompatible fluorescence-enhanced ZrO_2–CdTe quantum dots nanocomposite for in vitro cell imaging. Nanotechnology **22**, 155604 (2011)
8. Pinaud, F., Clarke, S., Sittner, A., Dahan, M.: Probing cellular events, one quantum dot at a time. Nat. Meth. **7**, 275–285 (2010)
9. Zrazhevskiy, P., Sena, M., Gao, X.H.: Designing multifunctional quantum dots for bioimaging, detection, and drug delivery. Chem. Soc. Rev. **39**, 4326–4354 (2010)
10. Yu, W.W., Chang, E., Drezek, R., Colvin, V.L.: Water-soluble quantum dots for biomedical applications. Biochem. Biophys. Res. Commun. **348**, 781–786 (2006)
11. Du, Y.P., Xu, B., Fu, T., Cai, M., Li, F., Zhang, Y., Wang, Q.B.: Near infrared photoluminescent Ag_2S quantum dots from a single source precursor. J. Am. Chem. Soc. **132**, 1470–1471 (2010)
12. Mansur, H.S.: Quantum dots and nanocomposites. WIREs Nanomed. Nanobiotechnol. **2**(2), 113–129 (2010). https://doi.org/10.1002/wnan

13. Vatanparast, M., Shariatinia, Z.: Computational studies on the doped graphene quantum dots as potential carriers in drug delivery systems for isoniazid drug. Struct. Chem. **29**, 1427–1448 (2018)
14. Wang, D., Chen, J.F., Dai, L.: Recent advances in graphene quantum dots for fluorescence bioimaging from cells through tissues to animals. Part Part Syst. Char. **5**, 515–523 (2015)
15. Neubeck, S., Ponomarenko, L.A., Freitag, F., Giesbers, A.J.M., Zeitler, U., Morozov, S.V., et al.: From one electron to one hole: quasiparticle counting in graphene quantum dots determined by electrochemical and plasma etching. Small **6**, 1469–1473 (2010)
16. Li, X., Rui, M., Song, J., Shen, Z., Zeng, H.: Carbon and graphene quantum dots for optoelectronic and energy devices: a review. Adv. Funct. Mater. **25**, 4929–4947 (2015)
17. Bacon, M., Bradley, S.J., Nann, T.: Graphene quantum dots. Part Part Syst. Char. **31**, 415–428 (2014)
18. Li, L., Wu, G., Yang, G., Peng, J., Zhao, J., Zhu, J.J.: Focusing on luminescent graphene quantum dots: current status and future perspectives. Nanoscale **5**, 4015–4039 (2013)
19. Sun, H., Wu, L., Wei, W., Qu, X.: Recent advances in graphene quantum dots for sensing. Mater. Today **16**, 433–442 (2013)
20. Qian, Z.S., Shan, X.Y., Chai, L.J., Chen, J.R., Feng, H.: Dual-colored graphene quantum dots-labeled nanoprobes/graphene oxide: functional carbon materials for respective and simultaneous detection of DNA and thrombin. Nanotechnology **25**, 415501 (2014)
21. Wang, Y., Zhang, L., Liang, R.P., Bai, J.M., Qiu, J.D.: Using graphene quantum dots as photoluminescent probes for protein kinase sensing. Anal. Chem. **85**, 9148–9155 (2013)
22. Zheng, L., Ye, D., Xiong, L.: Preparation of cobalt-tetraphenylporphyrin/reduced graphene oxide nanocomposite and its application on hydrogen peroxide biosensor. Anal. Chim. Acta **768**, 69–75 (2013)
23. Wang, Y., Wang, S., Ge, S., Wang, S., Yan, M., Zang, D., Yu, J.: Facile and sensitive paper-based chemiluminescence DNA biosensor using carbon dots dotted nanoporous gold signal amplification label. Anal. Meth. **5**, 1328–1336 (2013)
24. Zheng, X.T., Than, A., Ananthanaraya, A., Kim, D.H., Chen, P.: Graphene quantum dots as universal fluorophores and their use in revealing regulated trafficking of insulin receptors in adipocytes. ACS Nano **7**, 6278–6286 (2013)
25. Chen, J., Than, A., Li, N., Ananthanarayanan, A., Zheng, X., Xi, F., et al.: Sweet graphene quantum dots for imaging carbohydrate receptors in live cells. Flat Chem. **5**, 25–32 (2017)
26. Shen, J., Zhu, Y., Yang, X., Li, C.: Graphene quantum dots: emergent nanolights for bioimaging, sensors, catalysis and photovoltaic devices. Chem. Commun. **48**, 3686–3699 (2012)
27. Yang, W., Zhang, H., Lai, J., Peng, X., Hu, Y., Gu, W., et al.: Carbon dots with red-shifted photoluminescence by fluorine doping for optical bio-imaging. Carbon **128**, 78–85 (2018)
28. Kim, C.O., Hwang, S.W., Kim, S., Shin, D.H., Kang, S.S., Kim, J.M., et al.: High-performance graphene-quantum dot photodetectors. Sci. Rep. **4**, 5603 (2014)
29. Tran, V.T., Hur, S.H., Chung, J.S., Choi, W.M.: Ultraviolet light sensor based on graphene quantum dots/reduced graphene oxide hybrid film. Sensor Actuat. A Phys. **233**, 368–373 (2015)
30. Zhuang, S.D., Chen, Y., Zhang, W.C., Chen, Z., Wang, Z.L.: Humidity sensor and ultraviolet photodetector based on carrier trapping effect and negative photoconductivity in graphene quantum dots. Sci. Chin. Phys. Mech. **61**, 0142111 (2018)
31. Cui, B., Feng, X.T., Zhang, F., Liu, X.G., Yang, Y.Z., Jia, H.S., et al.: The use of carbon quantum dots as fluorescent materials in white LEDs. N Carbon Mater. **32**, 385–401 (2017)
32. Sudhagar, P., Herraiz-Cardona, I., Park, H., Song, T., Noh, S.H., Gimenez, S., et al.: Exploring graphene quantum dots/TiO_2 interface in photoelectrochemical reactions: solar to fuel conversion. Electrochim. Acta **187**, 249–255 (2016)
33. Zhu, Z., Ma, J., Wang, Z., Mu, C., Fan, Z., Du, L., et al.: Efficiency enhancement of perovskite solar cells through fast electron extraction: the role of graphene quantum dots. J. Am. Chem. Soc. **136** (2014)

34. Wang, X.Y., Narita, A., Mullen, K.: Precision synthesis versus bulk-scale fabrication of graphenes. Nat. Rev. Chem. **2**, 0100 (2017)
35. Wang, Z., Xu, C., Lu, Y., Chen, X., Yuan, H., Wei, G., Ye, G., Chen, J.: Fluorescence sensor array based on amino acid derived carbon dots for pattern-based detection of toxic metal ions. Sens. Actuators B Chem. **241**, 1324–1330 (2016)
36. Liu, H., Na, W., Liu, Z., Chen, X., Su, X.: A novel turn-on fluorescent strategy for sensing ascorbic acid using graphene quantum dots as fluorescent probe. Biosens. Bioelectron. **92**, 229–233 (2017)
37. Qian, Z.S., Shan, X.Y., Chai, L.J., Ma, J.J., Chen, J.R., Feng, H.A.: Universal fluorescence sensing strategy based on biocompatible graphene quantum dots and graphene oxide for the detection of DNA. Nanoscale **6**, 5671–5674 (2014)
38. Wang, W., Zhang, Y., Wang, Y.B.: Noncovalent interaction between graphene and aromatic molecule: structure, energy, and nature. J. Chem. Phys. **140**, 094302 (2014)
39. Xiao, S., Zhou, D., Luan, P., Gu, B., Feng, L., Fan, S., et al.: Graphene quantum dots conjugated neuroprotective peptide improve learning and memory capability. Biomaterials **106**, 98–110 (2016)
40. Radhakrishnan, S., Samanta, A., Sudeep, P.M., Maldonado, K.L., Mani, S.A., Acharya, G., et al.: Metal-free dual modal contrast agents based on fluorographene quantum dots. Part Part. Syst. Char. **34** (2017)
41. Markovic, Z.M., Ristic, B.Z., Arsikin, K.M., Klisic, D.G., Harhaji-Trajkovic, L.M., Todorovic-Markovic, B.M., et al.: Graphene quantum dots as autophagy-inducing photodynamic agents. Biomaterials **33**, 7084–7092 (2012)
42. Nurunnabi, M., Khatun, Z., Reeck, G.R., Lee, D.Y., Lee, Y.K.: Photoluminescent graphene nanoparticles for cancer phototherapy and imaging. ACS Appl. Mater. Interfaces **6**, 12413–12421 (2014)
43. Wang, X., Sun, X., Lao, J., He, H., Cheng, T., Wang, M., et al.: Multifunctional graphene quantum dots for simultaneous targeted cellular imaging and drug delivery. Colloids Surf. B Biointerfaces **122**, 638–644 (2014)
44. Zou, X., Liu, M., Wu, J., Ajayan, P.M., Li, J., Liu, B., et al.: How nitrogen-doped graphene quantum dots catalyze electroreduction of CO_2 to hydrocarbons and oxygenates. ACS Catal. **7**, 6245–6250 (2017)
45. Fernández-García, J.M., Evans, P.J., Medina Rivero, S., Fernández, I., García-Fresnadillo, D., Perles, J., et al.: Extended corannulene-based nanographenes: selective formation of negative curvature. J Amfluorescent materials in white LEDs. N Carbon Mater. **32**, 385–401 (2018)
46. Li, Q., Zhang, S., Dai, L., Li, L.S.: Nitrogen-doped colloidal graphene quantum dots and their size-dependent electrocatalytic activity for the oxygen reduction reaction. J. Am. Chem. Soc. **134**, 18932–18935 (2012)
47. Zhuo, S., Shao, M., Lee, S.T.: Upconversion and down conversion fluorescent graphene quantum dots: ultrasonic preparation and photocatalysis. ACS Nano **6**, 1059–1064 (2012)
48. Ge, J., Minhuan, L., Zhou, B., Liu, W., Guo, L., Wang, H.: A graphene quantum dot photodynamic therapy agent with high singlet oxygen generation. Nat. Commun. **5**, 4596–603 (2014)
49. Zhu, S., Zhang, J., Liu, X., Li, B., Wang, X., Tang, S.: Graphene quantum dots with controllable surface oxidation, tunable fluorescence and up-conversion emission. RSC Adv. **2**, 2717–2720 (2012)
50. Achadu, O.J., Uddin, I., Nyokong, T.: Fluorescence behavior of nanoconjugates of graphene quantum dots and zinc phthalocyanines. J. Photochem. Photobiol. Chem. **317**, 12–25 (2016)
51. Gonçalves, L.M., Ribeiro, H.A., Mendes, A.M., deZea, B.V.: Dye-sensitized solar cells: a safe bet for the future. Energy Environ. Sci. **1**, 655–667 (2008)
52. Tian, P., Tang, L., Teng, K.S., Lau, S.P.: Graphene quantum dots from chemistry to applications. Mater. Today Chem. **10**, 221–258 (2018)
53. Kadish, K.M., Smith, K.M., Guilard, R.: Handbook of Porphyrin Science with Applications to Chemistry, Physics, Materials Science, Engineering, Biology and Medicine. World Scientific, Singapore (2010)

54. Imahori, H., Hayashi, S., Hayashi, H., Oguro, A., Eu, S., Umeyama, T.: Effects of porphyrin substituents and adsorption conditions on photovoltaic properties of porphyrin-sensitized TiO_2 cells. J. Phys. Chem. C **113**, 18406–18413 (2009)
55. Imahori, H., Umeyama, T., Ito, S.: Large aromatic molecules as potential sensitizers for highly efficient dye-sensitized solar cells. Acc. Chem. Res. **42**, 1809–1818 (2009)
56. Umeyama, T., Hanaoka, T., Yamada, H., Namura, Y., Mizuno, S., Ohara, T., et al.: Exclusive occurrence of photoinduced energy transfer and switching of its direction by rectangular _ -extension of nanographenes. Chem. Sci. **10**, 6642–6650 (2019)
57. Gao, G., Jiang, Y.W., Jia, H.R., Yang, J., Wu, F.G.: On-off-on fluorescent nanosensor for Fe^{3+} detection and cancer/normal cell differentiation via silicon-doped carbon quantum dots. Carbon **134**, 232–243 (2018)
58. Gao, F., Yang, C.L., Wang, M.S.: Ma XG (2018) Computational studies on the absorption enhancement of nanocomposites of tetraphenylporphyrin and graphene quantum dot as sensitizers in solar cell. J. Mater. Sci. **53**, 5140–5150 (2018)
59. Karachevtsev, V.A., Stepanian, S.G., Karachevtsev, M.V., Adamowicz, L.: Graphene induced molecular flattening of meso-5,10,15,20-tetraphenyl porphyrin: DFT calculations and molecular dynamics simulations. Comp. Theor. Chem. **1133**, 1–6 (2018)
60. Georgakilas, V., Tiwari, J.N., Kemp, K.C., Perman, J.A., Bourlinos, A.B., Kim, K.S.: Noncovalent functionalization of graphene and graphene oxide for energy materials, biosensing, catalytic, and biomedical applications. Chem. Rev. **116**, 5464–5519 (2016)
61. Zhang, Z., Huang, H., Yang, X., Zang, L.: Tailoring electronic properties of graphene by stacking with aromatic molecules. J. Phys. Chem. Lett. **2897–905**,(2011). https://doi.org/10.1021/jz201273r
62. Zarudnev, E.S., Stepanian, S.G., Adamowicz, L., Karachevtsev, V.A.: Noncovalent interaction of graphene with heterocyclic compounds: benzene, imidazole, tetracene and imidazophenazines. Chem PhysChem **17**, 1204–1212 (2016)
63. Karachevtsev, V.A., Stepanian, S.G., Glamazda, A.Y., Karachevtsev, M.V., Eremenko, V.V., Lytvyn, O.S.: Noncovalent interaction of single-walled carbon nanotubes with1-pyrenebutanoic acid *d succinimide ester and glucoseoxidase*. J. Phys. Chem. C **115**, 21072–21082 (2011)
64. Ishida, Y., Masui, D., Shimada, T., Tachibana, H., Inoue, H.: The mechanism of the porphyrin spectral shift on inorganic nanosheets: the molecular flattening induced by the strong host−guest interaction due to the "size-matching rule." J. Phys. Chem. C **116**, 7879–7885 (2012)
65. Shawkat, M.A., Parida, M.R., Alarousu, E., Mohammed, O.F.: Ultrafast electron injection at the cationic porphyrin–graphene interface assisted by molecular flattening. Chem. Commun. **50**, 10452–10455 (2014)
66. Sakthinathan, S., Kubendhiran, S., Chen, S.M., Tamizhdurai, P.: Reduced graphene oxide/gold tetraphenyl porphyrin (RGO/Au–TPP) nanocomposite as an ultrasensitive amperometric sensor for environmentally toxic hydrazine. RSC Adv. **6**, 56375–83 (2016)
67. Hou, H.L., Dasler, D., Hauke, F., Hirsch, A.: Reductive functionalization of graphenides with nickel(II) porphyrin diazonium compounds. Phys. Status Solidi RRL **11**(1700306), 1–5 (2017)
68. Qi, Z.-L., Cheng, Y.-H., Xu, Z., Chen, M.-L.: Recent advances in porphyrin-based materials for metal ions detection. Int. J. Mol. Sci. **21**, 5839 (2020). https://doi.org/10.3390/ijms21165839
69. Novoselov, K.S., Fal'ko, V.I., Colombo, L., Gellert, P.R., Schwab, M.G., Kim, K.: A roadmap for graphene. Nature **490**, 192–200 (2012)
70. Dreyer, D.R., Park, S., Bielawski, C.W., Ruoff, R.S.: The chemistry of graphene oxide. Chem. Soc. Rev. **39**, 228–240 (2010)
71. Pei, S., Cheng, H.M.: The reduction of graphene oxide. Carbon **50**, 3210–3228 (2012)
72. Ball, M., Zhong, Y., Wu, Y., Schenck, C., Ng, F., Steigerwald, M.: Contorted polycyclic aromatics. Acc. Chem. Res. **48**, 267–276 (2015)

73. Wang, H., Yang, Y., Zhou, X., Li, R., Li, Z.: NiCo2S4/tryptophan-functionalized graphene quantum dot nanohybrids for high-performance supercapacitors. New J. Chem. **41**, 1110–1118 (2017)
74. Mandal, B., Sarkar, S., Sarkar, P.: Theoretical studies on understanding the feasibility of porphyrin sensitized graphene quantum dot solar cell. J. Phys. Chem. C **119**, 3400–3407 (2015)
75. Sharma, G., Kumar, A., Naushad, M., Kumar, A., Al-Muhtaseb, A.H., Dhiman, P., Ghfar, A.A., Stadler, F.J., Khan, M.R.: Photoremediation of toxic dye from aqueous environment using monometallic and bimetallic quantum dots-based nanocomposites. J. Clean. Prod. **172**, 2919–2930 (2018)
76. Singh, P., Shandilya, P., Raizada, P., Sudhaik, A., Rahmani-Sani, A., Bandegharaei, H.: A review on various strategies for enhancing photocatalytic activity of graphene-based nanocomposites for water purification. Arab. J. Chem. **13**, 3498–3520 (2020)
77. Reshma, V.G., Mohanan, P.V.: Quantum dots: applications and safety consequences. J. Lumin. **205**, 287–298 (2019)
78. Hu, Z.M., Fei, G.T., Zhang, L.D.: Synthesis and tunable emission of Ga_2S_3 quantumdots. Mater. Lett. **239**, 17–20 (2019)
79. Hagiwara, K., Horikoshi, S., Serpone, N.: Luminescent monodispersed carbon quantum dots by a microwave solvothermal method toward bioimaging applications. J. Photochem. Photobiol. A Chem. **415**, 113310 (2021)
80. Ding, H., Zhou, X.X., Zhang, Z.H., Zhao, Y.P., Wei, J.S., Xiong, H.M.: Large scale synthesis of full-color emissive carbon dots from a single carbon source by a solvent-free method. Nano Res. **15**, 3548–3555 (2021)
81. Yue, X.Y., Zhou, Z.J., Wu, Y.M., Li, Y., Li, J.C., Bai, Y.H., Wang, J.L.: Application progress of fluorescent carbon quantum dots in food analysis. Chin. J. Anal. Chem. **48**, 1288–1296 (2020)
82. Abd Rani, U., Yong Ng, L., Yin Ng, C., Mahmoudi, E.: A review of carbon quantum dots and their applications in waste water treatment. Adv. Colloid Interface Sci. **278**, 102124 (2020)
83. Rahal, M., Atassi, Y., Alghoraibi, I.: Quenching photoluminescence of carbon quantum dots for detecting and tracking the release of minocycline. J. Photochem. Photobiol. A Chem. **412**, 113257 (2021)
84. Wei Heng, Z., Chan Chong, W., Ling Pang, Y., Hoon Koo, C.: An overview of the recent advances of carbon quantum dots/metaloxides in the application of heterogeneous photocatalysis in photodegradation of pollutants towards visible-light and solarenergy exploitation. J. Environ. Chem. Eng. **9**, 105199 (2021)
85. Ju-Meng, W., Bi-Tao, L., Xin, Z., Chang-Chun, S.: One-pot synthesis of N, S co-doped photoluminescent carbon quantum dots for Hg^{2+} ion detection. New Carbon Mater. **33**, 333–340 (2018)
86. Cirone, J., Rahin Ahmed, S., Wood, P.C., Chen, A.: Green synthesis and electrochemical study of cobalt/graphene quantum dotsfor efficient water splitting. J. Phys. Chem. C **123**, 9183–9191 (2019)
87. Shin, J., Guo, J., Zhao, T., Guo, Z.: Functionalized carbon dots on graphene as outstanding non-metal bifunctional oxygen electrocatalyst. Small **15**, 1900296 (2019)
88. Chen, L., Yang, G., Wu, P., Cai, C.: Real-time fluorescence assay of alkaline phosphatase in living cells using boron-doped graphene quantum dots as fluorophores. Biosens. Bioelectron. **96**, 294–299 (2017)
89. Liu, L., Miao, Q., Liang, G.: Quantum dots as multifunctional materials for tumor imaging and therapy. Materials **6**, 483–499 (2013)
90. Gao, Y., Hou, F., Hu, S., Wu, B., Wang, Y., Zhang, H., Jiang, B., Fu, H.: Graphene quantum-dot-modified hexagonal tubular carbon nitride for visible-light photocatalytic hydrogen evolution. ChemCatChem **10**, 1330–1335 (2018)
91. Ahmad Nazri, N.A., HidayahAzeman, N., Luo, Y., Ashrif, A., Bakar, A.: Carbon quantum dots for optical sensor applications: a review. Opt. Laser Technol. **139**, 106928 (2021)

92. Kumar Barman, M., Patra, A.: Current status and prospects on chemical structure driven photoluminescence behaviour of carbon dots. J. Photochem. Photobiol. C **37**, 1–22 (2018)
93. Ying Lim, S., Shen, W., Gao, Z.: Carbon quantum dots and their applications. Chem. Soc. **44**, 362–381 (2015)
94. Yu, X., Liu, J., Yu, Y., Zuo, S., Li, B.: Preparation and visible light photocatalytic activity of carbon quantum dots/TiO_2 nanosheet composites. Carbon **68**, 718–724 (2014)
95. Zhu, J., Tang, Y., Wang, G., Mao, J., Liu, Z., Sun, T., Wang, M., Chen, D., Yang, Y., Li, J.: Green, rapid, and universal preparation approach of graphene quantum dots under ultraviolet irradiation. ACS Appl. Mater. Interfaces **9**, 14470–14477 (2017)
96. Zhang, H.Y., Wang, Y., Xiao, S., Wang, H., Wang, J.H., Feng, L.: Rapid detection of Cr(VI) ions based on cobalt(II)-doped carbon dots. Biosens. Bioelectron. **87**, 46–52 (2017)
97. Zhang, N., Zhang, L., Ruan, Y.F., Zhao, W.W., Xu, J.J., Chen, H.Y.: Quantum-dots-based photoelectrochemical bioanalysis highlighted with recent examples. Biosens. Bioelectron. **94**, 207–218 (2017)
98. Zhou, X., Gao, X., Song, F., Wang, C., Chu, F., Wu, S.: A sensing approach for dopamine determination by boronic acid functionalized molecularly imprinted graphene quantum dots composite. Appl. Surf. Sci. **423**, 810–816 (2017)
99. Gong, P., Wang, J., Hou, K., Yang, Z., Wang, Z., Liu, Z., Han, X., Yang, S.: Small but strong: the influence of fluorine atoms on formation and performance of graphene quantum dots using a gradient F-sacrifice strategy. Carbon **112**, 63–71 (2017)
100. Yuan, F., Ding, L., Li, Y., Li, X., Fan, L., Zhou, S., Fang, D., Yang, S.: Multicolor fluorescent graphene quantum dots colorimetrically responsive to all-pH and a wide temperature range. Nanoscale **7**, 11727–11733 (2015)
101. Madurani, K.A., Suprapto, S., Syahputra, M.Y., Puspita, I., Masudi, A., Rizqi, H.D., Hatta, A.M., Juniastuti, J., Lusida, M.I., Kurniawan, F.: Recent development of detection methods for controlling COVID-19 outbreak. J. Electrochem. Soc. **168**, 37511 (2021)
102. Li, Y., Ma, P., Tao, Q., Krause, H.J., Yang, S., Ding, G., Dong, H., Xie, X.: Magnetic graphene quantum dots facilitate closed-tube one-step detection of SARS-CoV-2 with ultra-low field NMR relaxometry. Sens. Actuators B Chem. **337**, 129786 (2021)
103. Yuan, H., Lin, J.H., Dong, Z.S., Chen, W.T., Chan, Y.K., Yeh, Y.C., Chang, H.T., Chen, C.F.: Detection of pathogens using graphene quantum dots and gold nanoclusters on paper-based analytical devices. Sens. Actuators B Chem. **363**, 131824 (2022)
104. Kuo, W.S., Chen, H.H., Chen, S.Y., Chang, C.Y., Chen, P.C., Hou, Y.I., Shao, Y.T., Kao, H.F., Lilian Hsu, C.L., ChenYC,: Graphene quantum dots with nitrogen-doped content dependence for highly efficient dual-modality photodynamic antimicrobial therapy and bioimaging. Biomaterials **120**, 185–194 (2017)
105. Feng, H., Qian, Z.: Functional carbon quantum dots: a versatile platform for chemosensing and biosensing. Chem. Rec. **18**, 491–505 (2018)
106. Chen, W., Li, D., Tian, L., Xiang, W., Wang, T., Hu, W., Hu, Y., Chen, S., Chen, J., Dai, Z.: Synthesis of graphene quantum dots from natural polymer starch for cell imaging. Green Chem. **20**, 4438–4442 (2018)
107. Wang, M., Fang, Z., Zhang, K., Fang, J., Qin, F., Zhang, Z., et al.: Synergistically enhanced activity of graphene quantum dots/graphene hydrogel composites: a novel all-carbon hybrid electrocatalyst for metal/air batteries. Nanoscale **8**, 11398–11402 (2016)
108. Li, C., Liu, W., Ren, Y., Sun, X., Pan, W., Wang, J.: The selectivity of the carboxylate groups terminated carbon dots switched by buffer solutions for the detection of multi-metal ions. Sens. Actuators B Chem. **240**, 941–948 (2017)
109. Vinod Kumar, V., Raman, T., Philip Anthony, S.: Fluorescent carbon quantum dots chemosensor for selective turn-on sensing of Zn^{2+} and turn-off sensing of Pb^{2+} in aqueous medium and Zebra fish egg. New J. Chem. **41**, 15157–15164 (2017)
110. Lin, Z., Chen, H., Lin, J.M.: Peroxide induced ultra-weak chemiluminescence and its application in analytical chemistry. Analyst **138**, 5182–5193 (2013)
111. Amjadi, M., Manzoori, J.L., Hallaj, T., Azizi, N.: Sulfur and nitrogen co-doped carbon quantum dots as the chemiluminescence probe for detection of Cu^{2+} ions. J. Lumin. **182**, 246–251 (2017)

112. Ali Shah, S.N., Dou, X., Khan, M., Uchiyama, K., Lin, J.M.: N-doped carbon dots/H_2O_2 chemiluminescence system for selective detection of Fe^{2+} ion in environmental samples. Talanta **196**, 370–375 (2019)
113. Amjadi, M., Hallaj, T.: Dramatic enhancement effect of carbon quantum dots on the chemiluminescence of Ru (bpy)32+–Ce (IV)reaction and application to the determination of 4-nitrophenol. J. Lumin. **171**, 202–207 (2016)
114. Nakano, K., Honda, T., Yamasaki, K., Tanaka, Y., Taniguchi, K., Ishimatsu, R., Imato, T.: Carbon quantum dots as fluorescent component in peroxyoxalate chemiluminescence for hydrogen peroxide determination. Bull. Chem. Soc. Jpn. **91**, 1128–1130 (2018)
115. Kumar, A., Chowdhuri, A.R., Laha, D., Mahto, T.K., Karmakar, P., Sahu, S.K.: Green synthesis of carbon dots from Ocimumsanctum for effective fluorescent sensing of Pb^{2+} ions and live cell imaging. Sens. Actuators B Chem. **242**, 679–686 (2017)
116. Shi, W., Wang, Q., Long, Y., Cheng, Z., Chen, S., Zheng, H., Huang, Y.: Carbon nanodots as peroxidase mimetics and their applications to glucose detection. Chem. Commun. **47**, 6695–6697 (2011)
117. Ashrafi Tafreshi, F., Fatahi, Z., Fatemeh Ghasemi, S., Taherian, A., Esfandiari, N.: Ultrasensitive fluorescent detection of pesticides in real sample by using green carbon dots. PLoS ONE **15**, e0230646 (2020)
118. Khaledian, S., Noroozi-Aghideh, A., Kahrizi, D., Moradi, S., Abdoli, M., Haji Ghasemalian, A., FoadHeidari, M.: Rapid detection of diazinon as an organophosphorus poison in real samples using fluorescence carbon dots. Inorg. Chem. Commun. **130**, 108676 (2021)
119. Ali Kamyabi, M., Moharramnezhad, M.: A novel cathodic electrochemiluminescent sensor based on CuS/carbon quantum dots/g-C3N4 nanosheets and boron nitride quantum dots for the sensitive detection of organophosphate pesticide. Microchem. J. **179**, 107421 (2022)
120. Duan, L., Du, X., Zhao, H., Sun, Y., Liu, W.: Sensitive and selective sensing system of metallothioneins based on carbon quantum dots and gold nanoparticles. Anal. Chim. Acta **1125**, 177–186 (2020)
121. Yang, Y., Fang, G., Wang, X., Zhang, F., Liu, J., Zheng, W., Wang, S.: Electrochemiluminescent graphene quantum dots enhanced by MoS2 as sensing platform: a novel molecularly imprinted electro chemiluminescence sensor for 2-methyl-4-chlorophenoxyacetic acid assay. Electrochim. Acta **228**, 107–113 (2017)
122. Sammi, H., Kukkar, D., Singh, J., Kukkar, P., Kaur, R., Kaur, H., Rawat, M., Singh, G., Kim, K.H.: Serendipity in solution–GQDs zeolitic imidazole frameworks nanocomposites for highly sensitive detection of sulfide ions. Sens. Actuators B Chem. **255**, 3047–3056 (2017)
123. KaanKoc, O., Uzer, A., Apak, R.: High quantum yield nitrogen-doped carbon quantum dot-based fluorescent probes for selective sensing of 2,4,6-trinitrotoluene. ACS Appl. Nano Mater. **5**, 5868–5881 (2022)
124. Kuo, W.S., Shen, X.C., Chang, C.Y., Kao, H.F., Lin, S.H., Wang, J.Y., Wu, P.C.: Multiplexed graphene quantum dots with excitation-wavelength- independent photoluminescence, as two-photon probes, and in ultraviolet−near infrared bioimaging. ACS Nano **14**, 11502–11509 (2020). https://doi.org/10.1021/acsnano.0c03915
125. Abbas, M., Shirkavand, B., Hadavand,: Fluorescent epoxy-graphene quantum dots nanocomposites: synthesis and study of properties. Polymer-Plast. Technol. Mater. **61**(2), 117–130 (2022). https://doi.org/10.1080/25740881.2021.1959929
126. Singh, N.S., Kulkarni, H., Pradhan, L., Bahadur, D.: A multifunctional biphasic suspension of mesoporous silica encapsulated with YVO4: Eu^{3+} and Fe_3O_4 nanoparticles: synergistic effect towards cancer therapy and imaging. Nanotechnology **24**, 065101 (2013)
127. Ye, F., Barrefelt, Å., Asem, H., Abedi-Valugerdi, M., El-Serafi, I., Saghafian, M., Abu-Salah, K., Alrokayan s, Muhammed M, Hassan M,: Biodegradable polymeric vesicles containing magnetic nanoparticles, quantum dots and anticancer drugs for drug delivery and imaging. Biomaterials **35**, 3885–3894 (2014)
128. Badruddoza, A.Z.M., Rahman, M.T., Ghosh, S., Hossain, M.Z., Shi, J., Hidajat, K., Uddin, M.S.: β-Cyclodextrin conjugated magnetic, fluorescent silica core-shell nanoparticles for biomedical applications. Carbohydr. Polym. **95**, 449–457 (2013)

129. Huynh, W.U., Dittmer, J.J., Alivisatos, A.P.: Science **295**, 2425 (2002)
130. Wang, F., Wang, J.A., Liu, X.G.: Direct evidence of a surface quenching effect on size-dependent luminescence of up conversion nanoparticles. Angew. Chem. In. Ed. **49**, 7456–7460 (2010)
131. Lee, J., Hwang, G., Hong, Y.S., Sim, T.: One step synthesis of quantum dot-magnetic nanoparticle heterodimers for dual modal imaging applications. Analyst **140**, 2864–2868 (2015)
132. Aleksey, E., Kuznetsov Fernando, K.A.S., Sahu, S., Liu, Y., Lewis, W.K., Guliants, E.A., Jafariyan, A., et al.: Carbon quantum dots and applications in photocatalytic energy conversion. ACS Appl. Mater. Interfaces **7**, 8363–8376 (2015)
133. Bin Chen, B., Xi Liu, Z., Yan Zou, H., Zhi Huang, C.: Highly selective detection of 2,4,6-trinitrophenol by using newly developed terbium-doped blue carbon dots. Analyst **141**, 2676–2681 (2016)
134. Chen, W., Lu, Z.S., Li, C.M.: Sensitive human interleukin 5 impedimetric sensor based on polypyrrole-pyrrolepropylic acid-gold nanocomposite. Anal. Chem. **80**, 8485–8492 (2008)
135. Chen, X., Gong, F., Cao, Z., Zou, W., Gu, T.: Highly cysteine-selective fluorescent nanoprobes based on ultrabright and directly synthesized carbon quantum dots. Anal. Bioanal. Chem. **410**, 2961–2970 (2018)
136. Costas-Mora, I., Romero, V., Lavilla, I., Bendicho, C.: In situ building of a nanoprobe based on fluorescent carbon dots for methylmercury detection. Anal. Chem. **86**, 4536–4543 (2014)
137. Dai, G., Wang, J., Zhao, Y., Zhu, Y., Ma, X., Mei, T., Li, J., Ma, P., Wang, X.: Dual-mode high sensitive detection of Fe(III) ions via fluorescent photonic crystal films based on co-assembly of silica colloids and carbon dots. Sci. Adv. Mater. **9**, 873–880 (2017)
138. Di Corato, R., Bigall, N.C., Ragusa, A., Dorfs, D., Genovese, A., Marotta, R., Manna, L., Pellegrino, T.: Multifunctional nanobeads based on quantum dots and magnetic nanoparticles: synthesis and cancer cell targeting and sorting. ACS Nano **5**, 1109–1121 (2011)
139. Li, L., Yan, X.: Colloidal graphene quantum dots. J. Phys. Chem. Lett. **1**, 2572–2576 (2010)
140. Li Liu, M., Bin Chen, B., Yang, T., Wang, J., Dong Liu, X., Zhi Huang, C.: One-pot carbonization synthesis of europium-doped carbon quantum dots for highly selective detection of tetracycline. Meth. Appl. Fluoresc. **5**, 015003 (2017)
141. Li, L.L., Diau, E.W.G.: Porphyrin-sensitized solar cells. Chem. Soc. Rev. **42**, 291–304 (2013)
142. Li, Y., Shu, H., Niu, X., Wang, J.: Electronic and optical properties of edge-functionalized graphene quantum dots and the underlying mechanism. J. Phys. Chem. C **119**, 24950–24957 (2015)
143. Liu, Y.M., Hou, H., Zhou, Y.Z., Zhao, X.J., Tang, C., Tan, Y.Z.: Nanographenes as electron-deficient cores of donor-acceptor systems. Nat. Commun. **9**, 1901 (2018)
144. Lu, Z.S., Hu, W.H., Bao, H.F., Qiao, Y., Li, C.M.: Interaction mechanisms of CdTe quantum dots with proteins possessing different isoelectric points. Med. Chem. Commun. **2**, 283–286 (2011)
145. Ponomarenko, L.A., Schedin, F., Katsnelson, M.I., Yang, R., Hill, E.W., Novoselov, K.S., et al.: Chaotic Dirac billiard in graphene quantum dots. Science **320**, 356–358 (2008)
146. Stepanian, S.G., Karachevtsev, M.V., Karachevtsev, V.A., Adamowicz, L.: Interactions of the Watson-Crick nucleic acid base pairs with carbon nanotubes and graphene: DFT and MP2 study. Chem. Phys. Lett. **610–611**, 186–191 (2014). https://doi.org/10.1016/j.cplett.2014.07.035
147. Wang, F., Han, Y., Lim, C.S., Lu, Y.H., Wang, J., Xu, J., Chen, H.Y., Zhang, C., Hong, M.H., Liu, X.G.: Direct evidence of a surface quenching effect on size-dependent luminescence of up conversion nanoparticles. Nature **463**, 1061–1065 (2010)
148. Wang, J., Huang, X., Xi, S., Lee, J.M., Wang, C., Du, Y.: Linkage effect in the heterogenization of cobalt complexes by doped graphene for electrocatalytic CO_2 reduction. Angew. Chem. Int. Ed. **58**, 2–10 (2019)
149. Wang, Y., Chen, G., Yang, M.X., Silber, G., Xing, S.X., Tan, L.H., Wang, F., Feng, Y.H., Liu, X.G., Li, S.Z., Chen, H.Y.: A systems approach towards the stoichiometry-controlled hetero-assembly of nanoparticles. Nat. Commun. **1**, 87 (2010)

150. Wei, J., Zhou, B., Gu, S., Lv, S., Zhou, Y., Liu, B.: Facile synthesis of fluorescent carbon nanodots from cornstalk and their application as a sensing platform for detection of Cu^{2+} ions. Sci. Adv. Mater. **9**, 901–906 (2017)
151. Wang, W., Liu, J., Yu., Xibin, Yang, G.: Transparent poly(methyl methacrylate)/ZnO nanocomposites based on KH570 surface modified ZnO quantum dots. J. Nanosci. Nanotechnol. **10**, 5196–5201 (2010)
152. Xu, Q., Wei, J., Wang, J., Liu, Y., Li, N., Chen, Y., Gao, C., Zhang, W., Sreenivasan Sreeprased, T.: Facile synthesis of copper doped carbon dots and their application as a "turn-off" fluorescent probe in the detection of Fe_3 ions. RSC Adv. **6**, 28745–28750 (2016)
153. Zhang, P., Bin, D., Wei, J.S., Niu, X.Q., Chen, X.B., Xia, Y.Y., Xiong, H.M.: Efficient oxygen electrocatalyst for Zn-air batteries:Carbon dots and Co9S8 nanoparticles in a N, S-codoped carbon matrix. ACS Appl. Mater. Interfaces **11**, 14085–14094 (2022). Chemosensors **10**, 367:22–29

Quantum Dots/Polymer Composites

Soumita Talukdar and Amit Gupta

Abstract Quantum dots (QD) are semiconductor nanoparticles which exhibit exciting quantum mechanical properties at a macro scale, such that they are often touted as the "macroscopic atoms". However, owing to their sizes, real world applications often need to cast them in a bulk object or anchor them to a substrate. Most common approaches include casting as a film or making composites. In this chapter we will discuss one such approach of casting the QD in a polymer matrix. This approach is flexible, easy to synthesize, and provides exceedingly useful applications in the field of analytical chemistry, photoluminescent materials, catalysis, and biologically useful materials.

Keywords Quantum dots · Polymer · Nanocomposite · Optoelectronics · Sensor

1 Introduction

Recent emergence of quantum dots (QD) as the magic material is being established for their unique physical properties, tunable surface properties, advanced synthetic routes and vast array of applications [1]. Moreover, colloidal-quantum dots are emerging to be a new branch of research domain as 2357 review work have been reported according to Scifinder®, since 2012 [2]. The applications ranges from opto-electronics, photocatalysis, fluorescent sensors, drug delivery, gamma scintillation, energy storage, solar cells, etc. [3–7]. Higher charge-carrier mobility, generation of multiple exciton ability, tunable surface and band structure properties make QDs an ideal candidate for photovoltaic applications [8]. Photoluminescence (PL) is the most dominant characteristics that is extensively studied [9]. The absorbance and emission characteristics of various polymers and their composites are shown in Table 1. In

S. Talukdar (✉)
GD Goenka University, Gurugram 122103, India
e-mail: soumita.talukdar@gdgu.org

A. Gupta
Aerospace Engineering and Mechanics, University of Minnesota, Minneapolis, USA

S. Thomas et al. (eds.), *Quantum Dots Based Nanocomposites*, Engineering Materials,
https://doi.org/10.1007/978-3-031-54779-9_6

addition, high surface-area to volume ratio and tunable redox potentials enable these QDs highly effective as catalysts for photo-induced redox reactions [10]. Varied sizes of QDs can be prepared by modification of parameters such as temperature, reaction time and synthetic routes, enabling desired applications [1]. Thereby enabling array of QDs excited at single wavelength with multiple well-separated emissions. However, this phenomenon may also lead to energy loss and emission re-absorption [11], phenomenon known as the inner filter effect. An interesting property of inner filter effect (IFE) was also observed in core–shell quantum dot structures, where the photo luminescence (PL) of smaller QD was self-absorbed by larger QDs exhibiting a red shift of photoluminescence. Therefore, these QD based nanocomposites can be considered as potential candidate for exploitation of IFE and develop spectroscopic displacement sensors [12].

Composites of such QDs, significantly improve the inadequacies of QDs, while restoring their original diverse properties. Due to the presence of diverse functional groups on the surface of QDs, they readily form composites with improved dispersibility [9]. Thus, properties such as optical absorptivity, photobleaching resistance, low toxicity, chemical stability, bio-compatibility and hydrophilicity can be highly alleviated in the QD/polymer composites. Development of polymer based QD materials proved to be a boon in various fields. Third generation solar cells, mainly comprised of inorganic–organic, polymer based nanocomposites [13]. Nanoparticles

Table 1 Absorbance and emission characteristics of various QD/polymer composites [9]

QD/polymer composite	Absorbance (nm)	Emission (nm)	Emission color	QY (%)	Ref.
PVA-N@C-dots	286, 355	~460	Cyan	44	[17]
N-doped CD in chitosan/PVA	360	436	Cyan	–	[18]
CD in polyethylene glycol	200–500	550	Yellow	14.86	[19]
Magnetite-MIPs	–	470	Blue	–	[20]
Cu-alginate	–	513	Green		[21]
C-dots/PVB film	353, 410, 500	550	Green–Blue, Orange–Red		[22]
CD polymer	200–500	550	Yellow	14.86	[23]
CDs@MIPs	250–300	450	Blue		[24]
WCDs@polystyrene		~590	Orange, Blue	10.7, 15.2	[25]
CD in polyethyleneimine	290, 340, 380	~565	Cyan	1.9–4	[26]
C-dots/PVA 360	282, 341	459	Green	8.64	[27]
CDs@PVA	~350	420–440	Blue		[28]
Poly(VPBA-AAm)	241	515	Blue		[29]
CQD/PAN nanofibers	314, 316, 318	530 560, 598, 660	Red, Green, Blue		[30]

(NP) have enabled modification and improvement in properties of polymer matrix [14]. NP reinforced polymer matrix exhibited enhanced mechanical, thermal and electrical properties which largely depends on the polymer-NP interface, concentration and uniformity of NP added. For specific and selective sensing probes, molecularly imprinted polymer (MIP) have gained much attention recently, due to the generation of specific-recognition sites or cavities during synthesis [15]. Conducting polymers on the other hand are extensively used as electrode materials due to their interactions capability such as H-bonding and π-π interactions, ion-exchange ability, hydrophobicity and presence of multi-functional groups [16].

With ever emerging applications and possibilities, QD/polymer composites have unlocked path for next generation materials. We are already witnessing breakthroughs of such materials in consumer electronics, sensors and bio-imaging applications. This chapter will discuss in detail about the nature, synthesis and applications of such composites.

2 Synthesis

Developing a highly efficient QD/polymer composite is very crucial. Fabrication of polymers with particles such as QDs, enables greater strength and stability. The QDs can also be stabilised by polymer encapsulation. Common techniques used for the synthesis of polymer nanocomposites are solution mixing, melt mixing, in situ particle processing, in-situ polymerization, template synthesis, grafting, electrohydrodynamic, etc., depending upon the types of QDs and polymers and their interactions [31]. There are mainly two synthetic approaches for synthesizing QDs, namely the (a) top-down and (b) bottom-up approach. The physio-chemical properties significantly depend on the type of precursor and the synthetic technique. The top-down approach is beneficial in case of large-scale production and oxygen-functionalised surface, whereas bottom-up approach facilitates the size, morphology and surface functionalised tailored synthesis [9]. The various synthetic methods used for preparing QD/polymers composites are discussed.

2.1 Solution Mixing

A magnetic and photoluminescent nanocarrier Fe_3O_4-HBPLCArg/QDs was developed by mixing dispersed CD solution in DMSO and N,N-dicyclohexylcarbodiimide (DCC), 4-dimethylaminopyridine (DMAP) with Fe_3O_4-HBPLC (hyperbranched poly L-lysine citramid) for 24 h [32]. QD dispersed piezoelectric polymer films were synthesized by mixing ZnO-D capped and stabilized by dopamine, in a mixture of acetone and DMF, with polyvinylidene fluoride-co-hexafluoropropylen (PVDF-HFP) copolymer, stirred for 12 h, followed by spin-coating over flexible electrode for 60 s at 1000 rpm. The hybrid film was then annealed for 1 h, at low temperature below 130 °C to avoid melt or delamination [33].

2.2 Template Synthesis

For synthesis of functionalised QD/polymer composites hard and soft template technique can be employed. Soft templates are easy to dissolve and economical whereas the hard templates used provides shape and size controlling ability and uniform dispersion [10]. In soft templates such as self-assembled biopolymers, water emulsions, soft carbon, surfactants or oil are commonly used. SnS_2 QDs/pg-C_3N_4/MWCNT (multi walled carbon nanotubes) surface was imprinted with 3-aminopropyltriethoxysilane (APTES) polymer resulting in Molecularly imprinted polymer on SnS_2 (MIP@ SnS_2) QDs/pg-C_3N_4/MWCNTs nanocomposite by sol–gel method. Indoxacarb (IDC) is used as soft template which can be easily dissolved by a solvent mixture of ethanol/acetonitrile (8:2 v/v)0 [34]. On the contrary, hard rigid structures are used as hard templates. Facilitates the shape and size tuning of nanocomposite synthesis. Hierarchical structure of $CsPbBr_3$ QDs embedded in di-block, amphiphilic copolymer polystyrene-poly2-vinyl pyridine (PS-b-P2VP, S2VP), was achieved using S2VP as micelle template, in which the pyridine group closely binds to $CsPbBr_3$ facilitating in-situ growth. The nanocomposite was synthesized by spin-coating of the mixture of $CsPbBr_3$ QD and S2PV polymer in toluene, followed by annealing under mild temperature [35]. HA-GQD-MIP nanocomposite probe was developed using Norfloxacin as template, which thereby create norfloxacin specific cavities in the synthesized MIP, shown in Fig. 1 [15].

2.3 Grafting

Various functionalised polymer coated CdSe QDs core–shell nanocomposite synthesis was reported by Zhu et al., a light-mediated polymerization, facilitating one-pot synthesis of QD/polymer nanocomposites [36]. CdSe QDs acting both as a catalyst as well as the inorganic core to the organic–inorganic core shell structure. This approach can be highly beneficial to synthesise and tune the polymer functionality and can be used for developing complex hybrid nanomaterials. Recently thiol-ene photo-polymerization have been reported to be highly efficient and instantaneous crosslinking method unaffected by ambient reaction environment. Here nanoparticle ligand act as the site from where the step-growth polymerization initiates resulting in crosslinked composites. Other photopolymerization was also reported initiated by UV light over QD site for efficient LED applications. Fabrication of Blue LED was also reported by blending red and green QD in appropriate ratio, with monomer, oligomer along with the photoinitiater [37]. By incorporation of QDs in polymer matrix, significant improvement of thermal stability was observed.

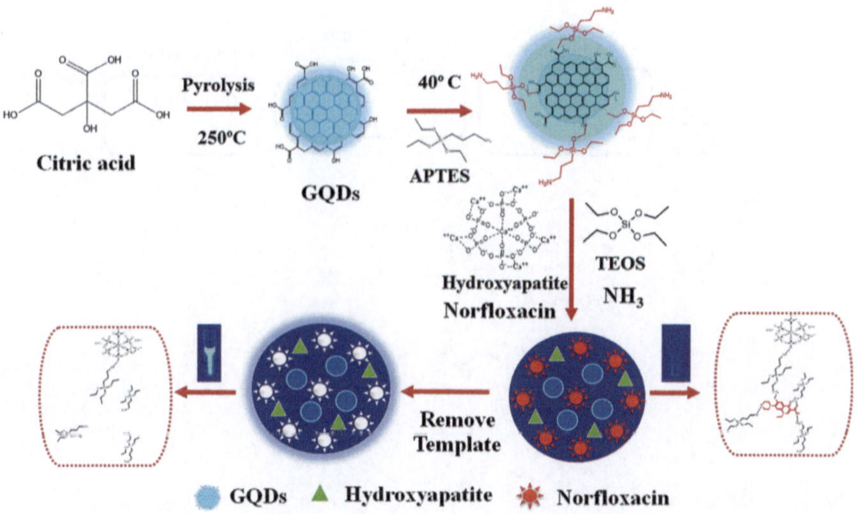

Fig. 1 Fabrication of HA-GQD-MIP nanocomposite using Norfloxacin template for Norfloxacin specific detection. Reproduced with permission from *Microchemical Journal, 2020, 158, 105,127* [15]

Thiol-ene crosslinking was also reported to be initiated thermally forming $CsPbBr_3$ QD-Polymer self-healing, elastic and luminescent composite [38]. Figure 2 also lists schematic for grafting QD with bidentate ligands [1].

2.4 Hydrothermal/Solvothermal

A photoluminescent nanocomposite was prepared using arginine derived hyperbranched polymer and glucose derived QD by simple hydrothermal method in a stainless steel autoclave at 180 °C for 6 h [39]. This nanocomposite was used as nano carrier in drug delivery. Green QDs such as ZnCdSeS/ZnS and Red QDs such ZnCdSe/ZnS encapsulated in polypropylene (PP) thermoplastic matrix, composite film exhibited color conversion property with high stability as compared to conventional PMMA film. Besides providing better protection from oxygen and moisture, PP polymer matrix also rendered a mild environment due to the absence of ester groups, as compared to PMMA matrix [40].

2.5 Electrochemical Method

A liquid phase synthesis of nanocomposite via electrohydrodynamic emulsification mediated by self-assembly approach was reported by Lee et al. [41] Here a

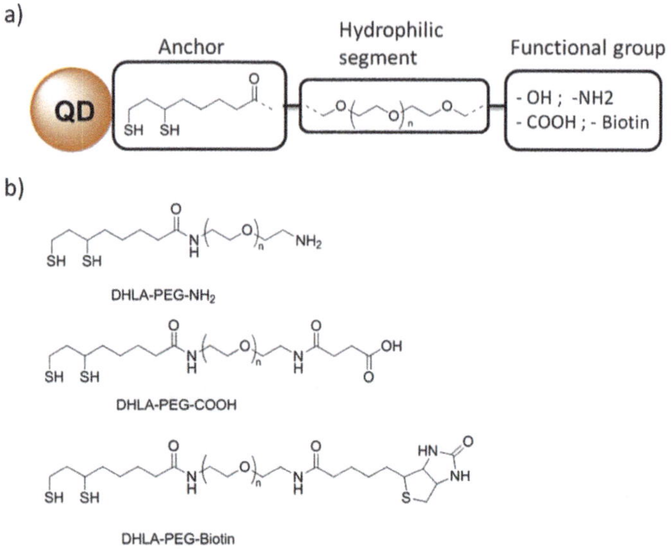

Fig. 2 **a** Schematic of grafting to the surface of QDs using bidentate ligands. **b** The end-functionalized polymers with suitable chemical groups for further derivatization, reproduced with permission from *Nanoscale, 2013, 5, 12,018*

non-conductive fluid chloroform containing poly(styrene-b-ethylene oxide) functionalised with amphiphilic block copolymer octadecylamine is being sprayed into a conductive fluid distilled water containing CdSe/ZnS QD, in between a metal needle and a counter electrode. The size of the emulsion droplets decreased with the increased voltage. On the other hand, polymerization of thionine was facilitated by amino-functionalised GQD (afGQD), to produce poly(thionine) modified afGQD nanocomposite electrochemically, due to their improved electro-conductivity [42].

2.6 Chemical Oxidation

Chemical oxidation is another simple in-situ synthesis of QD@polymer nanocomposite. Polyaniline@GO nanocomposite was synthesised by polymerization of aniline in presence of oxidising agent ammonium persulfate (APS) and acidic medium [16, 43].

2.7 *Wrapping*

Strategy to functionalise nanostructures using non-covalent interaction between organic polymers and inorganic QDs. A nano-bio catalyst graphene oxide/glucose oxidase was synthesised by wrapping enzyme molecules over graphene oxide surface via adsorption and crosslinking [44].

2.8 *Sol–gel Method*

One of the simplest, low cost and greener method of polymerization under mild conditions. Non-imprinted polymer NIP@SiCQDs and molecularly imprinted MIP@SiCQDs nanocomposite was reported to be synthesized using this method, coupled with template synthesis technique, using target analyte as template for synthesizing MIP@SiCQDs. SiCQDs were first functionalized by 3-aminopropyltriethoxysilane (APTES), then MIP and NIP nanocomposite was prepared by sol–gel method, using APTES monomer and tetraethoxysilane (TEOS) as cross-linker [45].

3 Reinforcements, Design, and Fabrication

3.1 *Surface Fabrication of QDs*

Formation of efficient and specific QD/polymer nanocomposites requires good dispersibility of QDs in the solvent with minimum aggregation. Good dispersibility can be commonly achieved by exchanging hydrophobic ligands with a chemical group with high affinity towards the binding QD. Amphiphilic ligands can also be coated over the QDs, the hydrophobic part easily binds to non-polar solvent, simultaneously the hydrophilic part renders the QD soluble in water. CdSe/ZnS QDs were coated with designed peptide comprising of a hydrophilic part for dispersion in water and an adhesive part, which binds to the QD [55] (see also Fig. 3).

Multidentate ligands also enhances the binding affinity with the QDs due to the presence of multiple binding sites, which thereby results in the QD/polymer interface stability [1] (also Fig. 2).

Results show that the chain length of the dispersant polymer also play a significant role in the dispersibility of the QDs due to effects of static and steric stabilization [31]. Silane modified ZnO QD-silicone nanocomposite exhibited slightly lower transmittance than that of pure silicone polymer, which suggests excellent dispersibility of QD over silicone matrix.

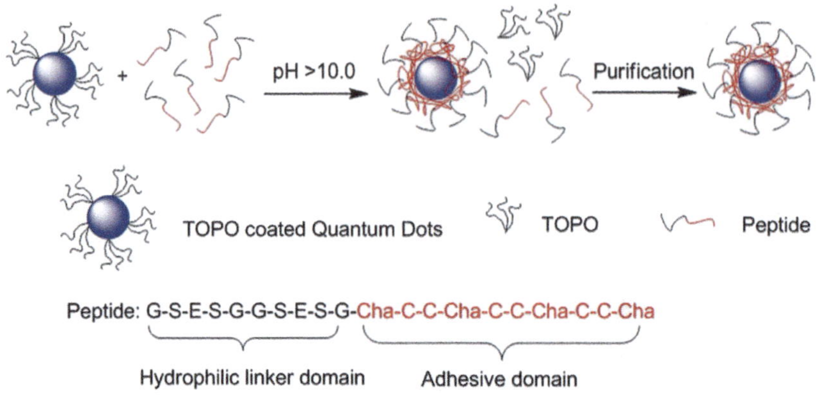

Fig. 3 CdSe/ZnS QDs with amphiphilic peptide ligand with an adhesive part to bind to QD and hydrophilic part to promote water dispersibility, reproduced with permission from, *Nanoscale, 2013, 5, 12,018–12,032*

3.2 Composite Fabrication

Polymer matrixes have wide range of applications and have replaced conventional materials such as ceramics, metals, alloys mostly due to their low cost of production and lightweight property. However, these polymer matrices have certain drawbacks such as thermal, chemical and photo instability [56]. Particle reinforced polymer matrix especially with nanoparticles or QDs are reported to be significantly beneficial in improving the stability of such polymer materials. Recently multi-scale composites attracted significant attention due to their advanced multi-functionality and improved electrical, mechanical, and optical properties. They are superior composites which are reinforced with nanoscale particles and macroscale fibres. The most common nanoparticle reinforced materials that are widely used are mainly carbon nanotubes (CNT), graphene, metal oxides, metals, etc. The key to effective reinforcements is uniform dispersion of the QDs into the polymer matrix and their stabilization from agglomeration. Ultra-sonication is most widely used mechanical technique for QD dispersion. Both ultra-sonication and ball-milling use shear forces for the dispersion of QDs which can also results in deformation of the original particles, generation of heat, defective structure, etc. Cryo-milling improved the thermal and electrical properties under lower temperatures.

To overcome the challenges of the mechanical techniques, low-cost chemical techniques can be used for improved dispersion by functionalization or tuning the surface of the QDs. Large bulky groups can be attached to these QD surface to create steric repulsion, thereby inhibiting aggregation. Functionalization of QDs plays a significant role in reinforcement of the QD/polymer composites. 2D structured graphene QDs with both alcoholic and acid groups resulted in tougher and stronger graphene QD-epoxy polymer composites [14]. Oxygen, aromatic rings and amine groups facilitate the bonding of the graphene QDs to the epoxy group which can be further associated with weak interactions such as van der Waal's interaction, hydrogen bonding, etc. Various nanoparticles which enabled reinforcement of polymer matrix are shown in Table 2 [57].

Dewetting of QD embedded in Poly(lauryl methacrylate) (PLMA) polymer films were reported to enhance fivefold thin film emission. The QD@PLMA composite was prepared by rotomixing followed by sonication. Thin films were then developed by spin-coating the composite at 30 s over silicon substrate. Controlled dewetting was then done by spraying with ethanol. Surface patterning was also done by using suitable mask and irradiating the uncovered surface with mercury lamp. This irregular film morphology led to increase in scattering of light which, thereby resulted in improved out-coupling [58]. Similarly, dewetting techniques is used for developing CdSe/ZnS core–shell structure in a polymer film for LED and Photovoltaic applications [59]. Highly efficient solar cell heterojunction was developed with high light to energy conversion, incorporating fullerene (C_{70}) covalently attached to CdSe QD blended over Poly(3-hexylthiophene-2,5-dyil) P3HT polymer.

Recently direct laser writing (DLW) technology exhibited promising tool for fabrication and processing QD@polymer nanocomposites [60]. β-carboxyethyl acrylates surface modified red CdSe/CdZnS/ZnS QDs and green CdSe/ZnS QDs was mixed with dipentaerythritol penta-acrylate (SR399) monomer resin. Polymerization was then initiated by femto second laser of 532 nm. It is a multi-photon polymerization initiated by high energy laser irradiating the QD@polymer nanocomposite. Firstly, the surface modification enabled the uniform distribution of QD into polymer matrix, secondly the size of the QD@polymer lines significantly depended on power P and velocity v.

3.3 Composite Stability

The stability of the prepared QD@polymer composite is of high significance for their optimum efficiency. Gray et al., studied the leaching of CdSe QDs from polyethylene polymer matrix in presence of common acids present in food, such as acetic acid, hydrochloric acid and citric acid. Significant leaching of QDs was observed after long term exposure to acetic acid as compared to citric which supressed the QD transfer, resulting in decrease in the PL intensity [61].

Table 2 QD reinforced polymer nanocomposites and their applications

QD	Polymer	Synthetic route	Applications	Ref.
CdSe/ZnS	Polypropylene	Mixing	–	[46]
Si capped with divinyl benzene	Polystyrene	Spin coating	Tunable refractive index	[47]
CdSe/ZnS	Polydimethyl siloxane	Ligand exchange/imprinting lithography	Photonic/optoelectronic	[44]
CdSe/CdS/ZnS	Poly(maleic acid-altoctadecene) in LDH	Drop-casting	Display/optical coatings	[48]
CdSe/ZnS	Dioctyloxybenzodithiophene-based polythiophene	Ligand exchange	Optoelectronic	[49]
PbS	Naphthalene diimide based n-type polymer	Thiol-ene	Photovoltaic	[50]
ZnCdS/ZnS QD/g-C_3N_4	Carboxymethyl Chitosan	Self-assembly	Mn^{2+} detection	[51]
CdSe/CdS	Poly(styrene-ethylene-butylene-styrene) (SEBS)	Drop casting	Displacement sensors	[12]
Ag (AMPSA)/ GO QDs	PANI	In-situ oxidative polymerization	Cr (IV) sensing	[52]
Hydroxyapetite-GQD	MIP (APTES/TEOS)	Template synthesis	Norfloxacin sensing in milk and chicken meat	[15]
GQDs@Fe_3O_4	MIP (APTES/TEOS)	Template synthesis	Ceftazidime sensing in milk	[53]
Mn–ZnS	MIP N,N′-Methylenebisacrylamide (MBA) and poly (N-isopropylacrylamide) (NIPAM) monomers	Template synthesis	Glyphosate sensing	[54]
GO	PANI	In-situ chemical-oxidative polymerization	Extraction of antibiotic tetracycline (TC)	[16]
SiCQDs	MIP (APTES/TEOS)	Sol–gel	Indoxacarb (IXC) sensing	[45]
afGQDs	Poly(thionine)	Electropolymerization	Hg^{2+} sensing	[42]

4 Applications

4.1 *Light Emitting Diodes and Displays*

Given the extraordinary optical properties of the quantum dots, and coupled with the fact that they can be easily tuned to entire visible spectra [62–65] unsurprisingly one of the first applications of QD-polymer nanocomposite is the high-efficiency Light Emitting Diodes (LEDs) [66]. Various QD-polymer composites have been demonstrated to have excellent yields and brightness. Usual construction of such an LED involves sandwiching a QD-polymer nanolayer between a hole transport layer for contact and conductivity (Fig. 4). These layers must be chosen carefully to ensure optimal hole-electron recombination, following which emission wavelengths can be tuned using the size and chemical constituent of the quantum dots.

Another way such composites are used in LEDs is to enhance their post hoc capabilities. e. g. using QD-polymer LED masks for white light emitting LEDs [67]. There the LED Poly(dimethylsiloxane) (PDMS) lens of a blue led was replaced with ZnCdSe/ZnSe-PDMS composites. ZnCdSe/ZnSe coreshell particles were tuned to emit in red-green region of visible spectra, rendering the LED white. There are increasing strides being made in miniaturizing these LEDs enough to develop a QD-LED based displays [62].

While, QD-LED based displays have still not found their way in the commercial market, QD-composites based displays have matured enough to be launched in consumer commercial space. Samsung electronics launched their "QLED" series of displays, which uses an composite-QD layer as a contrast enhancer on top of the LCD-LED layer [68], following which several commercial vendors have released similar displays (ViewSonic [69]). The current generation of displays use the composites-QD as a passive fluorescent layer.

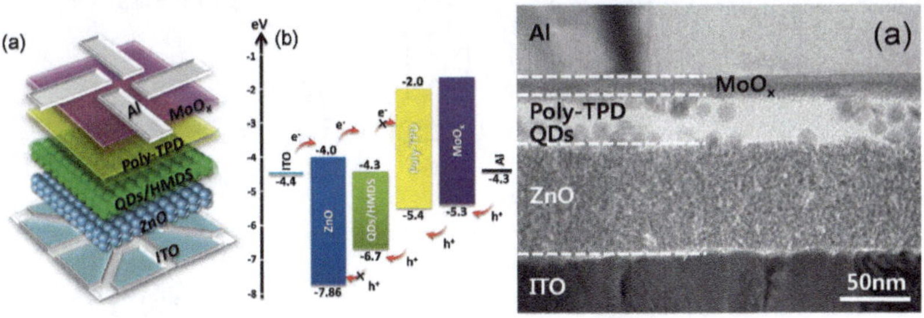

Fig. 4 QD-polymer nanocomposite LED constituting layer. Here QD-polymer composite can be seen sandwiched between ZnO and Al layers, which act as hole and electron transport respectively. Reproduced with permission from *J. Mater. Chem. C, 2017, 5, 522–526* [66]

4.2 Sensors

It was long hypothesized that due to higher optical sensitivity/variability of the quantum dots, they can be used in various sensors by the primary means of distinguishing energy flow [70]. This is primarily exploited in FRET based biological QD tags but has also resulted in large number of composite based sensors.

Simplest of them can be attributed to change-of-environment sensors, where the optical properties of QDs changes with mechanical or chemical changes in the composite film. Example of these are CdSe-polymer inkjet printed films for detecting explosives [71], measuring strength of sea-water damaged Glass Fibre Reinforced Polymer by fluorescent measurements of infused 3-(trimethoxysilyl)-1-propanethiol (MPS) modified CdS nanoparticles, and tertracycline sensor based on fluorescence quenching of B, N doped carbon dots—PVA hydrogel composite [72]. Other interesting examples includes “reverse” core–shell composite of polypyrrole/graphene quantum dots, where the polymer core is coated with graphene QDs [73]. This is somewhat different from conventional composite materials, where a polymer matrix is used to embed the active QD particle. The authors demonstrated that when these polymer graphene QD hybrids can detect Dopamine in external environments by red-shifting fluorescent emission. The change-of-environment need not come solely form direct interaction with an external chemical stimulus but can also be a result of deformation in the polymer-composite [22]. Demonstrated mechanical sensing property of carbon dots/ polyvinyl butyral composite, where the fluorescence of the c-dot/ polyvinyl butyral (PVB) film changed with the applied deformation; by measuring the relative fluorescence luminescence they were able to characterize the stress/strain of the film (Fig. 5).

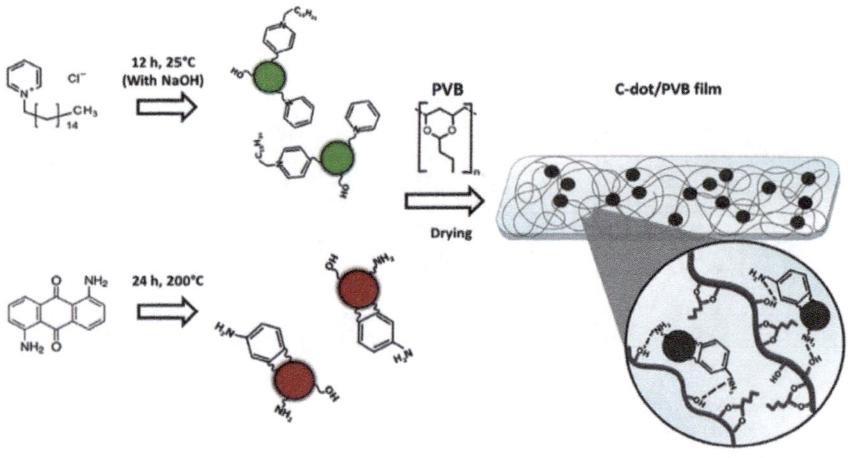

Fig. 5 Synthesis of composite film susceptible to fluorescent changes on applied strain. Reproduced from *Carbon, 2019, 152, 363–371* [22]

While most popular, optical sensing is not the only method to detect the change of chemical environment. Several reports exist, for example, detecting Metobromuron herbicide in aqueous media using gold-nitrogen doped carbon quantum dots with silver core–shell polymer composite [74]. Using polypyrrole/graphene quantum dots composition [75] reported electrochemical sensing of bisphenol A.

QD based sensors are not limited to the change-in-environment sensing. While the above sensing applications rely on external interaction with the composites, a more flexible approach is the intrinsic sensing using the inner filter effect [76, 77].

One of the most interesting example of exploiting the inner filter effect was given by [12] where the polymer composites were used to yield the microscopic z-height measuring sensor where the height of the sensor can be measured simple by measuring the red-shift in emission wavelength. This was demonstrated to measure small movements on a chicken lung during using optical measurements.

4.3 Photovoltaics

N. C. Greenham reported photovoltaic effect in a CdSe—MEH-PPV (poly(2-methoxy, 5-(2'-ethyl)-hexyloxy-p-phenylenevinylene) QD-polymer composite [78]. Interestingly here the polymer matrix is not just inter-spacer as we saw in the LED applications, but rather MEH-PPV acts as a hole conducting polymer which takes away the photogenerated holes away from the semiconducting QD. This was among the first reports of QD photovoltaics using such conducting polymer matrix, and such photovoltaics are called bulk heterojunction photovoltaics [79]. Further reports from the same group soon after used nanorods like wurtzite CdSe nanodots in poly(3-hexylthiophene) matrix and reported much higher yields [80, 81]. It was suggested that the rod-like shape might improve the hole conductance across the QD-polymer junction, hence resulting in higher efficiency of the photo-voltaic. Different iterations of it include using an additional TiO_2 layer for electron transport, which increases the efficiency even further [82]. Other alternative approach developed was to use electron conducting polymer matrix instead.

Following the above discoveries more optically active polymer matrix was considered and it was reported that narrow bandgap polymer, poly[{2,5-bis(2-hexyldecyl)-2,3,5,6-tetrahydro-3,6-dioxopyrrolo[3,4-c]pyrrole-1,4-diyl}-*alt*-{[2,2′-(1,4-phenylene)bis-thiophene]-5,5′-diyl}], (PDPPTPT), when mixed with PbS can be used to yield much more efficient solar cells [83]. Several other similar small band-gap polymers have been since reported PDTPBT [84], P3HT [84], PDTPQx [85] and si-PCPDTBT [86] to work excellently with the quantum dots. It was reported that instead of bulk-heterojunction, if polymer-QD interaction was mediated by a small far-IR active molecule, the efficiency of the solar cell can be improved even further [79].

As an alternative to polymer matrix or small molecule junction, surface functionalized CdSe QDs were also studied, where the surface of the QD was functionalized using extremely bulky molecule, such as C_{70}. It provides the right balance between bulk-heterojunction and small molecule interface [87].

4.4 Catalysis

The fact that quantum dots are such an efficient photoactive material, is not only exploited in photovoltaics, but also in photo-degradation/photo-catalysis as well. One of the most widely sought after application is the photo-active water splitting. Several reports exist which demonstrate the capability of the QD composites in splitting water and generating hydrogen or simply as a possible fuel cell, e. g. [88] reported a B-doped graphene QD composite for Zn-air battery, and water splitting application. CdS/polymeric g-C_3N_4 composites were reported to have hydrogen evolution rate of about 4.494 mmol h^{-1} g^{-1} under irradiation of simple 300 W xenon lamp [89]. However it seems the majority of community prefers QD-Nanostructure composites over polymer composites for water splitting [90]. Which is understandable as majority of polymeric composites does not favor high water miscibility.

Other than water splitting another emerging use of QD composite photoactivity seems to be emerging in the form of water purification. Under irradiation QD composites can degrade small molecules. Reports of photodegradation of Rhodamine B [91], Methylene blue [92], Ciprofloxacin [93], napthelene and pyrene [94] etc. are aplenty [95].

New exciting application of QD catalysed reactions include polymerization using CdSe QD composite as an photoelectron transfer agent for initiating and catalysing electron transfer reversible addition—fragmentation chain transfer catalysis [7].

4.5 Biomedical

Owing to high quantum yield and narrow excitation spectra of the QDs they have already found strong footing in microscopy fluorescent tags. Gao et. al. [96] used amphiphilic triblock copolymer (hydrophilic polymethacrylic segments, and two hydrophobic polybutylacrylate and polyethylacrylate segments) with TOPO protected CdSe/ZnO coreshell quantum dots. They reported enhanced cancer cell targeting in live mice. They also used prostate binding antibodies, tagged with above quantum dots for active targeting of the cancer cells. A PVP coated BiS QD were used as CT scan enhancing aid. $CsPbBr_3$ Perovskite Quantum Dots with polystyrene-*block*-poly(ethylene-*ran*-butylene)-*block*-polystyrene (PS-PEB-PS) and poly(ethylene glycol)-*block*-poly(propylene glycol)-*block*-poly(ethylene glycol (PEG-PPG-PEG) polymer were used for tracking triple-negative MDA-MB-231 breast tumor-derived exosomes [97]. PEG/CdSe composites were also used for nontoxic live imaging of a Xenopus (frog) embryo [98]. Similar applications have been reported in all aspects of qualitative and quantitative biological imaging.

QD nanocomposites are also making huge inroads in bioimaging and targeted photodynamic therapy. QD nanocomposites are also used as an active agent in photodynamic therapies in mice [99], as an anti-bacterial agent for agricultural applications [100], and very recently as theragnostic probe for COVID detection [101, 102]. PEGylated MoS_2 QDs (MoS_2-PEG) were used for their pH responsive properties and developed a composite that could be used to deliver anti-cancer drug such as doxorubicin to U251 cell lines [103]. Hu et al. developed branched polyethyleneimine (bPEI) based multifunctional photoluminescent carbon dots materials that had quantum efficiencies as high as 54.3%, and were used for gene delivery applications in MCF7 cells [104]. Abdullah-Al-Nahain et al. prepared dopamine hydrochloride conjugated hyaluronic acid (HA), anchored to a graphene quantum dot (GQD) and injected it to tumour-bearing balb/c female mice having overexpressed CD44 receptor [105]. The tumour tissue showed bright fluorescence. These biocompatible materials were successfully sued for drug delivery and in vivo cellular imaging.

There have also been reports of completely novel applications of QD composites, e. g. $Cd_xZn_{1-x}S/ZnS$ core–shell QD—polymer composites can be used as a scintillation gamma ray detectors 5, use of carbon dots in tracing counterfeit currency [106], and as a therapeutic agent in neuro-degenerative diseases [107]. This shows that QD nanocomposites might still have undiscovered applications.

5 Conclusion

In can be concluded that QD and polymer exhibits synergistic effects, enhancing the strength and properties of both via composite formation. Different synthetic routes exhibited various functionalised and well-dispersed QD@polymer composites, among which grafting and template synthesis being the most used. Wrapping technique can be adapted for non-covalent QD-polymer interaction. Uniform QD dispersion can also be achieved using simple bulky ligands to enable steric repulsion which inhibits aggregation. For high photo-luminescent composite fabrication using dewetting can be done to increase surface irregularity, thereby increasing the scattering.

These physical properties of QD-Polymer nanocomposites, when coupled with the optical properties of the QD, results in materials which can be applied to problems ranging from LEDs and white light generation, to bioimaging and photodynamic therapy. QD composites have already made inroads in higher contrast commercial LED displays, and optical microscopy tags. They also seem to be promising in the field of photovoltaics, photocatalysis, water purification, hydrogen/oxygen generation, and chemical sensing.

References

1. Tomczak, N., Liu, R., Vancso, J.G.: Polymer-coated quantum dots. *Nanoscale* **5**, 12018 (2013)
2. Owen, J., Brus, L.: Chemical synthesis and luminescence applications of colloidal semiconductor quantum dots. J. Am. Chem. Soc. **139**, 10939–10943 (2017)
3. Chaitong, N., Chansud, N., Orachorn, N., Limbut, W., Bunkoed, O.: A magnetic nanocomposite optosensing probe based on porous graphene, selective polymer and quantum dots for the detection of cefoperazone in milk. Microchem. J. **171**, 106838 (2021)
4. Siva, V., et al.: Gel combustion synthesized NiMoO4 anchored polymer nanocomposites as a flexible electrode material for solid state asymmetric supercapacitors. Int. J. Hydrog. Energy **48**, 18856–18870 (2023)
5. Liu, C., et al.: Transparent ultra-high-loading quantum dot/polymer nanocomposite monolith for gamma scintillation. ACS Nano **11**, 6422–6430 (2017)
6. Lu, H., Ho, S.-J., Chen, H.-S., Yang, A.C.-M.: Interactions of excited charges with substrate fields in CdSe/ZnS quantum dots embedded in nano polymer films: implications for light-emitting and photovoltaic devices. ACS Appl. Nano Mater. **4**, 2108–2117 (2021)
7. McClelland, K.P., Clemons, T.D., Stupp, S.I., Weiss, E.A.: Semiconductor quantum dots are efficient and recyclable photocatalysts for aqueous PET-RAFT polymerization. ACS Macro Lett. **9**, 7–13 (2020)
8. Xie, Q., et al.: Parameters in planar quantum dot-polymer solar cell: Tuned by QD Eg, ligand exchange and fabrication process. Org. Electron. **69**, 1–6 (2019)
9. Ahirwar, R.C., et al.: Progression of quantum dots confined polymeric systems for sensorics. Polymers **15**, 405 (2023)
10. Yu, C.-H., et al.: Design strategies, surface functionalization, and environmental remediation potentialities of polymer-functionalized nanocomposites. Chemosphere **306**, 135656 (2022)
11. Heo, H.S., et al.: Fabrication of cross-linked quantum dot/colorless-polyimide nanocomposites for highly efficient and long-term stable light-emitting diodes. Chem. Mater. **34**, 6958–6967 (2022)
12. Koc, M.A., et al.: Characterizing photon reabsorption in quantum dot-polymer composites for use as displacement sensors. ACS Nano **11**, 2075–2084 (2017)
13. Nazim, M., Kim, B., Lee, S., Min, B.K., Kim, J.H.: UV-Curable Polymer–QD Flexible films as the downconversion layer for improved performance of Cu(In, Ga)Se2 Solar Cells. Energy Fuels **34**, 14581–14590 (2020)
14. Gobi, N., Vijayakumar, D., Keles, O., Erogbogbo, F.: Infusion of graphene quantum dots to create stronger, tougher, and brighter polymer composites. ACS Omega **2**, 4356–4362 (2017)
15. Bunkoed, O., Donkhampa, P., Nurerk, P.: A nanocomposite optosensor of hydroxyapatite and graphene quantum dots embedded within highly specific polymer for norfloxacin detection. Microchem. J. **158**, 105127 (2020)
16. Sereshti, H., Karami, F., Nouri, N., Farahani, A.: Electrochemically controlled solid phase microextraction based on a conductive polyaniline-graphene oxide nanocomposite for extraction of tetracyclines in milk and water. J. Sci. Food Agric. **101**, 2304–2311 (2021)
17. Kumar, V.B., Sahu, A.K., Mohsin, A.S.M., Li, X., Gedanken, A.: Refractive-Index Tuning of Highly Fluorescent Carbon Dots. ACS Appl. Mater. Interfaces **9**, 28930–28938 (2017)
18. Kurt, S.B., Sahiner, N.: Chitosan based fibers embedding carbon dots with anti-bacterial and fluorescent properties. Polym. Compos. **42**, 872–880 (2021)
19. Fernandes, D., Heslop, K.A., Kelarakis, A., Krysmann, M.J., Estevez, L.: In situ generation of carbon dots within a polymer matrix. Polymer **188**, 122159 (2020)
20. Wang, M., et al.: A magnetic and carbon dot based molecularly imprinted composite for fluorometric detection of 2,4,6-trinitrophenol. Microchim. Acta **186**, 86 (2019)
21. Roucan, M., Kielmann, M., Connon, S.J., Bernhard, S.S.R., Senge, M.O.: Conformational control of nonplanar free base porphyrins: towards bifunctional catalysts of tunable basicity. Chem. Commun. **54**, 26–29 (2018)
22. Shauloff, N., Bhattacharya, S., Jelinek, R.: Elastic carbon dot/polymer films for fluorescent tensile sensing and mechano-optical tuning. Carbon **152**, 363–371 (2019)

23. Zheng, X., Ding, G., Wang, H., Cui, G., Zhang, P.: One-step hydrothermal synthesis of carbon dots-polymer composites with solid-state photoluminescence. Mater. Lett. **238**, 22–25 (2019)
24. Xu, X., et al.: Carbon dots coated with molecularly imprinted polymers: A facile bioprobe for fluorescent determination of caffeic acid. J. Colloid Interface Sci. **529**, 568–574 (2018)
25. Wang, C., Hu, T., Chen, Y., Xu, Y., Song, Q.: Polymer-Assisted self-assembly of multicolor carbon dots as solid-state phosphors for fabrication of warm, high-quality, and temperature-responsive white-light-emitting devices. ACS Appl. Mater. Interfaces **11**, 22332–22338 (2019)
26. Bhattacharya, S., et al.: Fluorescent self-healing carbon dot/polymer gels. ACS Nano **13**, 1433–1442 (2019). https://doi.org/10.1021/acsnano.8b07087
27. Carvalho, J., et al.: Hydrothermal synthesis to water-stable luminescent carbon dots from acerola fruit for photoluminescent composites preparation and its application as sensors. Mater. Res. **22**, e20180920 (2019)
28. Tian, Z., et al.: Multilevel data encryption using thermal-treatment controlled room temperature phosphorescence of carbon dot/polyvinylalcohol composites. Adv. Sci. **5**, 1800795 (2018)
29. Wang, H., Yi, J., Yu, Y., Zhou, S.: NIR upconversion fluorescence glucose sensing and glucose-responsive insulin release of carbon dot-immobilized hybrid microgels at physiological pH. Nanoscale **9**, 509–516 (2017)
30. Safaei, B., Youssefi, M., Rezaei, B., Irannejad, N.: Synthesis and properties of photoluminescent carbon quantum dot/polyacrylonitrile composite nanofibers. Smart Sci. **6**, 117–124 (2018)
31. Fu, S., Sun, Z., Huang, P., Li, Y., Hu, N.: Some basic aspects of polymer nanocomposites: A critical review. Nano Mater. Sci. **1**, 2–30 (2019)
32. Aslani, R., Namazi, H.: Fabrication of a new photoluminescent and pH-responsive nanocomposite based on a hyperbranched polymer prepared from amino acid for targeted drug delivery applications. Int. J. Pharm. **636**, 122804 (2023)
33. Fu, H., et al.: Quantum dot hybridization of piezoelectric polymer films for non-transfer integration of flexible biomechanical energy harvesters. ACS Appl. Mater. Interfaces **14**, 29934–29944 (2022)
34. Khosropour, H., Saboohi, M., Keramat, M., Rezaei, B., Ensafi, A.A.: Electrochemical molecularly imprinted polymer sensor for ultrasensitive indoxacarb detection by tin disulfide quantum dots/carbon nitride/multiwalled carbon nanotubes as a nanocomposite. Sens. Actuators B Chem. **385**, 133652 (2023)
35. Jiang, Q., et al.: $CsPbBr_3$ perovskite quantum dots embedded in polystyrene-poly2-vinyl pyridine copolymer for robust and light-tunable memristors. ACS Appl. Nano Mater. **6**, 8655–8667 (2023)
36. Pan, Z., et al.: A mannosylated PEI–CPP hybrid for TRAIL gene targeting delivery for colorectal cancer therapy. Polym. Chem. **8**, 5275–5285 (2017)
37. Chou, C.-C., Wang, T.-L., Chen, W.-J., Yang, C.-H.: Quantum dot-acrylic acrylate oligomer hybrid films for stable white light-emitting diodes. ACS Omega **4**, 3234–3243 (2019)
38. Hsu, Y.-C., et al.: Fabrication of Intrinsic, Elastic, Self-healing, and Luminescent $CsPbBr_3$ Quantum dot-polymer composites via thiol-ene cross-linking. ACS Appl. Polym. Mater. **4**, 8987–8995 (2022)
39. Aslani, R., Namazi, H.: Synthesis of a new polymer from arginine for the preparation of antioxidant, pH-sensitive, and photoluminescence nanocomposite as a cancer drugs carrier. J. Ind. Eng. Chem. **112**, 335–347 (2022)
40. Fang, F., et al.: Thermally processed quantum-dot polypropylene composite color converter film for displays. ACS Appl. Mater. Interfaces **14**, 31160–31169 (2022)
41. Ho Lee, K., Ireland, M., Miller, B.L., Wyslouzil, B.E., Winter, J.O.: Synthesis of polymer nanoparticles via electrohydrodynamic emulsification-mediated self-assembly. J. Colloid Interface Sci. **586**, 445–456 (2021)
42. Tian, B., et al.: Ultrasensitive determination of mercury ions using a glassy carbon electrode modified with nanocomposites consisting of conductive polymer and amino-functionalized graphene quantum dots. Microchim. Acta **187**, 210 (2020)

43. Tang, S.-J., et al.: Polymerization of aniline under various concentrations of APS and HCl. Polym. J. **43**, 667–675 (2011)
44. Kim, T.S., et al.: Antifouling membranes employing a 2D planar nanobiocatalyst of crosslinked glucose oxidase aggregates wrapping extra-large graphene oxide. Chem. Eng. J. **424**, 130343 (2021)
45. Shirani, M.P., Rezaei, B., Ensafi, A.A., Ramezani, M.: Development of an eco-friendly fluorescence nanosensor based on molecularly imprinted polymer on silica-carbon quantum dot for the rapid indoxacarb detection. Food Chem. **339**, 127920 (2021)
46. Bobrovsky, A., et al.: Quantum dot–polymer composites based on nanoporous polypropylene films with different draw ratios. Eur. Polym. J. **82**, 93–101 (2016)
47. Choi, J.-K., Dung, M.X., Jeong, H.-D.: Novel synthesis of covalently linked silicon quantum dot–polystyrene hybrid materials: Silicon quantum dot–polystyrene polymers of tunable refractive index. Mater. Chem. Phys. **148**, 463–472 (2014)
48. Cho, S., Kwag, J., Jeong, S., Baek, Y., Kim, S.: Highly Fluorescent and stable quantum dot-polymer-layered double hydroxide composites. Chem. Mater. **25**, 1071–1077 (2013)
49. Han, Y.D., et al.: Quantum dot and π-conjugated molecule hybrids: nanoscale luminescence and application to photoresponsive molecular electronics. NPG Asia Mater. **6**, e103–e103 (2014)
50. Yoshida, K., Chang, J.-F., Chueh, C.-C., Higashihara, T.: Hybridization of an n-type semiconducting polymer with PbS quantum dots and their photovoltaic investigation. Polym. J. **54**, 323–333 (2022)
51. Wang, S., Liu, J., Huang, Y., Yang, N.: Selective sensing of Mn2+ based on ZnCdS/ZnS QD/carboxymethyl chitosan/g-C3N4 nanosheet nanocomposite film by the conformational regulation of polymer chain. Appl. Surf. Sci. **530**, 147255 (2020)
52. Shokry, A., Khalil, M.M.A., Ibrahim, H., Soliman, M., Ebrahim, S.: Highly luminescent ternary nanocomposite of polyaniline, silver nanoparticles and graphene oxide quantum dots. Sci. Rep. **9**, 16984 (2019)
53. Bunkoed, O., Raksawong, P., Chaowana, R., Nurerk, P.: A nanocomposite probe of graphene quantum dots and magnetite nanoparticles embedded in a selective polymer for the enrichment and detection of ceftazidime. Talanta **218**, 121168 (2020)
54. Sawetwong, P., Chairam, S., Jarujamrus, P., Amatatongchai, M.: Enhanced selectivity and sensitivity for colorimetric determination of glyphosate using Mn–ZnS quantum dot embedded molecularly imprinted polymers combined with a 3D-microfluidic paper-based analytical device. Talanta **225**, 122077 (2021)
55. Pinaud, F., King, D., Moore, H.-P., Weiss, S.: Bioactivation and cell targeting of semiconductor CdSe/ZnS nanocrystals with phytochelatin-related peptides. J. Am. Chem. Soc. **126**, 6115–6123 (2004)
56. Mittal, G., Rhee, K.Y., Mišković-Stanković, V., Hui, D.: Reinforcements in multi-scale polymer composites: Processing, properties, and applications. Compos. Part B Eng. **138**, 122–139 (2018)
57. Karki, S., Gohain, M.B., Yadav, D., Ingole, P.G.: Nanocomposite and bio-nanocomposite polymeric materials/membranes development in energy and medical sector: A review. Int. J. Biol. Macromol. **193**, 2121–2139 (2021)
58. Geldmeier, J., et al.: Dewetting-induced photoluminescent enhancement of poly(lauryl methacrylate)/quantum dot thin films. Langmuir **33**, 14325–14331 (2017)
59. Yoon, C., Yang, K.P., Kim, J., Shin, K., Lee, K.: Fabrication of highly transparent and luminescent quantum dot/polymer nanocomposite for light emitting diode using amphiphilic polymer-modified quantum dots. Chem. Eng. J. **382**, 122792 (2020)
60. Jue, J., Gan, Z., Luo, Z., Li, K.: Direct laser writing of functional QD–polymer structure with high resolution. Materials **16**, 2456 (2023)
61. Gray, P.J., et al.: Influence of different acids on the transport of CdSe quantum dots from polymer nanocomposites to food simulants. Environ. Sci. Technol. **52**, 9468–9477 (2018)
62. Choi, M.K., Yang, J., Hyeon, T., Kim, D.-H.: Flexible quantum dot light-emitting diodes for next-generation displays. Npj Flex. Electron. **2**, 1–14 (2018)

63. Lee, J., Sundar, V.C., Heine, J.R., Bawendi, M.G., Jensen, K.F.: Full color emission from II–VI semiconductor quantum dot-polymer composites. Adv. Mater. **12**, 1102–1105 (2000)
64. Yang, M., et al.: Flexible quantum dot light emitting diodes based on copper nanowires/poly(P-Phenylene Benzobisoxazole) composite electrode. J. Phys. Conf. Ser. **1637**, 012066 (2020)
65. Yoon, S.H., et al.: Flexible quantum dot light-emitting diodes without sacrificing optical and electrical performance. Appl. Surf. Sci. **566**, 150614 (2021)
66. Fu, Y., Kim, D., Moon, H., Yang, H., Chae, H.: Hexamethyldisilazane-mediated, full-solution-processed inverted quantum dot-light-emitting diodes. J. Mater. Chem. C **5**, 522–526 (2017)
67. Onal, A., et al.: High-performance white light-emitting diodes over 150 lm/W using near-unity-emitting quantum dots in a liquid matrix. ACS Photonics **9**, 1304–1314 (2022)
68. Xiang, C., Cao, W., Yang, Y., Qian, L., Yan, X.: The dawn of QLED for the FPD Industry. Inf. Disp. **34**, 14–17 (2018)
69. Kan, D.: What is quantum dot technology? ViewSonic Library. (2021). https://www.viewsonic.com/library/tech/explained/what-is-quantum-dot-technology/
70. Willard, D.M., Mutschler, T., Yu, M., Jung, J., Van Orden, A.: Directing energy flow through quantum dots: towards nanoscale sensing. Anal. Bioanal. Chem. **384**, 564–571 (2006)
71. Bright, C.J., Nallon, E.C., Polcha, M.P., Schnee, V.P.: Quantum dot and polymer composite cross-reactive array for chemical vapor detection. Anal. Chem. **87**, 12270–12275 (2015)
72. Fang, Y., et al.: Monitoring of seawater immersion degradation in glass fibre reinforced polymer composites using quantum dots. Compos. Part B Eng. **112**, 93–102 (2017)
73. Zhou, X., et al.: Dopamine fluorescent sensors based on polypyrrole/graphene quantum dots core/shell hybrids. Biosens. Bioelectron. **64**, 404–410 (2015)
74. Kıran, T.R., Yola, M.L., Atar, N.: Electrochemical sensor based on au@nitrogen-doped carbon quantum dots@Ag core-shell composite including molecular imprinted polymer for metobromuron recognition. J. Electrochem. Soc. **166**, H691 (2019)
75. Tan, F., et al.: An electrochemical sensor based on molecularly imprinted polypyrrole/graphene quantum dots composite for detection of bisphenol A in water samples. Sens. Actuators B Chem. **233**, 599–606 (2016)
76. Ceresa, L. et al.: On the origin and correction for inner filter effects in fluorescence. Part II: secondary inner filter effect -the proper use of front-face configuration for highly absorbing and scattering samples. Methods Appl. Fluoresc. **9**, 035005 (2021)
77. Kimball, J., et al.: On the origin and correction for inner filter effects in fluorescence Part I: primary inner filter effect-the proper approach for sample absorbance correction. Methods Appl. Fluoresc. **8**, 033002 (2020)
78. Greenham, N.C., Peng, X., Alivisatos, A.P.: Charge separation and transport in conjugated-polymer/semiconductor-nanocrystal composites studied by photoluminescence quenching and photoconductivity. Phys. Rev. B **54**, 17628–17637 (1996)
79. Baek, S.-W., et al.: Efficient hybrid colloidal quantum dot/organic solar cells mediated by near-infrared sensitizing small molecules. Nat. Energy **4**, 969–976 (2019)
80. Huynh, W.U., Peng, X., Alivisatos, A.P.: CdSe Nanocrystal Rods/Poly(3-hexylthiophene) composite photovoltaic devices. Adv. Mater. **11**, 923–927 (1999)
81. Liu, Y., et al.: Selective synthesis of wurtzite cdse nanorods and zinc blend CdSe nanocrystals through a convenient solvothermal route. J. Nanoparticle Res. **9**, 745–752 (2007)
82. Arango, A.C., Carter, S.A., Brock, P.J.: Charge transfer in photovoltaics consisting of interpenetrating networks of conjugated polymer and TiO2 nanoparticles. Appl. Phys. Lett. **74**, 1698–1700 (1999)
83. Piliego, C., et al.: Charge separation dynamics in a narrow band gap polymer–PbS nanocrystal blend for efficient hybrid solar cells. J. Mater. Chem. **22**, 24411–24416 (2012)
84. Seo, J., et al.: Enhancement of the photovoltaic performance in PbS nanocrystal:P3HT hybrid composite devices by post-treatment-driven ligand exchange. Nanotechnology **20**, 095202 (2009)
85. Noone, K.M., et al.: Broadband absorbing bulk heterojunction photovoltaics using low-bandgap solution-processed quantum dots. Nano Lett. **10**, 2635–2639 (2010)

86. Lu, H., Joy, J., Gaspar, R.L., Bradforth, S.E., Brutchey, R.L.: Iodide-passivated colloidal PBS nanocrystals leading to highly efficient polymer: nanocrystal hybrid solar cells. Chem. Mater. **28**, 1897–1906 (2016)
87. Albero, J., Riente, P., Clifford, J.N., Pericàs, M.A., Palomares, E.: Improving CdSe quantum dot/polymer solar cell efficiency through the covalent functionalization of quantum dots: implications in the device recombination kinetics. J. Phys. Chem. C **117**, 13374–13381 (2013)
88. Tam, T.V., et al.: Novel graphene hydrogel/b-doped graphene quantum dots composites as trifunctional electrocatalysts for Zn−air batteries and overall water splitting. Adv. Energy Mater. **9**, 1900945 (2019)
89. Ge, L., et al.: Synthesis and efficient visible light photocatalytic hydrogen evolution of polymeric g-C3N4 coupled with CdS quantum dots. J. Phys. Chem. C **116**, 13708–13714 (2012)
90. Kandi, D., Martha, S., Parida, K.M.: Quantum dots as enhancer in photocatalytic hydrogen evolution: A review. Int. J. Hydrog. Energy **42**, 9467–9481 (2017)
91. Vattikuti, S.V.P., Reddy, P.A.K., Shim, J., Byon, C.: Visible-Light-Driven Photocatalytic Activity of SnO2–ZnO Quantum Dots Anchored on g-C3N4 nanosheets for photocatalytic pollutant degradation and H_2 Production. ACS Omega **3**, 7587–7602 (2018)
92. Huang, H., Leung, D.Y.C., Kwong, P.C.W., Xiong, J., Zhang, L.: Enhanced photocatalytic degradation of methylene blue under vacuum ultraviolet irradiation. Catal. Today **201**, 189–194 (2013)
93. Rashid, J., et al.: Butterfly cluster like lamellar BiOBr/TiO2 nanocomposite for enhanced sunlight photocatalytic mineralization of aqueous ciprofloxacin. Sci. Total. Environ. **665**, 668–677 (2019)
94. Yang, X., et al.: Insight into the highly efficient degradation of PAHs in water over graphene oxide/Ag3PO4 composites under visible light irradiation. Chem. Eng. J. **334**, 355–376 (2018)
95. Ahmad, N., et al.: Visible light-conducting polymer nanocomposites as efficient photocatalysts for the treatment of organic pollutants in wastewater. J. Environ. Manage. **295**, 113362 (2021)
96. Gao, X., Cui, Y., Levenson, R.M., Chung, L.W.K., Nie, S.: In vivo cancer targeting and imaging with semiconductor quantum dots. Nat. Biotechnol. **22**, 969–976 (2004)
97. Pramanik, A., et al.: Water-soluble and bright luminescent cesium-lead-bromide perovskite quantum dot-polymer composites for tumor-derived exosome imaging. ACS Appl. Bio Mater. **2**, 5872–5879 (2019)
98. Dubertret, B., et al.: In Vivo imaging of quantum dots encapsulated in phospholipid micelles. Science **298**, 1759–1762 (2002)
99. Sharker, S.M., Do, M.: Nanoscale carbon-polymer dots for theranostics and biomedical exploration. J. Nanotheranostics **2**, 118–130 (2021)
100. Schwartz, S.H., Hendrix, B., Hoffer, P., Sanders, R.A., Zheng, W.: Carbon dots for efficient small interfering RNA delivery and gene silencing in plants. Plant Physiol. **184**, 647–657 (2020)
101. Adam, G.O., Sharker, S.M., Ryu, J.H.: Emerging biomedical applications of carbon dot and polymer composite materials. Appl. Sci. **12**, 10565 (2022)
102. Belza, J., Opletalová, A., Poláková, K.: Carbon dots for virus detection and therapy. Microchim. Acta **188**, 430 (2021)
103. Liu, L., et al.: PEGylated MoS2 quantum dots for traceable and pH-responsive chemotherapeutic drug delivery. Colloids Surf. B Biointerfaces **185**, 110590 (2020)
104. Hu, L., et al.: Multifunctional carbon dots with high quantum yield for imaging and gene delivery. Carbon **67**, 508–513 (2014)
105. Abdullah-Al-Nahain, et al.: Target delivery and cell imaging using hyaluronic acid-functionalized graphene quantum dots. Mol. Pharm. **10**, 3736–3744 (2013)
106. Guo, J., et al.: Green synthesis of carbon dots toward anti-counterfeiting. ACS Sustain. Chem. Eng. **8**, 1566–1572 (2020)
107. Ann Mathew, S. et al.: Luminescent chitosan/carbon dots as an effective nano-drug carrier for neurodegenerative diseases. RSC Adv. **10**, 24386–24396 (2020)

Quantum Dots@Metal–Organic Frameworks Composites

Amanpreet Kaur Jassal and Priyanka Kajal

Abstract The development and design of novel quantum dots@metal organic frameworks (QDs@MOFs) composites have been performed to address various societal issues such as energy crises, bacterial infections, the sensing of harmful gases or substances, etc. Combining the advantages of tunable characteristics, unique structural features, and advanced physiochemical properties of MOFs with quantum confinement in quantum dots, these hybrid materials can become a potential candidate for gas storage, catalysis, chemical sensing, and electrochemical energy storage applications. However, MOFs alone possess certain limitations, such as limited electrical conductivity, fixed pore size, and poor stability. These limitations can be overcome by the smart integration of QDs into MOF structures. The resulted host-guest structure shows a wide range of synergetic properties as the host (MOFs) framework suggests appropriate associations to the guest (QDs) for increasing stability, long carrier lifetime, enhanced, higher photoluminescence quantum yield and synergic properties. Thus, this chapter delivers a perception about the recent breakthroughs in the field of QDs@MOFs, their historical evolution, structural features, various synthesis strategies, and progress towards their utilization in numerous applications, including optoelectronics, biomedical, electrocatalysis, and energy storage. Along with this, it also addresses conclusion and future prospects of QDs@MOFs hybrid materials towards their implementation in energy and environmental applications.

Keywords Quantum dots · Composites · Photoluminescence · Metal–organic frameworks · Catalysis

A. K. Jassal
Department of Chemistry, UGC Sponsored-Centre of Advance Studies-II, Guru Nanak Dev University, Amritsar, Punjab 143005, India

P. Kajal (✉)
Ananyavijaya Consultancy LLP, Bengaluru, Karnataka 562107, India
e-mail: priyanka.kajal45@gmail.com

S. Thomas et al. (eds.), *Quantum Dots Based Nanocomposites*, Engineering Materials,
https://doi.org/10.1007/978-3-031-54779-9_7

List of abbreviations

QDs@MOFs	Quantum dots@metal organic frameworks
ZIF	Zeolite imidazole frameworks
LEDs	Light emitting diodes
PeQDs	Perovskite QDs
$Ca(OH)_2$	Calcium hydroxide
PbO_2	Lead oxide
PbS	Lead sulfide
CdS	Cadmium sulfide
CdSe	Cadmium selenide
Cos	Composites
MBE	Molecular-beam epitaxy
CdTe	Cadmium telluride
GQDs	Graphene quantum dots
HRTEM	High-resolution transmission electron microscopy
BTC	1,3,5-Benzenetricarboxylic acid
TiO_2	Titanium dioxide
ZnO	Zinc oxide
$Pb(Ac)_2$	Lead acetate
HPLC	High performance liquid chromatography
DA	Dopamine
CDs	Carbon dots
BPA	Bisphenol A
CQDs	Carbon quantum dots
ECL	Electrochemiluminescence
TPA	Terephthalic acid
NiS	Nickel sulfide
AFM1	Aflatoxin M1
MoS_2	Molybdenum disulfide
CCQDs	Chiral carbon QDs
TDPAT	2,4,6-Tris (3,5-dicarboxylphenylamino)-1,3,5-triazine
DNA	Deoxyribonucleic acid

1 Introduction

Metal-organic frameworks (MOFs) are enthralling materials which exhibit adjustable surface functionalities, tunable porosity, and their usefulness in hybrid or composite structural composition based applications [1]. These functional materials are designed by a combination of definite metal nodes with organic linkers (Fig. 1), resulting in materials with tunable pore size, an overall framework, structural stability, and expanded physicochemical properties [2–5]. Thus, this results in their utilization in various applications, for example, electrocatalysis, gas storage or separation, drug delivery, chemical and physical sensing, etc. [6]. Among these, metal ions with coordinatively unsaturated active sites and luminescent features in the counter parts of MOFs such as MIL, UiO, ZIF (zeolite imidazole frameworks), PCN, HKUST (Cu-MOF), and IR-MOF, etc. behave as functionally dynamic sites for selective sensing applications [7, 8]. Sustainability in the pores of MOFs acts as an ordinary territory for guest molecules, which results in improving sensing properties due to their better interaction with guest molecules [9]. These rigid MOFs exhibit higher thermal stability with durable structural support for fluorescent guests. Due to the reduction in the non-radiative easing of organic linkers in MOFs, the origin of strong fluorescence emission has received substantial attention for various applications. Most recently, there have been some reports of buckling of the overall straight structure of MOFs due to the uniaxial pressure on them. Resulting MOFs show quantum behaviour at lower temperature (T) due to the superposition of degenerated bucking states and open a new pathway towards strain caused quantum states. At $T = 0\,°C$, three strained buckling phases (normal, parabuckling, and ferrobuckling) are observed in MOFs due to the quantum phase transitions [10].

Despite being so advantageous, there are certain limitations to MOFs that hinder their direct utilization in various applications. These can be overcome by uniting them with other materials such as highly conductive polymers, quantum dots (QDs) or other semiconductive materials to enhance their properties. Thus, QDs are classified as a class of nanomaterials that exhibit three dimensional confinement, i.e., classified as zero dimensional (0D) particles with dimensions shorter than the Bohr exciton radius [11]. These are generally synthesized using materials from elements of groups I–III–VI [12], II–VI [13], III–V [14–16], and IV [17] and exhibit various properties with disparities in their sizes ranging from 2 to 10 nm [18]. Based on their bandgaps, the materials emit a wide range of colors upon absorption of white light [19, 20]. As a result, these materials find applications in photovoltaics, light emitting diodes (LEDs), energy storage, and targeted drug delivery, etc. [17, 21, 22]. QDs (size < 10 nm), exhibit quantum effects including inelastic electron tunnelling, quantum tunnelling, Fermi-Dirac distribution, etc. [23]. However, these materials also exhibit certain limitations, such as the easy agglomeration pertaining to fluorescence quenching, which limits their utility in a various of applications. Significant efforts have been invested to overcome the hurdle of passivation by utilizing supplementary semiconductor films with an appropriate bandgap and embedding QDs inside a wide range of materials (e.g., organic polymers or micelles, etc.). Though all the reported methods are time consuming and can lead to undesirable products with unexpected behaviour and properties. Thus, it has been attempted to encapsulate the semiconductor nanomaterials with MOFs to get QDs with multifunctional functionality and adaptable composite architectures [24]. Incarceration of purely inorganic perovskite QDs (PeQDs) embedded inside MOFs e.g., UiO-67-4,4′-bipyridyl and MIL-101(Cr) in order to synthesize PeQDs@MOFs provides a wide range of advantages such as electronic structure, band gap tuning, thermal stability, chemical reactivity, and ultrafast scintillator application [25, 26]. The aim of this chapter is to discuss the synthetic approaches for QDs@MOFs composites

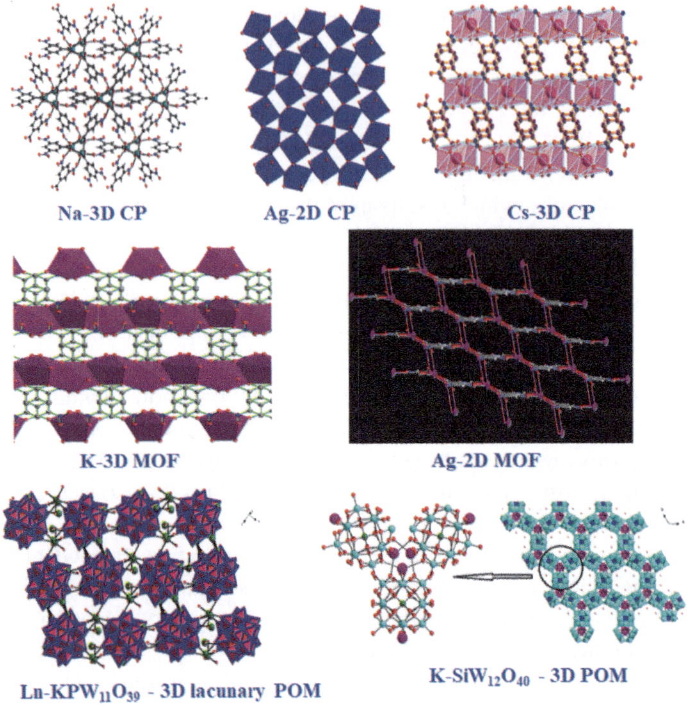

Fig. 1 Examples of earlier reported coordination polymers (Na-3D CP, Ag-2D CP), MOFs (K-3D MOF and Ag-2D MOF) and Polyoxometalates (Ln-$KPW_{11}O_{39}$—3D lacunary POM, K-$SiW_{12}O_{40}$—3D POM), where CP = coordination polymer, POM = polyoxometalates, reproduced with permission from the References [3–5]

and their advancements over MOFs and QDs. Additionally, the discussion related to the structural properties, activity, and selectivity of composites towards various catalytic reactions and other applications have been discussed in details.

2 Historical Evaluation of MOFs, QDs and Composites (Cos)

According to literature, the fields of zeolites, solid state, and coordination chemistry came into reality before the origin of MOFs. Thus, there is a wide history related to the investigation of coordination compounds for numerous applications. Since 1964, curiosity about porous materials such as Prussian blue and Hofmann has increased due to their adjustable adsorption properties [27]. Subsequently, in early 1990s, the first report about MOFs was published followed by more than twenty thousand examples of MOFs have been reported in the last thirty years. These materials are important due to their extreme porosity with 90% free volume inside the framework and broad internal specific surface areas (more than 6000 m^2/g) [28]. Hoskins and Robson reported basic information related to the future benefits of MOFs in their articles from 1989 to 1990. After their findings, many researchers all over the world have published reports related to the construction of an extensive range of highly ordered crystalline compounds, conceivably utilizing structure-templating agents and post-synthetic modification approaches. Yaghi et al. in 1995 [29] first coined the popular term MOF. In 1997, Kitagawa and co-workers reported a 3D MOF that shows excellent gas sorption properties [30]. Later, the MOF-518 and HKUST-119 came into the attention in year 1999 and in 2002, Férey et al. reported the permeable and flexible MOFs (MIL-47, MIL-53, and MIL-88) [31]. In continuation of the progress in MOFs, the idea of isoreticular chemistry originated in 2002 for simple Zn dicarboxylates and mixed-ligand based MOFs and then spread to other compounds. In the same year, the MOFs family has expanded to include imidazolate-based compounds, currently known as zeolite imidazole frameworks (ZIFs). Apart from the ZIF family, plenty of examples of MOFs, including UiO-66 (Zr), MIL-125 (Ti), MIL-101 (Cr), MIL-100 (Fe), and, CAU-1 (Al) etc. have been reported by researchers along with their usefulness in various applications [32].

In addition to this, the history of QDs is over 2000 years old; where naturally occurring, cost effective substances such as calcium hydroxide ($Ca(OH)_2$), lead oxide (PbO_2), and water were utilized for the synthesis of lead sulfide (PbS) QDs [33]. Then, these were utilized by Roman and Greek civilizations for applications such as hair dyes [34]. Later in the 20th century, Cadmium sulfide (CdS) and Cadmium selenide (CdSe) QDs were utilized in the preparation of the colourful glasses by controlling their size in silicate glasses. Such doped glasses found their application in optical filters in 1932 Following this from 1991, great deal of interest moved towards establishing the synthesis approach towards the growth of QDs. Thus, this two—era's of history of QDs, ranging from ancient hair dye applications to scientific advancements, shows an increasing trend in the applications of these materials from lab to their commercialization [34]. Depending on the QDs synthesis techniques, the two major classes are categorized as *top-down* and *bottom-up* techniques. *Top-down* technique includes the division of the main material into smaller subunits by using various processes (e.g., ion implantation, and X-ray lithography etc.) [35] and is also known as thinning down technique. Whereas the *bottom-up* technique results in the growth of QDs from the precursor solution. This is the most utilized technique, which includes the self-assembly of the QDs in the solution over time [36, 37]. Depending on the growth conditions, bottom-up techniques include two major synthesis methods: (a) wet chemical method, (b) vapor phase method. As the name suggests, *wet chemical* technique involves the utilization of the precipitation technique for the synthesis of QDs. Depending on solute, solvent, and growth conditions, nucleation can be classified as homogeneous and heterogeneous nucleation. Homogeneous nucleation on one side arises from the alliance of the solute atoms/molecules and reaches to a critical size without any solid interface. This includes techniques such as sol-gel, hot solution decomposition, microwave, and microemulsion techniques [38]. Major advantage of this technique includes the lower synthesis cost, along with challenges such as toxicity to humans and environment. Thus *biogenic approach* which is the extended version of the wet technique and offers advantages in terms of bio-compatibility and lower toxicity [39, 40]. The vapor phase method, also known as the physical approach, in the synthesis of QDs involves the atom-by-atom deposition of the QDs using techniques such as molecular-beam epitaxy (MBE), sputtering, etc. This process of nucleation of particles in the vapor phase to form QDs results in advantages such as simple and controlled fabrication technique, but a major challenge is the higher cost.

Along with advantages, QDs suffer from various challenges, including agglomeration. Thus, for better utilization of their properties, they are generally used in combination with other materials, such as composites (Cos). Cos are types of materials that are designed such that the advantages of both components can be effectively utilized, avoiding the shortcoming of both [23, 41]. QD@MOF Cos were first introduced in 2010 by Feng et al. [42], where they utilized ZnO QDs in combination with the MOF-5. In this report, the main emphasis was on the evaluation of the optical properties of Cos. Additionally, many other reports published in the same year 2010 which were later described by Lopamudra Giri et al. [24] in 2022. In this study, a wide range of synthesis techniques, i.e., bottle-ship, ship-bottle, photochemical deposition, sol-gel, template assisted method, drop casting, in-situ consecutive chemical bath deposition, physical fusion, photodecomposition, direct deposition, spray coating, simple dispersion, pyrolysis, calcination, direct surface functionalization, intercalation, co-electrospinning followed by carbonization and hydrothermal treatment, thermal induction, bi-sacrificial templates etc. have been discussed [43]. On the other side, highly porous MOFs with a large surface area deliver a thrilling podium to create a superlative atmosphere for QDs loading and prevent the chances of aggregation. Simultaneously, QDs develop the physicochemical behaviour of MOF materials. The combination of both QDs and MOFs results in good dispersibility, resulting in overall stability. In this situation, QDs@MOFs Cos have piqued curiosity and then unlocked an advanced avenue for various applications These hybrid substances have captivating properties, such as good mechanical or thermal stability, outstanding photoluminescence, excellent biocompatibility, and simplicity of functionalization [26]. A list of advantages offered by the QD@MOF Cos includes:

- *Improved stability*: The MOF matrix protects the QDs from degradation due to environmental factors including moisture, temperature, and oxidation, leading to their higher stability.
- *Tunable porosity*: The porosity of the MOF can be controlled as per the QDs used for better stability of the QDs in the matrix. This also improves the performance of the Cos for applications such as sensing, catalysis, etc.
- *Controlled interfaces*: The interface between QD and MOF can be precisely controlled, leading to controlled charge transfer processes between both QD and MOFs.

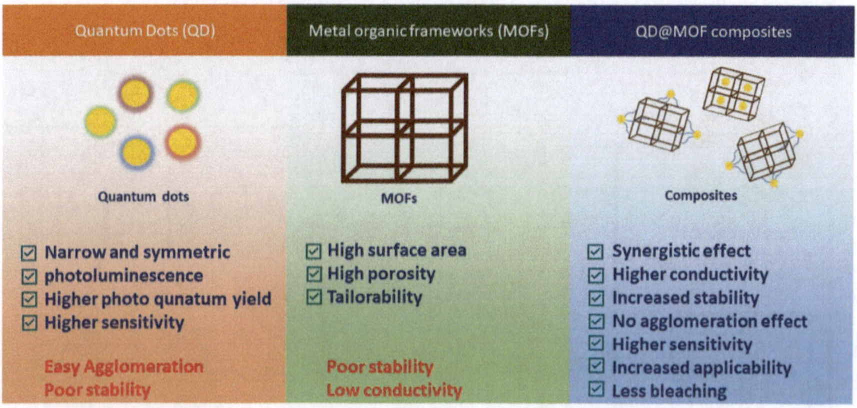

Fig. 2 Illustration of the characteristics of the QDs, MOFs and their composites

- *Enhanced photoluminescence (PL)*: The integration of QD and MOF results in increment in light emission, with only QDs showing improvement in light emitting efficiency by incorporation of both.
- *Improved properties*: Standalone QDs are mostly utilized for LEDs, and light absorption applications, while MOFs generally offer traits such as porosity, functionality, etc. Thus, combining both as Cos results in more versatile applications than standalone.

Thus, these enhanced features by combining both MOFs and QDs open pathways for more advanced possibilities in photonics, sensing, and catalysis applications. Figure 2 illustrates the characteristics of QDs, MOFs and their composites; indicating that combination of both can bring new possibilities due to their synergetic effects. Thus, the goal of this book chapter is to take a deep dive into the world of QD-MOF Cos, their synthesis techniques, material selection depending on application.

3 Synthesis Techniques of QD@MOF Cos

As stated above there are various reported techniques for the fabrication of the QD@MOF Cos. For the synthesis of Cos pre-synthesized QDs and MOFs are utilized in various ways as shown in Fig. 3 depending on the MOFs, QDs properties and required applications.

Initially, photochemical deposition and direct surface functionalisation of semiconductor QDs with MOFs were the two major techniques utilised for the preparation of the QDs@MOFs Cos [44, 45]. The construction of QD@MOF Cos by photochemical deposition requires two significant strategies, which are *in-situ* development of QDs and their accumulation on the exterior sides of MOFs based on proper photoreduction potential. Direct surface functionalization includes the immobilization of

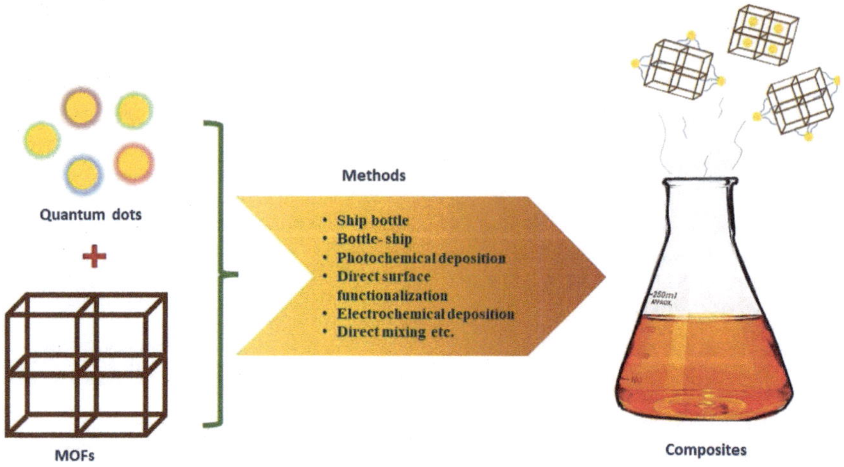

Fig. 3 Methods of the production of QD@MOF Cos [43]

QDs on MOFs by suitable ligands. There are different capping agents used to change the surface of QDs and directly make contact with respective MOFs by coordinating the association between the overlaying agents of QDs and the centre metal ions of MOFs. This synthetic approach has some advantages in that the morphology of MOFs; with proper controlled shape and size QDs.

In addition to these two methods, the fabrication of ZIF-8@CQDs can be done by in-situ fabrication involving the direct mixing of all precursors at room temperature [46]. It has been observed that the resulting ZIF-8@CQDs composite is an example of perfect luminescent MOFs-derived sensory materials because high quantum yields containing luminescent guest materials have been introduced inside the MOF channels.

There is other two well-constructed approaches i.e., '*ship-bottle*' and '*bottle-ship*' which find application of QDs inside- outside of MOFs pores. Among these approaches, the *ship-bottle* method as the name suggests focuses on fitting of the QDs into the pores of the MOFs. This immobilisation of small nanoparticle precursors/ QDs into pores leaves less tendency for them to pierce through the pores of MOF, followed by further treatments to transform them into the required practical assembly. For the synthesis of QD@MOF Cos, the QDs precursors are loaded into the MOF by heating, hydrogenation, and reduction reactions [44]. Consequently, in the *bottle-ship* method, the QDs@MOFs Cos encloses the guest material (QDs) with QD size > MOF pore size; retaining it outside the lattice structure. There have been numerous reports showing the applicability of the ship-bottle approach, i.e., H. Kitagawa and co-workers summarized palladium-based nanocubes with HKUST-1 (an example of Cu^{2+} ion-based MOF) without any obvious organic linker molecules [47]. It has experienced an upsurge in the ability for hydrogen storage through modification of the outer surface of palladium atoms with MOF. Consequently, the formation of direct contact between MOF and nanoparticles is anticipated to become fascinating for QDs and their luminescent properties are vulnerable to the surface conditions. It has been explored the prospect of employing MOFs as novel classes of surface shielding materials for fluorescent QDs. ZIF-8 MOF utilized to cover CdSe/CdS core/shell QDs via a direct connection between the two species in the absence of other bonding constituents. They have aimed to synthesize Cos which would bring the cooperative effects of both tunable MOFs and semiconductor QDs. Oleylamine of the QDs has switched with pyridine, and then the pyridine-plugged QDs undergo in-situ ligand exchange with 2-methylimidazole of ZIF-8. Further using a controlled concentration of ZIF-8, monodispersed growth of ZIF-8 on the outer layer of CdSe/CdS core/shell QDs is obtained. Fully merged QDs into the ZIF-8 framework show excellent photoluminescence properties, with photoluminescence quantum yield >40% as compared to the untreated QDs [45]. The major obstacle to the utilization of the ship-bottle technique is the lack of control over the inner surface, shape, and size of guest molecules within the MOF structure [48]. Thus, ship-bottle method could be disadvantageous for applications where the shape of QDs is predominantly important; i.e., utilization of the gold nanoparticles for biomedical applications. On the flip side, for the bottle-ship method, a stable framework is required under optimised conditions to construct the functional material or nanoparticles on the outer part of MOFs.

4 Structural Features and Properties of MOFs@QDs Cos

It has been proven from literature that the methods of preparation for the MOFs@QDs Cos play crucial roles in controlling their physical properties for various applications. In earlier reports, the injection reaction protocol was utilised for the synthesis of ultra-small (~2 nm) CdSe QDs in the confined cavities of MIL-101(Al)-NH_2 [49]. Due to the mismatch in the size of QDs and the pore size of MOF, the morphologies of the resultant QD@MOF Cos can be varied and are categorised into three sub categories. In the former case, QDs are completely encapsulated by the frameworks of MOFs. The evolution and orientation of QDs are limited or negligible. So, the QDs cannot be mobilized in the MOF matrix in any order to get agglomerated. For the fabrication of luminescent semiconductor QDs, MOFs can act as efficient stabilizing agents because of their insulating nature and sometimes semiconducting properties, as well as the versatility of the designed materials [45]. In most of the QD@MOF Cos, there are some direct bonding interactions between the QD and MOF-like polymers, which are required to exploit the synergistic effects of both QDs and MOFs. A dropwise mixing of MOFs based solution into QDs solution, an easy detachment of pyridine derivatives, and the growth of MOFs on the outer surface of QDs has been observed. The resultant composite has been characterized based on photoluminescence spectra. The complete shielding of MOFs on the outer layer of QD has established by the impassiveness of the QDs@MOFs to the solid-state photoluminescence quenching tests [45]. In an earlier report, the construction of the core-shell assembly of CdTe@ZIF-8 (CdTe = Cadmium telluride) was mentioned, in which various numbers of CdTe QDs were orderly dispersed in the ZIF-8 framework [50].

Additionally, the encapsulation of graphene quantum dots (GQDs) inside the pores of ZIF-8 framework shows the robust connections between Zn^{2+} ions of ZIF-8 and −COOH groups of GQDs [51]. Such interactions are forming zinc-carboxyl bonds which have endorsed the association of GQDs and ZIF-8 and withdrawn the chances of GQDs aggregation in ZIF-8. In the second case, some QDs are surrounded by the MOF matrix, although others are unprotected from the MOF's surface. The probable cause is that some QDs are larger in size than the pores of MOFs, and sometimes the amount of QDs is high as compared to the voids of MOFs. Consequently, all the QDs cannot enter the pores of MOFs and some QDs are exposed on the surface of MOF. He et al. captured CdS QDs inside the pores of MIL-101, where the excess CdS QDs had aggregated on the surface of MIL-101 [52]. Similarly, Gan et al. fabricated the CdSe QDs@UiO-66 in which the CdSe QDs have been encapsulated by the UiO-66 framework [53]. It has been shown from high-resolution transmission electron microscopy (HRTEM) images that CdSe QDs have grown both internal and external surfaces of the UiO-66. The particle size of CdSe QDs on the outer surface was bigger than those inside the MOFs. The third case shows that QDs are affixed to the external surface of MOFs, so their growth is not constrained and leads to excessive accumulation on the MOFs. This concept is supported by the earlier report of Li and co-workers, in which they fabricated $Cu_3(BTC)_2$ (BTC = 1,3,5-benzenetricarboxylic acid) with a core-shell structure. The titanium dioxide (TiO_2) shell with a thickness ~210 nm has bounded the central portion to form a porous and rough texture on $Cu_3(BTC)_2$ [54]. Similarly, Xu et al. have immobilized ZnO (zinc oxide) QDs with ~10 nm particle size on the UiO-66 (Zr-MOF) with a fixed pore size the MOF. Consequently, the ZnO QDs did not enter the Zr-MOF matrix but rather agglomerated on the surface of Zr-MOF, which is established by the TEM images [55]. Direct impregnation of GQDs with ~3.1 nm particle size into a mesoporous porphyrin Zr-MOF has increased electrical conductivity and electrochemical sensing activity by simply transferring charge from GQDs to porphyrin linkers of Zr-MOF. The resultant composite holds half the porosity compared to the parent MOF and contains 100-fold advanced electrical conductivity. The utilization of porphyrin linkers in Zr-MOF as highly catalytic active components for the synthesized GQD@MOF composite to display improved electrochemical sensing activities towards nitrite in aqueous medium compared to both GQD and the parent MOF [56]. The in-situ confined progress of ultrasmall PeQDs in the UiO-66 MOF matrix has performed to get their quantum confinement effect. These composite materials display gifted optical and optoelectronic properties for photovoltaic applications by regulating the particle size from 6.4 to 3.3 nm with variation in the concentration of lead acetate ($Pb(Ac)_2$). The reduction in the size of PeQDs results in a blue shift in their PL emission spectra from 521 to 486 nm due to the enhanced quantum confinement effect. Thus, the UiO-66 matrix has endowed the surface passivation effect, and ultrasmall PeQDs have exhibited a 43.3% photoluminescence quantum yield with a large photoluminescence lifetime of 100 ns [57]. In another case, $AgInS_2/GaS_x$ core/shell QDs encapsulated by indium-fumarate MOFs led to an enhancement in structure stability with a slight transition in band-edge and a 12.6% photoluminescence quantum yield; which will be conserved after one week due to the enhanced stability of encapsulated QDs in MOF [57].

5 Applications of QDs@MOFs Cos

After the comparison of previously reported porous inorganic substances with the MOFs based composite, it has been proposed that there is diversity in the chemical compositions of MOFs, which allows a wide range of structure features and functionalities to be tuned by using different building blocks. These Cos exhibit features such as ultrahigh specific surface areas, easy tunability, and micro-porosity, resulting in enhanced structural properties leading to their utilization in various applications (e.g., photocatalysis, gas storage, sensing, energy storage, targeted drug delivery, and antibacterial activities) [44]. Thus, this section of the chapter primarily recapitulates various applications of QD@MOFs in detail. Moreover, it also describes the opportunities related to the QDs@MOFs Cos and briefly discusses various challenges associated with Cos during synthesis along with their stability issues.

5.1 Sensing

Luminescence based sensing is one of the major breakthroughs in sensing applications using QD@MOFs [58] because of the higher quantum yield achieved by integrating QDs and MOFs. A broad range of sensing applications using these materials include Cu^{2+} ion sensing, visual detection of glucose, dopamine detection, etc., where these materials are utilized as fluorescent sensing probes or electrodes. Along with this, these Cos offer a wide range of advantages, such as a facile or cost-effective synthetic approach, environmental friendliness, ultrahigh PL stability, lower toxicity, and excellent biocompatibility. Various sensing applications can be further described as follows.

5.2 Detection of Dopamine

Dopamine (DA) is a type of neurotransmitter in the human central nervous system that contains catecholamine. It generally takes part in numerous biological activities, such as central nervous system, hormonal, and cardiovascular system. The deficiency of dopamine may lead to poor regulation of muscular movement, resulting in the distraction of muscles and leading to Parkinson's disease. To date, plentiful methods have been developed by chemists to regulate the level of dopamine, including UV spectrophotometer, electrochemical analysis, high performance liquid chromatography (HPLC) and various fluorescent probes. However, these methods have some shortcomings, such as lethal substances and organic ligands, etc., which would confine their applications, difficult instrumental techniques, and complicated sample pre-treatment procedures. To overcome these difficulties, it is extremely required to grow eco-friendly, facile, and commercial approaches to regulate the amount of dopamine. However, there are limited reports for determining the quantity of dopamine by utilizing carbon dots (CDs)@MOFs Cos. In one of the reports, it has been observed that CDs with high quantum yield have been embedded within the voids of ZIF-8 matric for the facile assembly of CDs@MOFs (Fig. 4) which can behave as fluorescent probes for smooth recognition of the dopamine concentration. In CDs@MOFs, the synergistic effect between CDs and MOF matrix shows remarkable results. It has provided a habitual situation to sensibly capture dopamine into the MOF pores and harvest the bonding interactions between dopamine and the basic spot of ZIF-8 (Fig. 4a). Dopamine has been effortlessly oxidised to harvest dopamine-quinone and captivated on ZIF-8 by the weak π···π interactions with phenyl rings of ZIF-8 MOF, resulting in fluorescent quenching of CDs (Fig. 4b). CDs have been utilized as fluorescent signal probes to transmute chemical signals of contacts between CDs and dopamine. The sensing performance of CDs for dopamine is uniformly improved compared to earlier reported signals, with a wide range of concentration (0.1–200 μM) and a low detection limit of 16.64 nM [46].

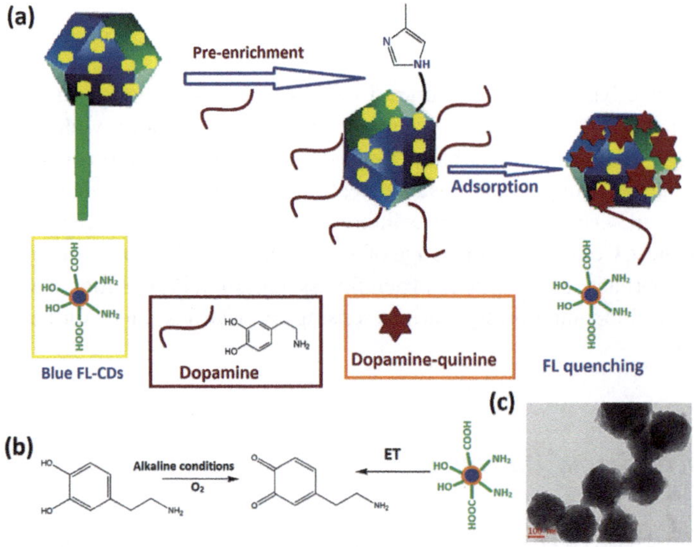

Fig. 4 **a** Schematic representation of the CDs@ZIF-8 synthesis and their application in the recognition of dopamine; **b** possible mechanism for the detection of dopamine; **c** HRTEM image of the CDs@ZIF-8/DA-quinone, reproduced with permission from the reference [46]

5.3 *Sensing of Endocrine-Disrupting Chemicals*

There are a series of dangerous endocrine-disrupting chemical substances (e.g., dioxins, triethyl lead, bisphenol A, vinclozolin, and phthalates etc.) that may result in numerous health issues, especially for the reproductive systems and transgenerational diseases in adults. From them, bisphenol A (BPA) is known as the endocrine-disrupting substance. It is a predominant substance due to its use as a raw material for manufacturing rigid foams, thermoplastic organic polymers, and large-scale industrial coatings. Therefore, there is a need to grow sensors or probes for the initial discovery of endocrine-disrupting chemicals [59].

5.4 *Detection of Ascorbic Acid*

Lately, a dual-emission ratiometric sensor such as CdTe QDs@UiO-66-NH_2 has been constructed and used for the sensing of ascorbic acid [60]. Ascorbic acid is a vital micronutrient and antioxidant that is used by various biosynthetic enzymes in several physiological and pathological processes. Ascorbic acid can also be oxidized to synthesize dehydroascorbic acid and remove oxidative stress from living organisms. The amount of ascorbic acid greater than 110 μM, creates major health issues such as diarrhoea, urinary stones, and stomach cramps. Consequently, quantitative analysis of ascorbic acid is the key method for primary recognition of associated health diseases. Lately, fluorescence sensing has attracted major attention to determine the concentration of ascorbic acid because of its rapid response time and better sensitivity. Though, in earlier reports, the fluorescent probes provided only a single emission feature and were blocked by numerous interferences, including instrumental parameters, environmental response, photobleaching, and the concentration around probes or sensors. Therefore, a stable fluorescent sensor, including both CdTe QDs and UiO-66-NH_2 MOF is required to detect the amount of ascorbic acid and the pristine MOF displays only a single emission peak. The simple mixing of MOF, i.e., UiO-66-NH_2 and CdTe QDs, includes the electron transfer from CdTe QDs to MOF UiO-66 (Fig. 5), resulting in an intrinsic dual emission peak [60]. Besides, the fluorescence intensity of CdTe QDs@UiO-66-NH_2 MOF can be transformed by ascorbic acid, the emission band of UiO66-NH_2 MOF is reduced, though the emission band intensity of CdTe QD is amplified after adding ascorbic acid. In conclusion, the CdTe@UiO-66-NH_2 behaves as an excellent fluorescence sensor to show virtuous sensitivity towards ascorbic acid.

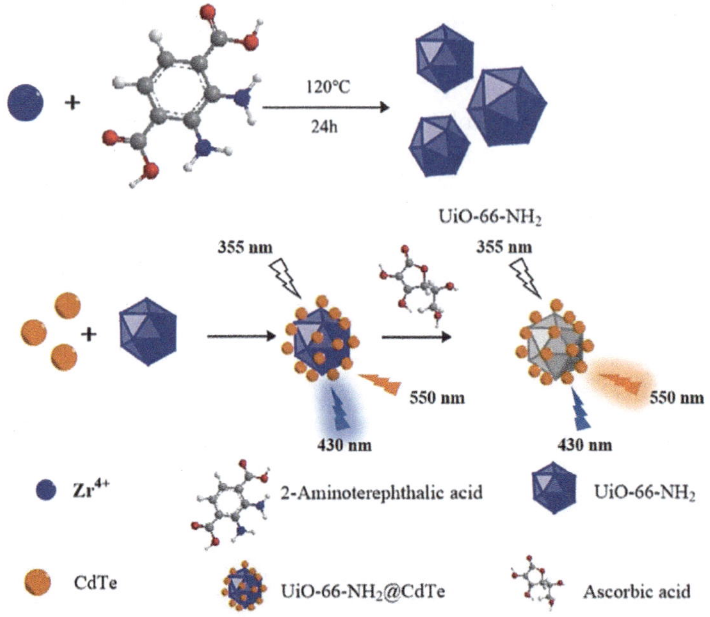

Fig. 5 Diagrammatic representation of the construction of UiO-66-NH$_2$@CdTe, reproduced with permission from the reference [60]

5.5 Highly Sensitive Detection of 4-nitrophenol

For the manufacturing of dyes, pharmaceuticals, and leather products, 4-nitrophenol is known as a common intermediate that comes out in all reactions. Nevertheless, 4-nitrophenol is also predicted to be a common lethal toxin that causes extreme harm to the human health and creates many environmental issues. Therefore, selective detection of 4-nitrophenol is of prodigious importance to save human health and protect the environment. Smartly impregnating CQDs (carbon quantum dots) within the voids of Zr-MOFs leads to the formation of fluorescence-functionalized Cos by applying a post-synthetic alteration approach. The resultant CQDs@Zr-MOF act as fluorescent probes for the determination of toxic substances [61]. Hence, a rapid, facile, and relative sensitive approach has been developed to synthesize NH$_2$-CQDs@UiO-66 MOF fluorescence sensor probe to detect the traces of 4-nitrophenol with 3.5 nM as a low detection limit. In order to expand the selectivity and sensitivity of fluorescence sensors, either UiO-66 MOF has been utilized to enrich the target molecules or fluorescent NH$_2$-CQDs have been used directly to sense the synergistic effect and interactions between UiO-66 MOF and targeted 4-nitrophenol [62].

5.6 *Selective Gossypol Detection*

Gossypol is known as a category of polyphenolic toxins that are naturally occurring within the mallow family. This substance is primarily present in the cottonseeds and can be easily removed from cotton-based plants. There is a fear related to the protection of cottonseed plant-based products (e.g., animal feed and cottonseed oil) from gossypol. Major issues related to gossypol are that its excretion from human body is very slow, and ingesting the food items with extreme gossypol accumulates toxic substances and shows adverse health effects [63]. In reports, multiple interpenetrating MOFs with channel-size-dependent behaviour easily encapsulate the PeQDs, resulting in a PeQDs@MOFs composite that is useful for the selective detection of gossypol and in host-guest chemistry [63].

5.7 *Targeted Drug Delivery*

Conjugated MOFs and QDs based Cos are useful for target drug delivery applications and bioimaging of cancer. All-inorganic PeQDs of $CsPbBr_3$ captured within the voids of ZIF-8 to form $CsPbBr_3$ QD@ZIF-8 MOF Cos by utilizing an in-situ chemical reduction approach. The $CsPbBr_3$ QD@ZIF-8 MOF Cos further tried for an advanced technique, i.e., surface-enhanced Raman scattering, in which a charge transfer is noted within the composite, and indicating the potential application of the composite for surface-enhanced Raman scattering detection. The adsorption behaviour of 4-mercaptopyridine over $CsPbBr_3$ QD@ZIF-8 MOF has been evaluated by utilizing typical surface-enhanced Raman scattering spectral bands [64].

5.8 *Energy Storage Applications*

In earlier reports, GQDs@MOFs have been used to derive porous carbon for electrode material with enhanced semiconducting behaviour for supercapacitors with ideal electric double layer capacitor (EDLC) behaviour and lithium/sodium-ion batteries [65]. In another report, a self-sustaining suprastructure has been fabricated by sequentially loading NH_2-CDs and $CsPbBr_3$ PeQDs into the hierarchical ZIF-8 for electrochemiluminescence (ECL) bioanalysis [66]. The resultant PeQDs@ZIF-8 MOF composite has enhanced structural stability because ZIF-8 contains a protective layer and enhances charge transfer from CDs to MOFs. There is also an efficiently self-enhanced ECL emitter shown between PeQDs and NH_2-CDs in a compact structure with the confinement effect. Thus, the ternary Cos (NH_2-CDs@PeQDs@ZIF-8) is gifted with excellent stability and better ECL efficiency [64]. For the first time, this ECL probing approach is used on ternary Cos (used as an electrode substrate) to examine the T4 polynucleotide kinase activity [66].

5.9 Electrocatalysis

From earlier reports, utmost MOFs are electrically insulating in nature, which limits their practical utility in electrochemical (Fig. 6) and optoelectronic applications. Though various synthetic or post-synthetic approaches have been established to project electrically conducting MOFs with tunable pores and structure stability [67]. The main example of such modifications is the introduction of numerous guest (QDs) molecules to reduce the charge transfer between donor–acceptor in respective MOFs. The most common approach to reducing the charge transport in Zr-MOFs is to utilize redox hopping under different electrochemical situations. Generally, the slow redox hopping in Zr-MOFs frequently limits the electrochemical activities [68]. In an earlier report of GQD@PCN-222, GQDs have an extra negative conduction band as compared to the porphyrinic linkers of Zr-MOFs. Typically, the charge transfer between donor and acceptor has been recorded from electron-releasing GQDs to electron-accepting porphyrinic linkers, resulting in heightened electrical conductivity and improved electrochemical sensing towards different nitrites [56].

In another report, NiS/Ni_3S_2 QDs@NiFe-terephthalic acid (TPA) nanosheets (~5 nm) were prepared by an *in-situ* two step hydrothermal method and used for water splitting applications [26]. These nanosheets show only 219 mV overpotential (current density = 10 mA/cm^2 for 60 h). With increasing the amount of sulphur-containing agent, the Ni_3S_2@NiFe-TPA displayed excellent HER performance. During complete water electrolysis, the electrolyzer has been accumulated with N/S/QDs@NiFe-TPA and Ni_3S_2@NiFe-TPA as an anodic electrode. The cathodic electrode displays 1.66 V a low cell voltage at 10 mA/cm^2 current density, and retained the structure of Cos after 60 h. The admirable electrocatalytic properties of these Cos are because of their synergistic effect between Nickel sulfide (NiS) and MOFs, which can offer quick ion or electron transportation, reachable active sites, and appropriate O_2 and H_2 release channels.

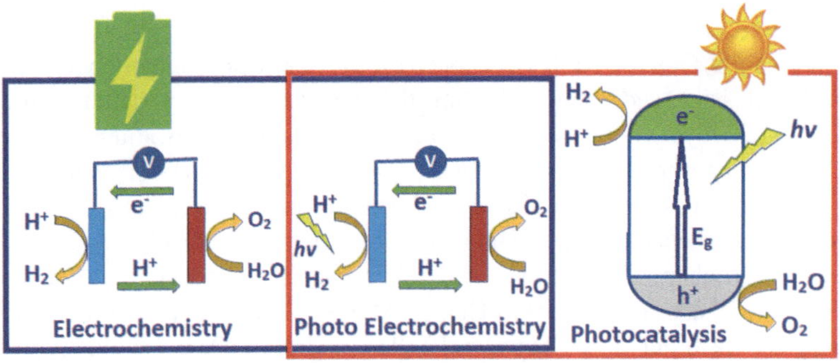

Fig. 6 Schematic representation of mechanisms of electrochemical, photo electrochemical and photocatalytic processes

5.9.1 Electrochemical Biosensor

Molybdenum disulfide (MoS_2) QDs@MOFs Cos can behave as electrochemical biosensors for the recognition of Aflatoxin M1 (AFM1) [69] and their working principle is as shown in Fig. 6. AFM1 are known as hepatotoxic metabolites, which are carcinogenic and root contrary human health issues such as diminished immune response, liver functioning issues, and numerous infections. The electrochemical biosensor for the active recognition of AFM1 has been designed by the modification of screen-printed carbon-based electrodes with MoS_2 QDs and UiO-66-NH_2 and decorated with AFM1. The developed biosensor has been validated according to its performance in recognising AFM1 in spiked milk samples. In another report, a Cu-TDPAT@CCQDs/NiF derived electrode (where, TDPAT = 2,4,6-tris (3,5-dicarboxylphenylamino)-1,3,5-triazine) has been prepared by electrodepositing MOF and chiral carbon QDs (CCQDs) on the nickel foam (NiF). This electrode can be used for the electrochemical enantio-recognition of tyrosine isomers from racemic mixtures with high selectivity and sensitivity [70].

5.10 *Antibacterial Activity*

New GQDs modified with NH_2-MIL-125 are efficient for antibacterial activity. Bacterial infection threatens human health, and an innovative photocatalyst, GQDs@NH_2-MIL-125 (Fig. 7), has been used for its photocatalytic disinfection and antibacterial properties. Active substances have destroyed the cell membranes and deoxyribonucleic acid (DNA), completely killing the bacterial infection. The chelation in NH_2-MIL-125 MOFs may diminish the polarity of metals, thus increasing the lipophilicity and enabling the NH_2-MIL-125 to breach the wall of a bacterial cell for killing the bacteria [71].

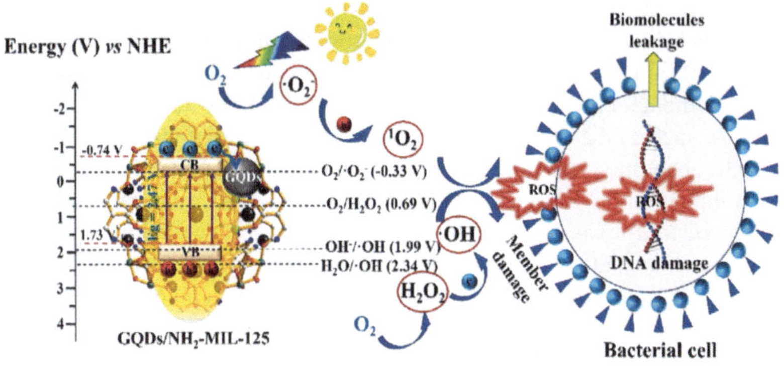

Fig. 7 Proposed mechanism of GQDs/NH_2-MIL-125 (UiO-66 MOF), reproduced with permission from the reference [71]

5.11 Photocatalysis

In current years, environmental pollution and the depletion of energy resources have become major problems for society. Photocatalysis (Fig. 6) is an effective technique to produce energy from renewable sources and is helpful for environmental remediation. Therefore, the emerging field of research is transiting towards the development of novel photocatalysts with their utilization for energy generation and storage. Nevertheless, there are a few drawbacks to single-phase photocatalysts, such as poor tunability, a wide bandgap, easy agglomeration in solution, a higher photogenic charge carrier recombination rate, slow photogenic electron transmission, and responding only to ultraviolet light. These limitations may lead to low photon energy utilization proficiency and low photocatalytic activity. To solve these issues, a combination of MOFs and semiconductor QDs has been reported to make composites that show porosity, exhibit large surface area. In these Cos, QDs are distributed inside the pores to decrease the rate of aggregation and enhance their light interaction capacity. Due to unique structural features of MOFs (e.g., NH2-MIL-125), they can combine with distinct semiconductor substances, such as GQDs with enhanced visible light harvesting ability and effectively endorse in photogenerated electrons transmission and recombination, thus enhancing the photocatalytic performance [43, 44]. However, GQD@MOFs Cos have some limitations in the field of photocatalysis [71, 72], and can be managed by using various synthetic approaches. The applications of QD@MOFs Cos in photocatalytic CO_2 reduction to CH_4, photocatalytic water splitting [72], photocatalytic reduction of Cr^{6+}, degradation of organic dyes, and oxidation of nitric oxide have been described in detail.

5.12 Photocatalytic Activity

5.12.1 Photocatalytic Hydrogen Evolution

Confined lead-free PeQDs@MOFs are used for proton reduction due to visible light interaction [25]. The semiconductor material MoS_2 has been considered an auspicious supplement for the noble-metal photocatalyst [72]. But this material has certain limitations, such as the fact that it can easily agglomerate in the solution, showing weak light absorption capability and poor photocatalytic performance. Therefore, MoS_2 is united with MOFs to produce a MoS_2@MOFs Cos because MOFs with ultra-high surface areas, and organized voids afford a favourable place for MoS_2 molecules to load and prevent their aggregation. MoS_2 QDs@graphene@TiO_2 has been designed by a facile one-step solvothermal route with a noteworthy growth in the photocatalytic activities due to the amplified charge separation and absorbance rate under visible-light [72]. MoS_2QDs@graphene@UiO-66-NH_2 is an excellent candidate for the photocatalytic hydrogen evolution reactions (Fig. 8a) due to the virtuous dispersal ability of the resultant composite. Sometimes, the excess QDs

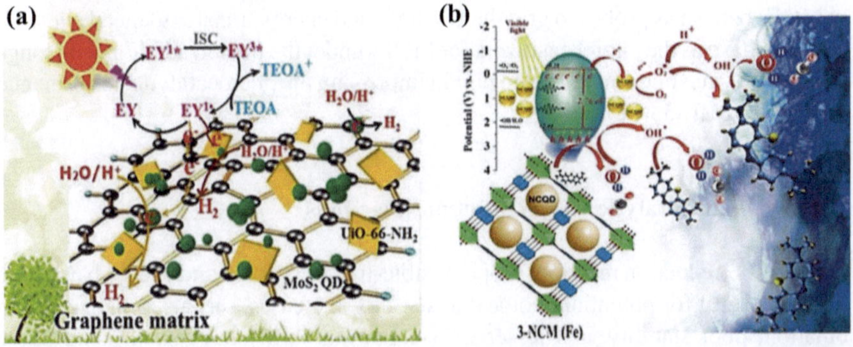

Fig. 8 **a** Possible mechanism for photocatalytic H_2 production, using MoS_2 QDs/UiO-66-NH_2/G under the irradiation of visible light, reproduced with permission from Elsevier [72], **b** proposed mechanism for methylene blue removal, using of 3-NCM(Fe), reproduced with permission from the reference [73]

occupy all the voids of MOFs and even block them, leading to reduced porosity and surface area of QD@MOF. Still, the interactions between QDs and MOFs have maintained enhanced photocatalytic activity [43].

Taking another example, amino acid-based CQDs@MIL-53 (Fe) behave as an extremely effective photoreactor, showing visible-driven photocatalytic activity for methylene blue photodegradation (Fig. 8b). The amino acid-based CQDs heightened the migration of photogenerated electrons and holes. Along with this, holes and radicals of hydroxyl groups are more projecting than the radicals of superoxide throughout the degradation process [73].

5.12.2 Photocatalytic Cr^{6+} Reduction

Major industrial water wastes include typical heavy metal contaminants such as hexavalent chromium (Cr^{6+}) ions [74]. Cr^{6+} is known as an enormously toxic compound to living organisms and causes carcinogenic problems in humans. A major issue with Cr^{6+} is that it is exceedingly soluble in water [75]. Photocatalytic reduction is observed as an efficient strategy to reduce Cr^{6+} contamination to Cr^{3+} by utilizing the semiconductor as a photocatalyst. The resultant Cr^{3+} is environment friendly and critical to the plants and human health [74, 76]. Temporarily, Cr^{3+} can be simply precipitated to form $Cr(OH)_3$ and detached in a neutral or alkaline medium. In recent reports, limited photocatalysts such as ZnO, TiO_2, WO_3 and CdS have been projected to reduce Cr^{6+}. Though these photocatalysts suffer from low catalytic activity due to objectionable factors, such as photocorrosion, severe photogenerated carrier recombination, and UV-limited activation. A facile generation of CQDs@MOF (MOF = MIL-53, MIL-125, UiO-66 (Zr), and MIL-68) behaves as confined electron acceptors for the enhanced Cr^{6+} reduction [76]. Active metal sites in MOFs serve as light absorbers and charge generators, whereas the organic ligands in MOFs behave as probes to grip the UV light and energy transfer to metal sites. The CQDs behave as the photosensitizers for MOFs under the light irradiation with longer wavelength, and play an important part in improving the photocatalytic performances of CQDs@MIL-53(Fe) [76].

5.12.3 Photocatalytic CO_2 Reduction Reactions

For the CO_2 reduction reaction, proper visible-light responses and energy band structure are useful for potential photocatalysts. In some cases, serious radiative recombination, poor stability, and lesser CO_2 capturing ability have been noted due to the low conversion efficiency. To solve these issues, designing a semiconducting material@MOFs composite as a photocatalyst is required with an improved CO_2 reduction process. In recent reports, a facile *in-situ* procedure has been applied to prepare $CsPbBr_3$@Zn/Co ZIFs for efficient photocatalytic CO_2 reduction [77]. The $CsPbBr_3$@Zn-ZIF or Co-ZIF Cos show enhanced moisture stability, charge separation efficiency, and CO_2 capturing ability. Upon irradiation with visible light, $CsPbBr_3$ QDs shift from ground to the excited state to generate pairs of electron-holes that are easily detached at the catalytic active metal centres in ZIFs, resulting in enhanced CO_2 reduction performance.

In another report, PeQDs@PCN-221 is synthesized by a consecutive deposition method, where PeQDs are captured by the Fe-porphyrin-based PCN-221 and capably used for the photocatalytic reduction of CO_2–CO and CH_4, using water as a sacrificial reductant [78]. The growth of an effective artificial photosynthesis depends on economical photosensitizers and the earth-abundant photocatalysts, to reduce carbon dioxide into feedstocks or fuel. In some reports, CQDs-sensitized semiconductor photocatalytic systems have limitations i.e., low efficiency and high rate of energy consumption due to maximum number of barriers. These limitations are responsible for the kinetically favourable reduction of water rather than carbon dioxide in carbonated water. In earlier reports, significant efforts have been made to hunt for novel CQDs that may overcome the limitations. Rational design and formation of GQDs@Zn-MOFs systems have been done to use them as photocatalysts for effective CO_2 reduction under visible light [79]. Compared with previous GQDs, the GQD@Zn-Bim-His-1 composite exhibits boosted photocatalytic activity for CO_2 conversion (20.9 and 3.7 μ mol h^{-1} g^{-1} are the formation rates for CH_4 and CO, respectively). The synergistic effect between rich active sites in imperfect Zn-Bim-His-1 and GQDs encouraged the photogenerated electron–hole separation in photocatalytic conversion [79].

5.12.4 Photocatalytic Removal of NO

Nitrogen oxides (NO_x) are known as poisonous gases, and nitric oxide and nitrogen dioxide, mainly cause air pollution. These gases create harmful effects on human health and cause severe environmental damage. Photocatalysis is an auspicious technique and can be used to remove NO at the ppb level. This process can be done with less consumption of energy and operates at normal temperature and pressure conditions [80, 81]. In recent studies, the 2.5 vol% N-doped CQDs@MIL-125(Ti) MOF composite, synthesized via a facile route at room temperature, showed 49% removal efficiency for NO under enhanced visible light [82]. The MIL-125 is showing an energy band gap of 3.6 eV, responds to UV light relative to visible light, thus suffering from fast electron and hole recombination. Consequently, this material cannot be used directly for the effective removal of NO. There is a crucial need to overcome the limitations of MIL-125 [83, 84]. Electrochemical impedance (EIS) information exhibited that the efficiency for charge carrier separation of N-doped CQDs@MOF is suggestively higher than the MIL-125. As additional information, the Ti^{3+}–Ti^{4+} in MIL-125 behaves as an active centre for the photocatalytic NO exclusion. There are two probable pathways for the migration of electrons. First is the transfer of electrons from N-doped CQDs to the Ti^{3+}–Ti^{4+} centre in MIL-125 due to the photoinduced electron transfer property of N doped CQDs. Second is the absorption of UV light originated from N doped CQDs by an organic linker, i.e., terephthalic acid, trailed by the transfer of electrons to active metal sites for the photocatalytic removal of NO.

5.12.5 Enhanced Light Harvesting

In many reports, it has been mentioned that QDs@MOFs Cos can be used as photocatalysts to produce energy and for inhaled light harvesting applications [43, 85]. The designed QDs@MOFs always hold carbonaceous materials to evade the chances of particle agglomeration and barricade the limitations of the light harvesting process. The functionalized porphyrin-based Zr-MOFs with CdSe or ZnS core/shell QDs are the best examples as photocatalysts for enhanced light harvesting by the transfer of energy from QDs to MOFs. In this case, nearly 80% energy has transferred from QDs to MOFs, as measured by time-resolved emission studies. There is a wide absorption band related to the QDs observed in the visible region, which suggests the better coverage of the solar spectrum by CdSe or ZnS@MOF. There is about 50% enhancement observed in the number of harvested photons harvested with QDs coated MOFs [86]. Thus, depending on this Table 1 shows a list of Cos, their type, synthesis techniques and potential applications [87–99].

6 Conclusion and Future Perspectives

This chapter has presented an overview of advancements in these Cos, synthetic approaches, diversified structural features, unique properties, and outstanding applications. MOFs have enormous structural features and chemical stability, which makes them attractive substances for constructing composites with semiconductor-like behaviour containing QDs. Sequentially, QDs have desirable light-harvesting

Table 1 Showing information related to synthetic approaches for QD@MOFs Cos, and their applications

S. No	Application	MOF	QD	QD@MOF composite	Methods of synthesis	Refs.
1	Hydrogen production	MIL-101/ MOF-5	CdS	CdS@MIL-101/MOF-5	Ship-bottle	[52]
2	Hydrogen production	ZAVA	CdS	CdS@ZAVA- MOF	Bottle-ship	[87]
3	Hydrogen production	UiO-66-NH_2	MoS_2	MoS_2 QDs/UiO-66-NH_2/G	Solvothermal	[72]
4	Hydrogen production	ZIF-8	ZnCdS	ZnCdS@ ZIF-8	Template method	[88]
5	CO_2 reduction	PCN-221(Fex)	$CH_3NH_3PbI_3$	$CH_3NH_3PbI_3$@PCN-221(Fex)	Encapsulating technique	[78]
6	Solid state lightning	ZIF-8	CdSe/CdS	CdSe/CdS-PVP@ZIF-8	Solution process	[89]
7	Temperature and aqueous heavy metal detection	MOF-5	$CH_3NH_3PbBr_3$	$CH_3NH_3PbBr_3$@MOF-5	Solvothermal	[90]
8	Chemical sensing	ZIF-8	ZnS:Eu	ZnS:Eu @ZIF-8	One-pot method	[91]
9	Chemical sensing	Zn-MOF-8	Ag	Zn-MOF-8@Ag	Sonochemical technique	[92]
10	Chemical sensing	ZIF-8	GaN	GaN@ZIF-8	Solution process	[93]
11	Chemical sensing (oxidase additives)	ZIF-8	CdTe	CdTe@ZIF-8	Solution process	[94]
12	Supercapacitor	ZIF-8	SnO_2	SnO_2 QDs@ ZIF-8	In-situ epoxide precipitation method	[95]
13	Photocatalytic and photovoltaic	NTU-9	CdTe	CdTe@ NTU-9	In-situ	[96]
14	Photovoltaic	Eu-MOF	CdTe	CdTe QD/Eu-MOF	Solution process	[97]
15	Light harvesting	IRMOF-3	CdSe/CdS/ZnS	CdSe/CdS/ZnS QDs @ IRMOF-3	Solution process	[98]
16	Visual recognition	ZIF-8	CdTe	CDs/QDs@ZIF-8	Solution process	[99]

capacity, optoelectronic, and fluorescent properties. The fashioned Cos have gained approval because of their improved stability, remarkable physio-chemical/opto-electrical features, and broader range of applications such as electrocatalysis, light harvesting, bioimaging, energy storage, etc. MOFs play numerous roles in the stabilization, dispersion of QDs and the modulation of electron-hole recombination by the charge separation. Along with this MOFs also enhance the coordinatively unsaturated metal sites for the accommodation of chemically active species in QDs. MOFs with tunable pores can behave as encapsulating agents to construct well-organized semiconductor particles without any chances of agglomeration. Recently, QDs@MOFs composites have been involved in the areas of energy production, storage, and environment remediation, such as green H_2 production, CO_2 reduction to produce methane, degradation of organic dyes, Cr^{6+} reduction, and NO oxidation etc.

Despite the advantages of QDs@MOFs in various applications, there are still many challenges exist. A major concern in the synthesis of these composites is the control over the morphology and size of semiconductor QDs. It is also vital to stabilize the dispersion ability of the QDs in the solution containing MOF components to get all the profits from both the properties of MOFs and QDs. In recent approaches to fashioning the QD@MOF composites, there has been a random and uncontrolled growth of QDs in the MOF matrix. The controlled synthesis of QDs@MOFs with advanced surface properties is a frightening task. A precise ratio of all precursors, experimental conditions directly affect the size of QDs and placement of organic linkers in MOFs. Then, the progress of QDs@MOFs composite can be accomplished by proper association between the functional groups of MOFs and QDs. Although, major developments have been made in the area of QDs@MOFs-based photocatalysis. There are still a lot of efforts to be made for the industrial-scale production of these type of photocatalysts. Several kinds of QDs have poisonous compositions, especially inorganic QDs contain heavy metal atoms that can cause environmental pollution and human health issues. It is essential to develop low-toxic QDs@MOFs composites from environment friendly precursors such as carbon-based QDs and encapsulate them in MOFs, which have great potential in biomedical applications and lesser environmental risk in other applications. It is also required to encompass the applications of Cos, such as their utilization for removal of organic pollutants such as pesticides and antibiotics as well as detection of heavy metals. Thus, in conclusion, these Cos are a developing class of materials which shows an optimistic future with their great potential for high-impact contributions towards numerous applied research areas.

Acknowledgements Dr. Priyanka Kajal (Senior scientist) at Ananya Vijaya consultancy LLP Bangalore, India is thankful to the company for providing infrastructure, facilities, and financial assistance for carrying out study effectively. Dr. Amanpreet Kaur Jassal, Assistant Professor at Guru Nanak Dev University located in Amritsar, Punjab, India wholeheartedly thankful to her University and especially Department of Chemistry for providing top-notch infrastructure and cutting-edge facilities.

References

1. Chen, Z., Kirlikovali, K., Shi, L., Farha, O.K.: Rational design of stable functional metal-organic frameworks. Mater. Horiz. (2023). https://doi.org/10.1039/D3MH00541K
2. Linnane, E., Haddad, S., Melle, F., Mei, Z., Fairen-Jimenez, D.: The uptake of metal–organic frameworks: a journey into the cell. Chem. Soc. Rev. **51**, 6065–6086 (2022). https://doi.org/10.1039/D0CS01414A
3. Jassal, A.K., Sran, B.S., Mandal, K., Mukhopadhyay, P., Hundal, G.: Role reversal of the carboxylate group from coordination to hydrogen bonding only, in structurally diverse metal-2-amino,5-nitro-benzoates: a first report, Cryst. Growth Des. **18**, 4737–4748 (2018). https://doi.org/10.1021/acs.cgd.8b00771
4. Jassal, A.K., Sharma, S., Hundal, G., Hundal, M.S.: Structural diversity, thermal studies, and luminescent properties of metal complexes of Dinitrobenzoates: a single crystal to single crystal transformation from dimeric to polymeric complex of copper(II). Cryst. Growth Des. **15**(1), 79–93 (2015). https://doi.org/10.1021/cg500883w
5. Kaur, A., Hundal, G., Hundal, A.S.: Spontaneous resolution upon crystallization of 3D, chiral inorganic networks assembled from achiral, polyoxometallate units and metal ions. Cryst. Growth Des. **13**, 3996–4001 (2013). https://doi.org/10.1021/cg4007197
6. Aggarwal, V., Solanki, S., Malhotra, B.D.: Applications of metal–organic framework-based bioelectrodes. Chem. Sci. **13**, 8727–8743 (2022). https://doi.org/10.1039/D2SC03441G
7. Introduction to metal-organic frameworks. Chem. Rev. **112**, 673–674 (2012). https://doi.org/10.1021/cr300014x
8. Li, X., Wang, Z., Wang, L.: Metal-organic framework-based materials for solar water splitting. Small Sci. **1**, 2000074 (2021). https://doi.org/10.1002/smsc.202000074
9. Gutiérrez, M., Zhang, Y., Tan, J.-C.: Confinement of luminescent guests in metal-organic frameworks: understanding pathways from synthesis and multimodal characterization to potential applications of LG@MOF systems. Chem. Rev. **122**, 10438–10483 (2022). https://doi.org/10.1021/acs.chemrev.1c00980
10. Geilhufe, R.M.: Quantum buckling in metal-organic framework materials. Nano Lett. **21**, 10341–10345 (2021). https://doi.org/10.1021/acs.nanolett.1c03579
11. Chan, W.C.W., Maxwell, D.J., Gao, X., Bailey, R.E., Han, M., Nie, S.: Luminescent quantum dots for multiplexed biological detection and imaging. Curr. Opin. Biotechnol. **13**, 40–46 (2002). https://doi.org/10.1016/s0958-1669(02)00282-3
12. Yang, L., Zhang, S., Xu, B., Jiang, J., Cai, B., Lv, X., Zou, Y., Fan, Z., Yang, H., Zeng, H.: I–III–VI quantum dots and derivatives: design, synthesis, and properties for light-emitting diodes. Nano Lett. **23**, 2443–2453 (2023). https://doi.org/10.1021/acs.nanolett.2c03138
13. Al-Douri, Y., Khan, M.M., Jennings, J.R.: Synthesis and optical properties of II–VI semiconductor quantum dots: a review. J. Mater. Sci. Mater. Electron. **34**, 993 (2023). https://doi.org/10.1007/s10854-023-10435-5
14. Mittelstädt, A., Schliwa, A., Klenovský, P.: Modeling electronic and optical properties of III–V quantum dots—Selected recent developments. Light Sci Appl. **11**, 17 (2022). https://doi.org/10.1038/s41377-021-00700-9
15. Kim, Y., Chang, J.H., Choi, H., Kim, Y.-H., Bae, W.K., Jeong, S.: III–V colloidal nanocrystals: control of covalent surfaces. Chem. Sci. **11**, 913–922 (2020). https://doi.org/10.1039/C9SC04290C
16. Zhao, T., Oh, N., Jishkariani, D., Zhang, M., Wang, H., Li, N., Lee, J.D., Zeng, C., Muduli, M., Choi, H.-J., Su, D., Murray, C.B., Kagan, C.R.: General synthetic route to high-quality colloidal III–V semiconductor quantum dots based on pnictogen chlorides. J. Am. Chem. Soc. **141**, 15145–15152 (2019). https://doi.org/10.1021/jacs.9b06652
17. Cotta, M.A.: Quantum dots and their applications: what lies ahead? ACS Appl. Nano Mater. **3**, 4920–4924 (2020). https://doi.org/10.1021/acsanm.0c01386
18. Krizkova, S., Dostalova, S., Michalek, P., Nejdl, L., Kominkova, M., Milosavljevic, V., Moulick, A., Vaculovicova, M., Kopel, P., Adam, V., Kizek, R.: SDS-PAGE as a tool for hydrodynamic

diameter-dependent separation of quantum dots. Chromatographia **78**, 785–793 (2015). https://doi.org/10.1007/s10337-015-2893-z
19. GhoshMitra, S., Diercks, D.R., Mills, N.C., Hynds, D.L., Ghosh, S.: Excellent biocompatibility of semiconductor quantum dots encased in multifunctional poly(N-isopropylacrylamide) nanoreservoirs and nuclear specific labeling of growing neurons. Appl. Phys. Lett. **98**, 103702 (2011). https://doi.org/10.1063/1.3562036
20. Bakalova, R., Ohba, H., Zhelev, Z., Ishikawa, M., Baba, Y.: Quantum dots as photosensitizers? Nat. Biotechnol. **22**, 1360–1361 (2004). https://doi.org/10.1038/nbt1104-1360
21. Duan, L., Hu, L., Guan, X., Lin, C.-H., Chu, D., Huang, S., Liu, X., Yuan, J., Wu, T.: Quantum dots for photovoltaics: a tale of two materials. Adv. Energy Mater. **11**, 2100354 (2021). https://doi.org/10.1002/aenm.202100354
22. Hvam, J.M.: Optoelectronic properties and applications of quantum dots. In: Optical Properties of Materials and Their Applications, pp. 503–536. Wiley (2019). https://doi.org/10.1002/9781119506003.ch17
23. Jeevanandam, J., Balu, S.K., Andra, S., Danquah, M.K., Vidyavathi, M., Muthalagu, M.: Quantum dots synthesis and application. In: Contemporary Nanomaterials in Material Engineering Applications, pp. 229–265. Springer, Cham (2021). https://doi.org/10.1007/978-3-030-62761-4_9
24. Giri, L., Rout, S.R., Varma, R.S., Otyepka, M., Jayaramulu, K., Dandela, R.: Recent advancements in metal–organic frameworks integrating quantum dots (QDs@MOF) and their potential applications. Nanotechnol. Rev. **11**, 1947–1976 (2022). https://doi.org/10.1515/ntrev-2022-0118
25. Wang, Y.-Y., Ji, X.-Y., Yu, M., Tao, J.: Confining lead-free perovskite quantum dots in metal–organic frameworks for visible light-driven proton reduction. Mater. Chem. Front. **5**, 7796–7807 (2021). https://doi.org/10.1039/D1QM01006A
26. Lin, J., Zhou, H., Amin, R.S., Fetohi, A.E., El-Khatib, K.M., Wang, C., Guo, L., Wang, Y.: In situ formation of nickel sulfide quantum dots embedded into a two-dimensional metal–organic framework for water splitting. Inorg. Chem. Front. **10**, 1294–1304 (2023). https://doi.org/10.1039/D2QI02279F
27. Stock, N., Biswas, S.: Synthesis of metal-organic frameworks (MOFs): routes to various MOF topologies, morphologies, and composites. Chem. Rev. **112**, 933–969 (2012). https://doi.org/10.1021/cr200304e
28. Zhu, Q.-L., Xu, Q.: Metal–organic framework composites. Chem. Soc. Rev. **43**, 5468–5512 (2014). https://doi.org/10.1039/C3CS60472A
29. Yaghi, O.M., Li, G., Li, H.: Selective binding and removal of guests in a microporous metal–organic framework. Nature **378**, 703–706 (1995). https://doi.org/10.1038/378703a0
30. Kitagawa, S., Kitaura, R., Noro, S.N.: Functional porous coordination polymers. **43**, 2334–2375 (2004). https://doi.org/10.1002/anie.200300610
31. Serre, C.: Gérard Férey (1941–2017) **56**, 14802 (2017). https://doi.org/10.1002/anie.201709508
32. Jiao, L., Seow, J.Y.R., Skinner, W.S., Wang, Z.U., Jiang, H.-L.: Metal–organic frameworks: structures and functional applications. Mater. Today **27**, 43–68 (2019). https://doi.org/10.1016/j.mattod.2018.10.038
33. Manikandan, A., Chen, Y.-Z., Shen, C.-C., Sher, C.-W., Kuo, H.-C., Chueh, Y.-L.: A critical review on two-dimensional quantum dots (2D QDs): from synthesis toward applications in energy and optoelectronics. Prog. Quantum Electron. **68**, 100226 (2019). https://doi.org/10.1016/j.pquantelec.2019.100226
34. Bera, D., Qian, L., Tseng, T.-K., Holloway, P.H.: Quantum dots and their multimodal applications: a review. Materials (Basel) **3**, 2260–2345 (2010). https://doi.org/10.3390/ma3042260
35. Mohamed, W.A.A., El-Gawad, H.A., Mekkey, S., Galal, H., Handal, H., Mousa, H., Labib, A.: Quantum dots synthetization and future prospect applications. Nanotechnol. Rev. **10**, 1926–1940 (2021). https://doi.org/10.1515/ntrev-2021-0118

36. Liu, R., Wu, D., Feng, X., Müllen, K.: Bottom-up fabrication of photoluminescent graphene quantum dots with uniform morphology. J. Am. Chem. Soc. **133**, 15221–15223 (2011). https://doi.org/10.1021/ja204953k
37. Crista, D.M.A., Esteves da Silva, J.C.G., Pinto da Silva, L.: Evaluation of different bottom-up routes for the fabrication of carbon dots. Nanomaterials **10**, 1316 (2020). https://doi.org/10.3390/nano10071316
38. Valizadeh, A., Mikaeili, H., Samiei, M., Farkhani, S.M., Zarghami, N., Kouhi, M., Akbarzadeh, A., Davaran, S.: Quantum dots: synthesis, bioapplications, and toxicity. Nanosc. Res. Lett. **7**, 480 (2012). https://doi.org/10.1186/1556-276X-7-480
39. Jeevika, A., Alagarsamy, G., Celestina, J.J.: Biogenic synthesis of carbon quantum dots from garlic peel bio-waste for use as a fluorescent probe for sensing of quercetin. Luminescence **37**, 1991–2001 (2022). https://doi.org/10.1002/bio.4381
40. Singh, N., Qutub, S., Khashab, N.M.: Biocompatibility and biodegradability of metal organic frameworks for biomedical applications. J. Mater. Chem. B **9**, 5925–5934 (2021). https://doi.org/10.1039/D1TB01044A
41. Xue, Y., Zheng, S., Xue, H., Pang, H.: Metal–organic framework composites and their electrochemical applications. J. Mater. Chem. A **7**, 7301–7327 (2019). https://doi.org/10.1039/C8TA12178H
42. Feng, P.L., Perry, J.J.I., Nikodemski, S., Jacobs, B.W., Meek, S.T., Allendorf, M.D.: Assessing the purity of metal–organic frameworks using photoluminescence: MOF-5, ZnO quantum dots, and framework decomposition. J. Am. Chem. Soc. **132**, 15487–15489 (2010). https://doi.org/10.1021/ja1065625
43. Wu, T., Liu, X., Liu, Y., Cheng, M., Liu, Z., Zeng, G., Shao, B., Liang, Q., Zhang, W., He, Q., Zhang, W.: Application of QD-MOF composites for photocatalysis: energy production and environmental remediation. Coord. Chem. Rev. **403**, 213097 (2020). https://doi.org/10.1016/j.ccr.2019.213097
44. Aguilera-Sigalat, J., Bradshaw, D.: Synthesis and applications of metal-organic framework–quantum dot (QD@MOF) composites. Coord. Chem. Rev. **307**, 267–291 (2016). https://doi.org/10.1016/j.ccr.2015.08.004
45. Kumagai, K., Uematsu, T., Torimoto, T., Kuwabata, S.: Direct surface modification of semiconductor quantum dots with metal–organic frameworks. Cryst. Eng. Comm. **21**, 5568–5577 (2019). https://doi.org/10.1039/C9CE00769E
46. Xie, S., Li, X., Wang, L., Zhu, F., Zhao, X., Yuan, T., Liu, Q., Chen, X.: High quantum-yield carbon dots embedded metal-organic frameworks for selective and sensitive detection of dopamine. Microchem. J. **160**, 105718 (2021). https://doi.org/10.1016/j.microc.2020.105718
47. Chen, Y., Sakata, O., Nanba, Y., Kumara, L.S.R., Yang, A., Song, C., Koyama, M., Li, G., Kobayashi, H., Kitagawa, H.: Electronic origin of hydrogen storage in MOF-covered palladium nanocubes investigated by synchrotron X-rays. Commun. Chem. **1**, 1–8 (2018). https://doi.org/10.1038/s42004-018-0058-3
48. Chen, L., Xu, Q.: Metal-organic framework composites for catalysis. Matter **1**, 57–89 (2019). https://doi.org/10.1016/j.matt.2019.05.018
49. Wakaoka, T., Hirai, K., Murayama, K., Takano, Y., Takagi, H., Furukawa, S., Kitagawa, S.: Confined synthesis of CdSe quantum dots in the pores of metal–organic frameworks. J. Mater. Chem. C **2**, 7173–7175 (2014). https://doi.org/10.1039/C4TC01136H
50. Zhang, W., Arramel, A., Wong, P.K.J., Zhang, L., Zheng, J., Zhang, W., Zhang, H., Yan, X., Qi, J., Li, J.: Core–shell hybrid zeolitic imidazolate framework-derived hierarchical carbon for capacitive deionization. J. Mater. Chem. A **8**, 14653–14660 (2020). https://doi.org/10.1039/D0TA05709F
51. Biswal, B.P., Shinde, D.B., Pillai, V.K., Banerjee, R.: Stabilization of graphene quantum dots (GQDs) by encapsulation inside zeolitic imidazolate framework nanocrystals for photoluminescence tuning. Nanoscale **5**, 10556–10561 (2013). https://doi.org/10.1039/C3NR03511E

52. He, J., Yan, Z., Wang, J., Xie, J., Jiang, L., Shi, Y., Yuan, F., Yu, F., Sun, Y.: Significantly enhanced photocatalytic hydrogen evolution under visible light over CdS embedded on metal–organic frameworks. Chem. Commun. **49**, 6761–6763 (2013). https://doi.org/10.1039/C3CC43218A
53. Gan, H., Wang, Z., Li, H., Wang, Y., Sun, L., Li, Y.: CdSe QDs@UIO-66 composite with enhanced photocatalytic activity towards RhB degradation under visible-light irradiation. RSC Adv. **6**, 5192–5197 (2016). https://doi.org/10.1039/C5RA23565K
54. Li, R., Hu, J., Deng, M., Wang, H., Wang, X., Hu, Y., Jiang, H.-L., Jiang, J., Zhang, Q., Xie, Y., Xiong, Y.: Integration of an inorganic semiconductor with a metal-organic framework: a platform for enhanced gaseous photocatalytic reactions. Adv. Mater. **26**, 4783–4788 (2014). https://doi.org/10.1002/adma.201400428
55. Xu, X.-Y., Yan, B.: Eu(III)-functionalized ZnO@MOF heterostructures: integration of pre-concentration and efficient charge transfer for the fabrication of a ppb-level sensing platform for volatile aldehyde gases in vehicles. J. Mater. Chem. A. **5**, 2215–2223 (2017). https://doi.org/10.1039/C6TA10019H
56. Chen, Y.-C., Chiang, W.-H., Kurniawan, D., Yeh, P.-C., Otake, K., Kung, C.-W.: Impregnation of graphene quantum dots into a metal-organic framework to render increased electrical conductivity and activity for electrochemical sensing. ACS Appl. Mater. Interfaces **11**, 35319–35326 (2019). https://doi.org/10.1021/acsami.9b11447
57. Ren, C., Li, Z., Huang, L., Xiong, X., Nie, Z., Yang, Y., Zhu, W., Yang, W., Wang, L.: Confinement of all-inorganic perovskite quantum dots assembled in metal–organic frameworks for ultrafast scintillator application. Nanoscale **14**, 4216–4224 (2022). https://doi.org/10.1039/D1NR08120A
58. Buso, D., Jasieniak, J., Lay, M.D.H., Schiavuta, P., Scopece, P., Laird, J., Amenitsch, H., Hill, A.J., Falcaro, P.: Highly luminescent metal-organic frameworks through quantum dot doping. Small **8**, 80–88 (2012). https://doi.org/10.1002/smll.201100710
59. Ajibade, P.A., Oloyede, S.O.: Synthesis of metal-organic frameworks quantum dots composites as sensors for endocrine-disrupting chemicals. Int. J. Mol. Sci. **23**, 7980 (2022). https://doi.org/10.3390/ijms23147980
60. Chen, L., Zheng, L., Wang, F., Yi, S., Liu, D., Huang, X., Chen, R., He, H.: A ratiometric fluorescence sensor based on metal-organic frameworks and quantum dots for detection of ascorbic acid. Opt. Mater. **121**, 111622 (2021). https://doi.org/10.1016/j.optmat.2021.111622
61. Chen, X., Gao, H., Yang, M., Xing, L., Dong, W., Li, A., Zheng, H., Wang, G.: Smart integration of carbon quantum dots in metal-organic frameworks for fluorescence-functionalized phase change materials. Energy Storage Mater. **18**, 349–355 (2019). https://doi.org/10.1016/j.ensm.2018.08.015
62. Yang, J.-M., Hu, X.-W., Liu, Y.-X., Zhang, W.: Fabrication of a carbon quantum dots-immobilized zirconium-based metal-organic framework composite fluorescence sensor for highly sensitive detection of 4-nitrophenol. Microporous Mesoporous Mater. **274**, 149–154 (2019). https://doi.org/10.1016/j.micromeso.2018.07.042
63. Jiang, X., Zhang, J., Fan, R., Zhou, X., Zhu, K., Yang, Y.: Multiple Interpenetrating metal-organic frameworks with channel-size-dependent behavior for selective gossypol detection and perovskite quantum dot encapsulation. ACS Appl. Mater. Interfaces **14**, 49945–49956 (2022). https://doi.org/10.1021/acsami.2c13610
64. Xie, Z., Li, X., Li, R., Lu, S., Zheng, W., Tu, D., Feng, Y., Chen, X.: In situ confined growth of ultrasmall perovskite quantum dots in metal–organic frameworks and their quantum confinement effect. Nanoscale **12**, 17113–17120 (2020). https://doi.org/10.1039/D0NR04741D
65. Yu, H., Zhu, W., Zhou, H., Liu, J., Yang, Z., Hu, X., Yuan, A.: Porous carbon derived from metal–organic framework@graphene quantum dots as electrode materials for supercapacitors and lithium-ion batteries. RSC Adv. **9**, 9577–9583 (2019). https://doi.org/10.1039/C9RA01488H

66. Cao, Y., Zhou, Y., Lin, Y., Zhu, J.-J.: Hierarchical metal-organic framework-confined $CsPbBr_3$ quantum dots and aminated carbon dots: a new self-sustaining suprastructure for electrochemiluminescence bioanalysis. Anal. Chem. **93**, 1818–1825 (2021). https://doi.org/10.1021/acs.analchem.0c04717
67. Usman, M., Zhu, Q.-L.: Metal–organic frameworks for electrocatalysis. In: Methods for Electrocatalysis, pp. 29–66. Springer, Cham (2020). https://doi.org/10.1007/978-3-030-27161-9_2
68. Bai, Y., Dou, Y., Xie, L.-H., Rutledge, W., Li, J.-R., Zhou, H.-C.: Zr-based metal–organic frameworks: design, synthesis, structure, and applications. Chem. Soc. Rev. **45**, 2327–2367 (2016). https://doi.org/10.1039/C5CS00837A
69. Kaur, G., Sharma, S., Singh, S., Bhardwaj, N., Deep, A.: Selective and sensitive electrochemical sensor for aflatoxin M1 with a molybdenum disulfide quantum dot/metal–organic framework nanocomposite. ACS Omega **7**, 17600–17608 (2022). https://doi.org/10.1021/acsomega.2c00126
70. Hou, Y., Liu, Z., Tong, L., Zhao, L., Kuang, X., Kuang, R., Ju, H.: One-step electrodeposition of the MOF@CCQDs/NiF electrode for chiral recognition of tyrosine isomers. Dalton Trans. **49**, 31–34 (2019). https://doi.org/10.1039/C9DT04354C
71. Yang, G., Wang, L., Zhang, C., Li, P., Du, H., Mao, Y., Qiu, M., Li, Q., Hao, D., Wang, Q.: Novel graphene quantum dots modified NH_2-MIL-125 photocatalytic composites for effective antibacterial property and mechanism insight. Sep. Purif. Technol. **312**, 123433 (2023). https://doi.org/10.1016/j.seppur.2023.123433
72. Hao, X., Jin, Z., Yang, H., Lu, G., Bi, Y.: Peculiar synergetic effect of MoS_2 quantum dots and graphene on metal-organic frameworks for photocatalytic hydrogen evolution. Appl. Catal. B **210**, 45–56 (2017). https://doi.org/10.1016/j.apcatb.2017.03.057
73. Rezaei, H., Zinatizadeh, A.A., Joshaghani, M., Zinadini, S.: Amino acid-based carbon quantum dot modified MIL-53 (Fe): investigation of its visible-driven photocatalytic activity provided by a highly efficient photoreactor. Mater. Sci. Semicond. Process. **154**, 107190 (2023). https://doi.org/10.1016/j.mssp.2022.107190
74. Ellis, A.S., Johnson, T.M., Bullen, T.D.: Chromium isotopes and the fate of hexavalent chromium in the environment. Science **295**, 2060–2062 (2002). https://doi.org/10.1126/science.1068368
75. Saha, B., Orvig, C.: Biosorbents for hexavalent chromium elimination from industrial and municipal effluents. Coord. Chem. Rev. **254**, 2959–2972 (2010). https://doi.org/10.1016/j.ccr.2010.06.005
76. Lin, R., Li, S., Wang, J., Xu, J., Xu, C., Wang, J., Li, C., Li, Z.: Facile generation of carbon quantum dots in MIL-53(Fe) particles as localized electron acceptors for enhancing their photocatalytic Cr(VI) reduction. Inorg. Chem. Front. **5**, 3170–3177 (2018). https://doi.org/10.1039/C8QI01164H
77. Kong, Z.-C., Liao, J.-F., Dong, Y.-J., Xu, Y.-F., Chen, H.-Y., Kuang, D.-B., Su, C.-Y.: Core@Shell CsPbBr 3@Zeolitic imidazolate framework nanocomposite for efficient photocatalytic CO2 reduction. ACS Energy Lett. **3**, 2656–2662 (2018). https://doi.org/10.1021/acsenergylett.8b01658
78. Wu, L.-Y., Mu, Y.-F., Guo, X.-X., Zhang, W., Zhang, Z.-M., Zhang, M., Lu, T.-B.: Encapsulating perovskite quantum dots in iron-based metal-organic frameworks (MOFs) for efficient photocatalytic CO2 reduction. Angew. Chem. Int. Ed. **58**, 9491–9495 (2019). https://doi.org/10.1002/anie.201904537
79. Wei, D., Tang, W., Gan, Y., Xu, X.: Graphene quantum dot-sensitized Zn-MOFs for efficient visible-light-driven carbon dioxide reduction. Catal Sci. Technol. **10**, 5666–5676 (2020). https://doi.org/10.1039/D0CY00842G
80. Jin, S., Dong, G., Luo, J., Ma, F., Wang, C.: Improved photocatalytic NO removal activity of SrTiO3 by using SrCO3 as a new co-catalyst. Appl. Catal. B **227**, 24–34 (2018). https://doi.org/10.1016/j.apcatb.2018.01.020
81. Dong, G., Zhao, L., Wu, X., Zhu, M., Wang, F.: Photocatalysis removing of NO based on modified carbon nitride: the effect of celestite mineral particles. Appl. Catal. B **245**, 459–468 (2019). https://doi.org/10.1016/j.apcatb.2019.01.013

82. Chen, M., Wei, X., Zhao, L., Huang, Y., Lee, S., Ho, W., Chen, K.: Novel N/carbon quantum dot modified MIL-125(Ti) composite for enhanced visible-light photocatalytic removal of NO. Ind. Eng. Chem. Res. **59**, 6470–6478 (2020). https://doi.org/10.1021/acs.iecr.9b06816
83. Basit, M.A., Rashid, M., Khan, T.F., Muhyuddin, M., Butt, S.: Simplistic thermal transformation of MIL-125 to TiO2 nano-coins and nano-diamonds for efficient quantum-dot sensitized solar cells. Mater. Sci. Semicond. Process. **104**, 104663 (2019). https://doi.org/10.1016/j.mssp.2019.104663
84. Guo, F., Guo, J.-H., Wang, P., Kang, Y.-S., Liu, Y., Zhao, J., Sun, W.-Y.: Facet-dependent photocatalytic hydrogen production of metal–organic framework NH_2-MIL-125(Ti). Chem. Sci. **10**, 4834–4838 (2019). https://doi.org/10.1039/C8SC05060K
85. Jin, S., Son, H.-J., Farha, O.K., Wiederrecht, G.P., Hupp, J.T.: Energy transfer from quantum dots to metal-organic frameworks for enhanced light harvesting. J. Am. Chem. Soc. **135**, 955–958 (2013). https://doi.org/10.1021/ja3097114
86. Yan, F., Wang, X., Wang, Y., Yi, C., Xu, M., Xu, J.: Sensing performance and mechanism of carbon dots encapsulated into metal–organic frameworks. Microchim. Acta **189**, 379 (2022). https://doi.org/10.1007/s00604-022-05481-5
87. Saha, S., Das, G., Thote, J., Banerjee, R.: Photocatalytic metal-organic framework from CdS quantum dot incubated luminescent metallohydrogel. J. Am. Chem. Soc. **136**, 14845–14851 (2014). https://doi.org/10.1021/ja509019k
88. Chen, J., Lv, S., Shen, Z., Tian, P., Chen, J., Li, Y.: Novel ZnCdS quantum dots engineering for enhanced visible-light-driven hydrogen evolution. ACS Sustainable Chem. Eng. **7**, 13805–13814 (2019). https://doi.org/10.1021/acssuschemeng.9b01632
89. Stone, A.E.B.S., Irgen-Gioro, S., López-Arteaga, R., Hupp, J.T., Weiss, E.A.: Encapsulating CdSe/CdS QDs in the MOF ZIF-8 enhances their photoluminescence quantum yields in the solid state. Chem. Mater. **34**, 1921–1929 (2022). https://doi.org/10.1021/acs.chemmater.1c04355
90. Zhang, D., Xu, Y., Liu, Q., Xia, Z.: Encapsulation of $CH_3NH_3PbBr_3$ perovskite quantum dots in MOF-5 microcrystals as a stable platform for temperature and aqueous heavy metal ion detection. Inorg. Chem. **57**, 4613–4619 (2018). https://doi.org/10.1021/acs.inorgchem.8b00355
91. Wang, Q., Chu, H., Zhang, J., Ma, W., Qin, S., Gao, L.: ZnS: Eu @ZIF-8: selective formation of ZnS: Eu QDs within a zinc methylimidazole framework for chemical sensing applications. Heliyon **9**, 16081 (2023). https://doi.org/10.1016/j.heliyon.2023.e16081
92. Rani, S., Sharma, B., Kapoor, S., Malhotra, R., Varma, R.S., Dilbaghi, N.: Construction of silver quantum dot immobilized Zn-MOF-8 composite for electrochemical sensing of 2,4-dinitrotoluene. Appl. Sci. **9**, 4952 (2019). https://doi.org/10.3390/app9224952
93. Esken, D., Turner, S., Wiktor, C., Kalidindi, S.B., Van Tendeloo, G., Fischer, R.A.: GaN@ZIF-8: selective formation of gallium nitride quantum dots inside a zinc methylimidazolate framework. J. Am. Chem. Soc. **133**, 16370–16373 (2011). https://doi.org/10.1021/ja207077u
94. Wang, K., Li, N., Zhang, J., Zhang, Z., Dang, F.: Size-selective QD@MOF core-shell nanocomposites for the highly sensitive monitoring of oxidase activities. Biosens. Bioelectron. **87**, 339–344 (2017). https://doi.org/10.1016/j.bios.2016.08.026
95. Gao, Y., Wu, J., Zhang, W., Tan, Y., Zhao, J., Tang, B.: The electrochemical performance of SnO2 quantum dots@zeolitic imidazolate frameworks-8 (ZIF-8) composite material for supercapacitors. Mater. Lett. **128**, 208–211 (2014). https://doi.org/10.1016/j.matlet.2014.04.175
96. Kaur, R., Rana, A., Singh, R.K., Chhabra, V.A., Kim, K.-H., Deep, A.: Efficient photocatalytic and photovoltaic applications with nanocomposites between CdTe QDs and an NTU-9 MOF. RSC Adv. **7**, 29015–29024 (2017). https://doi.org/10.1039/C7RA04125J
97. Kaur, R., Sharma, A.L., Kim, K.-H., Deep, A.: A novel CdTe/Eu-MOF photoanode for application in quantum dot-sensitized solar cell to improve power conversion efficiency. J. Ind. Eng. Chem. **53**, 77–81 (2017). https://doi.org/10.1016/j.jiec.2017.04.002
98. Kumagai, K., Uematsu, T., Torimoto, T., Kuwabata, S.: Photoluminescence enhancement by light harvesting of metal-organic frameworks surrounding semiconductor quantum dots. Chem. Mater. **33**, 1607–1617 (2021). https://doi.org/10.1021/acs.chemmater.0c03367

99. Yi, K., Zhang, L.: Embedding dual fluoroprobe in metal-organic frameworks for continuous visual recognition of Pb^{2+} and PO_4^{3-} via fluorescence "turn-off-on" response: agar test paper and fingerprint. J. Hazard. Mater. **389**, 122141 (2020). https://doi.org/10.1016/j.jhazmat.2020.122141

Quantum Dots/2D Composites

Jaya V. Gade, Neeta Gupta, Reena Negi Rawat, Sangeeta Sahu, and Bhawana Jain

Abstract This chapter explores the intriguing domain of quantum dots (QDs) and two-dimensional (2D) materials, considering their synergistic combination to create innovative composites. In this context, the combination or synthesis of 2D materials with low-dimensional nanostructures in intricately engineered frameworks presents a novel opportunity to effectively integrate their distinct features in a complementary way. The combination of 2D and 0D nanomaterials in nanocomposites offers a promising approach to leverage the unique properties of each constituent. Combining the surface property tuning, small molecule attachments, external stimuli sensitivity, robust environmental resistance, negligible interference to other analytes, and recyclability of these materials, the resulting hybrid nanocomposites can overcome the limitations associated with each component synergistically.

Keywords Quantum dots · 2D materials · Nanocomposites · Graphene · MXenes

---

J. V. Gade
Department of Analytical Chemistry, SNDT Women's University, Mumbai 400049, Maharashtra, India

N. Gupta
Department of Chemistry, Govt. E. Raghavendra Rao P.G. Science College, Bilaspur 495001, Chhattisgarh, India

R. N. Rawat · B. Jain
Department of Chemical Sciences, Siddhachalam Laboratory, Raipur 493221, Chhattisgarh, India

S. Sahu (✉)
Department of Chemistry, Bhilai Institute of Technology, Raipur 493661, Chhattisgarh, India
e-mail: dr.sangeetasahu@bitraipur.ac.in

S. Thomas et al. (eds.), *Quantum Dots Based Nanocomposites*, Engineering Materials,
https://doi.org/10.1007/978-3-031-54779-9_8

1 Introduction

Two-dimensional (2D) nanomaterials are a type of nanomaterial with unique capabilities due to their ultra-thin, planar structure. One of the most well-known examples is graphene, a single layer of carbon atoms organized in a hexagonal lattice. Since the discovery of graphene, there has been a substantial boost in material science research due to the effective exfoliation of single-layer transition metal dichalcogenides. Graphene demonstrates extraordinary electrical conductivity, mechanical strength, and thermal properties [1–3]. Its unique structure and exceptional attributes have paved the way for extensive research and various applications, from electronics to materials science. Besides graphene, there has been a growing interest in exploring other two-dimensional (2D) nanomaterials such as MXenes, including transition metal carbides, nitrides, and carbonitrides, borophene, hexagonal boron nitride (h-BN), and phosphorene [4–12]. These materials possess various electrical, optical, and mechanical properties, rendering them appropriate for multiple applications. 2D nanomaterials are utilised across diverse domains, including electronics wherein they serve as integral components for transistors and sensors, optoelectronics wherein they facilitate the development of photodetectors and light-emitting devices, as well as energy storage wherever they function as electrodes within batteries and supercapacitors [13–16]. The considerable surface area and distinctive electrical characteristics of these materials play a significant role in enhancing their efficacy in catalytic reactions, sensing mechanisms, and biological implementations.

Recent advancements in the synthesis of nanomaterials and the research conducted on nanotechnology have showcased significant implications across all domains of science, engineering, and healthcare. Quantum Dots (QDs) are a significant category of nanoparticles widely utilized in many biological and industrial contexts [17]. The concept of QDs was initially formulated for both solid (glass crystals) and liquid states. The term 'Quantum' denotes a small and distinct unit of a physical attribute. The term of "quantum dots" is derived from the fundamental influence of quantum mechanics on the optical characteristics of nanocrystals [18]. QDs are recognised as a novel class of semiconductor inorganic crystals with dimensions on the nanometer scale. According to the technical definition, they can be described as tiny crystals that contain a variable number of electrons. These electrons occupy distinct quantum states and possess electronic properties that lie between those of bulk materials and discrete fundamental particles. Graphene has several limitations, including limited light absorption, rapid recombination of photo-generated carriers, ease of aggregation, and low dispersion. These shortages make it difficult to use these two materials in future optoelectronic devices [19]. Due to hydrogen bonding or van der Waals interactions, MXenes are prone to self-restacking, resulting in a considerable loss of electroactive area as well as ion and electrolyte inaccessibility [20]. The composites of 2D MXene and QDs offer a promising approach to leverage the unique properties of each constituent. This design enables the tuning of surface properties, attachment of small molecules, sensitivity to external stimuli, resistance to environmental factors, minimal interference with other analytes, and recyclability. Consequently, the hybrid nanocomposites effectively address the limitations associated with the individual components. The enhanced performance of QDs/2D materials hybrids can be attributed to the induced synergy between the constituent materials. The combination of 2D material and 0D nanomaterials in nanocomposites allows for the effective integration of the unique benefits offered by each component. This includes the ability to modify surface properties, attach small molecules, respond to external stimuli, withstand harsh environmental conditions, avoid interference with other substances, and be easily recycled. By combining these advantages, the hybrid nanocomposites overcome the limitations of each individual constituent in a synergistic manner.

2 Fundamentals of QDs

QDs are nanoscale zero dimensional particles with unique and favourable optical and electrical characteristics due to their quantum confinement phenomena. These tiny structures, typically measuring between 2 and 10 nm in diameter, can be regulated to generate distinct wavelengths of light through size modulation [21–24]. QDs such as graphene quantum dots (GQDs) or carbon quantum dots (CQDs) display advantageous properties, such as high luminescence, broad emission spectra, and resilience against photo bleaching [25–29]. These attributes render them highly beneficial in domains such as medicine for targeted therapy, diagnostics, imaging, energy-efficient screens, and solar cells for increasing light absorption [30, 31]. Furthermore, the ability to functionalize QDs with specific surface coatings allows for their incorporation into a wide range of applications, from biological markers to sensors [32]. While the technology evolves, researchers are looking for new methods to use QDs' unique features to advance materials science, electronics, and medical diagnostics [3, 32–37]. This chapter gives an overview of the latest developments and highlights research results for next-generation hetero structures of QDs/2D Composites.

3 Different Types of Quantum Dots/2D Composites

Graphene, a single sheet of hexagonally organised carbon atoms, offers remarkable electrical, thermal, and mechanical properties [38–42]. A synergetic relationship occurs when QDs, particularly semiconductor QDs such as CdSe or PbS, are integrated. The strong connection between graphene and QDs improves charge transport resulting in photoluminescence and electrical conductivity. Graphene has several limitations, including limited light absorption, rapid recombination of photogenerated carriers, ease of aggregation, and low dispersion. These shortages make using these materials in future optoelectronic devices difficult. Since the synthesis of QDs/Graphene composite materials in 2010, the synergistic effects of the composite materials have overcome the disadvantages of QDs and graphene [43, 44]. Within this hybrid arrangement, the QDs function as light-absorbing substances, generating

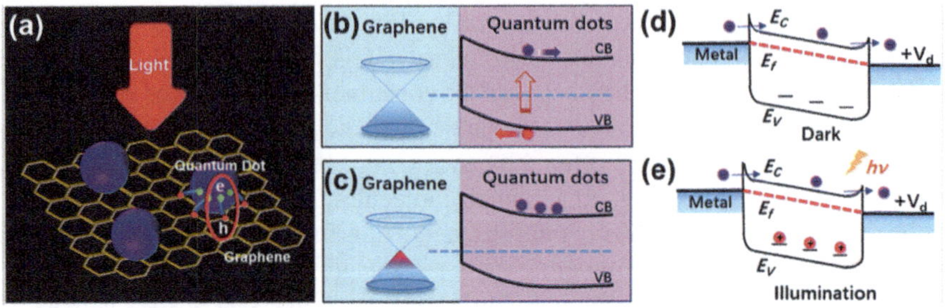

Fig. 1 **a** A schematic representation of the QDs/Graphene composite and light-generated electron–hole pairs. Energy level diagrams of the QDs/Graphene interface demonstrating **b** electron–hole pair production in QDs under illumination and **c** trapped carrier doping in graphene. Diagram illustrating the photogating effect **d** in the dark with external bias and **e** in the light [19]

electron–hole pairs when exposed to light. Subsequently, the carriers can be promptly transferred to graphene, resulting in photocurrent generation. This process facilitates the effective separation of electrons and holes, hence extending the lifespan of the carriers. The photo-generated carriers in QDs can be efficiently transported to the surface of graphene sheets, leading to significant optical amplification, exceptional optical sensitivity, and favourable photo-catalytic capabilities as shown in Fig. 1a–c [19]. The fundamental principle of the QDs/Graphene hybrid system is the charge transfer mechanism. The hybrid exhibits synergetic and cooperative effects, which can lead to the emergence of new phenomena and physical features.

Zhao and colleague utilised GO as a framework to grow oxidized SnO_2 QDs[45]. The functional groups of GO facilitated the oxidation of Sn^{2+} ions, while GO served as a template to enhance selective heterogeneous nucleation. Consequently, SnO_2 nanoparticles were formed and developed on the surface of GO. The SnO_2 QDs integrated with GO composite was synthesized by the hydrothermal method. The TEM image of SnO_2 QDs embedded in GO, exhibiting excellent dispersion and a significant amount of loading. Also, it was discovered that SnO_2 QDs tend to grow in the [001] direction, with (110) surfaces serving as oxygen bridging sites on the SnO_2 (110) plane. Liu et al. synthesised the CdTe QDs–TiO_2–graphene hybrid using a universal hydrothermal technique [46]. The CdTe QDs were synthesised and grown directly on the composite surface of reduced graphene oxide (rGO) and titanium dioxide (TiO_2). The reduced fluorescence in the generated hybrid implies that electrons transfer from CdTe QDs to TiO_2 and graphene.

The ZnO QDs/graphene heterojunction photodetectors are created by printing QDs directly on the channel of a graphene field-effect transistor (GFET). This combination of ZnO QDs' strong quantum confinement and graphene's high charge mobility results in exceptional quantum efficiency (or photoconductive gain) for detecting ultraviolet (UV) light while being blind to visible light. The crucial factor for achieving excellent performance is the presence of a clean van der Waals interface, which enables an effective transfer of electric charge from ZnOQDs to graphene when exposed to UV light. Gong et al. reported a robust ZnO QDs surface activation process and demonstrated that by systematically engineering the ZnO QDs surface, a transition from zero to extraordinarily high photo responsivity of 9.9×10^8 A/W and a photoconductive gain of 3.6×10^9 in ZnO QDs/GFET hetero junction photodetectors (Fig. 2) can be achieved [47]. The lithium-sulfur battery is highly regarded as a leading candidate for the future generation of energy storage devices, primarily because of its cost-effectiveness and exceptionally high energy density. Nevertheless, the uncontrolled development of Li dendrites and the shuttle effect significantly impede its practical utilization. To address both critical difficulties simultaneously, Yu et al.reported a novel Mo_2C quantum dots (MQDs) attached N-doped graphene functionalized interlayers (MQD@NG). The DFT simulations demonstrate that Mo_2C QDs have a lithiophilic characteristic that allows them to build a fast electrolyte diffusion channel for Li^+ transportation and accomplish uniform Li deposition [48]. Additionally, the polar Mo_2C quantum dots have a high affinity for the chemical adsorption of polysulfides. The lithium symmetric cells with MQD@NG interlayers successfully accomplished dendrite-free lithium deposition over 1600 h, even when subjected to a high current density of 10 mA cm^{-2}. The Li–S cells demonstrate a notable capacity of 1230 mAh g^{-1} at 0.2 C and maintain a remarkably stable cycling performance for over 400 cycles without any noticeable decrease in capacity. This study provided a novel perspective on the development and production of high-capacity safe lithium-metal batteries.

The use of ultratransparent electrodes has garnered significant interest in the fields of optoelectronics and energy technology. Nevertheless, achieving a harmonious equilibrium between energy storage capacity and transparency continues to pose a significant challenge. Yuan and coworkers presented a novel approach using a precisely controlled femtosecond laser to synthesise MXene quantum dots (MQDs) and attach them uniformly to laser reduced graphene oxide (LRGO) [49]. The resulting material exhibits outstanding electrochemical capacitance and extremely

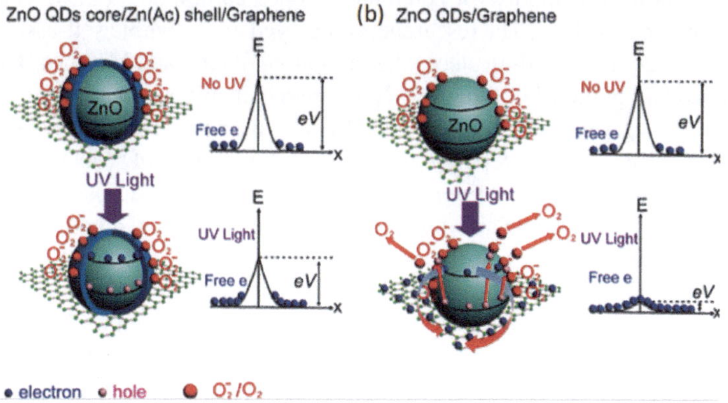

Fig. 2 Schematics of ZnO QDs/GFETs before (**a**) and after (**b**) removing the charge-transfer inhibiting Zn(Ac) shell from a ZnO QD. The influence of the Zn(Ac) shell on the energy band diagram and charge transfer to graphene is also shown [47]

high transparency. An analysis and observation of the synthesis process's mechanism and plasma dynamics occur simultaneously. The inclusion of the unique MQDs in LRGO significantly enhances the electrode's specific surface area by virtue of their nanoscale dimensions and the presence of additional edge states. The MQD/LRGO supercapacitor exhibits exceptional flexibility and durability. It has an ultrahigh energy density of 2.04×10^{-3}mWh cm^{-2}, a long cycle life with 97.6% capacity retention after 12,000 cycles, and an outstanding capacitance of 10.42 mF cm^{-2}. It also offers high transparency, a transmittance above 90%, and excellent overall performance. Moreover, this technique provides a way to fabricate nanostructured composite electrode materials and utilize quantum capacitance phenomena for energy storage.

As two-dimensional substrates for a wide variety of applications, it is anticipated that MXenes, which are both electrically conductive and flexible, would compete with graphene and, in some instances, even outperform itself [50–53]. As a result of hydrogen bonding or van der Waals interactions, MXenes have a tendency to self-restack, which results in a considerable loss of electroactive area as well as inaccessibility to ions and electrolytes [54]. Incorporating spacers, which reside on the surfaces of MXenes, can prevent the MXenes from self-restacking and safeguard their electroactive sites from being sacrificed. This is an intriguing and beneficial concept. In this context, hybridization or composite synthesis of two-dimensional MXenes with low-dimensional nanostructures in carefully designed offers an innovative way to combine their distinctive features in a manner that is complementary to one another [20]. By designing hybrid nanocomposites consisting of two-dimensional MXenes and zero-dimensional QDs, it is possible to effectively combine the benefits of both individual constituents. These benefits include surface property tuning, small molecule attachments, sensitivity to external stimuli, robust environmental resistance, negligible inferences to other analytes, and recyclability. This allows the hybrid nanocomposites to overcome the shortcomings of each individual constituent. Zou et al. described the use of heteroatom-doped carbon dots (CDs) as affordable conductive spacers [55]. The researchers employed a Coulombic self-assembly technique to create free-standing electrodes composed of CDs and MXenes. The fabrication procedure of the CDs/$Ti_3C_2T_x$MXene composite films is depicted in Fig. 3. The CDs have a negative charge, while the MXenes have a positive charge. The resulting electrodes exhibit unique topologies with a combination of 0D (dot) and 2D (sheet) motifs. The selected CDs, varying in particle sizes and heteroatoms, effectively regulated the interlayer spacing and influenced the composition and quantity of the hybrid films. CDs, as intercalators, efficiently reduce the self-restacking of MXene layers and enhance the interlayer distance. This is beneficial for the transport of electrolyte ions and the utilization of the surface active site. A unique design was built employing SnS_2 quantum dots (QDs) and N-doped $Ti_3C_2T_x$ in a 0D/2D heterostructure [56]. The growth process involved solvent spatial confinement and was characterised by its innovative nature. The nucleation and growth of SnS2 nanoparticles were effectively controlled by N-Methyl-2-pyrrolidone (NMP) during the hydrothermal process, leading to the consistent growth of SnS_2 quantum dots with a size of around

3 nm on the MXene matrix. In addition, the generation of NH_3 during the decomposition of the sulphur precursor resulted in the formation of N-doped $Ti_3C_2T_x$, which significantly enhances the transport of Na^+ ions at the interface. The application of this design model leads to the creation of an electrode made of SnS2 QDs and Ti_3C_2, which exhibits exceptional skills for storing sodium. These abilities include a high specific capacity and excellent cycling stability. The synthesis of black phosphorus quantum dots (BPQDs)/Ti_3C_2nanocomposites was achieved by Meng et al. through interfacial assembly [57]. These composites exhibit excellent dispersion of BPQDs that are chemically bonded to MXenes. BPQD/Ti_3C_2 composites are produced by combining BPQDs with $Ti_3C_2T_x$ in NMP by chemical bonding interactions, aided by sonication and stirring. The transmission electron microscopy (TEM) images showed that $Ti_3C_2T_x$ had a morphology consisting of only a few layers, resulting in a high specific surface area. Additionally, the BPQDs were uniformly distributed and had an average diameter of around 4 nm. The BPQDs that are anchored on the Ti_3C_2 nanosheets imparted with increased conductivity and relieved stress upon cycling, which enables a high-capacity and reliable energy storage.

Guo et al. constructed an arrangement of Sb single atoms and QDs on $Ti_3C_2T_x$ aerogels (Fig. 4) for use in potassium-ion batteries (PIBs) [58]. The DFT simulations revealed that the presence of atomically distributed Sb significantly enhances the charge transfer between Sb QDs and MXenes, resulting in improved adsorption performance of K at the interface. Furthermore, the MXene-based aerogel, which possesses a high number of surface functional groups and defects, offers a large number of anchoring sites. This characteristic enhances the structural stability and facilitates efficient electron transport in the composite. Desirable potassium reservoirs require a large loading of Sb, approximately 60.3 wt %, along with short

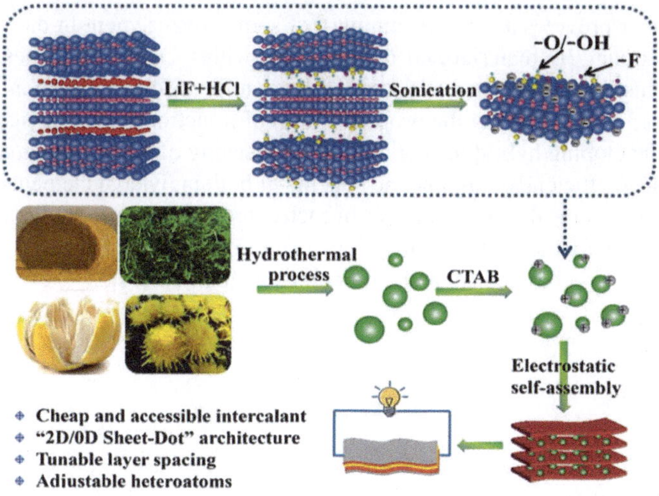

Fig. 3 Schematic of CDs/$Ti_3C_2T_x$MXene composite film fabrication [55]

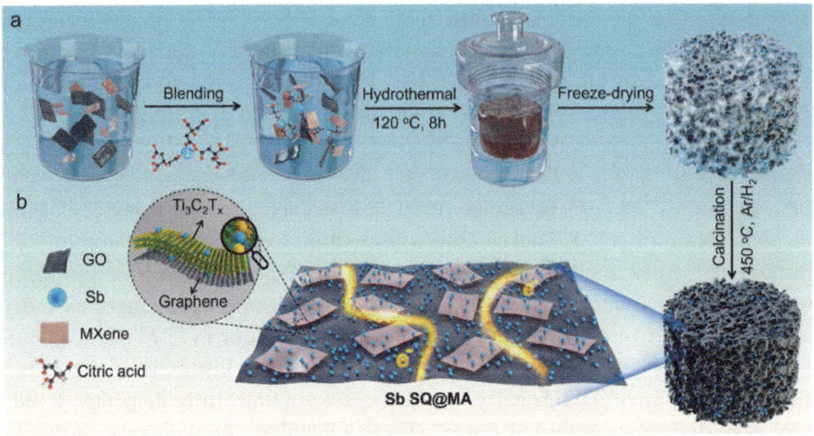

Fig. 4 Pictorial depiction of synthetic methods and structural properties of the Sb SQ@MA composite [58]

ionic transport channels. The combination of these qualities effectively improves the speed and cycling ability of the Sb SQ@MA electrodes in PIBs. This study has presented an illuminating method to customize the interface behavior of hetero structured electrodes for electrochemical purposes.

4 Conclusion

This chapter provides a concise summary of some advancement in the synthesis of multifunctional 2D material and their hybrids with QDs nanoparticles. Chemical components and constructed structures subject to precise manipulation are significant factors in determining the properties of multifunctional QDs/2D nanocomposites. By developing hybrid nanocomposites consisting of QDs/2D nanomaterials, it is possible to efficiently combine the benefits of both individual elements and synergistically overcome the disadvantages of each constituent. Future studies can serve as the foundation for the commercialization of QDs/2D nanocomposites for the purpose of applications in biomedical and environmental applications.

References

1. Seabra, A.B., et al.: Nanotoxicity of graphene and graphene oxide. Chem. Res. Toxicol. **27**(2), 159–168 (2014)
2. Razaq, A., et al.: Review on graphene-, graphene oxide-, reduced graphene oxide-based flexible composites: from fabrication to applications. Materials **15**(3), 1012 (2022)

3. Zhang, F., et al.: Recent advances on graphene: synthesis, properties and applications. Compos. A Appl. Sci. Manuf. **160**, 107051 (2022)
4. Das, P., Rosenkranz, A., Ganguly, S.: MXene Nanocomposites: Design, Fabrication, and Shielding Applications. CRC Press (2023)
5. Das, P., Srinivasan, S., Rajabzadeh, A.R.: Electromagnetic interference shielding behavior of MXenes: theoretical and experimental perspectives. In: MXene Nanocomposites, pp. 137–152. CRC Press (2023)
6. Ahmed, S.R., et al.: Electrochemical assisted enhanced nanozymatic activity of functionalized borophene for H2O2 and tetracycline detection. Biosens. Bioelectron. 115857 (2023)
7. Ganguly, S., et al.: Mussel-inspired polynorepinephrine/MXene-based magnetic nanohybrid for electromagnetic interference shielding in X-band and strain-sensing performance. Langmuir **38**(12), 3936–3950 (2022)
8. Das, P., et al.: Carbon-dots-initiated photopolymerization: an in situ synthetic approach for MXene/poly (norepinephrine)/copper hybrid and its application for mitigating water pollution. ACS Appl. Mater. Interfaces **13**(26), 31038–31050 (2021)
9. Caldwell, J.D., et al.: Photonics with hexagonal boron nitride. Nat. Rev. Mater. **4**(8), 552–567 (2019)
10. Naclerio, A.E., Kidambi, P.R.: A review of scalable hexagonal boron nitride (h-BN) synthesis for present and future applications. Adv. Mater. **35**(6), 2207374 (2023)
11. Idumah, C.I.: Recently emerging trends in flame retardancy of phosphorene polymeric nanocomposites and applications. J. Anal. Appl. Pyrolysis 105855 (2023)
12. Chaudhary, V., et al.: Phosphorene—an emerging two-dimensional material: recent advances in synthesis, functionalization, and applications. 2D Materials **9**(3), 032001 (2022)
13. Shanmugam, V., et al.: A review of the synthesis, properties, and applications of 2D materials. Part. Part. Syst. Charact. **39**(6), 2200031 (2022)
14. Marian, M., et al.: Roadmap for 2D materials in biotribological/biomedical applications–a review. Adv. Coll. Interface. Sci. **307**, 102747 (2022)
15. An, J., et al.: Perspectives of 2D materials for optoelectronic integration. Adv. Func. Mater. **32**(14), 2110119 (2022)
16. Chen, X., et al.: The marriage of two-dimensional materials and phase change materials for energy storage, conversion and applications. EnergyChem **4**(2), 100071 (2022)
17. Shukla, S.: Recent developments in biomedical applications of quantum dots. Adv Mater Rev **1**(1), 2–12 (2014)
18. Reshma, V., Mohanan, P.: Quantum dots: applications and safety consequences. J. Lumin. **205**, 287–298 (2019)
19. Tong, L., et al.: Recent progress in the preparation and application of quantum dots/graphene composite materials. RSC Adv. **7**(76), 47999–48018 (2017)
20. Das, P., et al.: MXene/0D nanocomposite architectures: Design, properties and emerging applications. Mater. Today Nano **24**, 100428 (2023)
21. Das, P., et al.: Waste-derived sustainable fluorescent nanocarbon-coated breathable functional fabric for antioxidant and antimicrobial applications. ACS Appl. Mater. Interfaces (2023)
22. Rajabzadeh, A.R., et al.: Magnetic Quantum Dots for Bioimaging. CRC Press (2023)
23. Ahmed, S.R., et al.: Graphdiyne quantum dots for H2O2 and dopamine detection. ACS Appl. Nano Mater. **6**(10), 8434–8443 (2023)
24. Das, P., et al.: Naturally derived carbon dots in situ confined self-healing and breathable hydrogel monolith for anomalous diffusion-driven phytomedicine release. ACS Appl. Bio Mater. **5**(12), 5617–5633 (2022)
25. Maity, P.P., et al.: Capra cartilage-derived peptide delivery via carbon nano-dots for cartilage regeneration. Front. Bioeng. Biotechnol. **11** (2023)
26. Lim, S.Y., Shen, W., Gao, Z.: Carbon quantum dots and their applications. Chem. Soc. Rev. **44**(1), 362–381 (2015)
27. Bacon, M., Bradley, S.J., Nann, T.: Graphene quantum dots. Part. Part. Syst. Charact. **31**(4), 415–428 (2014)

28. Das, P., et al.: Carbon dot biopolymer-based flexible functional films for antioxidant and food monitoring applications. ACS Appl. Polym. Mater. **4**(12), 9323–9340 (2022)
29. Das, P., et al.: Optical Properties of Quantum Dots. In: Quantum Dots and Polymer Nanocomposites, pp. 69–85. CRC Press (2022)
30. Jamieson, T., et al.: Biological applications of quantum dots. Biomaterials **28**(31), 4717–4732 (2007)
31. Das, P., et al.: Tailor made magnetic nanolights: fabrication to cancer theranostics applications. Nanoscale Adv. **3**(24), 6762–6796 (2021)
32. Ganguly, S., et al.: Advancement in science and technology of carbon dot-polymer hybrid composites: a review. Funct. Compos. Struct. **1**(2), 022001 (2019)
33. Das, P., et al.: Graphene based emergent nanolights: a short review on the synthesis, properties and application. Res. Chem. Intermed. **45**, 3823–3853 (2019)
34. Das, P., et al.: Carbon dots for heavy-metal sensing, pH-sensitive cargo delivery, and antibacterial applications. ACS Appl. Nano Mater. **3**(12), 11777–11790 (2020)
35. Das, P., et al.: Heteroatom doped blue luminescent carbon dots as a nano-probe for targeted cell labeling and anticancer drug delivery vehicle. Mater. Chem. Phys. **237**, 121860 (2019)
36. Ganguly, S., et al.: Microwave assisted green synthesis of Zwitterionic photolumenescent N-doped carbon dots: An efficient 'on-off'chemosensor for tracer Cr (+ 6) considering the inner filter effect and nano drug-delivery vector. Colloids Surf. A Physicochem. Eng. **579**, 123604 (2019)
37. Das, P., et al.: Converting waste Allium sativum peel to nitrogen and sulphur co-doped photoluminescence carbon dots for solar conversion, cell labeling, and photobleaching diligences: A path from discarded waste to value-added products. J. Photochem. Photobiol. B Biol. **197**, 111545 (2019)
38. Das, P., et al.: Acoustic green synthesis of graphene-gallium nanoparticles and PEDOT: PSS hybrid coating for textile to mitigate electromagnetic radiation pollution. ACS Appl. Nano Mater. **5**(1), 1644–1655 (2022)
39. Ganguly, S., et al.: Photopolymerized thin coating of polypyrrole/graphene nanofiber/iron oxide onto nonpolar plastic for flexible electromagnetic radiation shielding, strain sensing, and non-contact heating applications. Adv. Mater. Interfaces **8**(23), 2101255 (2021)
40. Ganguly, S., et al.: Green reduced graphene oxide toughened semi-IPN monolith hydrogel as dual responsive drug release system: rheological, physicomechanical, and electrical evaluations. J. Phys. Chem. B **122**(29), 7201–7218 (2018)
41. Ganguly, S., et al.: Mechanically robust dual responsive water dispersible-graphene based conductive elastomeric hydrogel for tunable pulsatile drug release. Ultrason. Sonochem. **42**, 212–227 (2018)
42. Ganguly, S., et al.: Sonochemical green reduction to prepare Ag nanoparticles decorated graphene sheets for catalytic performance and antibacterial application. Ultrason. Sonochem. **39**, 577–588 (2017)
43. Geng, X., et al.: Aqueous-processable noncovalent chemically converted graphene–quantum dot composites for flexible and transparent optoelectronic films. Adv. Mater. **22**(5), 638–642 (2010)
44. Guo, C.X., et al.: Layered graphene/quantum dots for photovoltaic devices. Angew. Chem. Int. Ed. **49**(17), 3014–3017 (2010)
45. Zhao, K., et al.: SnO2 Quantum dots@ graphene oxide as a high-rate and long-life anode material for lithium-ion batteries. Small **12**(5), 588–594 (2016)
46. Liu, J., Li, X.: Hydrothermal synthesis of CdTe quantum dots–TiO2–graphene hybrid. Phys. Lett. A **378**(4), 405–407 (2014)
47. Gong, M., et al.: All-printable ZnO quantum dots/graphene van der Waals heterostructures for ultrasensitive detection of ultraviolet light. ACS Nano **11**(4), 4114–4123 (2017)
48. Yu, B., et al.: Mo2C quantum dots@ graphene functionalized separator toward high-current-density lithium metal anodes for ultrastable Li-S batteries. Chem. Eng. J. **399**, 125837 (2020)
49. Yuan, Y., et al.: Ultrafast shaped laser induced synthesis of MXene quantum dots/graphene for transparent supercapacitors. Adv. Mater. **34**(12), 2110013 (2022)

50. Ding, L., et al.: A two-dimensional lamellar membrane: MXene nanosheet stacks. Angew. Chem. Int. Ed. **56**(7), 1825–1829 (2017)
51. Chen, X., et al.: Two-dimensional Ti 3 C 2 MXene-based nanostructures for emerging optoelectronic applications. Mater. Horiz. **8**(11), 2929–2963 (2021)
52. Ouyang, W., et al.: Improved photoelectric performance of UV photodetector based on ZnO nanoparticle-decorated BiOCl nanosheet arrays onto PDMS substrate: the heterojunction and Ti3C2Tx MXene conduction layer. Adv. Electron. Mater. **6**(6), 2000168 (2020)
53. Song, W., et al.: Self-powered MXene/GaN van der Waals heterojunction ultraviolet photodiodes with superhigh efficiency and stable current outputs. Adv. Mater. **33**(27), 2101059 (2021)
54. Luo, J., et al.: Sn4+ ion decorated highly conductive Ti3C2 MXene: promising lithium-ion anodes with enhanced volumetric capacity and cyclic performance. ACS Nano **10**(2), 2491–2499 (2016)
55. Li, L., et al.: Carbon dot-regulated 2D MXene films with high volumetric capacitance. Ind. Eng. Chem. Res. **59**(31), 13969–13978 (2020)
56. Ding, J., et al.: Integrating SnS2 quantum dots with nitrogen-doped Ti3C2T x MXene nanosheets for robust sodium storage performance. ACS Appl. Energy Mater. **4**(1), 846–854 (2021)
57. Meng, R., et al.: Black phosphorus quantum dot/Ti3C2 MXene nanosheet composites for efficient electrochemical lithium/sodium-ion storage. Adv. Energy Mater. **8**(26), 1801514 (2018)
58. Guo, X., et al.: MXene-based aerogel anchored with antimony single atoms and quantum dots for high-performance potassium-ion batteries. Nano Lett. **22**(3), 1225–1232 (2022)

Quantum Dots/One-Dimensional (1D) Composites

Supriya Rana, Vishwajit M. Gaikwad, and Krishna K. Yadav

Abstract Quantum dots are tiny semiconductor structures that have unique electronic and optical properties due to their small size and quantum confinement effects. In recent years, there has been increasing interest in combining quantum dots with one-dimensional materials, such as nanowires and carbon nanotubes, to create novel composite materials with enhanced properties. These composites have potential applications in a variety of fields, including optoelectronics, sensing, and energy conversion. The integration of quantum dots with one-dimensional (1D) materials can lead to improvements in charge transfer, light absorption, emission, and tunability of electronic and optical properties. The current chapter provides an overview of recent research on quantum dot/1D material composites and their potential applications.

Keywords Quantum dots · Composite · Semiconductor · 1D · Quantum confinement

1 Introduction

The unique properties and unconventional Physics at low dimensions significantly attract scientists' attention in this field. The low-dimensional materials can be categorized into 0D named Quantum dots (QDs), one-dimensional (1D) such as rods, nanotubes and two-dimensional (2D) such as nanosheets [1]. With diameters typically in the nanometer range, QDs exhibit size-dependent characteristics

S. Rana
Institute of Nano Science and Technology, Knowledge City, Sector 81, Mohali, India
e-mail: supriyarana8@gmail.com

V. M. Gaikwad
Departmet of Physics, Amolakchand Mahavidyalaya, Yavatmal, India
e-mail: vish9823@gmail.com

K. K. Yadav (✉)
Department of Chemical Engineering, Sami Shamoon College of Engineering, 8410802 Beer-Sheva, Israel
e-mail: Krishnay447@gmail.com; krishphy25@gmail.com

S. Thomas et al. (eds.), *Quantum Dots Based Nanocomposites*, Engineering Materials,
https://doi.org/10.1007/978-3-031-54779-9_9

such as tunable fluorescence, high photostability, and high quantum yield, making them highly desirable for a wide range of applications in optoelectronics, sensing, and biomedicine [2–4]. By integrating dissimilar materials into a 0D template, heterostructures with varying chemical compositions along the length of 1D materials can be created [5–7]. A typical contact between different dimension metals can be attached in the way shown in Fig. 1 [8, 9]. The interaction between 0D-1D components in these heterostructures results in significant changes in their properties due to variable electronic states, control of charge transport, and the emergence of new physical and chemical properties at their interface not observed in single-component systems [10]. These composites offer several advantages over individual QDs or 1D materials, including improved charge transport, enhanced light absorption and emission, and tunability of electronic and optical properties. Moreover, incorporating QDs into 1D materials can improve their compatibility with biological systems, making them suitable for bioimaging and biosensing applications. In the case of 1D materials, their confined size and high aspect ratio are responsible for their extraordinary properties as compared to their counterpart (2D and 3D). The synthesis of QD/1D material composites involves a combination of bottom-up and top-down approaches, where QDs are synthesized separately and then integrated into the 1D material matrix. Various techniques, such as solution-phase synthesis, vapour-phase deposition, and electrochemical methods, have been developed for the synthesis of these composites. Molecular beam epitaxy (MBE) or metal–organic chemical vapour deposition (MOCVD) growth techniques are also used for synthesising high-quality heterostructure materials; however, this technique requires little mismatch between the two interfaces which should be patterned therefore, only selected heterostructures can be prepared using this technique [11]. In recent years, there have been significant advancements in the synthesis and applications of QD/1D material composites, offering unique properties and promising potential for future research. One of the key advantages of QD/1D material composites is their improved charge transport properties. In individual QDs, electron–hole pairs can become trapped or recombine before they can contribute to charge transport. However, when combined with 1D materials, the 1D material can serve as an efficient pathway for charge transport, facilitating more efficient charge extraction and transport. Additionally, QD/1D material composites can exhibit enhanced light absorption and emission properties. The 1D material can serve as a scaffold for the QDs, increasing surface area and light scattering, which can result in higher photon conversion efficiency in applications such as solar cells and light emitting diodes (LEDs). Moreover, the confinement of charge carriers in both the QDs and 1D materials can lead to improved emission properties, including higher quantum yields and improved colour purity. Another notable advantage of QD/1D material composites is their potential for compatibility with biological systems, making them suitable for bioimaging and biosensing applications. This is particularly important in the field of biomedicine, where materials are needed to interact effectively with biological systems. Additionally, QD/1D composites have been shown to exhibit higher photon conversion efficiencies compared to individual QDs or 1D materials in various types of solar cells. Furthermore, these composites have potential applications in LEDs, where their improved emission properties

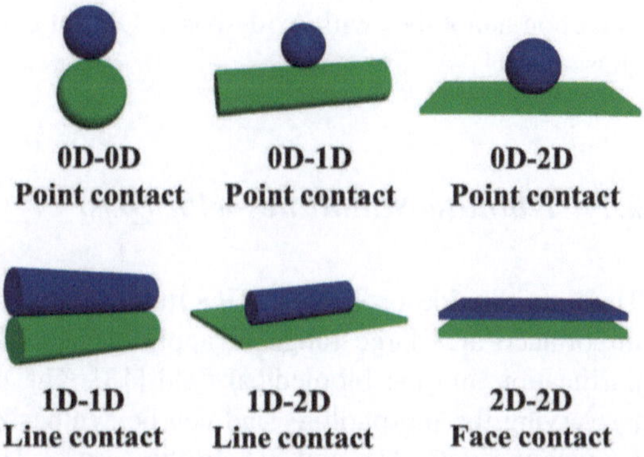

Fig. 1 Schematic diagram of possible combination of MDHs. Reprinted with the permission of ref [8]

can lead to higher efficiency and improved colour purity. Briefly, QD/1D material composites offer unique properties and advantages over individual QDs or 1D materials, including improved charge transport, enhanced light absorption and emission, improved compatibility with biological systems, and potential applications in solar cells, LEDs, and photocatalysis. The synthesis and applications of QD/1D material composites have seen significant advancements in recent years, and their potential for future research and applications is highly promising. 1D materials can be synthesized in different fashions such as single (CNTs) and multi-walled carbon nanotubes (MWCNTs), ultrafine nanorods, nanowires, fibres, pillar-like structures and so on [12–16]. Apart from 1D nanostructures, 0D materials can be also synthesized in mainly two types such as conventional 0D materials and core–shell-like quantum dots [17].

In the current chapter, several modern approaches have been discussed to fabricate quantum dot/1D material composites/heterostructures. Novel QDs/1D heterostructures by incorporating 1D components of titanate nanotubes, metal oxides and metal sulfides, respectively have been overviewed for the first time. The role of 1D structures in tuning local atomic structures to facilitate pathways for effective charge transport is discussed for novel QDs/1D heterostructures. Interface engineering for the manipulation of physical and chemical properties of the composite system is well explained owing to their modern technological applications.

2 Synthesis Route of QD/1D

Various methods have been adopted for the QDs/1D heterostructures synthesis and the approach for synthesis influences the properties of heterostructures. In the following section, we will provide an overview of various synthesis routes of QDs/1D heterostructures. The following sections are classified based on types of different 1D structures and their heterostructures irrespective of quantum dots structures such as carbon nanotubes with oxide-based QDs or carbon nanotubes with sulfide-based QDs.

2.1 Titanate Nanotubes with QDs

Titanium dioxide or Titania (TiO_2) are highly abundant materials and show their importance in a large range of applications such as catalysis, electronics, water purification and the biomedical field [18]. The properties of titania can be tuned by varying the morphology and can be synthesized in various morphologies such as quantum dots, 1D, and 2D. In the case of 1D morphology, it has been found that protonated titania named titanate nanotubes increase the range of application as well as enhance the existing properties [19]. Titanate can be synthesized via various methods including electrochemical synthesis, microwave irradiation, the hydrothermal method and the sol–gel template approach, where hydrothermal is most popular method due to its large-scale synthesis route as well as economical [20]. Various titanate materials such as protonated titanate nanotubes and sodium titanate nanotubes have been synthesised. Titanates have a hollow structure with accurate inner and outer diameters having a large surface area [18]. Titanates may find application in the fields of catalysis, gas sensing, water purification, energy storage and generation and many others [21]. Titanates are utilized as catalytic support with active species such as metals, oxides, and sulfides.

Recently, Titanate has been used as support for bismuth oxide quantum dots and utilized as the removal of toxic uranium and tetracycline. Sun et al. synthesized the Bi_2O_3 QDs heterostructures with titanates using a hydrothermal route [22]. Pure titanate powder was first synthesized followed by attachment of Bi_2O_3 QDs over titanates using hydrothermal route anchored titanate has been used using hydrothermal route. For the anchoring, different amounts of $Bi(NO_3)_3 \cdot 5H_2O$ were placed into ethylene glycol followed by the addition of ethanol, which is double to ethylene glycol and stirred continuously for 30 min. Afterwards, the desired amount of prepared titanate powder was added to the above solution and stirred for 60 min. The resulting mixture was placed into PolyPhenyLene (PPL)-lined stainless-steel autoclave and heated at 160 °C for 6 h. After completion of the reaction obtained samples were washed and centrifuged with deionized water and absolute ethanol. The end products after being centrifuged were dried in an oven at 60 °C for 12 h. It is interesting to note that this method provides a better attachment of Bi_2O_3 QDs with 1D titanate nanotube. The synthesized heterostructures were characterized using various microscopic and spectroscopic techniques. The transmission electron microscopy (TEM) imaging showed the formation of titanate nanotubes where Bi_2O_3 quantum dots were uniformly distributed over titanate nanotubes. It has been also found that the titanate structures have a very low impact with the addition of different amounts of QDs. The high resolution transmission electron microscopy (HRTEM) study also reveals the attachment of Bi_2O_3 over titanate nanotubes. TEM analysis shows that Bi_2O_3 has an average 2.5 nm diameter, which can be defined as 0D and these 0D QDs are firmly and uniformly attached to nanotubes and formed 0D/1D heterostructures. The Bi bonded with Ti surface Ti–O–Bi bonds, which will enhance the specific surface area and active site. The removal of uranium was used using an adsorption study and found that it could be 347.39 mg/g. It has been revealed that cation exchange between Na^+ ions and uranium cation is responsible for the adsorption of uranium. The complexation and cation- π interaction of U(VI) is responsible for synergistic adsorption. The synthesized 0D/1D heterostructure efficiently and economically removes the radioactive pollutants and antibiotics from hospital wastewater.

Shi et al. synthesized sodium titanate ($Na_2Ti_3O_7$) nanotube necklaces decorated with bismuth oxyhalide quantum dots (BiOBr) using a hydrothermal route [23]. The BiOBr QDs attachment over $Na_2Ti_3O_7$ necklaces forms a heterogeneous interface between BiOBr (QDs) and $Na_2Ti_3O_7$ (1D). In brief first, 1 g of titania powder was dispersed in aqueous NaOII and heated in Teflon hydrothermal at 120 °C for 24 h followed by washing and centrifugation till 8 pH with water. In the present process, pH plays a very crucial role in each step and the synthesized $Na_2Ti_3O_7$ powder was further dispersed in water and adjusted the pH to 11 by the addition of NH_3H_2O. Thereafter, a certain amount of $Bi(NO_3)_3{\cdot}5H_2O$ and NaBr were added to the above solution mixture and heated at 160 °C for 20 h hydrothermally. The different ratios of BiOBr/$Na_2Ti_3O_7$ composite were obtained after hydrothermal and washed with water and ethanol followed by centrifugation. The scheme for BiOBr/$Na_2Ti_3O_7$ nanocomposites synthesis has been shown in Fig. 2. Different concentrations of BiOBr were deposited over $Na_2Ti_3O_7$. Inductively coupled plasma atomic emission spectroscopy (ICP-AES) studies were performed for the actual loading of BiOBr, which were 4.7, 9.1 and 13.8 wt % and named BOR/NT-0.5, BOR/NT-1.0 and BOR/NT-1.5, respectively. The spectroscopic and microscopic analysis has been performed for the analysis.

The FTIR study reveals the Bi-O bonding (stretching vibration) as well as the stretching vibration of O–Br and O–H bending vibration of surface adsorbed water in BiOBr. In the case of $Na_2Ti_3O_7$, O–H bond and C–H deformation bond were found $Na_2Ti_3O_7$. The Ti–O and Ti–O–Ti vibration bands were also found in $Na_2Ti_3O_7$.

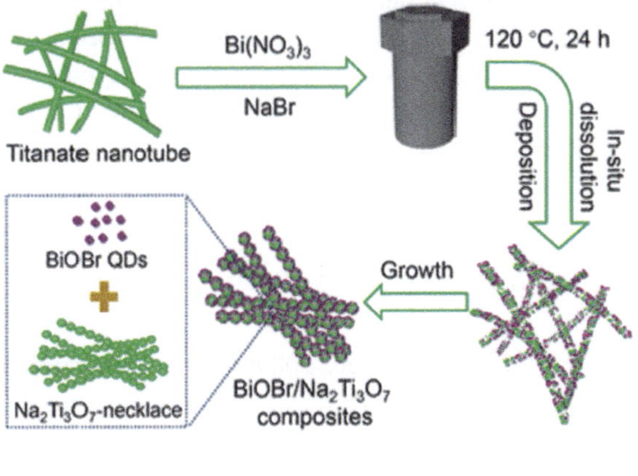

Fig. 2 Schematic representation for BiOBr/$Na_2Ti_3O_7$ necklace composite. Reprinted with the permission of ref [23]

In the case of BiOBr/$Na_2Ti_3O_7$ heterostructures, the existence of few overlapped peaks with somehow shifts unsurprisingly validates the presence of BiOBr/$Na_2Ti_3O_7$ heterostructures. The TEM and HRTEM micrographs show the formation of unique necklace-like morphology interwoven into a network structure. Apart from that it was seen that 2–4 nm QDs of BiOBr are attached to the surface of $Na_2Ti_3O_7$ supports. The synthesized heterogeneous interfaces between BiOBr/$Na_2Ti_3O_7$ decrease electron–hole pair recombination rate and improve the population of oxygen vacancies for capturing electrons. The synthesized BiOBr/$Na_2Ti_3O_7$ interface allows photo-oxidation of benzyl alcohol to benzaldehyde with 100% conversion and > 99% benzaldehyde. The present composite shows substantially higher than previously reported photocatalysts. Apart from these, g-C_3N_4 quantum dots with protonated titanate were developed using a hydrothermal route and utilized for improved photocatalytic hydrogen evolution activity [24, 25]. Out of these results, it has been found that titanates were synthesized using the hydrothermal route and till now the decoration of titanate with various types of QDs was prepared using the hydrothermal route and utilized for catalysis work.

2.2 Metal Oxide 1D Heterostructures with QD

QDs are sometimes referred to as artificial atoms and show improved intrinsic properties as compared to their bulk materials. For instance, when the size of bulk materials is confined to QDs size the bandgap of materials increases due to the widening of the position of the highest occupied molecular orbital (HOMO) and the lowest unoccupied molecular orbital (LUMO), which leads to size-dependent fluorescent materials and absorption properties [26, 27]. Apart from that various other applications were also explored for this type of heterostructure such as solar energy conversion, and catalysis [28, 29]. Chade et al. reported the synthesis of Bi_2O_3 quantum dots (QDs) over $BiVO_4$ nanofibers by direct heat treatment of as-spun fibres. The nanofibers of $BiVO_4$ have 400–700 nm in length where 5–15 nm QDs of Bi_2O_3 were decorated uniformly. The Bi_2O_3/$BiVO_4$ heterostructure shows higher photocatalytic performance than pure $BiVO_4$ nanofibers, which might be due to the enhancement in charge recombination. Lots of work has been done for oxide-based heterostructure synthesis. In another work, Lan et al. synthesized Fe_3O_4 nanowires decorated by CdTe quantum dots [30]. The bifunctional CdTe/Fe_3O_4 heterostructures show enhanced magnetic resonance imaging.

2.3 Carbon-Based Nanotubes/Nanocage Anchored with QDs

Apart from the 1D titanates, other 1D tube-like structures have also been synthesized as a support for various oxides and sulfide quantum dots. From various studies, carbon nanotubes (CNTs) are used as 1D supports. Liang et al. synthesized CNTs with CdS

QDs using a hydrothermal route followed by calcination [31]. Typically, aqueous melamine pH was altered to make it unity by adding phosphorous acid and followed by transferring the mixture into a Teflon-lined stainless autoclave and heated at 180 °C for 10 h and as prepared off-white precipitates after centrifuge used as precursors and calcined in a nitrogen atmosphere for 4 h and denoted as phosphorus-doped carbon nitride tubes (P-CNT). For the synthesis of phosphors-free CNT, the sample obtained after hydrothermal was used at least 30 times with water till the complete removal of phosphors and calcined in N_2 atmosphere in an earlier way. For the synthesis of CdS QDs oil batch method was utilized. $Cd(CH_3COO)_2.2H_2O$ dispersion was made in ethanol using sonication and dropwise addition of thioacetamide (TAA) was carried out during the heating of the oil bath followed by vigorous stirring. The final mixture was centrifuged and washed with deionized water and ethanol several times and dried for further use. The anchoring of CdS QDs over CNTs was done using the in-situ oil bath method. During the synthesis of CdS QDs, a specific amount of synthesized CNTs were inserted after the sonication and followed the previous steps. The morphological study reveals the formation of the well-defined hexagonal tube with smooth and hollow morphology obtained in the case of CNTs, whereas CNTs play a crucial role in anchoring CdS quantum dots over 1D tubes. The energy dispersive spectroscopy shows the presence of C, N, P, Cd and S elements, indicating the successful formation of CdS/P-CNT. The synthesized 0D/1D structures are utilized for photocatalytic H_2 production. It has been seen that there is an almost negligible amount was generated with CNTs, whereas when P-atom incorporation provides a small amount of H_2, it was ~4.5 times higher than P-CNTs in the case of CdS QDs. It is interesting to note that CdS QDs/P-CNT boosted hydrogen production due to their heterojunction, which was responsible for the proton reduction reaction. The production of hydrogen was greatly affected by the loading of CdS QDs. With optimized CdS QDs/P-CNT, a 31.6 times higher amount of hydrogen can be obtained, which is due to the very slow recombination rate of generated electron–hole pairs. However, further incorporation of CdS QDs results in lower hydrogen production due to the shielding effect. Due to the very strong electronic interaction between P-CNT and CdS QDs are the responsible for continuous production of hydrogen with the same amount of up to 15 h reaction under the light.

In another work by Yuan et al. in 2022 synthesized caged-like 1D carbon structures followed by their decoration with vanadium nitride quantum dots and utilized them for sodium-ion capacitors [32]. It has been seen that electrochemical energy storage emerges as one of the significant sustainable energy sources. Lithium-ion and carbon-based materials show their importance in the energy storage sector [33–35]. However, nowadays, sodium ion hybrid capacitors (SIHCs) show their importance in this field due to their high energy density and high output power but the kinetics between anode and cathode limit their performance. Despite being a valid approach, overcoming the obstacle of improving kinetics compatibility between anode and cathode through rational construction of anode materials with pseudocapacitive advantages for fast charge storage has proven to be a significant challenge until now. Concerning the above facts, Yuan et al. developed an advanced solid-state intercalation hybrid capacitors device by coupling freestanding vanadium nitride quantum dots embedded in nitrogen and fluorine co-doped 1D carbon nanofibers as the anode and corresponding carbon nanofibers as the cathode [32]. For the synthesis of QDs of vanadium nitride over CNTs, the first PVA dispersed was made in hot water followed by the addition of 2 mmol NH_4VO_3 and 3 mL of 60 wt % PTFE solution in PVA dispersion and stirred for another 10 h. The as-prepared dispersion was placed in an ink panel for electro span which was attached to a nozzle connecting a high-voltage power supply and aluminium foil was rolled up on the drum collector to collect nanofibers. The flow rate applied voltage and distance between collector to nozzle was optimized to get good electro-span fibre. The obtained nanofibers film was heated for 6 h at 250 °C then nitridation was performed in NH_3 atmosphere at 700 °C for 3 h and yielded the vanadium nitride QDs heterostructures with CNTs. The utilization of quantum dots confined within 1D nanocages has proven to be effective in overcoming the limitations of sluggish mass and electron transport, resulting in improved kinetics for the anode. Specifically, carbon nanocages co-doped with nitrogen and fluorine, which possess a highly conductive network and abundant active sites, demonstrate remarkable performance in terms of reversible sodium ion storage and anion adsorption/desorption. Consequently, the hybrid device exhibits superior electrochemical properties, including enhanced energy density, power density, and long-term cyclability. This study elucidates a promising strategy to achieve optimal kinetics and performance in energy storage systems. Few other works have been studied where metal oxides, perovskites and graphene quantum dots were decorated over CNTs and utilized for energy and sensor applications [36–39].

2.4 *Metal Sulfide 1D Structured with Metal QDs*

The synthesis of semiconductor-based materials plays a very crucial role in various applications such as electronics, energy, catalysis, and photocatalysis [40–42]. In semiconductors cadmium sulfide (CdS, $E_g = 2.4$ eV) emerges as an excellent choice for visible photocatalysis and their catalysis properties can be altered by varying the shape of CdS. Earlier, a few reports showed that 1D CdS nanorods will be an excellent choice over another shape for photocatalysis such as charge separation along the axial direction are very well whereas a high aspect ratio can significantly improve the light scattering provided along [43, 44]. As they have several important properties still there will be a scope for improving the performance of the 1D CdS by making heterostructures with metal quantum dots and the high surface area of CdS will provide a better environment for fabrication. By considering this, Chava et al. first time reported the growth of Bi quantum dots (Bi QDs) over the surface of CdS nanorods and utilized them for the photocatalytic degradation of the tetracycline [45]. Briefly, CdS nanorods were synthesized using the solvothermal approach in ethylenediamine at 180°C and for anchoring the Bi quantum dots over it, the synthesized CdS nanorods were sonicated in ethylene glycol followed by the addition of bismuth nitrate pentahydrate precursors and again the solvothermal process was carried out for 2 h at 160 °C. After the reaction samples (CdS/Bi) were obtained

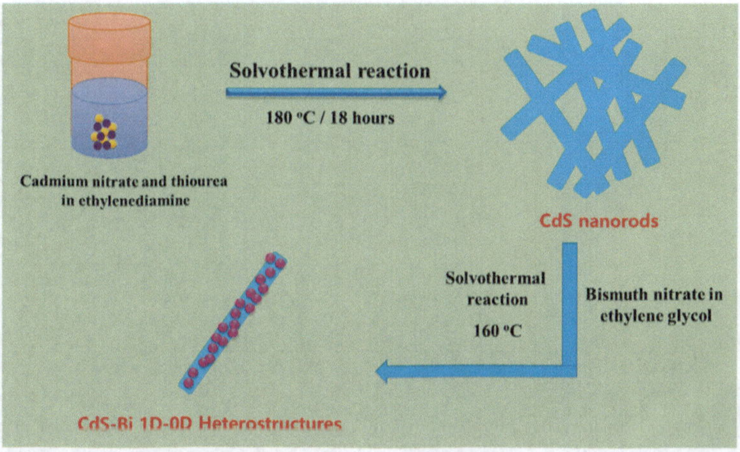

Fig. 3 Synthesis and schematic of Bi nanodots over CdS nanorods. Reprinted with the permission of ref [45]

by washing with ethanol followed by drying. Based on the amount of Bi-precursors used the obtained samples were named CBi-2, CBi-3 and CBi-5. A typical synthesis approach has been reported in Fig. 3.

The main confirmation for heterostructure analysis was performed using the (high-resolution transmission electron microscopy) HRTEM study. From Fig. 4 one can see the HR-TEM images, lattice fringes and SAED patterns of CBi-4 sample (highest amount of Bi). Figure 4a reveals the rods of CdS whereas the rough surface was due to the growth of Bi QDs. From the STEM-EDS elemental mapping, one can see that the Bi-quantum dots were uniformly distributed over the CdS nanorods and confirm the CdS/Bi.

The as-fabricated (CBi-4) heterostructures show the highest photocatalytic activity of tetracycline degradation (90%), which might be due to the synergistic effect of CdS/Bi heterojunction interfaces and surface plasmon resonance (SPR) effect. The fabrication of CdS/Bi plasmonic heterostructures improves the generation of photoexcited charge carriers and decreases the recombination rate of charge.

In another work, Jha et al. synthesized 0D/1D Cu_2O quantum dots sensitized CdS nanorods heterojunction for visible light H_2 generation [46]. For the synthesis of CdS nanorods, Cd and S precursors were inserted into ethylenediamine followed by heating at 180 °C for 48 h solvothermal and yellow precipitate of CdS NRs after reaction can collect via washing repeatedly with deionized (DI) water and ethanol followed by drying at 60 °C for 12 h. For the growth of Cu_2O QD over CdS nanorods, 0.05 M Cu_2SO_4 was dissolved into 10 ml of DI water followed by addition into the above CdS solvothermal precursors and similar steps were used. Different molarity of Cu_2SO_4 was used for making wt % of Cu_2O over CdS and named as x–Cu_2O/CdS (x is the wt % of Cu_2O). A schematic synthesis representation has been shown in Fig. 5.

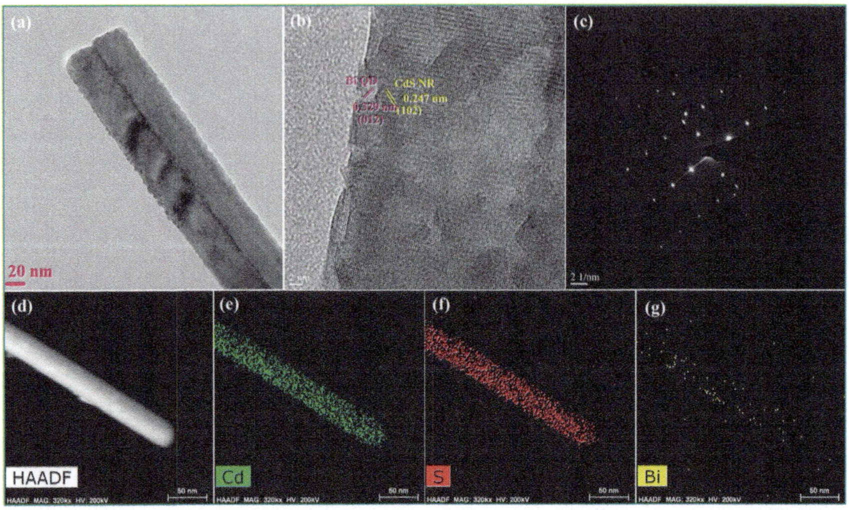

Fig. 4 **a** High-resolution TEM image, **b** inter-planar spacings, **c** SAED patterns and **d–g** STEM-EDS mapping profiles of CBi-4 photocatalyst sample. Reprinted with the permission of ref [45]

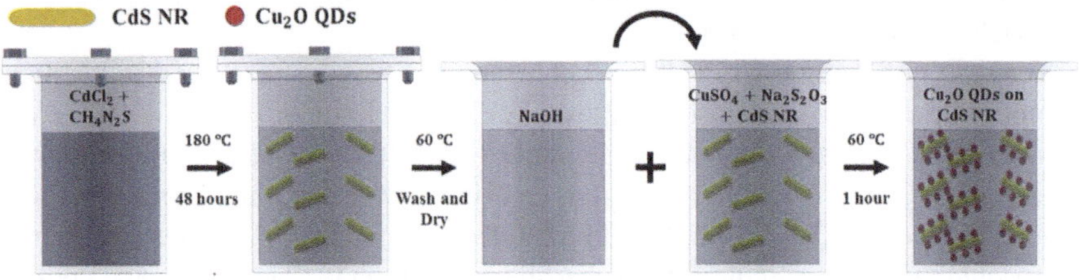

Fig. 5 A schematic diagram describing the synthesis of CdS NRs and Cu_2O QDs/CdS NRs. Reprinted with the permission of ref [46]

The TEM and HRTEM study were utilized for the analysis of Cu_2O/CdS heterostructure (Fig. 6). TEM of 20-Cu_2O–CdS study reveals the formation of nanorods of CdS with Cu_2O QD of size ~4–5 nm (Fig. 6). Figure 6c shows the HR-TEM image demonstrating the incorporation of Cu_2O QDs over the CdS NRs. The lattice fringes also confirmed the formation of Cu_2O/CdS heterostructure (Fig. 6d–e). Figure 6f displays the SAED pattern of Cu_2O QDs/CdS NRs, which confirms the polycrystalline nature of the heterostructure, and the interface between Cu_2O QDs and CdS NRs is abrupt.

The photocatalytic hydrogen evolution (PCHE) test was carried out in the presence of visible light and 20 wt% loaded Cu_2O/CdS shows a high rate of H_2 evolution (130.3 μmol/h, 18 times higher than pure Cu_2O QDs and 5.1 times higher than CdS NRs). The improved H_2 generation over heterostructure arises due to the increased lifetime of photogenerated electrons to 1.2×10^{-8} s.

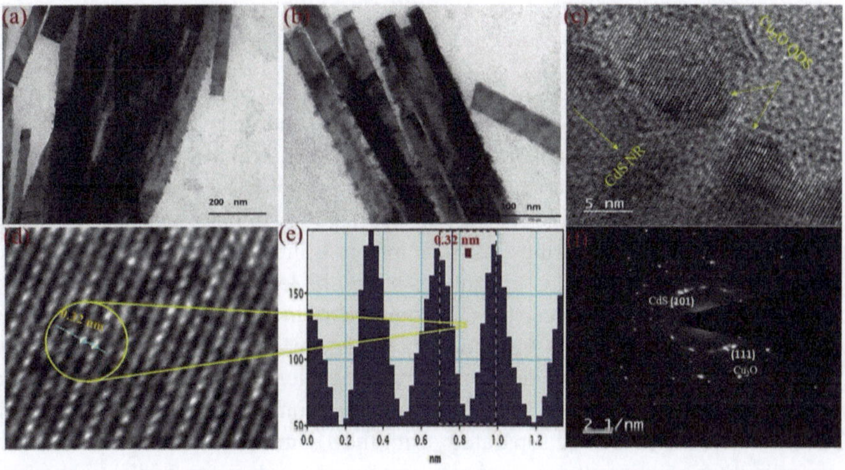

Fig. 6 The TEM image of 20 wt% Cu_2O QDs deposited on CdS NRs (**a–c**), Inverse FFT of Cu_2O QDs (**d**), line profile spectra of Cu_2O QDs (**e**), SAED pattern of 20 Cu_2O QDs/CdS NRs (**f**). Reprinted with the permission of ref [46]

3 Future Perspectives

The role of interfaces between the QD/1D heterostructures is yet to be defined. Moreover, there are no clear theoretical models that can guess the behaviour of these systems. Density functional theory (DFT) calculations predict energy levels of QDs shift with different chemical ligands and the binding modes of linker molecules can affect QD coverage and aggregation which have a further impact on charge separation, recombination, and transport properties. Thus, the study of the effect of local nature on these interfaces is important to control the performance of practically possible QD/1D composite systems. The structural and electronic nature of the interfaces would be needed for QD/1D heterostructures with high spatial and temporal resolution. In future, charge transfer at and across composite interfaces in real-time should be needed. A possible way of finding the reason for simultaneously enhanced absorption with optimized light harvesting efficiencies is a big deal. A critical issue remains to build up protocols for reproducibly generating ‘pure’, homogeneous, crystalline and uniformly well-characterized heterostructures which would be useful for practical commercial device applications. The potential toxicity of QD/1D heterostructures concerns the possible cytotoxicity. It motivated the search for green and sustainable alternatives for the reliable production of individual nanoscale constituent components and heterostructures. In summary, an important future step would be taken towards the spatial integration, stabilization, and ultimately, optimization of these QD/1D composite materials within more sophisticated, realistic device scenarios.

4 Conclusions

The unique properties of QD/1D heterostructures are highly promising as compared to their quantum dots (QDs) or one-dimensional (1D) materials. QDs, which are zero-dimensional materials with nanometer-sized diameters, exhibit size-dependent characteristics such as tunable fluorescence, high photostability, and high quantum yield, making them highly desirable for various applications in optoelectronics, sensing, and biomedicine. By integrating dissimilar materials into a 1D template, heterostructures with varying chemical compositions along the length of 1D materials can be created. The interaction between 0D–1D components in these heterostructures results in significant changes in their properties, including variable electronic states, control of charge transport, and the emergence of new physical and chemical properties at their interface. The synthesis of QD/1D material composites involves combining QDs, which are synthesized separately, with the 1D material matrix. Various techniques such as solution-phase synthesis, vapour-phase deposition, electrochemical methods, molecular beam epitaxy (MBE), or metal–organic chemical vapour deposition (MOCVD) have been developed for the synthesis of these composites. These techniques offer control over the composition and arrangement of the heterostructures, allowing for the customization of their properties. One of the key advantages of QD/1D material composites is tuning the charge transport properties. The 1D material serves as an efficient pathway for charge transport, facilitating more efficient charge extraction and transport compared to individual QDs. The 1D material acts as a scaffold for the QDs, increasing surface area and light scattering, thereby enhancing the photon conversion efficiency in applications like solar cells and LEDs. The confinement of charge carriers in both the QDs and 1D materials can also lead to improved emission properties, including higher quantum yields and improved colour purity.

Acknowledgements KKY thanks SCE for the fellowship.

References

1. An, Y., Tian, Y., Wei, C., et al.: Dealloying: An effective method for scalable fabrication of 0D, 1D, 2D, 3D materials and its application in energy storage. Nano Today **37**, 101094 (2021). https://doi.org/10.1016/J.NANTOD.2021.101094
2. Lim, S.Y., Shen, W., Gao, Z.: Carbon quantum dots and their applications. Chem. Soc. Rev. **44**, 362–381 (2014). https://doi.org/10.1039/C4CS00269E
3. Rohilla, D., Kaur, N., Shanavas, A., Chaudhary, S.: Microwave mediated synthesis of dopamine functionalized copper sulphide nanoparticles: An effective catalyst for visible light driven degradation of methlyene blue dye. Chemosphere **277**, 130202 (2021). https://doi.org/10.1016/j.chemosphere.2021.130202
4. Kaur, G., Kaur, V., Kaur, N., et al.: Design of Silica@Au hybrid nanostars for enhanced SERS and photothermal effect. Chem Phys Chem (2023). https://doi.org/10.1002/cphc.202200809

5. Qin, J.K., Wang, C., Zhen, L., et al.: Van der Waals heterostructures with one-dimensional atomic crystals. Prog. Mater. Sci. **122**, 100856 (2021). https://doi.org/10.1016/J.PMATSCI.2021.100856
6. Wadhwa, R., Yadav, K.K., Goswami, T., et al.: Mechanistic insights for photoelectrochemical ethanol oxidation on black gold decorated monoclinic zirconia. ACS Appl. Mater. Interfaces **13**, 9942–9954 (2021). https://doi.org/10.1021/acsami.0c21010
7. Singh, H., Sunaina, Y.K.K., et al.: Tuning the bandgap of m-ZrO2 by incorporation of copper nanoparticles into visible region for the treatment of organic pollutants. Mater. Res. Bull. **123**, 110698 (2020). https://doi.org/10.1016/j.materresbull.2019.110698
8. Low, J., Cao, S., Yu, J., Wageh, S.: Two-dimensional layered composite photocatalysts. Chem. Commun. **50**, 10768–10777 (2014). https://doi.org/10.1039/C4CC02553A
9. Ahmad, W., Tareen, A.K., Khan, K., et al.: A review of the synthesis, fabrication, and recent advances in mixed dimensional heterostructures for optoelectronic devices applications. Appl. Mater. Today **30**, 101717 (2023). https://doi.org/10.1016/J.APMT.2022.101717
10. Li, B., Cao, Z., Wang, S., et al.: BiVO4 quantum dot-decorated BiPO4 nanorods 0D/1D heterojunction for enhanced visible-light-driven photocatalysis. Dalton Trans. **47**, 10288–10298 (2018). https://doi.org/10.1039/C8DT02402B
11. Zubko, P., Gariglio, S., Gabay, M., et al.: Interface physics in complex oxide heterostructures. Annu Rev Condens Matter Phys **2**, 141–165 (2011). https://doi.org/10.1146/ANNUREV-CONMATPHYS-062910-140445
12. Bai, F., Sun, Z., Wu, H., et al.: Porous one-dimensional nanostructures through confined cooperative self-assembly. Nano Lett. **11**, 5196–5200 (2011). https://doi.org/10.1021/NL203598N/SUPPL_FILE/NL203598N_SI_001.PDF
13. Yadav, K.K., Khan, N., Jha, M.: Application of surface modified carbon nanotubes in energy. ACS Symp. Ser. **1425**, 101–119 (2022). https://doi.org/10.1021/BK-2022-1425.CH005
14. Yadav, K.K., Arora, A., Jangra, S., Jha, M.: Chemically modified carbon nanotubes in 3D and 4D printing. Chem. Modif. Carbon Nanotub. Commer. Appl., 419–439 (2023). https://doi.org/10.1002/9783527838790.CH18
15. Wadhwa, R., Arora, A., Rana, S., Yadav, K.K.: Overview of advanced fiber materials. In: Aslam, J., Verma, C (eds.) Fiber materials: design, fabrication and applications, 1st ed. p. 1. Walter de Gruyter GmbH & Co KG
16. Wadhwa, R., Arora, A., Yadav, K.K.: Fiber materials: design, fabrication and applications. In: Aslam, J., Verma, C. (Eds.) Fiber materials: design, fabrication and applications, 1st ed. Walter de Gruyter GmbH & Co KG, p. 191
17. Sahu, A., Kumar, D.: Core-shell quantum dots: A review on classification, materials, application, and theoretical modeling. J. Alloys Compd. **924**, 166508 (2022). https://doi.org/10.1016/J.JALLCOM.2022.166508
18. Ameur, N., Bachir, R.: Study of 1D titanate-based materials–new modification of the synthesis procedure and surface properties-recent applications. ChemistrySelect **5**, 1164–1185 (2020). https://doi.org/10.1002/SLCT.201904539
19. Adeleye, A.T., John, K.I., Adeleye, P.G., et al.: One-dimensional titanate nanotube materials: heterogeneous solid catalysts for sustainable synthesis of biofuel precursors/value-added chemicals—a review. J. Mater. Sci. **56**, 18391 (2021). https://doi.org/10.1007/S10853-021-06473-1
20. Chen, X., Mao, S.S.: Titanium dioxide nanomaterials: Synthesis, properties, modifications and applications. Chem. Rev. **107**, 2891–2959 (2007). https://doi.org/10.1021/CR0500535/ASSET/CR0500535.FP.PNG_V03
21. Vazquez-Santos, M.B., Tartaj, P., Morales, E., Amarilla, J.M.: TiO_2 nanostructures as anode materials for Li/Na-Ion batteries. Chem. Rec. **18**, 1178–1191 (2018). https://doi.org/10.1002/TCR.201700103
22. Sun, S., Chen, M., Dao, X., et al.: 0D/1D Bi2O3@TNTs composites synthesized by the decoration of Bi2O3 quantum dots onto titanate nanotubes: Synergistic adsorption of U(VI) and tetracycline. Sep. Purif. Technol. **306**, 122537 (2023). https://doi.org/10.1016/J.SEPPUR.2022.122537

23. Shi, Q., Raza, A., Xu, L., Li, G.: Bismuth oxyhalide quantum dots modified sodium titanate necklaces with exceptional population of oxygen vacancies and photocatalytic activity. J. Colloid Interface Sci. **625**, 750–760 (2022). https://doi.org/10.1016/J.JCIS.2022.06.066
24. Nasir, M.S., Yang, G., Ayub, I., et al.: In situ decoration of g-C3N4 quantum dots on 1D branched TiO2 loaded with plasmonic Au nanoparticles and improved the photocatalytic hydrogen evolution activity. Appl. Surf. Sci. **519**, (2020). https://doi.org/10.1016/j.apsusc.2020.146208
25. Amiri, M., Dondapati, J., Quintal, J., Chen, A.: Sodium hexa-titanate nanowires modified with cobalt hydroxide quantum dots as an efficient and cost-effective electrocatalyst for hydrogen evolution in alkaline media. ACS Appl. Mater. Interfaces **14**, 40021–40030 (2022). https://doi.org/10.1021/ACSAMI.2C11310/SUPPL_FILE/AM2C11310_SI_001.PDF
26. Murray, C.B., Norris, D.J., Bawendi, M.G.: Synthesis and characterization of nearly monodisperse CdE (E = S, Se, Te) semiconductor nanocrystallites. J. Am. Chem. Soc. **115**, 8706–8715 (1993). https://doi.org/10.1021/JA00072A025/ASSET/JA00072A025.FP.PNG_V03
27. Alivisatos, A.P.: Semiconductor clusters, nanocrystals, and quantum dots. Science **271**, 933–937 (1996). https://doi.org/10.1126/SCIENCE.271.5251.933
28. Cao, S.W., Yin, Z., Barber, J., et al.: Preparation of Au-BiVO 4 heterogeneous nanostructures as highly efficient visible-light photocatalysts. ACS Appl. Mater. Interfaces **4**, 418–423 (2012). https://doi.org/10.1021/AM201481B
29. Lv, C., Chen, G., Sun, J., et al.: One-dimensional Bi_2O3 QD-decorated BiVO4 nanofibers: electrospinning synthesis, phase separation mechanism and enhanced photocatalytic performance. RSC Adv. **5**, 3767–3773 (2014). https://doi.org/10.1039/C4RA11065J
30. Lan, X., Cao, X., Qian, W., et al.: Long Fe_3O_4 nanowires decorated by CdTe quantum dots: Synthesis and magnetic–optical properties. J. Solid State Chem. **180**, 2340–2345 (2007). https://doi.org/10.1016/J.JSSC.2007.06.007
31. Liang, Q., Zhang, C., Xu, S., et al.: In situ growth of CdS quantum dots on phosphorus-doped carbon nitride hollow tubes as active 0D/1D heterostructures for photocatalytic hydrogen evolution. J. Colloid Interface Sci. **577**, 1–11 (2020). https://doi.org/10.1016/j.jcis.2020.05.053
32. Yuan, J., Qiu, M., Hu, X., et al.: Pseudocapacitive vanadium nitride quantum dots modified one-dimensional carbon cages enable highly kinetics-compatible sodium ion capacitors. ACS Nano **16**, 14807–14818 (2022). https://doi.org/10.1021/ACSNANO.2C05662/SUPPL_FILE/NN2C05662_SI_001.PDF
33. Yadav, K.K., Wadhwa, R., Khan, N., Jha, M.: Efficient metal-free supercapacitor based on graphene oxide derived from waste rice. Curr. Res. Green Sustain. Chem. **4**, 100075 (2021). https://doi.org/10.1016/j.crgsc.2021.100075
34. Yadav, K.K., Singh, H., Rana, S., et al.: Utilization of waste coir fibre architecture to synthesize porous graphene oxide and their derivatives: An efficient energy storage material. J Clean Prod., 124240 (2020). https://doi.org/10.1016/j.jclepro.2020.124240
35. Devi, M.M., Ankush, G.S.K., et al.: Energy efficient electrodes for lithium-ion batteries: Recovered and processed from spent primary batteries. J. Hazard. Mater. **384**, 121112 (2020). https://doi.org/10.1016/j.jhazmat.2019.121112
36. Shu, M., Zhang, Z., Dong, Z., Xu, J.: CsPbBr 3 perovskite quantum dots anchored on multi-walled carbon nanotube for efficient CO2 photoreduction. Carbon N Y **182**, 454–462 (2021). https://doi.org/10.1016/J.CARBON.2021.06.040
37. Liu, Z.Q., Cheng, H., Li, N., et al.: $ZnCo_2O_4$ quantum dots anchored on nitrogen-doped carbon nanotubes as reversible oxygen reduction/evolution electrocatalysts. Adv. Mater. **28**, 3777–3784 (2016). https://doi.org/10.1002/ADMA.201506197
38. Wang, Y.X., Rinawati, M., De, Z.J., et al.: Boron-doped graphene quantum dots anchored to carbon nanotubes as noble metal-free electrocatalysts of uric acid for a wearable sweat sensor. ACS Appl Nano Mater **5**, 11100–11110 (2022). https://doi.org/10.1021/ACSANM.2C02279/SUPPL_FILE/AN2C02279_SI_001.PDF
39. Xu, N., Wang, Y-D., Zhou, X-D., Qiao, J.: Ni-Fe oxide quantum dots anchored on carbon nanotubes as a catalysts for oxygen reduction/evolution reactions for metal-air batteries. ECS Meeting Abstracts MA2018–02, 1668 (2018). https://doi.org/10.1149/MA2018-02/48/1668

40. Yadav, K.K., Sunaina, S.M., et al.: Excellent field emission from ultrafine vertically aligned nanorods of NdB6 on silicon substrate. Appl. Surf. Sci. **526**, 146652 (2020). https://doi.org/10.1016/j.apsusc.2020.146652
41. Samanta, S., Battula, V.R., Sardana, N., Kailasam, K.: Solar driven photocatalytic hydrogen evolution using graphitic-carbon nitride/NSGQDs heterostructures. Appl. Surf Sci., **563**. https://doi.org/10.1016/j.apsusc.2021.150409
42. Sunaina, G.A.K., Mehta, S.K.: High performance ZnSe sensitized ZnO heterostructures for photo-detection applications. J Alloys Compd., **894**, (2022). https://doi.org/10.1016/j.jallcom.2021.162263
43. Chava, R.K., Son, N., Kang, M.: Controllable oxygen doping and sulfur vacancies in one dimensional CdS nanorods for boosted hydrogen evolution reaction. J. Alloys Compd. **873**, 159797 (2021). https://doi.org/10.1016/J.JALLCOM.2021.159797
44. Li, J.Y., Li, Y.H., Zhang, F., et al.: Visible-light-driven integrated organic synthesis and hydrogen evolution over 1D/2D CdS-Ti3C2Tx MXene composites. Appl. Catal. B **269**, 118783 (2020). https://doi.org/10.1016/J.APCATB.2020.118783
45. Chava, R.K., Son, N., Kang, M.: Bismuth quantum dots anchored one-dimensional CdS as plasmonic photocatalyst for pharmaceutical tetracycline hydrochloride pollutant degradation. Chemosphere **300**, 134570 (2022). https://doi.org/10.1016/J.CHEMOSPHERE.2022.134570
46. Raheman, A.R.S., Momin, B.M., Wilson, H.M., et al.: Optimal fabrication of 0D/1D Cu_2O quantum dots sensitized CdS nanorods heterojunction: Efficient photoredox catalyst for H_2 generation under visible light irradiation. J. Alloys Compd. **835**, 155262 (2020). https://doi.org/10.1016/J.JALLCOM.2020.15526

Quantum Dots/Bioconjugates

Amandeep Singh, Vandana Dhiman, Kamlesh Kumari, and Patit Paban Kundu

Abstract The quantum dots (QDs) are excellent nanocrystals of prodigious attention to the domain of medical diagnosis and therapeutics. Further, the bioconjugation of QDs shows various unique and worthful optical characteristics which highlight the importance of QDs nanomaterials to be used for very prognosticating fluorescent probes. In the past two decades, super-resolution imaging techniques with different principles have been invented to visualize biomolecules at nanometer scales. This chapter focused on the latest development in the research related to quantum dot bioconjugates fluorescent probe for the super-resolution imaging applications.

Keywords Fluorescent-probes · Bio-imaging · Semi-conductor · Cancer-treatment

1 Introduction

The fluorescent microscopy technology is extensively used in the biotic research. But, the optical diffraction limit restricts the spatial resolution of fluorescent microscopy. Since several years, super-resolution imaging methods with diverse mechanisms are being investigated to envision the biomolecules of nanometer measures. Majorly, the advancement of such techniques is based on the functionality and efficiency of the fluorescent probes. Specifically, the inherent characteristics of the fluorescent probe limit to the optimum imaging performance of the high-resolution

A. Singh (✉) · P. P. Kundu
Department of Polymer Science and Technology, University of Calcutta, Kolkata, India
e-mail: adsingh.chemical@gmail.com

V. Dhiman
Department of Endocrinology, PGIMER, Chandigarh, India

K. Kumari
Department of Chemical Engineering, SLIET Longowal, Punjab, India

P. P. Kundu
Department of Chemical Engineering, Indian Institute of Technology (IIT), Roorkee, India

S. Thomas et al. (eds.), *Quantum Dots Based Nanocomposites*, Engineering Materials,
https://doi.org/10.1007/978-3-031-54779-9_10

fluorescent nanoscopy. The recent progress in fluorescent probe bioconjugates for super-resolution imaging techniques includes the bioconjugates based on fluorescent proteins (FPs), organic dyes, quantum dots (QDs), carbon dots (CDs), up-conversion nanoparticles (UCNPs), aggregation induced emission (AIE) nanoparticles, and polymer dots (PDs) [1–9].

The hurdle of chemo/antibiotic resistance in the modern medicine domain remains pertinent and is yet to be overcome. Sensitivity of microorganisms regulates the choice of drugs to be used that eventually influenced the efficiency of disease management as well as prognosis for patient. Though, considering the adaptation procedure of discrete strains of microorganism, unrestrained applications of antibiotics will certainly lead to the maintenance of crisis of antibiotic resistance globally including the realization of a circle which declines the functional and anatomical results of treatment of inflammatory diseases [10].

This chapter describes the physicochemical properties of quantum dots as well as biological nanoconjugates as an option to be used as fluorescent probes cancer treatment. In this chapter, the main emphasis is on quantum dots based bioconjugates fluorescent probes for the high- or super-resolution florescent micro- or nanoscopy. Also, based on the properties of already existing probes and their compliance with present imaging techniques, an outlook for further improvement in fluorescent probes domain for super-resolution imaging is also provided.

2 QDs for Biomedical

The fluorescence emission is the widely explored characteristic of QDs by doping of transition elements in the core of nanocrystals as an approach to be used to impart novel value-added multimodal features to the QDs for biomedical uses. One of the utmost fascinating features of the quantum dots is the uniqueness of their emission spectra that can be delicately regulated, just by changing the shape, size, and material composition of the QDs [11]. Due to the recent progress in bioconjugation chemistry as well as accessibility of various existing constituents for the functionalization, interest is being increasing to combine quantum dots with several other components to develop multimodal composites materials.

Different types of quantum dots include colloidal quantum dots which are used for biosensing, magnetic quantum dots which enable magnetic cell separation, and fluorescent quantum dots which can be utilized for magnetic resonance imaging (MRI).

2.1 Semiconductor QDs

Semiconductor quantum dots emerge as an outstanding material in the arena of biomedical study. Morphologically, the QDs are spherical-shaped semiconductor nanocrystals having photoluminescent nature and own particle dimensions ranges between 2 and 10 nm. The size of a quantum dot is greater than that of characteristic atomic radius, however, adequately delivers electrons' and holes' quantum confinement in the all of three spatial magnitudes. Subsequently, semiconductor quantum dots, sometimes, are also called artificial atoms. The exceptional characteristics of semiconductor quantum dots include the quantum confined states' energies as well as wave functions those are prone to be altering by changing the size, shape, and composition of QDs.

2.2 *Metal-Doped QDs*

Metal-based nanoparticles are in the focus of attention due to their massive prospective in the nanotechnology [12–14]. The metal-doped quantum dots can be prepared using hydrothermal, solvothermal, microwave, reduction, and sonochemical techniques. These QDs were found to be excellent antifungal, antibacterial, bioimaging agents, and antioxidant, in nature. Also, these nanomaterials exhibit good photodynamic as well as photothermal therapy properties. Doped QDs not only potentially retain almost all of the above advantages, but also avoid the self-quenching problem due to their substantial ensemble Stokes shift. Various compounds, i.e., ZnS, ZnO, ZnSe, CdS, CdSe QDs etc. are employed as host lattices inside which some additional transition-metal ions or lanthanide ions, i.e., Mn^{2+}, Cu^{2+}, Co^{2+}, Ni^{2+}, Ag^{+}, Pb^{2+}, Cr^{3+}, Eu^{3+}, Tb^{3+}, Sm^{3+}, Er^{3+} etc. are integrated as dopant moieties which turns to upsurge the luminescent properties of metal-doped quantum dotes with advanced features [15].

2.3 *Carbon-Based QDs (C-dots)*

Carbon-based quantum dots are new generation carbon-based nanomaterial. Recently, the C-dots have fascinated wide research attention due to their miscellaneous physicochemical features as well as promising characteristics, i.e., good biocompatibility, exclusive optical nature, economic, eco friendliness, presence of plentiful functional groups ($-NH_2$, $-OH$, $-COOH$), electron mobility, and significant stability.

3 QD Bioconjugation

Methods for bioconjugation of quantum dots include covalent binding, non-covalent binding, or direct coupling of ligands to coat the surface of QDs. Quantum dots can be attached to DNA, Proteins, and numerous other biomolecules.

3.1 QDs-Proteins Bioconjugation

QD-protein bioconjugates are considered to be influential assemblies for the biosensing and bioimaging as fluorescent probes despite of strenuous procedure for their preparation. The fabrication of QD-proteins bioconjugation assembly takes lots of time and efforts because it involves many steps, i.e., synthesis of colloidal QDs, their solubilization, and thereafter functionalization of QD-proteins bioconjugates.

The attachment of the QDs to the various proteins is determined by the presence of functional groups existing on both the moieties, i.e., QD and respective protein. For instances, the presence of thiols functionalities onto the surface of QD, suggests the bioconjugation with maleimides present onto the protein molecules. Like-wise, carboxyls or N-hydroxysuccinimide (NHS) functionalities present onto surface of QD suggests the bioconjugation with lysine. However, the rarer amino acid, i.e., arginine, is considered for the site-specific bioconjugation. Widely used approaches for the bioconjugation of QDs with proteins are metal-affinity interaction, covalent coupling, as well as avidin–biotin interaction [16].

3.2 QDs-DNA Bioconjugation

The distinctive physicochemical and functional features of DNA have tendered their role as main building blocks. The QDs-DNA bioconjugates are prepared generally by fastening DNA's functional groups with QDs. In general, QDs-DNA bioconjugates are prepared using below-mentioned four different methods.

- Using 1-ethyl-3-(3-dimethylaminopropyl) carbodiimide (EDC) coupling chemistry to attach carboxyl or amine-functionalized DNA to the appropriate groups existing onto surface of QD. The bioconjugation between amine-containing biomolecules and NHS ester brings up the formation of a steady amide bond. In order to react amine with an NHS ester, both are mixed in a polar aprotic solvent, i.e., Dimethyl sulfoxide (DMSO), maintaining a pH level where the amine can get deprotonated.
- A direct attachment of thiolated DNA onto surface of quantum dots through dative thiol interactions. In general, thiol-based bioconjugation process involves the interactions with para-fluoro compounds, maleimides, alkenes, alkynes, halogenated isotactic polypropylene, and vinyl sulfones. Thiol-based bioconjugates are used for various biomedical applications such as to develop biomaterial for drug delivery, to create glycopolymers, and to create various biologically significant monomers, i.e., s-glucuronides.
- Polyhistidine peptide-based DNA assembly to the surface of quantum dots through metal-affinity coordination mechanisms.
- Biotinylated DNA linked to the streptavidin-coated quantum dots.

3.3 QD-Fluorophores Bioconjugation

QDs are widely being used for the energy 'donor' for Forster resonance energy transfer (FRET) pairs. Quantum dots act as FRET acceptors if they are coupled with various polymers capable to act as energy donors, i.e., Polyvinylcarbazole (PVK), Polyaniline (PANI), arylamine-based polyazomethine (TPAPAM), etc. [17]. The conjugates bonded with covalent interactions with the lanthanide derivatives show a greater Stoke's shift, extended fluorescence decay duration, and sharper emission. A huge range of luminescent and quenching compounds are available for FRET applications. Apart from the intrinsic fluorophores of different proteins, i.e., tryptophan and tyrosine; general organic fluorescence dyes, i.e., pyrene, coumarin (UV emission), fluorescein, rhodamine, cyanine (Vis to NIR emission) are extensively used as FRET acceptors as well as donors [18].

4 Applications as Fluorescent Probes

Cancer is fast raising fatal ailments that distress about 18.1 million individuals globally in 2018. Numerous conventional methods, i.e., surgery, radiation, chemotherapy, etc. are used as a conventional cure management for the patients. But, use of these methods bars several restrictions, i.e., development of multi-drug resistance by tumor cells, cytotoxicity (due to off-targeted action), poor intra-tumor localization, physical and psychological stresses, etc. Such restrictions have inspired the researchers to put the effort to develop patient-oriented precision treatment expending progressive drug delivery methods, i.e., liposomes, NPs, nanoconjugates, etc. [19]. Though, such drug transporters too experience various restrictions, i.e., deprived biocompatibility, smaller payload capacity, dripping of the encapsulated drug, and short-range stability. Therefore, the development of biomacromolecule-functionalized nanoconjugates is required to potentiate the anticancer action of healing agents for different cancers, i.e., lung, colorectal, ovarian, breast, and liver cancer. However, the investigators have shown the interest in the biofunctionalized nanoconjugates due to several benefits i.e. good biocompatibility, site specificity with good localization, greater entrapment along with durable stability and also slighter off-target toxicity. The advanced development in the field of biomacromolecule nanoconjugates shall surely inspire the further investigation to develop effective carriers for drug-delivery.

The colloidal quantum dots are considered to be good fluorescent probes having different optical properties far better than organic dyes for the purpose of microscopy imaging. Zhou et al. [20] has used hydrofluoric acid (HF) etching passivation method to develop small nonblinking core-multishell QD. This approach consists of total four phases. First; preparation of InP core, second; preparation of oil-phase core-multishell quantum dots using HF and trioctylphosphine etching passivation method, third; the increment in water solubility by modifying the QDs by 3-mercaptopropionic acid and forth is increasing the capability of bio targeting using the surface coupling technology. The results revealed that the synthesized non-blinking quantum dots show good fluorescence efficiency, decent photo-stability, significant Stokes shift (greater than 135 nm), and good bio-compatibility along with good targeting. The infectious keratitis causes the weakening of visual functioning which may leads to the monocular blindness with time. By putting the causing factors of eye infectious, inflammatory diseases, antibiotic resistance, and internal mutations of the pathogens together, it has been felt that highly effective alternatives are required to combat the diseases related to eye infectious. For this purpose, fluorescent semiconductor nanocrystals are considered to be potential candidates to treat infectious diseases. Olegovich et al. [21] has evaluated the anti-infectious potentiality of CdTe/Cd MPA 710 QD bioconjugates as well as levofloxacin to counter the *staphylococcal* corneal infection. The results show significant possibilities to consider synthesized QDs to fight against the infectious keratitis diseases.

Furthermore, Rahmani et al. [22] have synthesized QD-DNA conjugates using a single step ligand-exchange approach to overcome the existing problems in preparing stable, bright, and tiny-sized DNA-functionalized QDs. The parameters, i.e., ligand-to-nanoparticle molar ratio, pH, and solvent composition, were varied to synthesize highly photo-luminescent and water-soluble QD-DNA conjugates. The prepared QD-DNA bio-conjugates showed significant colloidal stability against the DNA-directed self-assembly. Furthermore, the fluorescence quenching was done using gold nanoparticles (AuNP) was demonstrated and found that these QD-AuNP dimers can be used as biosensors with excellent sensitivity.

In another study, Savage et al. [23] has developed N−, C− and N−/C-naphthyl substituted hydrophobic amino acid sidechains with Val-and Phe and analyzed them for their propensity for the in vitro self-assembly. The analysis showed the existence of bio-conjugates in both the phases; solution state and solid state. In order to induce the self-assembly, the prepared bio-conjugates are supposed to exposed to the sonication or temperature variations. The bio-conjugates were found to be soft gels of widespread fibrillar linkages of almost same morphology which depends upon the bioconjugate, organic solvent being used, and stimuli provided. Out of six, three formulations of bioconjugates named as 2-NAP-Val-OMe, 2-NAP-Phe-OMe, and Boc-Val-1-NAP were not found to be undergo for the gelation. The forth bio-conjugate (Boc-Phe-1-NAP) has formed opaque soft gels when non-polar solvents were used along with temperature stimuli (without sonication). Furthermore, the di-substituted bioconjugates; 2-NAP-Val-1-NAP and 2-NAP-Phe-1-NAP showed the gelation. Bioconjugate 2-NAP-Val-1-NAP turns to soft transparent gel in benzene solvent on temperature treatment; therefore this bioconjugate can be doped with QDs as well as with highly fluorescent organogel. These results show that the prepared bioconjugates assemble themselves into the ordered structures.

The *Pseudomonas aeruginosa* is considered as one of the dangerous pathogen having fatal characteristics which leads to infectious diseases in eyes. Ponomarev et al. [24] has demonstrated an investigation on anti-infectious activity of bioconjugates based on QDs namely KTCdTe/CdMPA710 and KTInP/ZnSe/ZnS650 in cooperation with the 3rd generation cephalosporin (Cefotaxin) against the hospital-originated *Pseudomonas aeruginosa* strains. The results of the study showed that the use of KTCdTe/CdMPA710 and KTInP/ZnSe/ZnS650 quantum dot conjugates considerably raises the ZZR.

Furthermore, Manoj et al. [25] have developed an approach to conjugate the InP/GaP/ZnS quantum dots using bovine serum albumin globular protein which create stable QD-BC. The fluorescence polarization analysis shows that the bound-protein holds chromophoric acceptors. A decrease in donor's emission intensity was observed when InP-BSA segment was excited in the presence of bound-chromophores with a simultaneous rise in acceptor's emission.

Dutta et al. [26] have developed an electrochemical immunosensor based on the nitrogen doped graphene QD (N-GQD) as well as single-walled carbon nanohorns (SWCNHs) for α-fetoprotein (AFP) detection which is a cancer biomarker as shown in Fig. 1. Therefore, to construct the immunosensor, nanocomposite material was decorated with N-GQD onto surface of SWCNHs. Subsequent N-GQD@SWCNHs hybrid structure acts as a good base for the immobilization of antibody (Anti-AFP) by carbodiimide reaction having good bioactivity and stability. Immunosensor was fabricated by consistently dispensing the bioconjugates (N-GQD@SWCNHs/Anti-AFP) dispersion onto surface of glassy carbon electrode, and consequently obstructing the residual active sites by bovine serum albumin (BSA) to avert the nonspecific adsorption. In optimum circumstances, the immunosensor showed a wide dynamic range from 0.001 to 200 ng/mL and a little detection limit of 0.25 pg/mL. Moreover, the sensor exhibited greater selectivity, appropriate constancy, and reproducibility. The assessment of AFP in the human serum suggests an exceptional recovery, i.e., 99.2 and 102.1%. Therefore, this study and approach showed a prospective role of nanocomposite-based sensor for the early medical screening of cancer biomarkers.

Salgado et al. [27] have prepared functionalized hybrid semiconductor bioconjugates using fluorescent QD capped by chitosan along with chemically modified with O-phospho-L-serine (OPS) which are biocompatible for human cell sources. Conjugation with OPS leading signaling molecule permits the preferential accumulation in human bone mesenchymal stromal cells (HBMSC). Chitosan shell with fluorescent CdS core was characterized using transmission electron microscope (TEM). Result showed the presence of small sized particles (2.3 nm) as well as a stable photoluminescence emission band. The in vitro biocompatibility results were found to be depended on the surface modification. Also, the efficiency of cells to uptake the NPs depends upon the nature of cells and source of tissue. However, the cellular uptake pathways used during the cell labeling with NPs remains free from the normal cellular biology i.e., adhesion, proliferation, osteogenic differentiation, gene expression, etc. Evaluation of uptake of bone cells exhibits a probable pathway by Caveolin-1 which controls the cell transduction in membrane. Researchers have claimed that the OPS-modified bioconjugated QD evidenced to be a consistent and well-established fluorescent bioprobe for the cell imaging and targeting research that may help to clarifying cellular mechanisms of particles in cytoplasmic membrane and osteogenic differentiation induction. Also, HBMSC's in vitro biocompatibility results showed that OPS-modified chitosan QDs possesses a forthcoming future in laboratory as well as pre-clinical uses, i.e., analysis of bio-images and *ex-vivo* cellular assessment of biomedical implants.

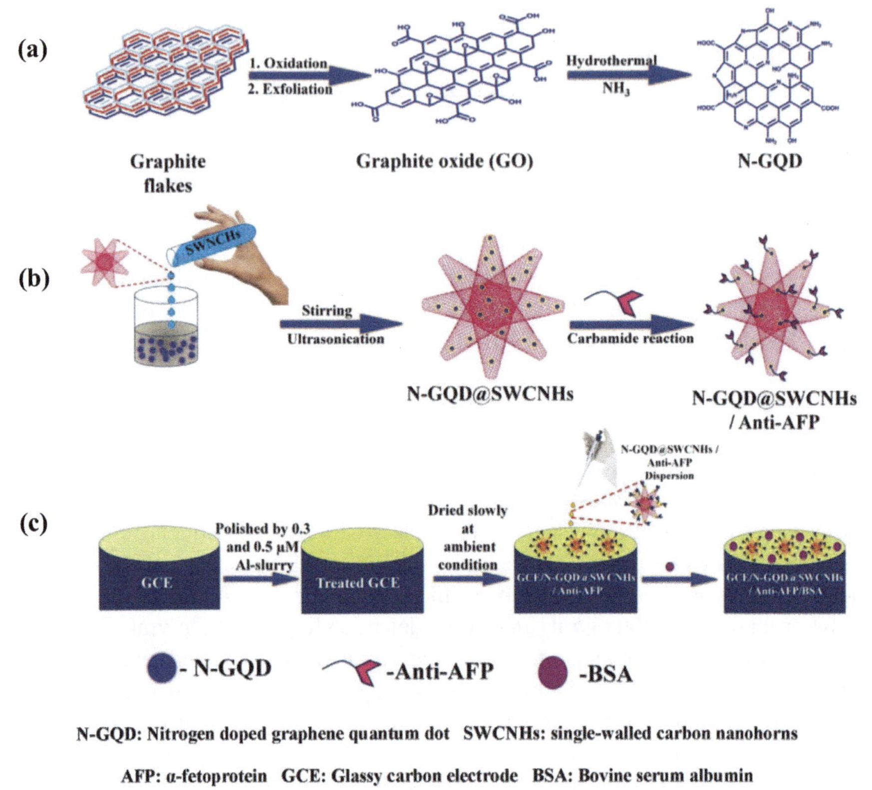

Fig. 1 Preparation of GO and N-GQD from graphite flakes (**a**), preparation of N-GQD@SWCNHs and N-GQD@SWCNHs/Anti-AFP (**b**), and fabrication procedure of immunosensor GCE/N-GQD@SWCNHs/Anti-AFP/BSA (**c**). Reproduced with permission from [26]

The applications of NP bioconjugates to regulate membrane ion channel's activity appeared as a new scope for the cells. Recently, a nanoparticle-based approach is proposed by Nag et al. [28] for the activation of channelrhodopsin C1V1 (ChR-C1V1) expressed in plasma membrane of HEK 293 T/17 cells. The hydrophilic CdSe/ZnS core–shell semiconductor QDs gets self-assembled towards ChR-C1V1 through the metal affinity-driven interaction between QD ZnS shell and N-terminal hexahistidine tag. Research revealed that the laser-induced opening of ChR-C1V1 channel in photoexcited QD activates the ChR-C1V1 channels which was established from a voltage-sensitive dye named bis-(1,3-diethylthiobarbituric acid) trimethine oxonol. Outcomes of the study shows the capability to control interface QDs with living cells for the ultimate activation of ChR membrane proteins.

Biocompatible methods to label the bacteria with fluorescent NPs are indispensible to develop bacterial-bioconjugates for various applications, i.e., biosensors, imaging, medicine, etc. Dong et al. [29] have reported that direct conjugation of carboxyl QDs with outer membrane of *Escherichia coli* through the surface-displayed binding peptides. Histidine-containing peptide (H6G9) was displayed at N-terminus of membrane-embedded circularly permuted outer membrane protein. Existence of binding peptide makes a milieu different from negatively charged *E. coli* surface and offers resilient binding affinity to carboxyl QDs. TEM results showed that E. coli-QD bioconjugates have higher loading densities immobilized onto cell surface even if loading concentration of QDs is very low (10 μg/mL) to decrease the exposure of cells. Such hybrid cells intensely fluoresce with all different colors of QDs with various emission wavelengths that could be simply imagined using fluorescence microscopy. *Escherichia coli*-QD bioconjugates were found to be highly feasible and preserved the capability to grow and split. In this research, authors claimed that a simple, direct, and highly efficient method can be used to label the bacteria using QDs deprived of considerably negotiating the vitality of cells.

Roy et al. [30] have presented an excited way for the targeted bioimaging of folate receptor (FR)-positive cancer cells by folic acid and MoS2 quantum dots. Water-soluble MoS2 quantum dotes (sized 4–5 nm) along with cysteine functionalization were prepared using a simple bottom-up hydrothermal approach. MoS2 QDs showed the blue emission with highest emission intensity at 444 nm. MoS2 QDs were found to be sensitive for folic acid to yield an active as well as stable nanofiber structure using supramolecular interaction that showed approx. 97% quenching of fluorescence as shown in Fig. 2. High sensitivity and selectivity of MoS2 QDs for folic acid make MoS2 QD-based nanoprobe a suitable applicant for folic acid-targeted imaging probes for in vivo investigation of folic acid-pretreated FR-overexpressed cancer cells. Confocal microscopy images showed that folic acid-pretreated B16F10 cancer cells exhibits greater population of darkened fluorescence as compared to untreated cancer cells as well as HEK-293 cells. Flow cytometry investigation exhibits the substantial variance in mean of fluorescence between the folic acid-pretreated and untreated B16F10 cancer cells. Therefore, such MoS_2 quantum dots-based nanoprobes could be used as potential material for pre-diagnosis of cancer using targeted bioimaging.

Efficient intraoperative identification of cancers involves the growth of bright, minimally-toxic, tumor-specific near-infrared (NIR) probes as contrast agents. Luminescent semiconductor QDs suggest numerous unique benefits for in vivo cellular imaging by giving bright and photostable fluorescent probes. Yakavets et al. [31] have offer the synthesis process of ZnCuInSe/ZnS core/shell QDs emitting in NIR (~750 nm) conjugated with NAVPNLRGDLQVLAQKVART (A20FMDV2) peptide in order to target αvβ6 integrin-rich head and neck squamous cell carcinoma (HNSCC). Result has showed the potential of these synthesized QDs from time-gated fluorescence imaging of stroma-rich 3D spheroids located onto the mm-thick tissue portions to mimic imaging conditions in tissues. Therefore, QD-A20 can be taken as highly capable nanoprobes for imaging-guided surgery as well as NIR bioimaging.

Since optical properties and conjugation capability with biomolecules are two advantageous factors of photoluminescent QDs, thus, are extensively being used in biomedicine domain. Though, the various derivatives of bioconjugation can impacts the interaction between nanoprobes and the targets. Also, the adaptation of size exclusion chromatography to separate the QDs and their derivatives is difficult because

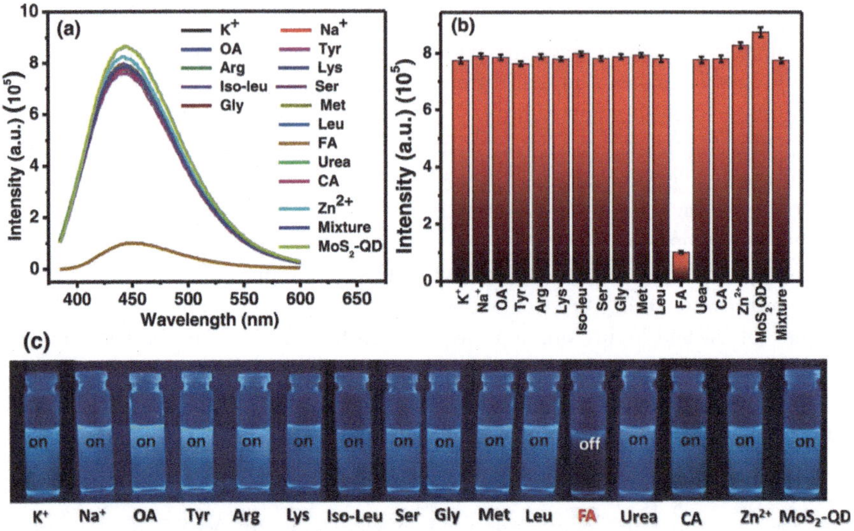

Fig. 2 **a** Emission spectra of MoS2 QDs in the presence of diverse potentially coexisting substances, **b** corresponding bar plot, and **c** photograph of emission images of MoS2 QD solutions in presence of different materials under 365 nm ultraviolet light irradiation. Reproduced with permission from [30]

of the variance in physical as well as chemical properties of biomolecules and NPs. In that regard, Linkov et al. [32] have presented a comprehensive procedure for size exclusion chromatography purification of water-soluble CdSe/ZnS QDs and QD-conjugates by choosing Sephadex resins of various porosities. Efficiency of SEC purification can also be measured by using the PEG derivatives and two types of small molecules named bisnetropsin and 4,5,9-trisubstituted acridine. The results showed that the use of common chromatographic media in group of Sephadex resins permits effective distillation of QD bioconjugates from the contaminants for their successive use for the bioimaging and diagnostics.

MicroRNAs are considered to be imperative post-transcriptional regulators and have great potential to work as noninvasive biomarkers for the disease diagnosis and biomedical research. Although, the multiplex and sensitive detection using microRNAs can expedite the accuracy as well as prompt clinical analysis, but the problem lies that the currently available techniques are generally conceded by taking organic dyes as signal probes, difficult enzymatic and chemical manipulations, and involving complex reaction mechanisms. Therefore, Ma et al. [33] have presented a siRNA-based self-assembled QD biosensor for the facile as well as instantaneous detection of multiple microRNAs. Mechanism wise, the attachment of microRNA targets to the respective QD nanoprobes causes the construction of siRNA duplexes. It further brings the spectrally resolved coding of microRNAs and also enables the association of QDs magnetic NPs bioconjugates in order to isolate and enrich the target microRNAs. However, the disassembled QDs are delicately sensed by the single-molecule detection method by facilitating the sensing of microRNAs at the level of single-particle. Interestingly, the biosensor recognizes only QDs as signal reporters that instantaneously detect several microRNAs from same specimen and accomplishes femtomolar level sensitivity as well as single-base misalliance selectivity deprived of the participation of the target labeling or/and amplification phases. The same can also be effectively used for the simultaneous detection of microRNAs present in the clinical serum specimen. The finding holds a great evidential supports for the microRNAs turn to be an excellent candidate for non-invasive biomarkers probe for the early disease diagnosis.

Fundamentals to get the most out of the optical properties of colloidal semiconductor QDs in the biological research includes the achievement of effective surface functionalization along with suitable bioconjugation approaches [34]. Since, the functionalization using dextran is greatly effective if used with some NPs materials, but, so far, has shown very restricted uses with QDs. Rees et al. [35] have reported the synthesis, characterization, and applications of dextran-functionalized quantum dots. Various methods for the dextran-based ligands were assessed, including the suitability with the colloidal at various pH values, general attachment with the proteins as well as cells, and the micro-injection to the cells and their viability assays. Different bioconjugation methods were used such as covalent coupling to make a facile pH sensor, attachment of polyhistidine-tagged peptides to quantum dots for the purpose of energy transfer-based proteolytic activity assays, as well as attachment with the tetrameric antibody complexes in order to permit a sandwich immunoassay, cell immunolabeling, and imaging. The outcomes of the study revealed that dextran ligands are extremely encouraging for functionalization of quantum dots and the design of ligands is adjustable to assist the desires of the applications.

Additionally, the circular economy concept's point of view should also be considered whenever a novel material id developed. The overall cost of the raw material as well as processing, their availability, sustainability towards end-use, and recyclability is must to take under consideration.

5 Conclusion

This chapter describes the physicochemical features of quantum dots and biological nanoconjugates for fluorescent probes. The chief highlighting of this chapter is on quantum dots based bioconjugates fluorescent probes for the high- or super-resolution florescent micro- or nanoscopy. Also, based on the properties of already existing probes and their compatibility with present imaging techniques, an outlook for further improvement in fluorescent probes domain for super-resolution imaging is also addressed. The conjugation process fascinates the attention of researchers as it enhances the existing qualities of medicinal products and imparts new properties. Conjugated peptides or proteins, when chemically linked, possess notable characteristics including altered pharmacokinetics, enhanced permeability and safety, and the ability to penetrate challenging areas such as the blood–brain barrier through transcytosis and intracellular delivery. Moreover, chemically-modified peptide/protein modifications for quantum dots/bioconjugates have a significant impact on the therapeutic domain, owing to their wide range of uses in several domains like oncology, ophthalmology, neurology, immunology, muscle diseases, endocrinology, and more.

6 Future Perspectives

Primary challenge associated with quantum dots/bioconjugates is their intricate nature. However, this problem can be overcome by carefully selecting and optimizing the peptide/protein, linker, and polymer payload. Moreover, it is necessary to carry out computational and theoretical investigations to substantiate the mechanisms of peptide or protein interactions with various polymers in order to investigate their stability and biological activity. Novel experimental and computational techniques will be employed to investigate the structural characteristics of both biological and synthetic constituents of a bioconjugate. This will facilitate a more streamlined understanding of the relationship between structure and function, enabling the fine-tuning of bioconjugates for specific applications.

References

1. Ahmed, S.R., et al.: Graphdiyne quantum dots for H_2O_2 and dopamine detection. ACS Appl. Nano Mater. (2023)
2. Das, P., et al.: Polymer composites for biosensors. In: Polymeric nanocomposite materials for sensor applications, pp. 323–342. Woodhead Publishing, (2023)
3. Optical properties of quantum dots. In: Quantum dots and polymer nanocomposites: synthesis, chemistry, and applications, p. 69. CRC Press, (2023)
4. Das, P, et al.: Optical properties of quantum dots. In: Quantum dots and polymer nanocomposites, pp. 69–85. CRC Press, (2022)
5. Das, P., et al.: Naturally derived carbon dots in situ confined self-healing and breathable hydrogel monolith for anomalous diffusion-driven phytomedicine release. ACS Appl. Bio Mater. (2022)
6. Das, P., et al.: Carbon dot biopolymer-based flexible functional films for antioxidant and food monitoring applications. ACS Appl. Polym. Mater. **4**(12), 9323–9340 (2022)
7. Das, P., et al.: Carbon-dots-initiated photopolymerization: An in situ synthetic approach for MXene/poly (norepinephrine)/copper hybrid and its application for mitigating water pollution. ACS Appl. Mater. & Interfaces **13**(26), 31038–31050 (2021)
8. Maruthapandi, M., et al.: Biocompatible N-doped carbon dots for the eradication of methicillin-resistant S. aureus (MRSA) and sensitive analysis for europium (III). Nano-Struct. & Nano-Objects **26**, 100724 (2021)
9. Singh, A., et al.: MXene-polymer nanocomposites for biomedical applications. MXene Nanocomposites: Des., Fabr., Shield. Appl., 197 (2023). https://doi.org/10.1201/9781003281511-10
10. Ponomarev, V.O., et al.: Evaluation of the ophthalmotoxic effect of quantum dots and bioconjugates based on them in terms of prospects for the treatment of resistant endophthalmitis. Experimental study (stage 1). Ophthalmology **18**(3), 476–487 (2021)
11. Zhao, Y.: A new era of metal–organic framework nanomaterials and applications. ACS Applied Nano Materials **3**(6), 4917–4919 (2020)

12. Singh, A., Kumari, K., Kundu, P.P.: Polyurethane nanocomposites for bone tissue engineering. In: Engineered nanomaterials for innovative therapies and biomedicine, pp. 373–40300. Springer International Publishing, Cham (2022). https://doi.org/10.1007/978-3-030-82918-6_15
13. Singh, A., et al.: Metal-based nanoparticles: synthesis and biomedical applications. nanoparticles reinforced metal nanocomposites: Mechanical performance and durability. Springer Nature Singapore, Singapore, pp. 365–408 (2023). https://doi.org/10.1007/978-981-19-9729-7_13
14. Singh, A., Kumari, K., Kundu, P.P.: Nanocellulose biocomposites for bone tissue engineering. Handbook of nanocelluloses: classification, properties, fabrication, and emerging applications. Springer International Publishing, Cham, pp. 1–51 (2021). https://doi.org/10.1007/978-3-030-62976-2_39-1
15. Wu, P., Yan, X.P.: Doped quantum dots for chemo/biosensing and bioimaging chem. Soc Rev **42,** 5489 (2013)
16. Xing, Y., et al.: Bioconjugated quantum dots for multiplexed and quantitative immunohistochemistry. Nat. Protoc. **2**(5), 1152–1165 (2007)
17. Fan, F., et al.: Conjugated polymer covalently modified graphene oxide quantum dots for ternary electronic memory devices. Nanoscale **9**(30), 10610–10618 (2017)
18. Geißler, D., Niko, H.: Lanthanide complexes in FRET applications. Curr. Inorg. Chem. (Discontin.) **1**(1), 17–35 (2011)
19. Shah, P., Shende, P.: Biomacromolecule-functionalized nanoparticle-based conjugates for potentiation of anticancer therapy. Curr. Cancer Drug Targets **22**(1), 31–48 (2022)
20. Zhou, L., et al.: Nonblinking Core–Multishell InP/ZnSe/ZnS quantum dot bioconjugates for super-resolution imaging. ACS Appl. Nano Mater., (2022)
21. Ponomarev, V.O., et al.: Evaluation of anti-infective activity of bioconjugates based on quantum dots CdTe/Cd MPA 710 and Levofloxacin against staphylococcal infection of the cornea. Experimental study. Ophthalmology **19**(4), 808–814 (2022)
22. Rahmani, P., et al.: One-step ligand-exchange method to produce quantum Dot–DNA conjugates for DNA-directed self-assembly. ACS Appl. Mater. & Interfaces **14**(42), 47359–47368 (2022)
23. Savage, P., et al.: Self-assembly of N-, C-and N-/C-terminated Val-and Phe-amino acid side chains of naphthalene. J. Mol. Struct. **1263**, 133116 (2022)
24. Ponomarev, V.O., et al.: Laboratory analysis of the anti-infectious activity of quantum dots and bioconjugates based on them in relation to potential ocular Pseudomonas aeruginosa infection. Experimental study (part 4). Ophthalmology **19**(2), 429–433 (2022)
25. Manoj, B., et al.: InP-Bovine serum albumin conjugates as energy transfer probes. J. Phys. Chem. B **126**(14), 2635–2646 (2022)
26. Dutta, K., et al.: Development of an efficient immunosensing platform by exploring single-walled carbon nanohorns (SWCNHs) and nitrogen doped graphene quantum dot (N-GQD) nanocomposite for early detection of cancer biomarker. ACS Biomater. Sci. & Eng. **7**(12), 5541–5554 (2021)
27. Salgado, C. L., et al.: Bioengineered fluorescent nanoprobe conjugates for tracking human bone cells: in vitro biocompatibility analysis. Materials **14**(16), 4422 (2021)
28. Nag, O.K., et al.: In situ self-assembly of quantum dots at the plasma membrane mediates energy transfer-based activation of channelrhodopsin. Part. & Part. Syst. Charact. **38**(7), 2100053 (2021)
29. Dong, H., et al.: Direct conjugation of fluorescent quantum dots with E. coli via surface-displayed histidine-containing peptides. Colloids Surf. B: Biointerfaces **203**, 111730 (2021)
30. Roy, S., et al.: Targeted bioimaging of cancer cells using free folic acid-sensitive molybdenum disulfide quantum dots through fluorescence “turn-off”. ACS Appl. Bio Mater. **4**(3), 2839–2849 (2021)
31. Yakavets, I., et al.: NIR imaging of the integrin-rich head and neck squamous cell carcinoma using ternary copper indium selenide/zinc sulfide-based quantum dots. Cancers **12**(12), 3727 (2020)

32. Linkov, P., et al.: Selection of the optimal chromatography medium for purification of quantum dots and their bioconjugates. Chem. Mater. **32**(21), 9078–9089 (2020)
33. Ma, F., Jiang, S., Zhang, C.-Y.: SiRNA-directed self-assembled quantum dot biosensor for simultaneous detection of multiple microRNAs at the single-particle level. Biosens. Bioelectron. **157**, 112177 (2020)
34. Singh, A., Kumari, K., Kundu, P.P.:Synthesis of biopolymer-polypeptide conjugates and their potential therapeutic interests. Phys. Sci. Rev. (0), (2023). https://doi.org/10.1515/psr-2022-0185
35. Rees, K., et al.: Dextran-functionalized semiconductor quantum dot bioconjugates for bioanalysis and imaging. Bioconjugate Chem. **31**(3), 861–874 (2020)

Cytotoxicity of Quantum Dots Based Nanocomposites

Konatham Teja Kumar Reddy, N. Venkatesan, M. V. Saranya, Jeevanandham Somasundaram, Jeetendra Kumar Gupta, Nagam Santhi Priya, Prakash Srichand Sukhramani, and Anasuya Patil

Abstract The present chapter provides an overview of the cytotoxic effects of nanocomposites containing quantum dots (QDs) in the context of their biomedical applications. Additionally, this chapter delves into the various approaches that have been proposed to counteract the potential risks associated with the use of QDs. In this discourse, we examine the physicochemical characteristics of QDs, the determinants that contribute to their cytotoxicity, and the surface functionalization approaches utilised to enhance their biocompatibility. The necessity for standardised protocols and regulatory frameworks to guarantee the safe utilisation of the aforementioned is also emphasised. The present chapter provides significant perspectives for

K. T. K. Reddy
Department of Pharmacy, University College of Technology, Osmania University, Amberpet, Hyderabad, Telangana 500007, India

N. Venkatesan
School of Pharmacy, Sri Balaji Vidyapeeth Deemed to Be University, Pillayarkuppam, Puducherry 607402, India

M. V. Saranya
Department of Pharmaceutics, JKKN College of Pharmacy, (Affiliated to the Tamilnadu Dr. MGR Medical University, Chennai), Kumarapalayam, Namakkal District, Tamil Nadu, India

J. Somasundaram
PPG College of Pharmacy, (Affiliated With the Tamil Nadu Dr. MGR. Medical University, Chennai), NH-209, Sathy Road, Saravanampatti Post, Coimbatore, Tamil Nadu 641035, India

J. K. Gupta
Institute of Pharmaceutical Research, GLA University Mathura, Mathura, Uttar Pradesh, India

N. S. Priya
Department of Pharmaceutics, Vignan Pharmacy College, Vadlamudi, Guntur, Andhra Pradesh 522213, India

P. S. Sukhramani
Department of Pharmaceutics, Veerayatan Institute of Pharmacy, Jakhania, Bhuj- Mandvi Road, Mandvi, Kutch, Gujarat 370460, India

A. Patil (✉)
Department of Pharmaceutics, K.L.E. College of Pharmacy, II Block Rajajinagar, Bengaluru, India
e-mail: anusuya.raghu@gmail.com

S. Thomas et al. (eds.), *Quantum Dots Based Nanocomposites*, Engineering Materials,
https://doi.org/10.1007/978-3-031-54779-9_11

researchers and clinicians who are engaged in the creation of secure and efficient nanocomposites based on QDs.

Keywords Cytotoxic effects · Quantum dots · Surface functionalization · Biocompatibility

1 Introduction

The scientific community has shown considerable interest in quantum dots (QDs) and nanocomposites as two distinct categories of nanomaterials, owing to their distinctive characteristics and potential utility in diverse domains. Quantum dots (QDs) are semiconductor nanocrystals with dimensions ranging from 2 to 10 nm [1–3]. Their fluorescence properties are determined by their size and composition, which enable them to emit light at different wavelengths. Quantum dots (QDs) have been utilised in various fields such as biological imaging, sensing, and photonics due to their exceptional photostability and adjustable fluorescence characteristics [4]. Nanocomposites are a type of composite material comprising a matrix material and one or more nanoscale components, such as nanoparticles, nanofibers, or nanotubes that are uniformly distributed throughout the matrix material [5, 6]. Nanocomposites have garnered significant attention due to their potential to augment the mechanical, electrical, or optical characteristics of the matrix material at the nanoscale level. This renders them a promising candidate for application in various domains, including aerospace, energy, and biomedical engineering [7]. The amalgamation of quantum dots (QDs) and nanocomposites has resulted in the emergence of innovative nanomaterials that exhibit superior characteristics. These include fluorescent nanocomposites that are suitable for bioimaging, as well as nanocomposite materials that can function as sensors for diverse applications [8].

Nanocomposites based on quantum dots (QDs) have become increasingly important due to their exceptional properties and potential applications [9]. It is important to note that the cytotoxicity of these materials should be considered when using them in biomedical applications. Studies have demonstrated that QDs can cause harm to cells both in laboratory settings and in living organisms [10]. The degree of toxicity is influenced by factors such as the size of the QDs, their surface chemistry, and the concentration at which they are present [11]. The cytotoxicity of QDs can be affected by the matrix material and manufacturing process when they are integrated into nanocomposites [11]. Hence, it is crucial to conduct a comprehensive examination of the cytotoxicity of nanocomposites based on quantum dots prior to their utilisation in biomedical applications [12]. In order to reduce the harmful effects of QDs based nanocomposites, different surface modifications have been investigated [13, 14]. One of the methods is to coat the QDs with biocompatible materials. Lowering the concentrations of QDs in the nanocomposites and optimising the manufacturing process are effective ways to reduce the cytotoxicity of the materials. Although QDs based nanocomposites have the potential to be cytotoxic, their distinct properties make them appealing for a range of biomedical applications. QDs-based nanocomposites have been extensively studied for their potential applications in targeted drug delivery, bioimaging, and tissue engineering. These materials have the potential to enable controlled drug release, high-resolution imaging, and improved mechanical and optical properties [15–17]. Furthermore, the significance of nanocomposites based on quantum dots in biomedical applications cannot be emphasised enough. It is important to carefully consider the cytotoxicity of these substances to ensure their safe use in vivo [6]. Ongoing research and development suggest that nanocomposites based on QDs have the potential to provide novel solutions to existing biomedical and environmental problems [6, 14, 18–22].

It is crucial to evaluate the cytotoxicity of nanocomposites containing quantum dots (QDs) to ensure their safety and effectiveness in biomedical applications. Quantum dots (QDs) are nanocrystals made of semiconductors that possess distinctive optical and electrical properties. These properties make them highly desirable for a wide range of applications, including bioimaging, drug delivery, tissue engineering, and environmental applications [23–28]. The potential cytotoxicity of these materials is a significant concern that requires attention before their application in biomedical fields. The toxicity of quantum dots (QDs) is influenced by several factors, including their size, surface charge, and concentration [29]. The cytotoxicity of QDs can also be influenced by the matrix material and the manufacturing process when they are incorporated into nanocomposites. It is crucial to evaluate the cytotoxicity of nanocomposites containing QDs by utilising both in vitro and in vivo models before implementing them in human applications. To assess the potential toxicity of nanocomposites, in vitro models involve testing them on cultured cells. These tests can assist in identifying the ideal concentration and duration of exposure for the nanocomposites, which would not result in any harmful impacts on the cells [30]. The evaluation of biocompatibility and safety of nanocomposites is conducted through in vivo models, which entail testing on animal subjects [31]. These tests can assist in identifying the possible harmful impacts of nanocomposites on various organs and tissues. Evaluating the cytotoxicity of nanocomposites made with QDs is crucial to verify their safety for use in biomedical applications. The potential applications of these could include targeted drug delivery, bioimaging, and tissue engineering. The creation of safe and efficient nanocomposites using QDs could provide inventive solutions to a range of biomedical issues and enhance human health.

2 Nanocomposites Based on Quantum Dots

The utilisation of nanocomposites that are founded on quantum dots (QDs) has garnered a growing amount of interest in contemporary times, owing to their exceptional optical and electronic characteristics, as well as their potential utility in diverse domains such as energy, electronics, and biomedicine. A nanocomposite refers to a type of material that merges two or more distinct components at the nanoscale to generate a hybrid material that exhibits enhanced properties.

The synthesis of nanocomposites utilising quantum dots (QDs) involves the amalgamation of QDs with diverse materials, including polymers, metals, and inorganic nanoparticles. The special properties and their applications are tabulated and summarized in Table 1. The nanocomposites exhibit distinctive characteristics that arise from the amalgamation of the properties of the quantum dots and the other constituent materials. These properties can be customised by manipulating the composition and structure of the nanocomposite. QD-polymer composites are among the most prevalent forms of nanocomposites that incorporate quantum dots. Quantum dots (QDs) have the capability to be incorporated into diverse polymer matrices, including polystyrene, poly(methyl methacrylate), and polyvinyl alcohol, resulting in the formation of QD-polymer nanocomposites. The nanocomposites have demonstrated superior optical characteristics, including heightened fluorescence intensity and enhanced stability in comparison to unbound quantum dots. The integration of QD-polymer nanocomposites exhibits promising prospects in various fields such as optical sensors, photovoltaic devices, and biological imaging.

QD-metal nanocomposites represent an additional category of nanocomposites that incorporate quantum dots. Quantum dots (QDs) have the potential to form nanocomposites with metals, including gold and silver. The nanocomposites exhibit distinctive plasmonic characteristics that render them suitable for employment in sensing and imaging applications. The utilisation of QD-metal nanocomposites has

Table 1 Various parameters for evaluating the role and features of QDs based nanocomposites

Categories	Quantum dot-based nanocomposites
Classification	
Composition	Core–shell, alloyed, heterostructure
Dimensionality	0D, 1D, 2D, 3D
Design	
Surface modification	Ligand exchange, surface functionalization
Morphology	Nanoparticles, nanowires, nanosheets, nanorods, nanotubes
Fabrication	
Synthesis	Chemical, physical, biological
Assembly	Self-assembly, Co-assembly, Template-assisted assembly
Applications	
Optoelectronics	LEDs, solar cells, display devices, photodetectors, photoelectrochemical cells
Biomedical	Imaging, drug delivery, biosensing, photodynamic therapy
Energy storage	Batteries, supercapacitors, fuel cells
Catalysis	Photocatalysis, electrocatalysis, biocatalysis
Cytotoxicity	
In vitro assays	MTT assay, LDH assay, ROS assay, cell viability assay
In vivo studies	Acute and chronic toxicity studies, biodistribution studies, histopathology, immunotoxicity studies

been demonstrated to manifest escalated catalytic efficacy in contrast to unbound QDs, thereby presenting a potential avenue for their application in photocatalytic contexts.

QD-inorganic nanoparticle nanocomposites can be generated by incorporating inorganic nanoparticles, such as silica and iron oxide, with quantum dots. The nanocomposites exhibit promising prospects for utilisation in drug delivery and magnetic resonance imaging. The utilisation of QD-inorganic nanoparticle nanocomposites has demonstrated enhanced biocompatibility and stability in contrast to unbound QDs, rendering them a propitious candidate for employment in biomedical domains. Quantum dot (QD) nanocomposites have been studied for their potential application in energy-related fields, including solar cells and light-emitting diodes (LEDs). Quantum dots (QDs) have the potential to be integrated into a range of materials, including perovskites and graphene, for the purpose of generating QD-energy nanocomposites. The nanocomposites have demonstrated superior light harvesting and charge transport characteristics in comparison to conventional materials, rendering them a viable candidate for deployment in future energy devices.

Nanocomposites based on quantum dots (QDs) exhibit distinctive characteristics that render them highly favourable for a diverse range of applications in the fields of energy, electronics, and biomedicine. The versatility and adaptability of QD-based nanocomposites can be attributed to the ability to modify their properties by manipulating the composition and structure of the nanocomposite. Additional investigation is required to comprehensively comprehend the potential of nanocomposites based on quantum dots and to enhance their characteristics for particular uses.

Nanocomposites that utilise quantum dots as their building blocks can be categorised according to various factors, such as their composition, dimensionality, surface modification, morphology, and methods of fabrication. Quantum dots of core–shell, alloyed, and heterostructure types can be produced in configurations of 0D, 1D, 2D, or 3D, accompanied by diverse surface modifications and morphologies. Various fabrication techniques can be employed, such as chemical, physical, or biological synthesis methods, in addition to self-assembly, co-assembly, and template-assisted assembly. Quantum dot-based nanocomposites have a wide range of potential applications, including but not limited to optoelectronics, biomedical, energy storage, and catalysis. The utilisation of quantum dots in optoelectronic devices such as light-emitting diodes, solar cells, display technologies, photodetectors, and photoelectrochemical cells is a prevalent area of research and development. Quantum dots have potential applications in the biomedical field, including but not limited to imaging, drug delivery, biosensing, and photodynamic therapy. Quantum dots have been employed in energy storage applications such as batteries, supercapacitors, and fuel cells. Additionally, catalysis applications have utilised quantum dots in photocatalysis, electrocatalysis, and biocatalysis. The evaluation of quantum dot-based nanocomposites' safety necessitates the conduct of cytotoxicity investigations. The toxicity of quantum dots on cultured cells can be assessed through in vitro assays, including but not limited to the MTT assay, LDH assay, ROS assay, and cell viability assay. Quantum dots' toxicity in living organisms can be evaluated through in vivo studies, including acute and chronic toxicity studies, biodistribution studies, histopathology, and immunotoxicity studies.

3 Cytotoxicity of Quantum Dots

Quantum dots (QDs) are diminutive particles, measuring in the nanometer range, that are composed of semiconducting materials. Their distinctive optical characteristics, such as size-dependent fluorescence and exceptional photostability, have rendered them extensively employed in biological investigations and medical imaging [32]. Notwithstanding their potential utility in biological and medical applications, concerns have been raised regarding their cytotoxicity, which may constrain their usage. Cytotoxicity pertains to the capacity of a substance to induce harm or mortality to cellular entities. QDs can undergo diverse mechanisms that lead to their toxicity, such as oxidative stress, cellular membrane disruption, and DNA damage [33]. The likelihood of cytotoxicity is contingent upon various factors, such as the dimensions, surface properties, and constitution of the quantum dots. The determination of cytotoxicity of quantum dots is significantly influenced by their size. QDs with a diameter smaller than 10 nm have been observed to exhibit greater toxicity compared to their larger counterparts. This is attributed to their higher probability of penetrating cell membranes and interacting with cellular components. Furthermore, the toxicity of quantum dots (QDs) can be significantly influenced by their surface chemistry. It has been demonstrated that the cytotoxicity of quantum dots (QDs) can be mitigated by coating them with hydrophilic molecules, such as polyethylene glycol (PEG), in contrast to QDs that are uncoated or coated with hydrophobic molecules [34]. The cytotoxicity of quantum dots (QDs) can be influenced by their composition. Cadmium-based quantum dots (QDs) are frequently employed owing to their robust fluorescence characteristics; however, they have demonstrated significant cytotoxicity towards cells. Cadmium is a heavy metal that possesses toxic properties and has the potential to induce oxidative stress and DNA damage [35]. Consequently, scholars have endeavoured to fabricate substitute quantum dot (QD) materials with reduced toxicity, including zinc-based QDs or QDs composed of non-toxic elements such as silicon. Several assays have been devised to evaluate the cytotoxicity of quantum dots (QDs). The aforementioned assays encompass techniques for evaluating cell viability, membrane integrity, reactive oxygen species (ROS) production, and DNA harm. Broadly speaking, the assays conducted have demonstrated that quantum dots (QDs) may exhibit cytotoxicity towards cells when present in elevated concentrations or in specific environments. The precise threshold for cytotoxicity may exhibit variability contingent upon the specific quantum dot (QD) variant and the particular cell type under investigation. Although QDs possess the potential for cytotoxicity, they remain a subject of interest for various biological and medical applications [36]. Scientists are currently engaged in the development of quantum dots that exhibit reduced toxicity, while retaining their distinctive optical characteristics.

Furthermore, endeavours are underway to enhance comprehension of the underlying mechanisms of quantum dot (QD) cytotoxicity [37]. This could facilitate the creation of safer QD materials and enhance their utility in biological and medical research. The matter of quantum dots' cytotoxicity is multifaceted and contingent upon various variables such as their dimensions, surface characteristics, and formation. Although quantum dots (QDs) possess the potential to induce cellular toxicity, they exhibit distinctive optical characteristics that render them highly advantageous for a diverse array of applications. Further investigation is required to enhance comprehension of the underlying mechanisms responsible for the cytotoxicity of quantum dots (QDs) and to formulate QD materials that are safer for application in biological and medical research.

4 Mechanisms of Cytotoxicity of Quantum Dots

Early reports talked about successes in single-molecule detection [38], single-cell tracking [39], and whole embryo or animal in vivo imaging. However, recent publications have brought up the need for a serious assessment of QDs' potential long-term cytotoxicity [33]. The research conducted on cadmium (Cd) chalcogenide materials, specifically those with CdS, CdSe, or CdTe cores, with or without a ZnS shell, has demonstrated that QDs exhibit cytotoxicity when present in micromolar concentrations in the growth medium [40]. Furthermore, the studies have revealed that the toxicity of QDs is positively correlated with the exposure of the core surface cadmium atoms to the surrounding medium. The investigation of the cytotoxicity of CdSe and CdSe/ZnS nanoparticles has been conducted with varying surface modifications, including mercaptopropionic acid coating, silanization, and polymer coating (Fig. 1). Quantitative values pertaining to the initiation of cytotoxic effects in culture media devoid of serum are provided for all instances. According to Kirchner et al. the viability of cells is reduced at elevated concentrations of nanoparticles, regardless of their type. Hence, it is plausible that diverse mechanisms of cytotoxicity could be operational (Fig. 2).

5 Factors Influencing Cytotoxicity of Quantum Dots

The potential for quantum dots (QDs) to exhibit cytotoxicity is contingent upon various factors, such as their size, shape, surface chemistry, composition, concentration, exposure duration, and the specific type of cell under consideration.

The dimension of quantum dots (QDs) is a crucial determinant in assessing their potential cytotoxicity. Quantum dots (QDs) that possess a diameter of less than 10 nm exhibit higher toxicity compared to their larger counterparts due to their ability to readily permeate cell membranes and interact with intracellular constituents. The diminutive dimensions of quantum dots (QDs) result in an augmented surface area

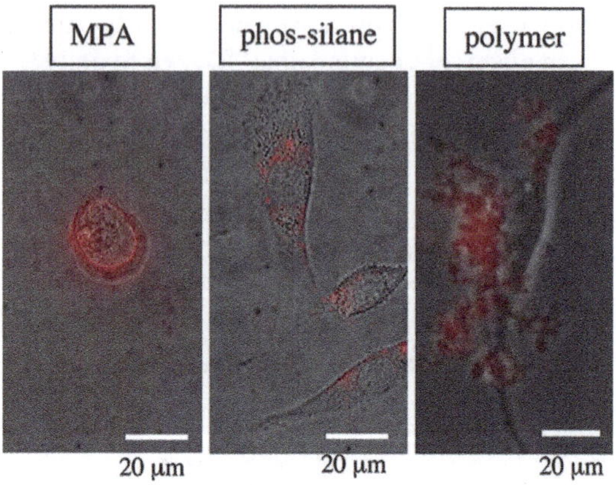

Fig. 1 Different type of nanoparticles and their cellular uptake. Reproduced with the permission from ref. [40] © 2005 American chemical society

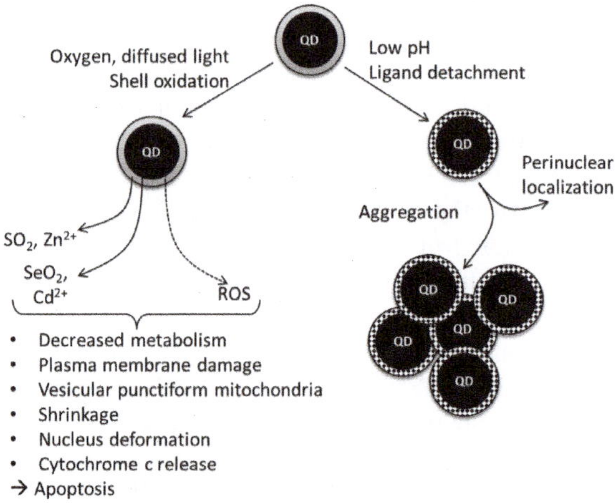

Fig. 2 Illustrations depicting the various mechanisms leading to the cytotoxicity of quantum dots

to volume ratio, thereby enhancing their reactivity and ability to generate reactive oxygen species (ROS). Furthermore, diminutive quantum dots exhibit a higher propensity to interact with proteins and nucleic acids, thereby inducing modifications in cellular activity and DNA impairment.

The cytotoxicity of quantum dots can also be influenced by their shape. The cytotoxicity of quantum dots (QDs) exhibiting distinct morphologies, including spherical, rod-shaped, and tetrapod-shaped, has been the subject of investigation. Several research studies have indicated that quantum dots (QDs) with a spherical shape exhibit lower levels of toxicity as compared to those with a rod or tetrapod morphology. One possible explanation for this phenomenon is that quantum dots (QDs) exhibiting a rod or tetrapod morphology possess a greater surface area, thereby facilitating heightened cellular interactions.

The cytotoxicity of quantum dots (QDs) is significantly influenced by their surface chemistry. The surface chemistry of QDs can be modified through the use of diverse molecules, including PEG, mercaptoundecanoic acid, or amines. Quantum dots (QDs) that are coated with hydrophilic molecules, such as polyethylene glycol (PEG), have demonstrated a decrease in cytotoxicity when compared to QDs that are uncoated or coated with hydrophobic molecules. The application of the coating has the potential to impede the aggregation of quantum dots (QDs) and subsequently reduce the probability of cellular interaction.

The cytotoxicity of quantum dots (QDs) can be influenced by their composition. Cadmium-based quantum dots (QDs) are frequently utilised in view of their robust fluorescence characteristics; however, they have been demonstrated to exhibit significant cytotoxicity towards cells. Cadmium is a heavy metal that possesses toxic properties and has the potential to induce oxidative stress and DNA damage. Consequently, scholars have devised substitute quantum dot (QD) substances, such as zinc-based QDs or QDs composed of innocuous elements like silicon, with the aim of mitigating toxicity.

The concentration of quantum dots (QDs) is a crucial element that significantly impacts their cytotoxicity. Elevated concentrations of quantum dots (QDs) may result in heightened toxicity due to the augmented probability of interaction with cellular constituents. At elevated concentrations, quantum dots (QDs) have been observed to induce apoptosis, resulting in cellular demise, through the disruption of cellular membranes.

The cytotoxicity of quantum dots can be influenced by the duration of exposure. Extended exposure to quantum dots (QDs) may result in heightened toxicity due to their potential to accumulate within cells and induce harm. If the concentration of QDs is sufficiently high, short-term exposure can result in toxicity. When evaluating the cytotoxicity of QDs, it is important to take into account the duration of exposure.

Cellular type plays a significant role in determining the cytotoxicity of quantum dots (QDs). Certain cellular varieties exhibit greater susceptibility to the toxicity of quantum dots than others. As an illustration, quantum dots (QDs) have demonstrated a greater cytotoxicity towards cancerous cells in comparison to healthy cells. Furthermore, the cellular attributes, such as their metabolic functionality and transporter expression, can impact their engagement with quantum dots and ascertain their cytotoxicity.

To summarise, the matter of QD cytotoxicity is multifaceted and can be impacted by various factors such as size, shape, surface chemistry, composition, concentration, exposure duration, and cellular classification. Comprehending these factors is of utmost importance for the secure and efficient utilisation of quantum dots (QDs) in the realm of biological and biomedical investigation. Additional investigation is required to gain a complete understanding of the mechanisms that underlie the cytotoxicity of quantum dots and to devise QD-based applications that are both safer and more efficacious.

6 In Vitro and in Vivo Assays for Evaluating Cytotoxicity of Quantum Dots

The evaluation of the cytotoxicity of quantum dots (QDs) is a crucial stage in their advancement for biomedical purposes (Table 2). Quantum dots (QDs) possess distinctive optical and electronic characteristics, rendering them highly desirable for a range of biomedical purposes, including but not limited to imaging, sensing, and drug delivery. Nevertheless, the utilisation of Quantum Dots (QDs) has been linked to probable toxicity owing to their diminutive dimensions, elevated surface area, and capability to produce reactive oxygen species (ROS). Hence, it is imperative to conduct a thorough evaluation of the cytotoxicity of quantum dots to ensure their safe and efficient utilisation as nanomaterials. The evaluation of QDs cytotoxicity is commonly performed through two primary assay types, namely in vitro assays and in vivo assays. Cell culture models are utilised for in vitro assays, whereas animal models are employed for in vivo assays. The selection of assay is contingent upon the particular research inquiry and developmental phase, as both in vitro and in vivo assays possess their respective merits and drawbacks. The simplicity, speed, and cost-effectiveness of in vitro assays make them a popular choice for the preliminary screening of quantum dot (QD) cytotoxicity. The MTT assay, LDH assay, ROS assay, and flow cytometry are among the frequently employed in vitro assays for evaluating the cytotoxicity of quantum dots.

The MTT assay is a widely used colorimetric technique that quantifies the metabolic activity of cells. The efficacy of the assay is predicated upon the capacity of living cells to convert the tetrazolium salt MTT, which is yellow in colour, to a purple formazan product. The quantity of formazan generated exhibits a direct correlation with the count of viable cells. The present assay is characterised by its straightforwardness and extensive application, owing to its capability to process a large number of samples and furnish numerical results. The MTT assay's reliability in determining cell viability may be compromised due to potential interference from QDs. The LDH assay is a biochemical technique utilised for quantifying the liberation of lactate dehydrogenase (LDH) from cells that have undergone damage. The assay evaluates cellular demise as opposed to vitality and is a straightforward and extensively employed technique. Nevertheless, the method's ability to differentiate between necrosis and apoptosis may be limited, and it could be susceptible to QD interference.

The quantification of reactive oxygen species (ROS) within cells is achieved through employment of the ROS assay. The aforementioned assay furnishes data pertaining to oxidative stress and plausible mechanisms underlying the cytotoxicity of quantum dots. The assessment of reactive oxygen species (ROS) may not provide a direct indication of cellular viability and could potentially be impacted by interference from quantum dots (QDs).

Flow cytometry is a robust technique that enables the quantification of cell viability, apoptosis, and cell cycle. Flow cytometry is a valuable tool for investigating various cellular parameters and can be employed to examine the impact of

Table 2 Various assays adopted to justify cytotoxicity of QDs based nanocomposites

Assay	Description	Advantages	Limitations
In vitro assays			
MTT assay	Measures metabolic activity of cells	Simple and widely used; High-throughput; Provides quantitative data	May not accurately reflect cell viability; Can be influenced by QD interference
LDH assay	Measures release of lactate dehydrogenase from damaged cells	Measures cell death rather than viability; Simple and widely used	May not distinguish between necrosis and apoptosis; Can be influenced by QD interference
ROS assay	Measures reactive oxygen species levels in cells	Provides information on oxidative stress and potential mechanism of cytotoxicity	May not directly reflect cell viability; Can be influenced by QD interference
Flow cytometry	Measures cell viability, apoptosis, and cell cycle	Provides detailed information on multiple cellular parameters	Requires specialized equipment and expertise
In vivo assays			
Histology	Examines tissue structure and cellular morphology	Provides information on tissue damage and inflammation	May not accurately reflect systemic toxicity; Requires tissue sampling
Biodistribution studies	Measures distribution and accumulation of QDs in tissues	Provides information on QD pharmacokinetics	May not accurately reflect toxicity; Requires specialized equipment and expertise
Blood chemistry analysis	Measures levels of enzymes, electrolytes, and other biomarkers in blood	Provides information on systemic toxicity and organ function	May not accurately reflect QD toxicity; Requires blood sampling
Behavioral assays	Measures changes in behavior or motor function	Provides information on potential neurological effects of QDs	May not directly reflect QD toxicity; Requires specialized equipment and expertise

quantum dots (QDs) on particular cell populations. Nevertheless, the utilisation of flow cytometry necessitates specialised equipment and expertise.

The utilisation of in vivo assays is crucial in the assessment of the systemic toxicity of quantum dots (QDs) and their capacity to accumulate in various tissues and organs. The assessment of quantum dot (QD) cytotoxicity in vivo typically involves a range of assays, including histology, biodistribution studies, blood chemistry analysis, and behavioural assays.

Histology pertains to the analysis of the structural composition of tissues and the morphology of cells. The utilisation of histological analysis enables the acquisition of valuable insights regarding tissue damage and inflammation, thereby facilitating the evaluation of the impact of QDs on various organs. Nevertheless, it should be noted that histology may not provide a precise representation of systemic toxicity and necessitates the collection of tissue samples.

Biodistribution investigations encompass the assessment of quantum dot (QD) dispersion and aggregation across diverse tissues and organs. Biodistribution investigations furnish insights into the pharmacokinetics of quantum dots (QDs) and serve as a means to evaluate the likelihood of QD accumulation in non-targeted anatomical sites. Notwithstanding, biodistribution investigations may not precisely mirror toxicity and necessitate specialised equipment and expertise.

The process of blood chemistry analysis entails the quantification of various biomolecules such as enzymes, electrolytes, and other biomarkers present in the blood. The analysis of blood chemistry offers valuable insights into the toxicity of a substance on the entire system and the functioning of organs. This method can be employed to evaluate the impact of quantum dots (QDs) on various organs. Notwithstanding, the toxicity of QD may not be precisely reflected by means of blood chemistry analysis and necessitates the collection of blood samples.

Behavioural assays encompass the quantification of alterations in animal conduct in reaction to exposure to QD. Behavioural assays have the capacity to furnish insights into the plausible impacts of quantum dots (QDs) on neurological function. Moreover, they can be employed to evaluate the likelihood of QDs to traverse the blood–brain barrier. Notwithstanding, behavioural assays are susceptible to various factors, including stress and environmental conditions.

The evaluation of quantum dots' cytotoxicity necessitates the utilisation of both in vitro and in vivo assays. The application of in vitro assays is advantageous for the preliminary assessment of cytotoxicity associated with QDs. Conversely, in vivo assays are indispensable for the comprehensive evaluation of systemic toxicity and the capacity of QDs to accumulate in diverse tissues and organs. The selection of an appropriate assay is contingent upon the particular research inquiry and developmental phase, and a comprehensive evaluation of quantum dot cytotoxicity may require the use of multiple assays.

7 Cytotoxicity of Quantum Dots Based Nanocomposites

The potential cytotoxicity of quantum dots (QDs) and QD-based nanocomposites has been a significant concern, particularly in the field of biomedical applications. The term cytotoxicity pertains to the capacity of a substance to induce harm or mortality to cells. The degree of cytotoxicity exhibited by QDs is predominantly influenced by factors such as their size, shape, surface chemistry, and composition. The diminutive dimensions of quantum dots (QDs) facilitate their cellular membrane penetration and subsequent interaction with intracellular constituents, potentially resulting in cytotoxicity. Research has demonstrated that quantum dots (QDs) have the potential to trigger oxidative stress, DNA harm, and apoptosis (a process of programmed cell death) within cellular structures. The potential for QDs to induce cell death is contingent upon the specific cellular population under investigation. As demonstrated by previous studies, quantum dots (QDs) have exhibited a greater propensity to elicit cytotoxic responses in malignant cells compared to their non-cancerous counterparts.

In order to address the issue of cytotoxicity associated with QDs, scholars have devised nanocomposites that are based on QDs and contain biocompatible substances, such as lipids and polymers. The utilisation of nanocomposites has been shown to enhance the stability, solubility, and biocompatibility of quantum dots (QDs), thereby rendering them more appropriate for employment in biomedical contexts. A research study was conducted to examine the cytotoxic properties of nanocomposites based on quantum dots (QDs), which were composed of zinc sulphide (ZnS)-coated QDs and polyethylene glycol (PEG). According to the research, the utilisation of PEG coating resulted in a notable decrease in the cytotoxic effects of QDs on human hepatoma cells, as indicated by the measurements of cell viability and lactate dehydrogenase (LDH) release. A further investigation was conducted to examine the cytotoxic properties of nanocomposites based on quantum dots (QDs), which were comprised of chitosan and hyaluronic acid coatings. According to the research, the application of chitosan coating resulted in enhanced stability and biocompatibility of quantum dots (QDs) in human breast cancer cells. The assessment was carried out by measuring cell viability and apoptosis. Nevertheless, the cytotoxicity associated with nanocomposites based on quantum dots remains a subject of ongoing investigation, and additional research is required to comprehensively comprehend the potential advantages and disadvantages of these substances. Furthermore, the comprehensive understanding of the enduring impacts of exposure to QD on human health remains incomplete, necessitating further investigation in this domain. The matter of cytotoxicity pertaining to QDs and QD-based nanocomposites is multifaceted and contingent upon several variables, such as dimensions, morphology, surface properties, and constitution. Although quantum dots (QDs) have demonstrated potential in various biomedical applications, it is crucial to meticulously assess and alleviate their potential cytotoxic impacts. Further investigation is required to comprehensively comprehend the potential advantages and drawbacks of biocompatible nanocomposites based on QDs, which could potentially mitigate the cytotoxicity of these substances.

8 Mechanisms of Cytotoxicity of Quantum Dots Based Nanocomposites

The matter of cytotoxicity pertaining to quantum dots (QDs) and QD-based nanocomposites is multifaceted and encompasses a range of mechanisms. The distinctive characteristics and diminutive dimensions of quantum dots (QDs) enable them to engage with intracellular constituents, thereby potentially engendering cytotoxic outcomes. The cytotoxicity of QDs and QD-based nanocomposites is attributed to several mechanisms, including oxidative stress, DNA damage, and apoptosis. The phenomenon of oxidative stress arises due to a state of disequilibrium between the generation of reactive oxygen species (ROS) and the capacity of cells to eliminate them (Fig. 3). Quantum dots (QDs) have the ability to produce reactive oxygen species (ROS) through various mechanisms such as surface defects, photoactivation, and electron transfer. The surplus of reactive oxygen species (ROS) has the potential to inflict harm upon the cellular constituents, including lipids, proteins, and DNA, thereby resulting in cytotoxic consequences. The cytotoxicity of quantum dots (QDs) and QD-based nanocomposites can be attributed to DNA damage as a mechanism. Quantum dots (QDs) have the capability to interact directly with DNA, which can result in the occurrence of single-strand breaks, double-strand breaks, and oxidative damage. DNA damage has the potential to trigger various cellular pathways, including the DNA damage response and p53 pathway, which may result in cell cycle arrest or apoptosis. The process of apoptosis is a predetermined cellular death pathway that is triggered as a result of internal or external cellular stressors. Quantum dots (QDs) have the ability to initiate the intrinsic apoptotic pathway through the disruption of the mitochondrial membrane potential, which results in the release of cytochrome c into the cytoplasm. Cytochrome c has the ability to initiate the activation of caspase enzymes, which in turn can induce the fragmentation of intracellular constituents, ultimately resulting in apoptosis.

The cytotoxicity of quantum dots (QDs) and QD-based nanocomposites is contingent upon the cell type that is subjected to them, in addition to the aforementioned mechanisms. As demonstrated by previous studies, quantum dots (QDs) have exhibited a greater propensity to elicit cytotoxic responses in malignant cells as compared to their non-malignant counterparts. The observed phenomenon could potentially be attributed to variances in cellular absorption, metabolic processes, and reactivity to oxidative stress. In order to address the issue of cytotoxicity associated with QDs and QD-based nanocomposites, scholars have devised biocompatible coatings, including polymers and lipids. The application of coatings on quantum dots (QDs) has been shown to enhance their stability, solubility, and biocompatibility, thereby mitigating their cytotoxicity. In general, the matter of cytotoxicity pertaining to quantum dots (QDs) and QD-based nanocomposites is multifaceted and encompasses a range of mechanisms. The diminutive dimensions and distinctive characteristics of quantum dots (QDs) may give rise to cellular oxidative stress, DNA harm, and programmed cell death. The creation of biocompatible coatings presents a potential solution for mitigating the cytotoxicity of said materials, thereby rendering them more appropriate for use in biomedical contexts.

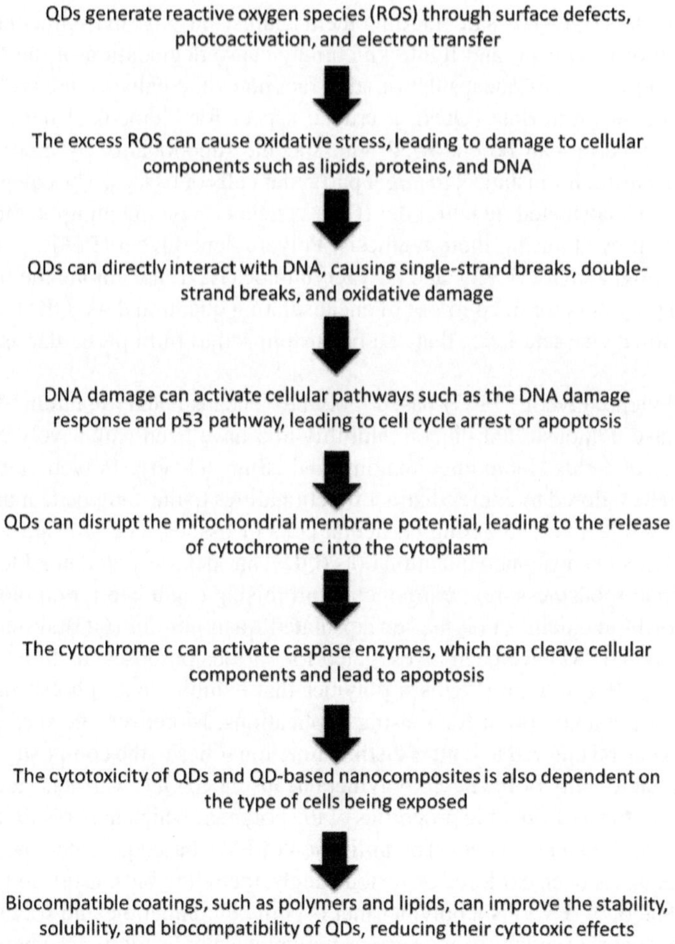

Fig. 3 Flow diagram of the consecutive steps of the mechanism of cytotoxicity of QDs based nanocomposites

9 Cytotoxicity for Polymer Encapsulated QDs Systems

Polymer-encapsulated quantum dot (QD) systems refer to composite materials that consist of QDs and a polymeric matrix that envelops and encapsulates the QDs. The polymeric matrix confers stability and safeguarding to the quantum dots, while simultaneously endowing distinctive characteristics to the composite material. The utilisation of polymer-encapsulated quantum dot (QD) systems has garnered considerable interest owing to their prospective applications in various fields including sensing, imaging, and drug delivery. The incorporation of quantum dots (QDs) into a polymer matrix through encapsulation confers numerous benefits compared to the utilisation of uncoated QDs. Encapsulation is a technique that enhances the stability of quantum dots (QDs) by safeguarding them against various environmental factors, including moisture, heat, and light, which may cause degradation of the QDs over time. The utilisation of encapsulation also facilitates the enhancement of biocompatibility of quantum dots (QDs), a crucial aspect for biomedical purposes. The composite material can be endowed with specific functionality by modifying the polymeric matrix, for instance, to target particular cells or tissues. The categorization of polymer encapsulated quantum dot (QD) systems is contingent upon the specific polymer employed during their synthesis. Polyethylene glycol (PEG), polystyrene (PS), polyvinyl alcohol (PVA), and polyacrylamide (PAA) are among the frequently employed polymers for the purpose of encapsulating quantum dots (QDs). Polymers exhibit distinct characteristics that can be customised to fulfil particular application demands.

Polyethylene glycol (PEG)-based polymer encapsulated quantum dot (QD) systems have demonstrated biocompatibility and have been extensively employed in biomedical fields, including imaging and drug delivery. Polyethylene glycol (PEG) can be tailored to confer distinct functionalities to the composite material, for instance, by selectively targeting particular cells or tissues. The utilisation of PEG-based polymer encapsulated quantum dots (QDs) has demonstrated notable biocompatibility and robustness, rendering them a promising candidate for employment in the field of biomedicine. Polymer-encapsulated quantum dot (QD) systems based on polystyrene (PS) have been investigated for various purposes, including sensing and imaging. Polystyrene (PS) is a polymer that exhibits hydrophobic properties, rendering it a viable option for sensing applications. Moreover, the surface chemistry of PS can be tailored to confer distinct functionalities to the composite material. The potential toxicity of PS-based polymer encapsulated QDs warrants careful evaluation due to the hydrophobic properties of the polymer, which may result in cellular uptake and subsequent toxicity. The utilisation of PVA-based polymer encapsulated QDs systems has been explored in various fields, including biosensors and imaging. Polyvinyl alcohol (PVA) is a polymer that is both biocompatible and water-soluble, rendering it appropriate for use in various biomedical applications. The optimisation of composite material properties can be achieved through the tuning of quantum dot size and surface chemistry. The utilisation of PAA-based polymer encapsulated QDs systems has been explored for various purposes, including biosensing and imaging. Polyacrylic acid (PAA) is a hydrophilic polymer that exhibits solubility in water and can be conveniently tailored to confer distinct attributes to the resulting composite material. The properties of the composite material can be adjusted by altering the surface chemistry of the PAA. The cytotoxicity mechanism of polymer-encapsulated quantum dot (QD) systems is contingent upon various factors, including the polymer type, QD size and surface chemistry, and composite material concentration. The deleterious effects of polymer encapsulated quantum dot (QD) systems on cells can be ascribed to multiple factors, such as oxidative stress, DNA damage, and apoptosis. It is imperative to conduct a thorough assessment of the toxicity of polymer-encapsulated quantum dot (QD) systems prior to their implementation in biomedical contexts, given the potential for significant variability in toxicity levels across different composite materials.

Polymeric matrices comprising natural polymers like chitosan, gelatin, and alginate have been employed as hosts for polymer nanocomposites containing quantum dots. Chitosan is a positively charged polysaccharide that originates from chitin, a readily available substance found in the shells of crustaceans. Gelatin is a biopolymer that is acquired through the process of partial hydrolysis of collagen, while alginate is a linear polysaccharide that originates from seaweed. The biocompatibility, biodegradability, and low toxicity of these polymers render them highly suitable for employment in biomedical contexts. The investigation of the cytotoxicity of polymer nanocomposites based on quantum dots in natural polymer systems has been conducted across multiple cell lines. Research findings indicate that QD nanocomposites based on chitosan demonstrate biocompatibility and low toxicity. The size of quantum dots (QDs) has been identified as a contributing factor to the cytotoxicity of chitosan-based QD nanocomposites, whereby larger QDs are more toxic than their smaller counterparts. Likewise, it has been demonstrated that QD nanocomposites based on gelatin manifest reduced cytotoxicity and are deemed appropriate for employment in the realm of biomedicine. Research has also explored nanocomposites consisting of alginate and quantum dots, which have demonstrated minimal cytotoxicity.

10 Strategies for Reducing Cytotoxicity of Quantum Dots Based Nanocomposites

Nanocomposites utilising quantum dots (QDs) have demonstrated significant promise for a range of biomedical purposes, including drug delivery and bioimaging. Nevertheless, the cytotoxicity of these substances remains a significant issue, impeding their extensive application. Various approaches have been investigated to address this concern, such as surface modification, encapsulation, and size and morphology optimisation. The objective of these methodologies is to mitigate the toxicity associated with nanocomposites based on quantum dots, while simultaneously preserving their intended characteristics.

11 Surface Functionalization

The implementation of surface functionalization has proven to be a viable strategy in mitigating the cytotoxic effects of nanocomposites that utilise quantum dots (QDs). Several techniques for surface functionalization have been devised to alter the surface characteristics of quantum dots (QDs) and enhance their biocompatibility. Several surface functionalization strategies have been utilised to mitigate the cytotoxicity of nanocomposites based on quantum dots (QDs).

The application of biocompatible polymers as a coating for quantum dots (QDs) has been demonstrated to effectively mitigate their cytotoxicity. Polyethylene glycol (PEG) is a commonly utilised biocompatible polymer for this purpose. The process of PEGylation enhances the hydrophilic nature of the surface of quantum dots (QDs), thereby impeding the binding of biomolecules and diminishing their interaction with cells, ultimately leading to an enhancement in biocompatibility.

The enhancement of biocompatibility of quantum dots (QDs) through surface modification with amino acids and peptides has been established. As an illustration, the surface conjugation of arginine onto quantum dots (QDs) has been shown to enhance their cellular uptake, whereas the conjugation of cysteine has been demonstrated to mitigate their cytotoxicity.

The process of conjugation with targeting ligands involves the binding of quantum dots (QDs) with specific targeting ligands, such as antibodies or aptamers. This conjugation process can enhance the specificity of QDs towards a particular cell type, thereby reducing their cytotoxicity. The utilisation of this methodology facilitates the targeted interaction of quantum dots (QDs) with specific cells, thereby reducing their interaction with non-targeted cells.

The utilisation of zwitterionic materials has been demonstrated to enhance the biocompatibility of nanocomposites based on quantum dots (QDs). These materials possess surface charges that are both positively and negatively charged. The surface charges of these materials confer upon them the ability to resist non-specific binding and minimise their interaction with cells.

The process of conjugation with antioxidant molecules has been observed to reduce the toxicity of quantum dots (QDs). This is achieved through the binding of QDs with antioxidant molecules such as glutathione or ascorbic acid, which effectively scavenge reactive oxygen species (ROS) and mitigate the oxidative stress induced by QDs.

The application of bioactive molecules for surface modification has been demonstrated to effectively mitigate the cytotoxicity of quantum dots (QDs). Specifically, enzymes or peptides have been utilised as bioactive molecules to modify the surface of QDs. The conjugation of catalase to quantum dots (QDs) has been observed to decrease the production of reactive oxygen species (ROS). Conversely, the conjugation of cell-penetrating peptides has been shown to augment the cellular uptake of QDs and mitigate their toxicity.

The utilisation of natural polymers, such as chitosan or alginate, for the purpose of surface modification of quantum dots (QDs) has been shown to enhance their biocompatibility and mitigate their toxicity. The innate biocompatibility and facile biodegradability of these natural polymers render them suitable contenders for the purpose of surface functionalization of quantum dots.

The biocompatibility of quantum dots (QDs) can be enhanced and their toxicity reduced through conjugation with non-toxic nanoparticles, such as gold or silver nanoparticles. Conjugated nanoparticles may be subjected to functionalization with targeting ligands, thereby augmenting their specificity towards a specific cell type.

12 Encapsulation of Quantum Dots

The process of encapsulating quantum dots (QDs) is a versatile approach that has the potential to enhance their biocompatibility and mitigate their cytotoxicity. The proposed methodology entails the incorporation of quantum dots (QDs) into a safeguarding matrix, resulting in the formation of core–shell architecture. The implementation of a protective matrix can serve as a physical impediment to hinder the release of hazardous quantum dots (QDs) into the surroundings. Additionally, it can enhance the durability and suitability of QDs for biological applications.

The utilisation of polymeric encapsulation is a widely employed approach in mitigating the cytotoxic effects of nanocomposites based on quantum dots. Polymeric materials, including PLGA, PEG, and PEI, are frequently employed for the encapsulation of QDs owing to their biocompatible nature, minimal toxicity, and adaptability. The utilisation of polymeric encapsulation presents a viable means of establishing a stable and safeguarding coating around quantum dots (QDs), thereby impeding their direct interaction with cells and reducing their potential for cytotoxicity.

The utilisation of lipid-based encapsulation has been implemented as an alternative strategy to mitigate the cytotoxic effects of nanocomposites based on quantum dots. The proposed methodology involves the encapsulation of quantum dots (QDs) within a lipid bilayer, resulting in the formation of a structure similar to a liposome. The lipid bilayer serves as a physical barrier that can hinder the direct interaction between quantum dots (QDs) and cells, thus mitigating their cytotoxic effects. The utilisation of lipid-based encapsulation presents an additional benefit of emulating the cellular biological membranes, thereby enhancing the cellular absorption and transportation of quantum dots (QDs).

The implementation of silica-based encapsulation has been utilised as a third approach in mitigating the cytotoxicity associated with nanocomposites based on quantum dots (QDs). The proposed methodology involves the integration of quantum dots (QDs) into a silica matrix, resulting in the formation of a core–shell architecture. The utilisation of a silica shell can serve as a physical impediment to impede the release of hazardous quantum dots (QDs) into the surroundings, augment their durability, and elevate their compatibility with biological systems. Studies have demonstrated that the utilisation of silica-based encapsulation can enhance the pharmacokinetic and biodistribution profiles of nanocomposites based on quantum dots (QDs), thereby rendering them more suitable for biomedical applications with improved safety.

The utilisation of protein-based encapsulation has been investigated as a potential strategy to mitigate the cytotoxic effects of nanocomposites based on quantum dots. The proposed methodology involves the encapsulation of quantum dots (QDs) within a protein shell, which may include Bovine Serum Albumin (BSA) or ferritin, resulting in the formation of core–shell architecture. The proteinaceous envelope can serve as a physical barricade, impeding the direct interaction of quantum dots with cellular entities and reducing their potential for cytotoxicity. The utilisation of protein-based encapsulation has demonstrated a notable enhancement in the biocompatibility and stability of nanocomposites based on quantum dots (QDs). This approach holds great potential as a safe method for the implementation of QDs in biomedical applications.

Various hybrid encapsulation strategies have been devised to mitigate the cytotoxicity of nanocomposites based on quantum dots (QDs) by integrating diverse encapsulation techniques. Core–shell architecture may be established through the process of enclosing quantum dots within a polymeric matrix, followed by the application of a lipid bilayer or silica shell coating. The amalgamation of various encapsulation techniques in hybrid encapsulation can yield synergistic advantages, resulting in enhanced efficacy and safety of nanocomposites based on QDs.

To summarise, encapsulation has shown potential as a viable approach to mitigate the cytotoxic effects of nanocomposites based on quantum dots (QDs). Several encapsulation techniques, including polymeric, lipid-based, silica-based, protein-based, and hybrid encapsulation, can be utilised to reduce the cytotoxicity of nanocomposites based on quantum dots (QDs) and enhance their safety for biomedical purposes. The aforementioned strategies provide flexible alternatives to enhance the biocompatibility and stability of nanocomposites based on quantum dots, thereby facilitating their secure and efficient implementation in diverse biomedical contexts.

13 Coating of Quantum Dots Based Nanocomposites

The application of a coating to quantum dots (QDs) represents a viable approach for mitigating the cytotoxic effects of nanocomposites based on QDs. The proposed methodology entails the application of a biocompatible substance onto the surface of quantum dots (QDs) in order to enhance their stability, biocompatibility, and mitigate their direct cellular interaction.

The implementation of biocompatible polymers, such as polyethylene glycol (PEG) or polyethyleneimine (PEI), to establish a hydrophilic layer on the surface of QDs, is a viable approach to mitigate the cytotoxicity of QDs-based nanocomposites. The utilisation of PEGylation or PEI coating has been observed to impede the direct interaction of quantum dots (QDs) with cells, while simultaneously enhancing their stability, biocompatibility, and mitigating their aggregation in biological surroundings. An additional approach to mitigate the cytotoxicity of nanocomposites based on quantum dots (QDs) involves the application of biomolecules, such as proteins or peptides, to generate a bioactive coating on the QD surface. The utilisation of bioconjugation has been observed to enhance the biocompatibility and specificity of nanocomposites based on quantum dots (QDs), thereby rendering them more suitable for targeted applications. Biomolecules may be chosen to selectively target particular receptors or tissues, thereby augmenting the cellular uptake of quantum dots (QDs) while simultaneously reducing their cytotoxicity. Inorganic coatings, namely silica or metal oxide coatings have been employed as a means to mitigate the cytotoxicity associated with nanocomposites based on quantum dots. The proposed methodology involves the application of a coating of inorganic material onto quantum dots (QDs), which serves as a protective layer. This layer is effective in preventing the release of hazardous QDs into the surroundings, while also enhancing their stability and biocompatibility. Studies have demonstrated that the application of coatings composed of silica or metal oxide can effectively mitigate the cytotoxicity associated with nanocomposites based on quantum dots (QDs), thereby enhancing their efficacy in the realm of biomedical applications. In addition, the functionalization of quantum dots (QDs) with diminutive molecules, such as enzyme inhibitors or antioxidants, can serve as a coating approach to mitigate their cytotoxic effects. The aforementioned molecules possess the ability to scavenge reactive oxygen species (ROS) or impede enzymes that participate in cellular toxicity. This leads to a decrease in the cytotoxicity of nanocomposites based on quantum dots (QDs) in biological settings.

The application of coating presents a promising approach towards mitigating the cytotoxic effects of nanocomposites based on quantum dots. Several coating techniques, including PEGylation, PEI coating, bioconjugation, inorganic coatings, and functionalization with small molecules, can be utilised to enhance the biocompatibility and stability of nanocomposites based on quantum dots (QDs) and mitigate their cytotoxicity. The implementation of these strategies presents a variety of adaptable solutions for enhancing the safety and effectiveness of nanocomposites based on quantum dots in various biomedical contexts.

14 Conclusion and Future Outlooks

To summarise, the potential cytotoxicity of nanocomposites utilising quantum dots (QDs) is a notable issue that requires resolution prior to their extensive implementation in the field of biomedicine. The distinctive physicochemical characteristics of quantum dots (QDs) that render them appealing for biomedical purposes also render them potentially perilous to both human health and the environment. It is imperative to possess a thorough comprehension of the mechanisms that underlie the cytotoxicity of nanocomposites based on quantum dots (QDs) in order to create QDs-based nanocomposites that are both safe and efficacious for biomedical purposes. Various approaches have been investigated to mitigate the cytotoxic effects of nanocomposites based on quantum dots (QDs), such as surface modification, encapsulation, and coating. The aforementioned methodologies endeavour to enhance the biocompatibility and stability of quantum dots (QDs), while concurrently mitigating their direct interaction with cells, thereby resulting in a reduction of their cytotoxicity. Nevertheless, additional research is imperative to ascertain the most effective coating approach for individual applications and to guarantee the enduring safety of nanocomposites based on quantum dots in vivo.

Prospective directions in this domain encompass the advancement of coating strategies that are both biocompatible and more efficient, the amalgamation of nanocomposites based on quantum dots with other sophisticated materials to augment their performance, and the exploration of their enduring safety and environmental implications. In addition, it is imperative to establish uniform procedures for assessing the cytotoxicity of nanocomposites based on quantum dots (QDs) and to create regulatory frameworks that ensure their safe application in clinical settings, in order to facilitate their successful integration into medical practise. In general, despite the existing obstacles in ensuring the secure and efficient utilisation of nanocomposites based on quantum dots (QDs), the emergence of inventive approaches to counteract their cytotoxicity presents encouraging prospects for their implementation in various biomedical domains such as diagnosis, imaging, therapy, and drug conveyance.

References

1. García de Arquer, F.P., et al.: Semiconductor quantum dots: Technological progress and future challenges. Science, **373**(6555), eaaz8541 (2021)
2. Gidwani, B., et al.: Quantum dots: Prospectives, toxicity, advances and applications. J. Drug Deliv. Sci. Technol. **61**, 102308 (2021)
3. Molaei, M.J.: Principles, mechanisms, and application of carbon quantum dots in sensors: a review. Anal. Methods **12**(10), 1266–1287 (2020)
4. Khan, M.E., Mohammad, A., Yoon, T.: State-of-the-art developments in carbon quantum dots (CQDs): Photo-catalysis, bio-imaging, and bio-sensing applications. Chemosphere, 134815 (2022)
5. Cao, J., et al.: Recent progress in NIR-II contrast agent for biological imaging. Front. Bioeng. Biotechnol. **7**, 487 (2020)
6. Ganguly, S., et al.: Microwave-synthesized polysaccharide-derived carbon dots as therapeutic cargoes and toughening agents for elastomeric gels. ACS Appl. Mater. & Interfaces **12**(46), 51940–51951 (2020)
7. Jacak, L., Hawrylak, P., Wojs, A.: Quantum dots. Springer Science & Business Media, (2013)
8. Bagher, A.M.: Quantum dots applications. Sens. & Transducers, **198**(3), 37 (2016)
9. Jamieson, T., et al.: Biological applications of quantum dots. Biomaterials **28**(31), 4717–4732 (2007)
10. Michler, P.: Single quantum dots: Fundamentals, applications and new concepts, vol. 90. Springer Science & Business Media, (2003)
11. Wang, Y., Hu, A.: Carbon quantum dots: synthesis, properties and applications. J. Mater. Chem. C **2**(34), 6921–6939 (2014)
12. Nozik, A.: Exciton multiplication and relaxation dynamics in quantum dots: applications to ultrahigh-efficiency solar photon conversion. Inorg. Chem. **44**(20), 6893–6899 (2005)
13. Shu, Y., et al.: Quantum dots for display applications. Angew. Chem. **132**(50), 22496–22507 (2020)
14. Das, P., et al.: Carbon dots for heavy-metal sensing, pH-sensitive cargo delivery, and antibacterial applications. ACS Appl. Nano Mater. **3**(12), 11777–11790 (2020)
15. Saravanan, A., et al.: Applications of N-doped carbon dots as antimicrobial agents, antibiotic carriers, and selective fluorescent probes for nitro explosives. ACS Appl. Bio Mater. **3**(11), 8023–8031 (2020)
16. Ganguly, S., et al.: Advancement in science and technology of carbon dot-polymer hybrid composites: a review. Funct. Compos. Struct. **1**(2), 022001 (2019)
17. Das, P., et al.: Heteroatom doped blue luminescent carbon dots as a nano-probe for targeted cell labeling and anticancer drug delivery vehicle. Mater. Chem. Phys. **237**, 121860 (2019)
18. Coe-Sullivan, S., et al.: Quantum dots for LED downconversion in display applications. ECS J. Solid State Sci. Technol. **2**(2), R3026 (2012)
19. Das, P., et al.: Tailor made magnetic nanolights: Fabrication to cancer theranostics applications. Nanoscale Adv. **3**(24), 6762–6796 (2021)

20. Das, P., et al.: Carbon-dots-initiated photopolymerization: An in situ synthetic approach for MXene/poly (norepinephrine)/copper hybrid and its application for mitigating water pollution. ACS Appl. Mater. & Interfaces **13**(26), 31038–31050 (2021)
21. Das, P., et al.: Immobilization of heteroatom-doped carbon dots onto nonpolar plastics for antifogging, antioxidant, and food monitoring applications. Langmuir **37**(11), 3508–3520 (2021)
22. Das, P., et al.: Biocompatible carbon dots derived from κ-carrageenan and phenyl boronic acid for dual modality sensing platform of sugar and its anti-diabetic drug release behavior. Int. J. Biol. Macromol. **132**, 316–329 (2019)
23. Wang, Z.M., *Self-assembled quantum dots*. Vol. 1. 2007: Springer Science & Business Media.
24. Das, P., et al.: MXene/0D nanocomposite architectures: Design, properties and emerging applications. Materials Today Nano **24**, 100428 (2023)
25. Das, P., et al.: Waste-derived sustainable fluorescent nanocarbon-coated breathable functional fabric for antioxidant and antimicrobial applications. ACS Appl. Mater. & Interfaces, (2023)
26. Rajabzadeh, A.R., et al.: Magnetic quantum dots for bioimaging. CRC Press, (2023)
27. Das, P., et al.: Naturally derived carbon dots in situ confined self-healing and breathable hydrogel monolith for anomalous diffusion-driven phytomedicine release. ACS Appl. Bio Mater. **5**(12), 5617–5633 (2022)
28. Das, P., et al.: Carbon dot biopolymer-based flexible functional films for antioxidant and food monitoring applications. ACS Appl. Polym. Mater. **4**(12), 9323–9340 (2022)
29. Zhang, L., et al.: Preparation of graphene quantum dots for bioimaging application. J. Nanosci. Nanotechnol. **12**(3), 2924–2928 (2012)
30. Vasudevan, D., et al.: Core–shell quantum dots: Properties and applications. J. Alloy. Compd. **636**, 395–404 (2015)
31. Chen, W., et al.: Synthesis and applications of graphene quantum dots: a review. Nanotechnol. Rev. **7**(2), 157–185 (2018)
32. Winnik, F.M., Maysinger, D.: Quantum dot cytotoxicity and ways to reduce it. Acc. Chem. Res. **46**(3), 672–680 (2013)
33. Derfus, A.M., Chan, W.C., Bhatia, S.N.: Probing the cytotoxicity of semiconductor quantum dots. Nano Lett. **4**(1), 11–18 (2004)
34. Priester, M.I., Ten Hagen, T.L.: Image guided drug delivery in nanosystem-based cancer therapies. Adv. Drug Deliv. Rev., 114621 (2022)
35. Flora, S., Mittal, M., Mehta, A.: Heavy metal induced oxidative stress & its possible reversal by chelation therapy. Indian J. Med. Res. **128**(4), 501–523 (2008)
36. Bradburne, C.E., et al.: Cytotoxicity of quantum dots used for in vitro cellular labeling: role of QD surface ligand, delivery modality, cell type, and direct comparison to organic fluorophores. Bioconjugate Chem. **24**(9), 1570–1583 (2013)
37. Kuo, T.-R., et al.: Studies of intracorneal distribution and cytotoxicity of quantum dots: risk assessment of eye exposure. Chem. Res. Toxicol. **24**(2), 253–261 (2011)
38. Dahan, M., et al.: Diffusion dynamics of glycine receptors revealed by single-quantum dot tracking. Science **302**(5644), 442–445 (2003)
39. Gao, X., et al.: In vivo cancer targeting and imaging with semiconductor quantum dots. Nat. Biotechnol. **22**(8), 969–976 (2004)
40. Kirchner, C., et al.: Cytotoxicity of colloidal CdSe and CdSe/ZnS nanoparticles. Nano Lett. **5**(2), 331–338 (2005)

Quantum Dots-Based Nanocomposites as Biosensors

Deeksha Thakur, Divya Hudda, Devendra Kumar, Yi-Ting Lai, and Prateek Sharma

Abstract Quantum dots (QD) are inorganic nanoparticles with unique optical and electronic properties, making them attractive candidates for biosensing applications. This chapter discusses the use of QD-based nanocomposites as biosensors for detecting a variety of biomolecules, including proteins, nucleic acids, and small molecules. QD can be functionalized with biomolecules such as antibodies, aptamers, and peptides, allowing for specific binding to target analytes. The resulting QD-biomolecule conjugates can then be incorporated into nanocomposites with other materials such as polymers, metals, or graphene, to enhance the sensitivity, stability, and selectivity of the biosensor. Various detection modalities, including fluorescence, electrochemistry, and surface-enhanced Raman spectroscopy, have been employed to detect the signal generated by QD-based biosensors. The performance of QD-based nanocomposite biosensors has been demonstrated in a wide range of applications, including environmental monitoring, clinical diagnostics, and food safety. However, challenges such as toxicity, stability, and reproducibility still need to be addressed for the practical application of QD-based nanocomposites as biosensors. Overall, the use of QD-based nanocomposites as biosensors shows great promise for the development of highly sensitive and specific biosensing platforms.

Keywords Quantum dots · Biosensors · 2D material · Nanocomposites · QDbased nanocomposites

D. Thakur · D. Hudda · D. Kumar · P. Sharma (✉)
Department of Applied Chemistry, Delhi Technological University, Shahbad Daulatpur, Main Bawana Road, Delhi 110042, India
e-mail: prateeksharma1688@gmail.com

Y.-T. Lai · P. Sharma
Department of Materials Engineering, Ming Chi University of Technology, New Taipei, Taiwan

S. Thomas et al. (eds.), *Quantum Dots Based Nanocomposites*, Engineering Materials,
https://doi.org/10.1007/978-3-031-54779-9_12

1 Introduction

Biosensors have become increasingly important in medical diagnostics due to their ability to provide rapid and accurate detection of various biological analytes. These analytes may include proteins, enzymes, nucleic acids, or metabolites, among others. Biosensors offer numerous advantages over traditional diagnostic methods, including faster detection times, higher sensitivity, specificity, and lower costs. One of the main benefits of biosensors is their ability to detect biomarkers indicative of various diseases, such as cancer, cardiovascular diseases, infectious diseases, and autoimmune disorders. Early detection of these biomarkers can lead to earlier diagnosis and treatment, significantly improving patient outcomes. Biosensors are also useful in monitoring disease progression and treatment efficacy. For example, biosensors can detect changes in the concentration of specific biomarkers over time, allowing clinicians to monitor the effectiveness of treatments and adjust them as needed.

In addition to their use in disease diagnosis and monitoring, biosensors are also being used for point-of-care testing, which enables rapid and on-site diagnosis without the need for specialized equipment or trained personnel. This is particularly important in resource-limited settings, where traditional laboratory-based diagnostic methods may not be feasible. Overall, biosensors have revolutionized medical diagnostics by providing rapid, accurate, and cost-effective detection of biological analytes. Their continued development and application are critical for improving patient outcomes and advancing personalized medicine. Biosensors are also important in environmental monitoring due to their ability to detect and quantify various pollutants and contaminants in air, water, and soil. These pollutants may include heavy metals, pesticides, industrial chemicals, and microbial contaminants, among others. Biosensors offer several advantages over traditional environmental monitoring methods, including their ability to provide real-time, on-site detection with high sensitivity, specificity, and selectivity. This enables rapid and efficient detection of pollutants, allowing for timely and effective interventions to prevent harm to the environment and human health.

Also, to their use in detecting pollutants, biosensors are also useful in monitoring biological processes in the environment, such as the activity of microorganisms and the degradation of pollutants. This can help in understanding the natural processes that occur in the environment and developing strategies for environmental remediation. Biosensors are also being used for the monitoring of food safety, agricultural practices, and the detection of harmful algal blooms in aquatic ecosystems. This allows for identifying potential hazards early on, preventing the spread of contaminated food, and minimizing environmental damage. Overall, biosensors play an important role in environmental monitoring by providing efficient, reliable, and cost-effective detection of pollutants and contaminants. Their continued development and application are critical for improving our understanding of the environment and mitigating the impact of pollutants on human health and the ecosystem.

Biosensors are analytical devices that can detect and measure specific biological or chemical substances in a sample. Nanomaterials have unique physical, chemical, and biological properties due to their small size and high surface area, which make them ideal for use in biosensors. Nanomaterials have been used to develop biosensors for a wide range of applications, including medical diagnostics, environmental monitoring, food safety, and drug discovery. For instance, biosensors using nanomaterials have been developed for the detection of glucose, cholesterol, and other biomarkers for various diseases such as diabetes and cancer.

Quantum dots (QD) are nanoscale semiconductors with unique physical and chemical properties that distinguish them from bulk materials. The QD possess the properties like size-tunability, high quantum yield, broad absorption spectrum, narrow emission, greater photostability, and high biocompatibility. The electronic and optical properties of QD strongly depend on their size, shape, and composition, which can be precisely controlled during synthesis. This allows for the creation of QD with specific properties that are tailored for various applications.

QD-based nanocomposites have tremendous potential for biosensing applications. Their unique optical properties, high sensitivity, and specificity make them ideal for the detection and analysis of biological molecules, cells, and pathogens.

The small size of QD allows them to penetrate the cell membrane, enabling the detection of intracellular targets with high sensitivity and specificity. The incorporation of QD into a matrix material can further enhance their sensing capabilities, providing a stable and biocompatible platform for biosensing.

One promising application of quantum dots-based nanocomposites in biosensing is the detection of biomolecules, such as proteins and nucleic acids. QD can be functionalized with specific ligands, such as antibodies or aptamers, which bind to the target biomolecule. The binding of the ligand to the target molecule can result in a change in the fluorescence properties of the quantum dot, providing a readout for the presence and concentration of the target biomolecule.

Along with biomolecule detection, quantum dots-based nanocomposites can also be used for the detection of cells and pathogens. The incorporation of QD into a matrix material can enable the fabrication of high-density arrays of sensing elements, providing multiplexed detection capabilities. The high sensitivity and specificity of QD can also enable the detection of low concentrations of cells or pathogens, making them ideal for early disease detection and monitoring. This chapter describes the QD-based nanocomposites for their magnificent properties and their best applications in biosensing in different sectors. The applications have been discussed in detail in further sections.

2 QD for Biosensing

QDs are a type of fluorescent nanomaterial that can emit various wavelengths and are typically composed of atoms from groups II–VI and III–V of the periodic table [1, 2]. Their optoelectronic characteristics are influenced by their shape and size, and their properties are intermediate between those of bulk semiconductors and those of discrete atoms or molecules [3, 4]. QDs are often capped with an inorganic layer to enhance their quantum yield and signal-to-noise ratio [5]. They have drawn much interest because of their distinctive electronic characteristics and tunable emission spectrum and can be utilized in many different devices and applications, such as solar cells, fluorescent probes, optical switches, and light sources.

QD can undergo modification by surface encapsulation with amphiphilic ligands, salinizing, and exchanging ligands, broadening their potential applications in sensors [6, 7]. QD can be used to create biosensors, allowing researchers to examine cell processes at the molecular level, potentially aiding in the diagnosis and treatment of diseases such as cancer 8. However, the utilization of QDs in biomedical applications is restricted due to the inclusion of toxic heavy metals, such as cadmium, making the development of a nontoxic and biocompatible nanomaterial highly vital [8]. Hybridizing QD with other substances can result in improved resistance to heat and chemicals, increased quantum efficiency, extended durations of excited state, and reduced harmful effects. Biosensors are devices that utilize biological interactions to confer molecular specificity to analytes, and QD can be used as active or passive sensors in high-resolution cellular imaging [9, 10].

3 Nanocomposites for Biosensing

a. Basics of nanocomposites and their properties

Nanocomposites (NCs) are a subset of hybrid materials that are created by combining two or more nano-sized entities under specific experimental conditions. These materials play a significant role in various scientific and technological fields. By optimizing the combination of matrix and filler, the properties of nanocomposites can be adjusted. For instance, conducting polymers (CPs) can act as the matrix and nanomaterials as the filler, resulting in hybrid inorganic–organic CP nanocomposites (CPNs) that possess excellent properties, making them ideal for electrical and biosensing platforms [11]. The properties of nanocomposites can be tailored by optimizing the experimental conditions and carefully selecting the matrix and filler materials. For instance, the addition of nanoparticles to a polymer matrix can improve its mechanical, thermal, and electrical properties. Similarly, the addition of carbon nanotubes to a metal matrix can enhance its strength, stiffness, and electrical conductivity. The tunability of nanocomposite properties makes them attractive for a wide range of applications, including electronics, aerospace, biomedical devices, energy storage, and biosensors [12].

Electrical sensing methods such as cyclic voltammetry, amperometry, potentiometry, capacitors, chemoresistance, and field effect transistors are well-suited for rapid detection of alterations in current, voltage, and resistance. Nanocomposites are superior to their counterparts due to their increased electrical breakdown strength during fabrication, higher melting temperature, magnetization, and charge capacity. Other benefits include modified electrical conductivity, corrosion resistance, dielectric, and semiconductor behavior. These properties make NCs ideal for use in solar cells, supercapacitors, biosensors, electrochemical sensors, and more [13, 14].

While the matrix phase is not directly used for assembling electrical sensor arrays due to limited linear range of response and interference by ionic and redox species, a solid contact phase—ion to electron transducer can be constructed, where the matrix acts as solid support and nanomaterial acts as an ion to electron transducer. Recent advancements in NCs-based electrical sensors have improved their limit of detection (LOD), and they are being used in practical applications with statistical analysis methods [15]. NCs are also useful in biosensing platforms such as tissue engineering, bio-electrodes, drug delivery, and biosensors to detect biological and synthetic molecules due to their nanoscale size and shape, biocompatibility, and biodegradability. Nanomaterials-based sensor arrays are in high demand for electrochemical sensors, biosensors, gas sensors, and immunosensors due to their exceptional performance in energy conversion performance, emission devices, integrated circuit-based nano-devices, medical devices, and more [16, 17].

Carbon nanomaterials, metal nanoparticles, and metal oxide nanoparticles can be embedded into conducting polymeric matrices to form CPNs with improved electrical, magnetic, thermoelectric, and biosensing properties. The incorporation of metal oxide nanoparticles into CPs has been extensively researched, with significant applications including supercapacitor electrodes, light-emitting diodes, transistors, electrochromic devices, electrochemical capacitors, photovoltaic cells, biosensors, gas sensors, actuators, and more [17].

The recent trends in nanocomposite synthesis by embedding carbon nanomaterials, metal and metal oxide nanoparticles into conducting polymers such as PANI, PPy, and PEDOT, and to examine their effect on the overall physical and chemical properties of CPNs. The article also highlights the utility of CPNs as biosensors, electrochemical sensors, gas sensors, therapeutic targeting sensors, and immunosensors [18].

b. **Types of nanocomposites used in biosensing**

QDs have special optical and electrical characteristics that make them valuable for a variety of applications, including biosensing. QDs are frequently used in nanocomposites along with other elements to enhance their functionality and compatibility with biological systems. As compared to QD, their nanocomposites display several novel traits and qualities as a result of the quantum confinement effect and edge effect.

1. Graphene-based QD

With a size range of 1–50 nm, graphene particles are stacked in a honeycomb shape to form GQD, a unique class of zero-dimensional semiconductor nanocrystals 19. GQD exhibits more advanced features compared to graphene due to the substantial change in physical size, including larger specific surface areas, more surface-active sites, and more accessible edges. Additionally, the functional groups included in GQD enable them to achieve superior solubility as well as stronger functionalization with organic, inorganic, or biological moieties for the development of electrochemical biosensors [20, 21]. Also, GQDs have a high photoluminescence quantum yield, which means that they emit strong fluorescent signals when excited by light. This property makes them useful for fluorescent imaging and detection applications. For example, Sun et al. developed an electrochemical immunosensor for the detection of heat shock protein 70(HSP70) by using polyaniline functionalized GQD. Under optimized conditions, the fabricated sensor exhibits a wide linear range of 0.0976-100ngmL^{-1} and LOD of 0.05ngmL^{-1} [22]. Potiphor's group fabricated an electrochemical biosensor based on gold nanoparticles/graphene quantum dots/graphene oxide film-modified three screen-printed electrodes (AuNPs/GQD/GO/SPCE) for the simultaneous detection of three different microRNAs (miRNA-21, miRNA-155, miRNA-210). Due to the synergistic effect of the modified electrode, the biosensor exhibits a wide linear range from 0.001pM-1000pM with LOD of 0.04fM, 0.33fm, and 0.28fM for miRNA-21, miRNA-155, miRNA-210 respectively [23].

2. Chalcogenide based QD

Chalcogenide QDs, which are the most important fluorescent probes in the VI group, are derived from metallic chalcogenides such as cadmium-selenium (CdSe), cadmium-titanium (CdTe), cadmium-sulfur (CdS), and zinc-sulfur (ZnS) 24. Due to their optical and electrochemical qualities, these QDs have attracted a lot of attention. With their superior water solubility, high stability, wide bandgap, and good fluorescence qualities, chalcogenide QDs have a variety of uses in in-vivo imaging, diagnostics, and optoelectronic devices with a 1–20 nm wavelength range [25]. Due to these outstanding properties, Chalcogenide-based QD are prospective candidates for future sensor devices. For example, Moazampour et al. used ZnS functionalized L-cysteine to design an electrochemical genosensor to detect miR-200a, an ovarian cancer biomarker. Under optimized conditions, the dynamic linear range and LOD of miR-200a were found to be 1.0fM–1.0μM and 8.4fM respectively [26].

3. MXene-based QD

MXenes are a family of two-dimensional (2D) transition metal carbides, nitrides, and carbonitrides with unique electronic, mechanical, and chemical properties. Regarding their 2D-QD, they exhibit high surface area, high biocompatibility, and excellent photoluminescence properties which makes them suitable for biosensing applications. For example, Jiang et al. and his colleagues reported a nitrogen-doped Ti3C2 MXene QD (N-Ti3C2 QD) based electro-chemiluminescent immunosensor for the detection of mucin 1(MUC1). In this work, N-Ti3C2 QD shows ECL emission peaks at around 612nm and higher quantum efficiency than Ti3C2 QD which can serve as an effective ECL emitter to fabricate ECL immunosensor [27]. Recently, a glucose electrochemical biosensor was constructed for the detection of glucose using PEDOT: PSS/Ti3C2/GQD modified SPCE. The synergistic effect and excellent electrical conductivity of the modified electrode enhance the electrochemical performance of the enzymatic biosensor, achieving a LOD of 65μM [28].

4. Doped QD

Doping technology has steadily been used to create unique QD with good functionalities and extraordinary features in order to enhance the photoluminescence capabilities of QD. In biosensing applications, doped QD can be used as fluorescent. Labels to detect and quantify biological analytes. For example, In 2020, Zang et al. and his colleagues synthesized boron-doped carbon QD(B-CQD) for the detection of amoxicillin antibiotics via fluorometry. The fabricated sensor exhibits a superior linear range and high stability with an LOD of 0.825μM [29].

4 Quantum Dots-Based Nanocomposites for Biosensing

Nanomaterials are conventional materials that have at least one dimension on the order of nanometres, while nanocomposites are combinations of two or more types of nanomaterials that result in a new integrated material system, still in the nano-range. Nanoparticles are a structural term of nanomaterials, describing particles with a size smaller than 100 nm, while QDs are nanoparticles with a size of less than 10 nm, and are characterized by the dominant quantum confinement effect [14]. These distinctions will be important to keep in mind when discussing the unique properties and potential applications of these materials in various fields.

Nanomaterials for biosensing possess distinct characteristics in electronics, optics, mechanics, and thermal properties, making them an emerging material for developing new analytical methods for many analyses. Nanoparticles like quantum dots, nanowires, nanotubes, nanorods, and nanofilms have attracted attention in recent years. These materials have excellent optical and electrical properties closely linked to their sizes [30]. Quantum dots, for instance, are nanoscale semiconductor materials, such as cadmium selenide (CdSe), InAs, CdS, GaN, InGeAS, CdTe, PbS, PbSe, and ZnS. The controllable size parameter combined with the effect of quantum restriction makes QD exhibit extraordinary optical and electrical properties 12. Consequently, QDs have become labeling materials for biosensors, optical probes for imaging, and biological detection. They offer high sensitivity and selectivity in detecting specific analytes and can significantly increase the density of electrochemical signals in detection systems [31]. QD-based sensors have proven to be highly suitable for developing highly accurate, rapid, and selective analytical instruments for detecting particular analytes [32].

Researchers are increasingly interested in the fields of electrochemistry and nanomaterials, as nanomaterials are considered ideal for creating sensors. QD has been used successfully for developing biosensors. We include the selected methods in Table 1 for identifying various targets with various quantum dots.QDs, in particular, exhibit extraordinary optical and electrical properties due to their controllable size parameter and the effect of quantum restriction [33]. Their luminescence color varies with alterations in their size, making them well-suited as fluorescent probes in medical diagnosis and imaging. However, the use of heavy metals such as CdSe, CdTe, and CdS in QD can lead to degradation and ion release, which is toxic under physiological conditions [34]. In the context of electrochemical sensors, there are both advantages and disadvantages associated with the use of quantum dots. The advantages include easy immobilization of bioreceptors, which enhances the possibility of the electroanalysis of multiple analytes, easy functionalization, and attractive electrochemical properties. On the other hand, the disadvantages include low stability, easy aggregation, and difficulty in controlling particle size [3, 4].

a. **Introduction to quantum dots-based nanocomposites**

This chapter discusses the use of QD in biosensors and highlights the importance of choosing the correct passivating agent to maintain their electronic properties and optical properties for real-life applications. Surface ligands such as metal–organic frameworks (MOFs), molecularly imprinted polymer (MIP), aptamers, multiwalled carbon nanotubes (MWCNTs), graphene quantum dots, and carbon QD can be used

Table 1 Selected QD-based electrochemical studies for different active compounds

QD type	Analyte	Linear range	Biomolecule	Ref.
Ni-doped CdTe	Pyrazinamide	2–100 μM	Plasma samples	[36]
CdTe	Dopamine	0.5–10 μM	Biological fluids	[68]
CdTe	Propafenone	0.003–7.0 μg mL $^{-1}$	Human serum	[69]
MoS2/CdTe	Tetracycline	0.1–1 μM	Milk samples	[70]
CdTe–Con A	Lipopolysaccharide and Serratia marcescens	10–90 fg/mL	–	[71]
CdSe	Urea	1–120 mM		[72]
CdTe	Salbutamol	6.27 × 10 − 8 to 2.09 × 10 − 7 M	Pig urine samples	[73]
Pd-doped CdTe	Diazinon	2.3–100 μM	Environmental water samples	[74]
CdS@MOF	Carcinoembryonic antigen	–	Human serum samples	[75]
CdTeS @SiO2	Folic acid	5–80 μM	Serum samples	[76]
α-FeOOH@CdS/Ag	17β-estradiol	0.01–10 pg. mL $^{-1}$		[77]
ZnCdS QD@MIP	Ascorbic acid	1–500 μM	Vitamin C tablets	[78]
MoS2/GQD	Caffeic acid	0.38–100 μM	Red wine samples	[79]
CdTe	Double-strand DNA	0.0874 μg mL $^{-1}$ and 20 μg mL $^{-1}$	Synthetic samples	[80]
Polymer CdTe/CdS	Glucose	0.2–5 mM	Human body fluids	[81]

to modify the surface of QD for various sensor applications [35]. The use of silica nanoparticles as encapsulating agents for QD, can increase the amount of QD on each target cell and improve the sensitivity and detection range of electrochemical immuno-biosensors. The use of transition metal oxides with variable valence states and doped transition metal ions such as Ni:CdTe is also explored as options for enhancing the sensitivity and selectivity of QD-based biosensors. Mesoporous silica nanoparticles are shown to provide higher loading sites and surface area for QD than solid silica [33].

QDs are being used as labeling materials for biosensors, which are devices that detect biological analytes such as proteins, DNA, and cells. QD are nanoscale particles that emit bright and stable fluorescent signals, making them an attractive component for optical probes used in cellular, tissue, or whole-body imaging and biological detection [33].

QD-based sensors offer high sensitivity and selectivity in detecting certain analytes in the chemical and biochemical sciences. QD has been integrated into electrochemical sensors to improve their selectivity and efficiency in detecting specific analytes. QDs can provide sharp and well-resolved voltammetry signals while also increasing the electrochemical signal density in the electrochemical detection system [36].

Sensors utilizing QDs are well suited for the development of highly selective, fast, and accurate tools for the detection of specific analytes in sensor technology. A review of the literature shows that QD-based sensors have been used to develop highly sensitive and specific biosensors for a wide range of applications, including medical diagnostics, environmental monitoring, and food safety. Figure 1 shows statistics on the number of articles published each year relating to nanocomposites based on QD for electrochemical sensing.

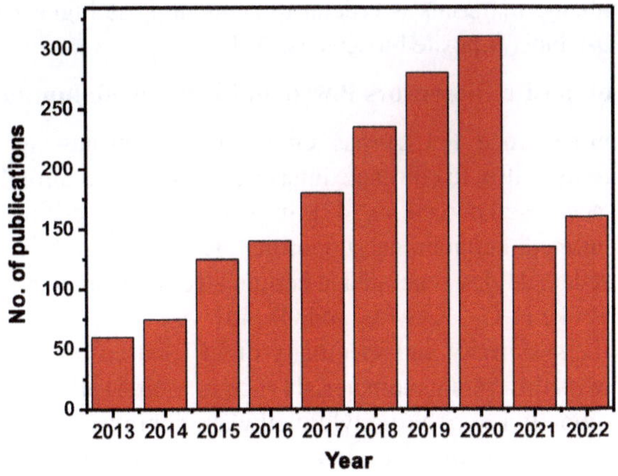

Fig. 1 Statistics of the number of publications per year related to quantum dot-based nanocomposites for electrochemical sensing

b. **Types of quantum dots-based nanocomposites for biosensing**

The popularity of QD as fluorescent biosensors is on the rise due to their robust photoluminescence, photostability, and sensitivity to changes in oxidative and acidic conditions. Additionally, QD can be modified to target specific biomolecules, allowing for the detection and measurement of various biomolecules in complex biological samples [37–39]. Moreover, QD has been integrated with other elements to boost their biosensing functionalities.

1. **QD-Nanocomposite Biosensors Based on Föster**

Föster resonance energy transfer (FRET) is a phenomenon where energy is transferred between a donor fluorophore and an acceptor fluorophore when they are in close proximity to each other. QDs are an ideal fluorescent donor due to their strong fluorescence, tunable narrow emission, and broad excitation wavelength range [40]. In a FRET system, when the donor emission matches the acceptor absorbance, energy is transferred from the excited donor to the acceptor, resulting in fluorescence emission of the acceptor fluorophore. FRET has been utilized in detecting dopamine by using a QD-dendrimer-AlexaFluor 488 labeled antibody, where the antigen–antibody reaction alters the distance between the donor and acceptor, leading to enhanced fluorescence emission [41]. DNA hybridization detection using molecular beacons is another well-known application of FRET. A nanocomposite consisting of QD-single-strand DNA and gold nanoparticles has been employed as a turn-on fluorescent sensor that activates to quantitatively measure target DNA. Quenchers such as carbon nanotubes, graphene, and graphene oxide have also been employed to fabricate nanomaterial-based biosensors. Dual-color sensors have been developed using QD-ruthenium complex nanocomposites for detection of double-strand DNA. The efficiency of FRET is influenced by two key factors which are the choice of the donor–acceptor pair and the distance between the donor and acceptor. Precisely designing the linkage materials is crucial to achieving the highest level of FRET efficiency in QD-nanocomposite biosensors [42].

2. **QD-Nanocomposite Biosensors Based on Electrochemiluminescence**

Electrochemiluminescence (ECL) has emerged as a promising technique for biosensing due to its ability to eliminate interference from excitation light sources in photoluminescence-based detections [43]. In ECL-based biosensing, QDs are electrically excited in media containing supporting electrolytes and ECL occurs when the oxidized or reduced QDs generated at certain electrochemical potentials interact with co-reactants like H_2O_2, $K_2S_2O_8$, and O_2 [44].

To enhance the sensitivity and specificity of ECL-based biosensors, QDs have been integrated with different nanomaterials such as multi-walled carbon nanotubes (CNTs), gold nanoparticles, poly(diallyldimethylammonium chloride) (PDDA), and graphene oxide to facilitate the electron injection process. For example, the combination of QD and CNTs has been shown to deliver a five-fold increase in ECL intensity and a positive shift of ECL starting voltage [45, 46]. Similarly, the use of CNT-QD-PDDA-gold nanoparticle composite films has been demonstrated as an immunosensor [47].

Interestingly, gold nanoparticles have been shown to both enhance and quench QD ECL through fluorescence resonance energy transfer (FRET), depending on the structure of the nanocomposite and the distance between QD and the gold nanoparticle. Gold nanoparticles can act as quenchers to eliminate QD fluorescence via FRET, but can also enhance ECL intensity when they are linked with QD through a hairpin-structured DNA molecule at close proximity [47].

Until now, the majority of the ECL biosensors utilizing QDs have been designed to identify biomolecules or bioreactions involving co-reactants actively involved in the ECL process. However, a universal ECL biosensor for the detection of various biomarkers using specially tailored QD nanocomposites still needs to be investigated.

3. QD-Nanocomposite Biosensors Based on Electrochemistry

The use of QD in nanocomposites for electrochemical sensors and photoelectrochemical solar cells has been demonstrated to have great potential for improving detection sensitivity and specificity. QDs are employed in electrochemical sensors as large-surface-area carriers to enhance enzyme loading and improve substrate detection in enzymatic reactions. Nanostructured composites incorporating QDs can also facilitate direct electron transfer between the redox centers of enzymes and the electrode, promoting direct electrochemistry, resulting in biosensors with good sensitivity and specificity [48, 49].

In photoelectrochemical solar cells, QDs are hybridized with other materials such as CNT, titanium dioxide (TiO2), and graphene to produce QD-based photoelectrochemical solar cells. When light irradiates the QD, it induces the dissociation of electrons and holes in the conduction and valence bands of QD, respectively. Based on the principle that electrons are injected into an electrode to produce a photocurrent when band energy is favorable and that holes are scavenged by suitable electrolytes to regenerate QD, QD-based photoelectrochemical sensors can be fabricated by measuring the change of photocurrent. This label-free approach is suitable for the detection of any antibody-antigen pair [50, 51].

As nanotechnology research continues to advance, the development and utilization of more energy-efficient and biocompatible QD nanocomposite are expected and used to enhance the sensitivity and specificity of diagnostic detection [49].

c. Functionalization of quantum dots-based nanocomposites for biosensing applications

QDs are semiconductor nanocrystals with unique optical and electronic properties, making them attractive for use in biosensing applications. However, to use QD for biosensing, they need to be functionalized with biomolecules such as antibodies, aptamers, or DNA strands that can specifically recognize and bind to target molecules.

Various methods, including covalent or non-covalent bonding, can be used to functionalize QD. Covalent bonding is the process by which biomolecules attach to the surface of QD by chemical reactions such as the formation of amide or thiol-ene bonds. On the other hand, non-covalent bonding involves the electrostatic interaction between positively charged QD and negatively charged biomolecules. The streptavidin–biotin system is a popular method for bioconjugation and one of the strongest noncovalent biological protein–ligand interactions known to date. For example, Rahul et al. synthesized poly-L-Lysine (PLL) functionalized graphene QD(GQD) for the detection of carcinoembryonic antigens (CEA). Prepared PLL-GQD showed improved fluorescence lifespan, high quantum yield, outstanding photostability, biocompatibility, and higher cellular uptake as compared to nonfunctionalized GQD [52]. Sopan et al. fabricate a biosensor for the detection of para-aminohippuric acid using poly-L-glutamic acid functionalized graphene QD(PLGA-fGQD) embedded on UiO-66-NH2 MOFs). In this work, GQD was synthesized from bovine serum albumin by using the hydrothermal method. The surface modification of GQD with PLGA not only enhances the optical properties of GQD but also results in high quantum yield and responsiveness [53]. Rahul et al. developed a fluorescent-based lung cancer biomarker using polyethyleneimine functionalized graphene quantum dots, enhancing the stability and fluorescence lifetime [54]. Aerzoo et al. synthesized a hybrid ink using silver nanoparticles and D-penicillamine functionalized graphene QD for the detection of CA 125 biomarkers in human plasma samples. The developed immunosensor exhibits a good performance with a linear range of 0.001-400U ml-1 and a limit of quantification of 0.001U mL^{-1} [55]. Mayank et al. fabricated a fluorescence-based immunosensor for ferritin detection using amine-functionalized graphene quantum dots. Under optimized conditions, the engineered fluorescent immunosensor shows a linear range of 10-4000ng mL^{-1} and LOD of 0.723ng mL^{-1} [56]. Chen et al. constructed a label-free surface plasmon resonance (SPR) aptasensor for the detection of N-gene SARS-CoV-2 by using niobium carbide MXene QD which results in high bio affinity towards aptamer and enhances SPR signal [57]. Yao et al. used sulfhydryl-functionalized CQD as a fluorescent probe to detect heavy metal ion Hg2 + in lake water and wastewater samples. Due to the production of nonfluorescent ground-state complexes and the electron transfer interaction between HS-CQD and Hg2+ , the resulting HS-CQD were able to detect mercury ions Hg2+ quickly and sensitively through fluorescence quenching. The fabricated sensor obtained a linear range from 0.45μM–2.1μM with LOD of 12nM [58].

In addition to biomolecule functionalization, QD can be incorporated into various nanocomposites for biosensing applications. For example, QD can be encapsulated within silica nanoparticles to improve their stability and biocompatibility. QD can also be incorporated into hydrogels, polymers, or liposomes to create nanocomposites with unique properties.

Overall, the functionalization of QD-based nanocomposites offers promising opportunities for the development of biosensors with high sensitivity and specificity, making them valuable tools for various applications.

d. **Applications of quantum dots-based nanocomposites in biosensing**

1. **In Pharmaceutical industry**

In 2018, Tang et al. developed an electrochemical sensor that used a glassy carbon electrode (GCE) with graphene quantum dots (GQDs) to measure hydroquinone (HQ) and catechol (CC). The electrochemical impedance spectra were used to describe this sensor's design through electrodeposition method. In the electrochemical measurement, the proposed GQD sensor showed excellent sensitivity, repeatability, and reliability with a detection limit of 0.08 μM in the range of 0.5 μM to 100 μM. In river water samples, simultaneous HQ and CC detection using a GQD/GC electrode was carried out with satisfactory recovery. In this study, the superior electrocatalytic and conductivity capabilities of the proposed sensor as well as its high precision, dependability, and repeatability in electrochemical measurement were used for HQ and CC [59].

Pasandideh-Nadamani and colleagues created a unique, incredibly sensitive, and extremely selective carbon paste electrode (CPE) in 2016 (33). Through the use of X-ray diffraction (XRD) and transmission electron microscopy (TEM) techniques, they were able to create CdS QD that were remarkably stable. Using an in-situ method and a thiosulfate precursor, CdS QD were produced. Without using any separation steps in the mixture, the electrochemical determination of p-aminophenol (PAP) and acetaminophen (Ac) was studied [60].

Algarra and colleagues have developed a modified glassy carbon electrode (GCE) with carbon QD(CQD) for the electrochemical measurement of dopamine and uric acid (34). The Hummers method was used to create CQD from graphite, and they were then examined using tools like the TEM microscope, XPS, Raman, solid-state NMR, and FTIR-ATR spectroscopies. The electrochemical analysis of both compounds revealed that the peak current in the CQD-GCE was significantly increased as compared to the bare glassy carbon electrode. The suggested sensor demonstrated good sensitivity using Linear Sweep Voltammetry (LSV). The lower limits of detection for uric acid and dopamine were determined to be 1.3 and 2.7 μM, respectively [60].

2. **In antibiotic detection**

Using magnetic nanoparticles (MNPs) and QD, Chen et al. created a suitable magnetic-QD (MNPs-SiO2-QD) material for the qualitative and quantitative detection of four antibiotics with a high percent recovery [61]. Meng et al. developed a novel technique that used fluorescent CdTe QD as a backdrop material to assess five quinolone antibiotic residues in food [62]. Li et al. developed a tetracycline (TC) detection method using GQD and Eu3+, and this system demonstrated outstanding potential for the detection of TCs in environmental and biological materials [63]. To identify tetracycline (TC) in serum samples, Zhang et al. created a novel fluorescent probe employing the hybrid QD/mesoporous silica/MIP (QD/MS/MIP) [64]. In order to identify enrofloxacin (ENR) in chicken muscle tissue, Chen et al. developed a competitive fluorescence-linked immunosorbent assay (c FLISA), which is appropriate for detecting veterinary medication residues [65]. To detect antibiotics, many QD types have been created and functionalized.

3. **QD and Nanocomposites in Cancer Detection and Therapy**

The use of molecular imaging with QD linked to biorecognition molecules is expected to play a significant role in cancer diagnosis, especially for the early detection of cancer lesions. QD can be used for cell tracking to study cell division and metastasis and can act as unique markers for tracking cancer cells during metastasis. Whole-body NIR optical imaging is a powerful technique that allows the observation of complex biological phenomena with minimum invasiveness [66]. QD emitting in the NIR region can be used as markers in sentinel lymph node mapping, a major procedure in cancer surgery, whereby the lymph node closest to the organ affected is monitored for the presence of roaming cancer cells. QD can also be conjugated with biomolecules such as peptides and antibodies to target tumors in vivo. Conjugation of QD with specific ligands enables the detection of cancer-related proteins and aids in diagnosis and imaging of breast cancer. The future of QD in cancer diagnosis and treatment lies in identifying metastatic cancer cells, quantitatively measuring the level of specific molecular targets, and providing biodynamic markers for target inhibition [67].

4. **Various biosensing application**

Pourghobadi et al. created TGA-capped CdTe QD to allow for the visual assessment of dopamine traces. As a result of the interaction between TGA-CdTe QD and dopamine, the fluorescence intensity of the samples was examined. A rise in dopamine concentration causes a sharp decline in fluorescence intensity. We created a simple, fluorescence-sensitive approach for the measurement of dopamine taking into account this tendency [68].

Hydrophilic QD can be employed as molecular probes to find minute levels of propafenone, claim Wang et al. Propafenone was able to mix with CdTe QD that had been treated with thioglycolic acid in the weak acid through electrostatic attraction and hydrogen bonding. The interaction of CdTe QD with propafenone has been investigated using spectral analysis, spectra, RRS, UV–vis absorption, and fluorescence. Propafenone concentrations of ppb (ng mL^{-1}) in serum samples can be found in less than 30 min using the RRS method [69].

Using Liang et al.'s CdTe and MoS2 QD, tetracycline was found in milk. First, under 365 nm excitation, MoS2 QD emits blue light at 433 nm, while CdTe QD emits yellow light at 573 nm. We created a fluorescent sensor with dual signals at 433 and 573 nm using MoS2 QD and CdTe QD. As tetracycline concentration is raised, MoS2/CdTe-based sensors exhibit a decrease in fluorescence intensity, and 573 nm is quenched more visibly than 433 nm [70].

As a successful new biosensor for detecting LPS at concentrations of fg/mL, TGA-capped conjugated CdTe QD was created. In the high-resolution transmission electron image of the CdTe QD, high crystalline lattice planes with an average size of 4–5 nm were visible. The adsorption of Con A onto CdTe QD surfaces was caused by the electrostatic attraction force between positively charged Con A species and negatively charged CdTe QD surfaces [70].

The fluorescent carbon nanoparticles known as carbon QD (CDs), which are less than 10 nm in size, have good biocompatibility, water solubility, and distinctive optoelectrical characteristics 15. Bioimaging, biochemical and chemical analyses, photocatalysis, and white light-emitting devices have all been effectively employed with them. The ECL behavior of water-soluble CDs has recently been described, and Zheng et al. have suggested the underlying mechanism. The usage of CDs as an ECL reagent may now have a hopeful future thanks to this research. Here, we explore the PL and ECL behaviors of Ru(bpy)32 + and CDs in solution for the first time, and we create CDs as nanoprobes for identifying Ru(bpy)32 + species 61.

Recent advances in quantum dots-based nanocomposites as biosensors

There have been several recent biosensing applications using QD-based nanocomposites. Here are some examples:

1. **Detection of cancer biomarkers**

QD-based nanocomposites have shown great potential in cancer biomarker detection due to their unique optical and electronic properties. To create QD-based nanocomposites for cancer biomarker detection, QDs are typically functionalized with a biomolecule such as an antibody or aptamer that specifically binds to the biomarker of interest. For example, Mehrnaz et al. developed an immunosensor for the detection of prostate cancer biomarkers in human serum using AuNP@MWCNTs-GQD. The surface of a glassy carbon electrode was coated with AuNPs@MWCNTs-GQD to act as a signal amplifier, and then an antibody against the prostate-specific antigen (PSA) was immobilized on the modified electrode. Under optimized conditions, the fabricated immunosensor determined the PSA concentration in a linear range of 1–10,000 $pgmL^{-1}$ and LOD of 0.48 $pgmL^{-1}$ [82]. Neeraj et al. synthesized GQD by electrochemically exfoliating waste dry batteries and fabricated an electrochemical sensor to detect CD44 cancer biomarkers. The fabricated biosensor shows high sensitivity with a LOD of 2.11fg/ml83. Qin et al. developed an electrochemiluminescence immunosensor for the detection of CA15-3, a breast cancer biosensor using AuNPs/CQD-PEI-GO composite. The developed immunosensor has high selectivity and reproducibility for the detection of CA15-384. Koushik et al. used single-walled carbon nanohorns and nitrogen-doped GQD nanocomposite for the detection of α-fetoprotein, a cancer biomarker. In this work, GCE was modified by N-GQD@SWCNHs and then anti-α-fetoprotein was immobilized. The incorporation of N-QD and SWCNHs enhances the conductivity and sensitivity of the fabricated immunosensor with a LOD of 0.25 $pgmL^{-1}$ 85 Overall, QD-based nanocomposites show great promise in cancer biomarker detection and could have significant impacts on early diagnosis and treatment of cancer.

2. **Detection of foodborne pathogens**

Foodborne infections continue to pose a serious threat to human health worldwide and cause significant economic losses. Foodborne pathogens come in a variety of forms, including bacteria, fungi, viruses, and parasites [86]. Additionally, E.coli, Salmonella enterica, Listeria monocytogenes, Campylobacter jejuni, Staphylococcus aureus, and Clostridium perfringens are the most prevalent foodborne pathogens. QD-based nanocomposites have shown promising results in detecting foodborne pathogens due to their unique optical and electronic properties. One advantage of using quantum dot-based nanocomposites for pathogen detection is their sensitivity as they can detect low levels of pathogens, even down to a single cell, which is important for ensuring the safety of food products. Additionally, they can provide rapid detection, which is crucial for preventing the spread of foodborne illness. For example, Wang et al. fabricates an electrochemical immunosensor for the detection of Staphylococcus aureus using Ag2S QD loaded with dendritic mesoporous silica nanospheres. With a detection limit of $2CFUml^{-1}$ and a wide linear range from $10CFUmL^{-1}$ to $108CFUmL^{-1}$, the fabricated immune biosensor shows high selectivity and stability in the electrochemical measurement [87]. In order to create the ECL system for smartphones, Li et al. used GQD/Ag nanoparticles (GQD/AgNPs). They successfully completed the real-time detection of E. coli, which could measure E. coli in the range of 10–107 $CFUmL^{-1}$ 88. With a low detection limit of 0.05 CFU/mL, GQD doped with nitrogen was combined with non-metallic, two-dimensional, hydrated defective tungsten oxide to create a photoelectrochemical aptamer biosensor for E. coli detection. The fabricated biosensor provides new opportunities to increase the efficiency of both disease prevention and diagnosis [89].

3. **Environmental monitoring**

QD are semiconductor nanoparticles that have unique optical and electronic properties, such as size-dependent fluorescence emission, high photostability, and tunable surface chemistry. These properties make QD ideal for environmental monitoring applications, as they can be used as sensors for detecting pollutants, heavy metals, pesticides, antibiotics, and other environmental contaminants. For example, Korram et al. develop a fluorescent sensor using cadmium telluride QD (CdTe QD) for the detection of various organophosphate pesticides including paraoxon, dichlorvos, malathion, and triazophos in apple and tomato juice [90]. By employing palladium ions as a dopant during the hydrothermal process, Najafi et al. reported Pd: CdTe QD in the presence of diazinon, and fluorescence intensity linearly decreases. Diazinon was discovered in environmental water using this unique nanoprobe at extremely high sensitivity and selectivity. Additionally, Pd: CdTe QD's excellent cytotoxicity assay demonstrates their enormous potential in biotechnology and medicine [91]. Roushani et al. [92] fabricate an aptasensor using AgNPs/GQD-N-S/AuNPs nanocomposite for the detection of streptomycin. The fabricated biosensor functions at a low concentration level of $pgmL^{-1}$ [92] Muthusankar et al. used N-CQD@Co_3O_4 /MWCNT blend nanocomposite as an electrode modifier for the simultaneous detection of anticancer drug flutamide and antibiotic nitrofurantoin. The fabricated electrochemical show a wide linear range of 0.05–590 μM and 0.05–1220 μM and LOD values were 0.0169 μM and 0.044 μM for flutamide and nitrofurantoin respectively [93]. Li et al. created an N-doped-CQD-GO modified GCE (NCQD-GO/GCE) sensor for the detection of Cd(II) and Pb(II). The fabricated sensor shows linear detection ranges of 100.357 nM–100.357 μM and 100 nM–50 μM for Cd(II) and Pb(II) respectively and corresponding LODs of 66.518 nm and 5.647 μM [94]. Pizarro et al. modified a GCE using GQD and nafion to create an eco-friendly sensor that can determine the concentrations of Cd(II) and Pb(II) in seafood [95].

Overall, these examples highlight the versatility and potential of QD-based nanocomposites for biosensing applications in various fields, including healthcare, food safety, and environmental monitoring.

Potential Future Developments in Quantum Dots-Based Nanocomposites for Biosensing

QD are nanoscale semiconductor particles that exhibit special optical and electrical characteristics, such as size-tunable fluorescence and high quantum yield. These characteristics have made QD a popular choice for biosensing applications. QD-based nanocomposites have been utilized in biosensing to detect a range of biomolecules with high sensitivity and specificity, including proteins, nucleic acids, and small molecules. Future advancements in QD-based nanocomposites for biosensing could include the following:

1. **Multiplexed detection**: Multiplexed detection involves the simultaneous detection of multiple analytes in a single sample. QD-based nanocomposites have the potential to be used for multiplexed biosensing, as they can be functionalized with different recognition molecules that bind specifically to different target analytes. Multiplexed detection using QD-based nanocomposites could enable the rapid and accurate diagnosis of complex diseases.
2. **High-throughput screening**: QD-based nanocomposites could be used for high-throughput screening of drugs and other compounds. By functionalizing QD with specific receptors or enzymes, it is possible to develop biosensors that can detect the activity of these molecules in a high-throughput manner.
3. **Imaging**: QD-based nanocomposites could be used for cellular and molecular imaging. QD can be conjugated with biomolecules, such as antibodies or peptides, to target specific cells or tissues. The fluorescence microscopy enables the visualization of specific cells or tissues.
4. **Portable biosensors**: QD-based nanocomposites have the potential to be used in portable biosensors for point-of-care testing. The small size of QD and their high sensitivity make them ideal for use in miniaturized biosensors that can be used in remote or resource-limited settings.
5. **Therapeutic applications**: QD-based nanocomposites could be used for therapeutic applications, such as targeted drug delivery. By functionalizing QD with specific targeting molecules, it is possible to deliver drugs specifically to the cells or tissues that need them, reducing side effects and improving efficacy.

Overall, QD-based nanocomposites hold great promise for the development of biosensors with improved sensitivity, specificity, and portability, as well as for therapeutic applications. Continued research in this area is likely to yield further exciting developments in the future.

5 Challenges and Future Prospects

(1) **Challenges and limitations of quantum dots-based nanocomposites as biosensors**

QD are nanocrystals that exhibit unique optical and electronic properties due to their size-dependent confinement effects. These properties make them attractive for use in biosensors, which are devices that, quantify biological analytes. However, there are several challenges and limitations associated with QD-based nanocomposites as biosensors.

1. **Biocompatibility**: One of the major challenges of using QD in biosensors is their potential toxicity to biological systems. Many QDs are made of heavy metals, such as cadmium, which can be toxic to cells. Therefore, it is important to ensure that QDs are coated with biocompatible materials to prevent them from interacting with biological systems in a harmful way.
2. **Stability**: QDs can be susceptible to degradation and aggregation, which can affect their optical properties and lead to inaccurate measurements. Therefore, it is important to ensure that QDs are stable in biological environments and that their optical properties are not affected by the presence of biological molecules.
3. **Sensitivity and selectivity**: QD-based biosensors must be highly sensitive and selective in order to detect low concentrations of biological analytes. However, QD can be prone to nonspecific binding, which can decrease their selectivity. Therefore, optimizing the surface chemistry of QD to enhance their selectivity and reduce nonspecific binding is important.
4. **Signal-to-noise-ratio**: QD-based biosensors must be able to detect signals in the presence of background noise, which can be challenging due to the small size of QD and the low concentration of analytes in biological samples.
5. **Cost**: QDs can be expensive to produce, which can limit their use in practical applications. Therefore, it is important to develop cost-effective methods for producing QD and incorporating them into biosensors.

In summary, while QD-based nanocomposites offer unique optical and electronic properties that make them attractive for biosensing applications, there are several challenges and limitations that must be addressed in order to ensure their effectiveness and safety in biological systems.

(2) **Future Prospects and potential impact of quantum dots-based Nanocomposites on Biosensing**

QD-based nanocomposites have the potential to revolutionize biosensing by enabling rapid and sensitive detection of biomolecules, pathogens, and other targets of interest. One of the key advantages of QD-based nanocomposites is their tunable emission spectrum, which allows for the detection of multiple targets simultaneously using a single probe. Additionally, QD exhibits high quantum yields and photostability, which enables long-term monitoring of biological processes and reduces the need for frequent sample preparation.

Another advantage of QD-based nanocomposites is their versatility in terms of surface functionalization. The surface of QD can be modified with a variety of biomolecules, such as antibodies, aptamers, and peptides, to selectively bind to target molecules. This enables the development of highly specific biosensors for a wide range of applications, including disease diagnosis, drug discovery, and environmental monitoring.

Furthermore, QD-based nanocomposites have the potential to enable real-time, in vivo monitoring of biological processes. QD can be incorporated into various biological systems, such as cells or tissues, and their unique optical properties can be used to monitor changes in the local environment, such as pH, temperature, and oxygen levels.

In conclusion, while there are still challenges to overcome, the development of QD-base biosensors holds great promise for a wide range of applications in healthcare, environmental monitoring, and beyond.

6 Conclusions

In conclusion, quantum dots-based nanocomposites have shown great potential as biosensors due to their unique optical and electronic properties. By combining QD with various materials such as metals, polymers, and biological molecules, it is possible to design highly sensitive and selective biosensors for the detection of various biomolecules, including proteins, DNA, and viruses. Moreover, the versatility of QD allows for the development of multiplexed biosensors capable of detecting multiple analytes simultaneously. The integration of QD into nanocomposites offers several advantages over conventional biosensors, including increased sensitivity, selectivity, and stability. Additionally, the use of nanocomposites allows for the development of biosensors with a wide range of optical and electronic properties, enabling the detection of biomolecules using a variety of sensing modalities. However, there are still challenges that need to be addressed to fully realize the potential of quantum dots-based nanocomposites as biosensors. These include issues related to toxicity, stability, and reproducibility. With continued research and development, these nanocomposites are likely to become increasingly important tools for the detection of biomolecules and the advancement of biomedical sciences.

References

1. Rajendiran, K., Zhao, Z., Pei, D.S., Fu, A.: Antimicrobial activity and mechanism of functionalized quantum dots. Polymers (Basel). **11**(10), (2019). https://doi.org/10.3390/polym11101670
2. Li, P., Liu, S., Cao, W., et al.: Low-toxicity carbon QDderived from gentamicin sulfate to combat antibiotic resistance and eradicate mature biofilms. Chem. Commun. **56**(15), 2316–2319 (2020). https://doi.org/10.1039/C9CC09223D

3. Xu, J., Zheng, J.: QDand Nanoclusters. Published online, Nano-inspired biosensors for protein assay with clinical applications (2019)
4. Zhao, M.X., Zhu, B.J.: The research and applications of QDas nano-carriers for targeted drug delivery and cancer therapy. Nanoscale Res. Lett. **11**(1), 207 (2016). https://doi.org/10.1186/s11671-016-1394-9
5. Amani-Ghadim, A.R., Arefi-Oskoui, S., Mahmoudi, R., et al.: Improving photocatalytic activity of the ZnS QD via lanthanide doping and photosensitizing with GO and g-C3N4 for degradation of an azo dye and bisphenol-A under visible light irradiation. Chemosphere **295**, 133917 (2022). https://doi.org/10.1016/j.chemosphere.2022.133917
6. Nann, T., Skinner, W.M.: QDfor electro-optic devices. ACS Nano **5**(7), 5291–5295 (2011). https://doi.org/10.1021/nn2022974
7. Roy, S., Tuinenga, C., Fungura, F., Dagtepe, P., Chikan, V., Jasinski, J.: Progress toward producing n-Type CdSe Quantum Dots: Tin and Indium Doped CdSe Quantum Dots. J. Phys. Chem. C **113**(30), 13008–13015 (2009). https://doi.org/10.1021/jp8113946
8. Fatima, B., Hussain, D., Bashir, S., et al.: Catalase immobilized antimonene QDused as an electrochemical biosensor for quantitative determination of H2O2 from CA-125 diagnosed ovarian cancer samples. Mater. Sci. Eng. C **117**, 111296 (2020). https://doi.org/10.1016/j.msec.2020.111296
9. Hamed, A.A., Saad, G.R., Abdelhamid, I.A., Elwahy, A.H.M., Abdel-Aziz, M.M., Elsabee, M.Z.: Chitosan Schiff bases-based polyelectrolyte complexes with graphene QDand their prospective biomedical applications. Int. J. Biol. Macromol. **208**, 1029–1045 (2022). https://doi.org/10.1016/j.ijbiomac.2022.03.199
10. Tufani, A., Qureshi, A., Niazi, J.H.: Iron oxide nanoparticles based magnetic luminescent QD(MQD) synthesis and biomedical/biological applications: A review. Mater. Sci. Eng. C **118**, 111545 (2021). https://doi.org/10.1016/j.msec.2020.111545
11. Senthil Kumar, P., Gunasundari, E.: Nanocomposites: recent trends and engineering applications. Nano Hybrids and Composites. **20**, 65–80 (2018). https://doi.org/10.4028/www.scientific.net/NHC.20.65
12. Bailey, E.J., Winey, K.I.: Dynamics of polymer segments, polymer chains, and nanoparticles in polymer nanocomposite melts: A review. Prog. Polym. Sci. **105**, 101242 (2020). https://doi.org/10.1016/j.progpolymsci.2020.101242
13. Meng, Z., Stolz, R., Mendecki, L., Mirica, K.: Electrically-transduced chemical sensors based on two-dimensional nanomaterials. Chem Rev. **119**, (2019). https://doi.org/10.1021/acs.chemrev.8b00311
14. Ahmad, R., Mahmoodi, T., Ahn, M.S., Hahn, Y.B.: Recent advances in nanowires-based field-effect transistors for biological sensor applications. Biosens Bioelectron. **100**, (2017). https://doi.org/10.1016/j.bios.2017.09.024
15. Fiorani, A., Merino, J.P., Zanut, A., et al.: Advanced carbon nanomaterials for electrochemiluminescent biosensor applications. Curr. Opin. Electrochem. **16**, 66–74 (2019). https://doi.org/10.1016/j.coelec.2019.04.018
16. Stephanie, R., Kim, M.W., Kim, S.H., Kim, J.K., Park, C.Y., Park, T.J.: Recent advances of bimetallic nanomaterials and its nanocomposites for biosensing applications. TrAC, Trends Anal. Chem. **135**, 116159 (2021). https://doi.org/10.1016/j.trac.2020.116159
17. Malhotra, B.D., Ali, M.A.: Nanomaterials in biosensors: fundamentals and applications. Nanomaterials for Biosensors. Published online 2018, pp. 1–74. https://doi.org/10.1016/B978-0-323-44923-6.00001-7
18. Mansur, H.S.: QDand nanocomposites. Wiley Interdiscip Rev Nanomed Nanobiotechnol. **2**(2), 113–129 (2010). https://doi.org/10.1002/wnan.78
19. Suginta, W., Khunkaewla, P., Schulte, A.: Electrochemical biosensor applications of polysaccharides chitin and chitosan. Chem. Rev. **113**(7), 5458–5479 (2013). https://doi.org/10.1021/cr300325r
20. Faridbod, F., Sanati, A.L.: Graphene QDin electrochemical sensors/biosensors. Curr. Anal. Chem. **15**(2), 103–123 (2018). https://doi.org/10.2174/1573411014666180319145506

21. Zhou, X., Zhang, Y., Wang, C., et al.: Photo-Fenton reaction of graphene oxide: A new strategy to prepare graphene QDfor DNA cleavage. ACS Nano **6**(8), 6592–6599 (2012). https://doi.org/10.1021/nn301629v
22. Sun, B., Wang, Y., Li, D., et al.: Development of a sensitive electrochemical immunosensor using polyaniline functionalized graphene QDfor detecting a depression marker. Mater. Sci. Eng. C. **111**, (2020). https://doi.org/10.1016/j.msec.2020.110797
23. Serhan, M., Sprowls, M., Jackemeyer, D., et al.: Total iron measurement in human serum with a smartphone. In: AIChE annual meeting, conference proceedings. vol 2019-November. American Institute of Chemical Engineers, (2019). https://doi.org/10.1039/x0xx00000x
24. Xu, X., Yang, Y., Jin, H., et al.: Fungal in situ assembly gives novel properties to CdSxSe1-x QDfor sensitive label-free detection of chloramphenicol. ACS Sustain Chem Eng. **8**(17), 6806–6814 (2020). https://doi.org/10.1021/acssuschemeng.0c01698
25. Sur, V.P., Kominkova, M., Buchtova, Z., Dolezelikova, K., Zitka, O., Moulick, A.: Cdse qd biosynthesis in yeast using tryptone-enriched media and their conjugation with a peptide hecate for bacterial detection and killing. Nanomaterials. **9**(10), (2019). https://doi.org/10.3390/nano9101463
26. Moazampour, M., Zare, H.R., Shekari, Z.: Femtomolar determination of an ovarian cancer biomarker (miR-200a) in blood plasma using a label free electrochemical biosensor based on l-cysteine functionalized ZnS quantum dots. Anal. Methods **13**(17), 2021–2029 (2021). https://doi.org/10.1039/d1ay00330e
27. Jiang, X., Wang, H., Shen, Y., Hu, N., Shi, W.: Nitrogen-doped Ti3C2 MXene QDas novel high-efficiency electrochemiluminescent emitters for sensitive mucin 1 detection. Sens Actuators B Chem. **350**, (2022). https://doi.org/10.1016/j.snb.2021.130891
28. Nashruddin, S.N.A., Abdullah, J., Mohammad Haniff, M.A.S., Mat Zaid, M.H., Choon, O.P., Mohd Razip Wee, M.F.: Label free glucose electrochemical biosensor based on poly(3,4-ethylenedioxy thiophene):Polystyrene sulfonate/titanium carbide/graphene quantum dots. Biosensors (Basel). **11**(8), (2021). https://doi.org/10.3390/bios11080267
29. Zhang, X., Ren, Y., Ji, Z., Fan, J.: Sensitive detection of amoxicillin in aqueous solution with novel fluorescent probes containing boron-doped carbon quantum dots. J Mol Liq. **311**, (2020). https://doi.org/10.1016/j.molliq.2020.113278
30. Fenzl, C., Hirsch, T., Baeumner, A.J.: Nanomaterials as versatile tools for signal amplification in (bio)analytical applications. TrAC, Trends Anal. Chem. **79**, 306–316 (2016). https://doi.org/10.1016/j.trac.2015.10.018
31. Wang, F., Hu, S.: Electrochemical sensors based on metal and semiconductor nanoparticles. Microchim. Acta **165**(1), 1–22 (2009). https://doi.org/10.1007/s00604-009-0136-4
32. Chomoucka, J., Drbohlavova, J., Masarik, M., et al.: Nanotechnologies for society. New designs and applications of nanosensors and nanobiosensors in medicine and environmental analysis. Int J Nanotechnol. **9**(8–9), 746–783 (2012). https://doi.org/10.1504/IJNT.2012.046752
33. Drbohlavova, J., Adam, V., Kizek, R., Hubalek, J.: QD-characterization, preparation and usage in biological systems. Int. J. Mol. Sci. **10**(2), 656–673 (2009). https://doi.org/10.3390/ijms10020656
34. Campuzano, S., Yáñez-Sedeño, P., Pingarrón, J.M.: Carbon dots and graphene QDin electrochemical biosensing. Nanomaterials. **9**(4), (2019). https://doi.org/10.3390/nano9040634
35. Wang, Y., Xu, K., Fan, L., Jiang, Y., Yue, Y., Jia, H.: B-Doped g-C3N4/Black TiO_2 Z-scheme nanocomposites for enhanced visible-light-driven photocatalytic performance. Nanomaterials. **13**(3). https://doi.org/10.3390/nano13030518
36. Safari, M.: recent advances in quantum dots-based biosensors. In: Thirumalai J (ed.) Quantum dots. IntechOpen, (2022). https://doi.org/10.5772/intechopen.108205
37. Li, X., Zhou, Y., Zheng, Z., et al.: Glucose biosensor based on nanocomposite Films of CdTe QDand glucose oxidase. Langmuir **25**, 6580–6586 (2009). https://doi.org/10.1021/la900066z
38. Wu, P., He, Y., Wang, H.F., Yan, X.P.: Conjugation of Glucose Oxidase onto Mn-Doped ZnS QDfor Phosphorescent Sensing of Glucose in Biological Fluids. Anal. Chem. **82**(4), 1427–1433 (2010). https://doi.org/10.1021/ac902531g

39. Soo Choi, H., Liu, W., Misra, P., et al.: Renal clearance of quantum dots. Nat. Biotechnol. **25**(10), 1165–1170 (2007). https://doi.org/10.1038/nbt1340
40. Sekar, R.B., Periasamy, A.: Fluorescence resonance energy transfer (FRET) microscopy imaging of live cell protein localizations. J. Cell Biol. **160**(5), 629–633 (2003). https://doi.org/10.1083/jcb.200210140
41. Dong, H., Gao, W., Yan, F., Ji, H., Ju, H.: Fluorescence resonance energy transfer between QDand graphene oxide for sensing biomolecules. Anal. Chem. **82**(13), 5511–5517 (2010). https://doi.org/10.1021/ac100852z
42. Lee, K., Kang, I.J.: Effects of dopamine concentration on energy transfer between dendrimer-QD and dye-labeled antibody. Ultramicroscopy **109**, 894–898 (2009). https://doi.org/10.1016/j.ultramic.2009.03.012
43. Forster, R.J., Bertoncello, P., Keyes, T.E.: Electrogenerated chemiluminescence. Annu Rev Anal Chem (Palo Alto Calif). **2**, 359–385 (2009). https://doi.org/10.1146/annurev-anchem-060908-155305
44. Myung, N., Ding, Z., Bard, A.: Electrogenerated chemiluminescence of CdSe nanocrystals. Nano Lett. **2**, (2022). https://doi.org/10.1021/nl0257824
45. Jie, G.F., Liu, P., Zhang, S.S.: Highly enhanced electrochemiluminescence of novel gold/silica/CdSe-CdS nanostructures for ultrasensitive immunoassay of protein tumor marker. Chem. Commun. **46**(8), 1323–1325 (2010). https://doi.org/10.1039/B919186K
46. Ding, S.N., Xu, J.J., Chen, H.Y.: Enhanced solid-state electrochemiluminescence of CdS nanocrystals composited with carbon nanotubes in H2O2 solution. Chem. Commun. **34**, 3631–3633 (2006). https://doi.org/10.1039/B606073K
47. Jie, G., Li, L., Chen, C., Xuan, J., Zhu, J.J.: Enhanced electrochemiluminescence of CdSe QDcomposited with CNTs and PDDA for sensitive immunoassay. Biosens. Bioelectron. **24**, 3352–3358 (2009). https://doi.org/10.1016/j.bios.2009.04.039
48. Guo, C.X., Sheng, Z.M., Shen, Y.Q., Dong, Z.L., Li, C.M.: Thin-Walled graphitic nanocages as a unique platform for amperometric glucose biosensor. ACS Appl. Mater. Interfaces **2**(9), 2481–2484 (2010). https://doi.org/10.1021/am100472j
49. Guo, C.X., Zheng, X.T., Lu, Z.S., Lou, X.W., Li, C.M.: Biointerface by cell growth on layered graphene-artificial peroxidase-protein nanostructure for in situ quantitative molecular detection. Adv. Mater. **22**(45), 5164–5167 (2010). https://doi.org/10.1002/adma.201001699
50. Bao, S.J., Li, C.M., Zang, J.F., Cui, X.Q., Qiao, Y., Guo, J.: New Nanostructured TiO2 for direct electrochemistry and glucose sensor applications. Adv. Funct. Mater. **18**(4), 591–599 (2008). https://doi.org/10.1002/adfm.200700728
51. Guo, C.X., Li, C.M.: Direct electron transfer of glucose oxidase and biosensing of glucose on hollow sphere-nanostructured conducting polymer/metal oxide composite. Phys. Chem. Chem. Phys. **12**(38), 12153–12159 (2010). https://doi.org/10.1039/C0CP00378F
52. Fabrication of Poly-l-lysine-functionalized graphene QDfor the label-free fluorescent-based detection of carcinoembryonic antigen
53. Nangare, S., Patil, S., Patil, A., Deshmukh, P., Patil, P.: Bovine serum albumin-derived poly-L-glutamic acid-functionalized graphene QDembedded UiO-66-NH2 MOFs as a fluorescence 'On-Off-On' magic gate for para-aminohippuric acid sensing. J Photochem Photobiol A Chem. **438**, (2023). https://doi.org/10.1016/j.jphotochem.2022.114532
54. Tade, R.S., Patil, P.O.: Biofabricated functionalized graphene QD(fGQD): unraveling its fluorescence sensing mechanism of human telomerase reverse transcriptase (hTERT) antigen and in vitro bioimaging application. Biomedical Materials (Bristol). **17**(5), (2022). https://doi.org/10.1088/1748-605X/ac84ba
55. Saadati, A., Hassanpour, S., Bahavarnia, F., Hasanzadeh, M.: A novel biosensor for the monitoring of ovarian cancer tumor protein CA 125 in untreated human plasma samples using a novel nano-ink: A new platform for efficient diagnosis of cancer using paper based microfluidic technology. Anal. Methods **12**(12), 1639–1649 (2020). https://doi.org/10.1039/d0ay00299b
56. Garg, M., Vishwakarma, N., Sharma, A.L., Singh, S.: Amine-Functionalized graphene QDfor fluorescence-based immunosensing of ferritin. ACS Appl Nano Mater. **4**(7), 7416–7425 (2021). https://doi.org/10.1021/acsanm.1c01398

57. Chen, R., Kan, L., Duan, F., et al.: Surface plasmon resonance aptasensor based on niobium carbide MXene QDfor nucleocapsid of SARS-CoV-2 detection. https://doi.org/10.1007/s00604-021-04974-z/Published
58. Yao, W., Hua, Y., Yan, Z., Wu, C., Zhou, F., Liu, Y.: Sulfhydryl functionalized carbon QDas a turn-off fluorescent probe for sensitive detection of Hg2+. RSC Adv. **11**(57), 36310–36318 (2021). https://doi.org/10.1039/d1ra06527k
59. Jing, T.: Simultaneous determination of hydroquinone and catechol using carbon glass electrode modified with graphene quantum dots. Int. J. Electrochem. Sci. **13**, 11250–11262 (2018). https://doi.org/10.20964/2018.11.52
60. Pasandideh-Nadamani, M., Omrani, A., Sadeghi-Maleki, M.R., Samadi-Maybodi, A.: Application of CdS QDmodified carbon paste electrode for monitoring the process of acetaminophen preparation. Anal. Biochem. **502**, 36–42 (2016). https://doi.org/10.1016/j.ab.2016.02.015
61. Chen, Y., Cao, Y., Ma, C., Zhu, J.J.: Carbon-based dots for electrochemiluminescence sensing. Mater Chem Front. **4**, (2019). https://doi.org/10.1039/C9QM00572B
62. Meng, H.L., Chen, G.H., Guo, X., Chen, P., Cai, Q.H., Tian, Y.F.: Determination of five quinolone antibiotic residues in foods by micellar electrokinetic capillary chromatography with quantum dot indirect laser-induced fluorescence. Anal. Bioanal. Chem. **406**(13), 3201–3208 (2014). https://doi.org/10.1007/s00216-014-7730-y
63. Li, W., Zhu, J., Xie, G., Ren, Y., Zheng, Y.Q.: Ratiometric system based on graphene QDand Eu3+ for selective detection of tetracyclines. Anal. Chim. Acta **1022**, 131–137 (2018). https://doi.org/10.1016/j.aca.2018.03.018
64. Zhang, L., Chen, L.: Fluorescence probe based on hybrid mesoporous silica/quantum dot/molecularly imprinted polymer for detection of tetracycline. ACS Appl. Mater. Interfaces **8**(25), 16248–16256 (2016). https://doi.org/10.1021/acsami.6b04381
65. Chen, J., Xu, F., Jiang, H., et al.: A novel quantum dot-based fluoroimmunoassay method for detection of Enrofloxacin residue in chicken muscle tissue. Food Chem. **113**(4), 1197–1201 (2009). https://doi.org/10.1016/j.foodchem.2008.08.006
66. Devi, S., Kumar, M., Tiwari, A., et al.: Quantum dots: An emerging approach for cancer therapy. Front Mater. **8**, (2022). https://doi.org/10.3389/fmats.2021.798440
67. Das, P., Maruthapandi, M., Saravanan, A., et al.: Carbon dots for heavy-metal sensing, pH-sensitive cargo delivery, and antibacterial applications. ACS Appl Nano Mater. **3**(12), 11777–11790 (2020). https://doi.org/10.1021/acsanm.0c02305
68. Pourghobadi, Z., Mirahmadpour, P., Zare, H.: Fluorescent biosensor for the selective determination of dopamine by TGA-capped CdTe QDin human plasma samples. Opt Mater (Amst). **84**, 757–762 (2018). https://doi.org/10.1016/j.optmat.2018.08.003
69. Wang, X., Liu, Z., Gao, P., Li, Y., Qu, X.: Surface functionalized QDas biosensor for highly selective and sensitive detection of ppb level of propafenone. Spectrochim Acta A Mol Biomol Spectrosc. **227**, 117709 (2020). https://doi.org/10.1016/j.saa.2019.117709
70. Liang, N., Hu, X., Li, W., et al.: A dual-signal fluorescent sensor based on MoS2 and CdTe QDfor tetracycline detection in milk. Food Chem. **378**, 132076 (2022). https://doi.org/10.1016/j.foodchem.2022.132076
71. Ebrahim, S., Reda, M., Hussien, A., Zayed, D.: CdTe QDas a novel biosensor for Serratia marcescens and Lipopolysaccharide. Spectrochim Acta A Mol Biomol Spectrosc. **150**, 212–219 (2015). https://doi.org/10.1016/j.saa.2015.05.042
72. Sajwan, R., Sharma, P., Lakshmi, G., Solanki, P.: Mercaptopropionic acid capped cadmium selenide QDbased urea sensor. Mater Lett. **305**, 130794 (2021). https://doi.org/10.1016/j.matlet.2021.130794
73. Zhang, J., Wei, Y., Qiu, S., Xiong, Y.: A highly selective and simple fluorescent probe for salbutamol detection based on thioglycolic acid-capped CdTe quantum dots. Spectrochim Acta A Mol Biomol Spectrosc. **247**, 119107 (2021). https://doi.org/10.1016/j.saa.2020.119107
74. Najafi, S., Safari, M., Amani, S., Mansouri, K., Shahlaei, M.: Preparation, characterization and cell cytotoxicity of Pd-doped CdTe QDand its application as a sensitive fluorescent nanoprobe. J. Mater. Sci.: Mater. Electronics. **30**, (2019). https://doi.org/10.1007/s10854-019-01792-1

75. Wei, X., Hua, Qiao, X., Fan, J., et al.: A label-free ECL aptasensor for sensitive detection of carcinoembryonic antigen based on CdS QD@MOF and TEOA@Au as bi-coreactants of Ru(bpy)32+. Microchem. J. **173**, 106910 (2021). https://doi.org/10.1016/j.microc.2021.106910
76. Yang, M., Wang, C., Yan, Y., et al.: Visual detection of folic acid based on silica coated CdTeS QDin serum samples. Mater. Res. Bull. **144**, 111509 (2021). https://doi.org/10.1016/j.materresbull.2021.111509
77. Liu, Y., Li, B., Yao, Y., et al.: An electrochemiluminescence sensor for 17β-estradiol detection based on resonance energy transfer in α-FeOOH@CdS/Ag NCs. Talanta **221**, 121479 (2021). https://doi.org/10.1016/j.talanta.2020.121479
78. Yang, M., Wang, C., Liu, E., Hu, X., Hao, H., Fan, J.: A novel ascorbic acid ratiometric fluorescent sensor based on ZnCdS QDembedded molecularly imprinted polymer and silica-coated CdTeS quantum dots. J. Mol. Liq. **337**, 116438 (2021). https://doi.org/10.1016/j.molliq.2021.116438
79. Vasilescu, I., Eremia, S.A. V., Kusko, M., Radoi, A., Vasile, E., Radu, G.L.: Molybdenum disulphide and graphene QDas electrode modifiers for laccase biosensor. Biosens Bioelectron. **75**, 232–237 (2016). https://doi.org/10.1016/j.bios.2015.08.051
80. Jamei, H.R., Rezaei, B., Ensafi, A.A.: Ultra-sensitive and selective electrochemical biosensor with aptamer recognition surface based on polymer QDand C(60)/MWCNTs—polyethylenimine nanocomposites for analysis of thrombin protein. Bioelectrochemistry **138**, 107701 (2021). https://doi.org/10.1016/j.bioelechem.2020.107701
81. Yu, M., Zhao, K., Zhu, X., et al.: Development of near-infrared ratiometric fluorescent probe based on cationic conjugated polymer and CdTe/CdS QD for label-free determination of glucose in human body fluids. Biosens. Bioelectron. **95**, 41–47 (2017). https://doi.org/10.1016/j.bios.2017.03.065
82. Ghanavati, M., Tadayon, F., Bagheri, H.: A novel label-free impedimetric immunosensor for sensitive detection of prostate specific antigen using Au nanoparticles/MWCNTs- graphene QDnanocomposite. Microchem. J. **159**, (2020). https://doi.org/10.1016/j.microc.2020.105301
83. Kumar, N., Yadav, S., Sadique, M.A., Khan, R.: Electrochemically exfoliated graphene QDBased biosensor for CD44 breast cancer biomarker. Biosensors (Basel). **12**(11), (2022). https://doi.org/10.3390/bios12110966
84. Qin, D., Jiang, X., Mo, G., Feng, J., Yu, C., Deng, B.: A Novel carbon QDsignal amplification strategy coupled with sandwich electrochemiluminescence immunosensor for the detection of ca15-3 in human serum. ACS Sens. **4**(2), 504–512 (2019). https://doi.org/10.1021/acssensors.8b01607
85. Dutta, K., De, S., Das, B., et al.: Development of an efficient immunosensing platform by exploring single-walled carbon nanohorns (SWCNHs) and nitrogen doped graphene quantum dot (N-GQD) nanocomposite for early detection of cancer biomarker. ACS Biomater. Sci. Eng. **7**(12), 5541–5554 (2021). https://doi.org/10.1021/acsbiomaterials.1c00753
86. Mei, Y., He, C., Zeng, W., et al.: Electrochemical biosensors for foodborne pathogens detection based on carbon nanomaterials: recent advances and challenges. Food Bioproc Tech. **15**(3), 498–513 (2022). https://doi.org/10.1007/s11947-022-02759-7
87. Wang, H., Yang, L., Tan, Y., Deng, W., Xie, Q.: Ag2S QDloaded dendritic mesoporous silica nanospheres as signal amplification labels for ultrasensitive electrochemical immuno-biosensor for Staphylococcus aureus. J. Electroanal. Chem. **919**, (2022). https://doi.org/10.1016/j.jelechem.2022.116496
88. Li, S., Liu, J., Chen, Z., et al.: Electrogenerated chemiluminescence on smartphone with graphene QDnanocomposites for Escherichia Coli detection. Sens Actuators B Chem. 297, (2019). https://doi.org/10.1016/j.snb.2019.126811
89. Jiang, D., Yang, C., Fan, Y., et al.: Ultra-sensitive photoelectrochemical aptamer biosensor for detecting E. coli O157:H7 based on nonmetallic plasmonic two-dimensional hydrated defective tungsten oxide nanosheets coupling with nitrogen-doped graphene QD(dWO3•H2O@N-GQD). Biosens Bioelectron. **183**, (2021). https://doi.org/10.1016/j.bios.2021.113214

90. Korram, J., Dewangan, L., Karbhal, I., et al.: CdTe QD-based inhibition and reactivation assay of acetylcholinesterase for the detection of organophosphorus pesticides. RSC Adv. **10**(41), 24190–24202 (2020). https://doi.org/10.1039/d0ra03055d
91. Najafi, S., Safari, M., Amani, S., Mansouri, K., Shahlaei, M.: Preparation, characterization and cell cytotoxicity of Pd-doped CdTe QDand its application as a sensitive fluorescent nanoprobe. J. Mater. Sci. Mater. Electron. **30**(15), 14233–14242 (2019). https://doi.org/10.1007/s10854-019-01792-1
92. Roushani, M., Ghanbari, K., Jafar, H.S.: Designing an electrochemical aptasensor based on immobilization of the aptamer onto nanocomposite for detection of the streptomycin antibiotic. Microchem. J. **141**, 96–103 (2018). https://doi.org/10.1016/j.microc.2018.05.016
93. Muthusankar, G., Devi, R.K., Gopu, G.: Nitrogen-doped carbon QDembedded Co3O4 with multiwall carbon nanotubes: An efficient probe for the simultaneous determination of anti-cancer and antibiotic drugs. Biosens. Bioelectron. **150**, 111947 (2020). https://doi.org/10.1016/j.bios.2019.111947
94. Li, L., Liu, D., Shi, A., You, T.: Simultaneous stripping determination of cadmium and lead ions based on the N-doped carbon quantum dots-graphene oxide hybrid. Sens Actuators B Chem. **255**, 1762–1770 (2018). https://doi.org/10.1016/j.snb.2017.08.190
95. Pizarro, J., Segura, R., Tapia, D., Navarro, F., Fuenzalida, F., Jesús Aguirre, M.: Inexpensive and green electrochemical sensor for the determination of Cd(II) and Pb(II) by square wave anodic stripping voltammetry in bivalve mollusks. Food Chem. **321**, (2020). https://doi.org/10.1016/j.foodchem.2020.126682

Quantum Dot-Based Nanocomposite as Metal Sensors

Bambesiwe M. May, Nkosingiphile Zikhalala, Olayemi J. Fakayode, Mokae F. Bambo, Andile Mkhohlakali, and Deogratius T. Maiga

Abstract The quest for searching new functional materials for efficient heavy metal sensors has motivated the development of productive hybrid materials in which the strengths and weaknesses of various materials are considered and the complimentary and synergetic functions of the individual materials are utilized for achieving better material performance. Quantum dots (QDs) are zero-dimensional nanomaterials with inherent optical and optoelectronic properties and thus are highly employed for heavy metal sensing. Recent advances in QD research have produced QDs-based nanocomposites by doping QDs with metals, conjugating QDs to molecules, polymers, magnetic materials, and biomolecules to improve hydrophilicity, and stability and enhance their sensing capabilities. This chapter describes the application of various QD-based nanocomposites as heavy metal sensors. The importance and involvement of materials such as metal-doped QD, QD-based polymeric, QD-based

B. M. May (✉) · O. J. Fakayode
Institute for Nanotechnology and Water Sustainability (iNanoWS), College of Science, Engineering, and Technology, University of South Africa, Florida Campus, 28 Pioneer Avenue, Roodepoort, Johannesburg 1709, South Africa
e-mail: bambesiwem@mintek.com

B. M. May · M. F. Bambo
DSI/Mintek Nanotechnology Innovation Centre, Advanced Materials Division, Mintek, 200 Malibongwe Drive, Randburg, Johannesburg, South Africa

N. Zikhalala
UNESCO-UNISA Africa Chair in Nanosciences & Nanotechnology Laboratory, College of Graduate Studies, University of South Africa, Muckleneuk Ridge, PO BOX 392, Pretoria 0002, South Africa

Nanosciences African Network-Material Research Department, iThemba LABS-National Research Foundation, Somerset West, PO BOX 722, Cape Town 7129, South Africa

A. Mkhohlakali
Analytical Chemistry Division, Mintek, 200 Malibongwe Drive, Randburg, Johannesburg, South Africa

D. T. Maiga
Measurement and Control Division, Mintek, 200 Malibongwe Drive, Randburg, Johannesburg, South Africa

S. Thomas et al. (eds.), *Quantum Dots Based Nanocomposites*, Engineering Materials,
https://doi.org/10.1007/978-3-031-54779-9_13

magnetic, QD-based metal–organic frameworks (MOFs), and QD-based bioconjugates are well described. The sensing mechanisms involved in the detection as well as challenges coupled with future projecions are also highlighted.

Keywords Quantum dots · Nanocomposites · Fluorescence · Metals · Sensing

List of Abbreviations

QD	Quantum dots
IFE	Inner filter effect
FRET	Fluorescence resonance energy transfer
PET	Photo-induced electron transfer
ACQ	Aggregation induced quenching
ACQ	Aggregation induced quenching
AIEE	Aggregation induced emission enhancement
AAS	Atomic absorption spectroscopy
ICP	Inductively coupled plasma
UV	Ultra-Violet
MOFs	Molecular organic frameworks
Er	Erbium
GQD	Graphene quantum dots
Eu	Europium
SET	Surface energy transfer
DET	Dexter energy transfer
PL	Photoluminescence
NIR	Near infrared
RBD	Rhodamine B derivative
PVDF-HFP	Poly (vinylidene fluoride-co-hexafluoropropylene)
TDDFT	Time-Dependent Density functional
LUMO	Least unoccupied molecular orbital
HOMO	Highest energy molecular orbital
CTSCD	Poly (β-cyclodextrin/chitosan
SQD	Sulphur quantum dot
MPA	Mercaptopropionic acid
ZAIS	Zn-doped $AgInS_2$
GSH	Glutathione
TEM	Transmission electron microscope
XPS	X-ray photoelectron spectroscopy
AES	Atomic emission spectroscopy
PSMA	Poly-(styrene-co-maleic anhydride
LOD	Limit of detection
CDs	Carbon dots
P-PEN	Phenolphthalin-grafted polyarylene ether nitrile

MNPs	Magnetic nanoparticles
NMs	Nanomaterials
EDC	1-ethyl-3-(3-dimethylaminopropyl) carbodiimide hydrochloride
CMCN	Carboxymethyl chitosan
ZIF	Zeolitic imidazolate framework
DNA	Deoxyribonucleic acid
TSPP	Meso-tetra (4-sulfonatophenyl) porphine dihydrochloride

1 Introduction

The high demand for resources has increased industrial and municipal activities, which has increased the occurrence of heavy metal ions pollution. Heavy metal ions are important environmental pollutants that caught the attention of environmental scientists due to their water solubility, small ionic state dimension, and non-degradability [1, 2]. Although some heavy metals (Zn, Cu, and Fe) are essential for human health they can be toxic at elevated concentrations. Non-essential metals such as Cd, Pb, and Hg are toxic even at trace levels. The effects of their presence in aquatic water bodies include bioaccumulation in aquatic plants and animals and a tainted food chain. In human bodies, the existence of toxic heavy metal ions affects body organs such as the kidney, brain, reproductive system, and heart leading to various organ disorders [3]. In addition, the detection of intracellular metal ions is important to understand biological functions and various medical diagnoses [4, 5]. Thus, developing sensitive techniques for the detection and quantification of heavy metal ions in both water environments and biological matrices is crucial. Various techniques have been developed for monitoring levels of heavy metal ions including the popular spectroscopic (Atomic absorption spectroscopy (AAS) and inductively coupled plasma (ICP), and spectrophotometric systems (ultra-violet (UV)–Visible) [6, 7] and the emerging sensors based on functional materials such as fluorescent organic materials, nanomaterials, molecular organic frameworks (MOF) [8, 9].

The pursuit for fabricating novel functional materials capable of efficiently detecting heavy metal has expanded the synthesis of hybrid materials whose strengths and weaknesses are considered and the complimentary and synergetic functions of the individual materials are utilized for realizing enhanced material performance. Over the years, nano-based materials have emerged in the field of heavy metal ion detection. Among the nano-detection systems, quantum dot (QD)-based fluorescent sensors have gained popularity due to their low detection limits and fast response time offerings. Quantum dots are zero-dimensional nanomaterials with inherent optical and optoelectronic properties and thus are highly employed for fluorescence bio-imaging, photo-thermal disease treatments, and energy processes [8, 10]. Semiconductor QDs such as II-VI, I-III-VI type [8, 11–15], and carbon-based QDs such as carbon QDs and graphene oxide QDs [16, 17] have widely applied fluorescence detection of heavy metals in water. However, despite these unique properties, some

QDs aggregate in water due to their extreme hydrophobicity, and thus their applications are limited in aqueous environments. To circumvent this challenge, efforts have been made to convert them to hydrophilic materials through surface functionalization with hydrophilic functional groups such as acid or amine groups [14, 18]. Alternatively, recent advances have produced QD-based nanocomposites to improve hydrophilicity, stability and enhance their sensitivity to heavy metals. The frequently reported QD-based nanocomposites for heavy metal ion sensing include metal-doped QDs, QD-based polymeric, QDs-molecular organic frameworks (QDs-MOFs), QD-based magnetic, and QD-based bioconjugate nanocomposites. The significance of this chapter is to provide information on QD-based nanocomposites and their applications as metal sensors in aqueous and biological environments. First, the various metal sensing mechanisms involved are discussed. The challenges associated with QD-based nanocomposites in metal sensing, conclusions, and future outlooks are also fully elucidated.

2 Metal Sensing Mechanisms in QD-Based Nanocomposites

QDs are excellent fluorescent sensors because they show a notable change in their fluorescence signal as a result of a specific interaction with a given molecule (metal ions), to detect and correlate the changes to the analyte concentration [19]. The interaction between QDs and the analyte results in either a turn-off [20] or a turn-on of the fluorescence signal of the nanoparticle due to the recombination processes of the electron–hole pair [21]. In as much as standalone QDs efficiently detect the presence of metal ions in various media such as water and biological samples, their drawback is the loss of fluorescence due to agglomeration when exposed to ambient conditions, moreover, they contribute to secondary pollution of the medium hence the need for their stabilization and immobilization in other materials to form QD-based nanocomposites [22]. Literature has shown that even composites of QDs obey the same mechanism. The mechanisms with which the fluorescence turns-off or on are known as the fluorescence quenching and fluorescence enhancing mechanisms, respectively.

2.1 *Fluorescence Quenching Mechanisms*

Fluorescence quenching occurs when the interaction between the QDs (fluorophore) and the analyte (quencher) results in a reduction of the fluorescence peak intensity and hence reduced photoluminescence (PL) properties [23]. Understanding these mechanisms is vital to avoid possible interferences and make selective strategies [24]. The most reported quenching mechanisms in QDs are static quenching, dynamic quenching, fluorescence resonance energy transfer (FRET), photo-induced electron transfer (PET), inner filter effect (IFE), aggregation-induced quenching (ACQ) and cation-exchange [25].

2.1.1 Static Quenching

Static quenching is a phenomenon that occurs when the interaction between the QD/fluorophore and the analyte/quencher forms a non-fluorescent, ground-state QD-quencher complex without emitting photons [26]. This mechanism could be linked to the formation of the ligand–metal complex [23]. Characteristics of static quenching are: (i) absorption spectra of the QDs become altered when the ground state complex is formed, (ii) an increase in temperature destabilizes the ground state complex and thereby reduces the efficiency of the static quenching, (iii) there is no significant difference between the QDs lifetime and that of ground state QDs-quencher complex [27].

2.1.2 Inner Filter Effect (IFE)

IFE is realized when the absorption spectrum of the metal ion overlaps with the emission spectra of QDs. This fluorescent decay mechanism is not a major quenching process because it does not involve radiative and non-radiative transitions in fluorescent measurements [23], but is caused by an attenuation of the excitation beam or absorption of the emitted radiation by an excess concentration of QDs or by the metal ion in the solution [28]. IFE cannot be categorized as a static or dynamic quenching process. The process can be confirmed when the PL intensity declines with the metal ion concentration but not the decay time. It has been reported to occur even when the distance between the QDs and the metal ion is more than 10 nm [29]. The process does not need energy transfer or chemical modification of the QDs [30].

Huynh et al. [31] produced erbium (Er)-doped graphene (G) QD nanocomposites via a one-pot hydrothermal method. The QD nanocomposite showed a good response to Fe^{3+} in the wide range of 0.01–140 μM with limit of detections (LODs) of 2.8 nM in PBS and 28 nM in human serum when excited at 360 nm and 730 nm for the down and up conversion respectively. The down conversion peak emerged at 440 nm and decreased proportionally with an increase in the concentration of Fe^{3+} (0–400 μm). The fluorescence lifetime decay of the Er-GQDs in the absence and the presence of Fe^{3+} was 7.4 ns signalling either a static or IFE quenching mechanism. However, due to the red-shift of the absorbance peak from 300 to 314 nm, the static mechanism was overruled because the red-shift suggests the formation of a new complex. Furthermore, the nanocomposite was found to be negatively charged while the metal ion was positive a phenomenon that allows electrostatic interaction between the nanocomposite and the Fe^{3+} resulting in static quenching of the fluorescence intensity. The excited electrons of hydroxyl groups on the surface of the Er-GQDs are transferred to the d orbitals of the Fe^{3+}.

Wang et al. [32] constructed a ratiometric fluorescence sensor using the fluorescence (467 and 616 nm) of nitrogen and cobalt (II) co-doped carbon dots (CDs) encapsulated in europium metal–organic frameworks (CDs@Eu-MOFs). The PL intensities of CDs@Eu-MOFs at 467 nm gradually declined with the increase of Cr(VI) concentrations from 0 to 500 μM, while the fluorescence intensity at 616 nm displayed insignificant changes. The observations implied that the adsorbed Cr(VI) blocks the excitation of composite and hence the fluorescence quenching at 467 nm, which indicated IFE quenching mechanism [33].

2.1.3 Dynamic Quenching Mechanism

During the dynamic quenching mechanism, the excited state of QDs return to the ground state by collision with the quencher via energy or charge transfer. The common examples of this type of mechanism include Forster resonance energy transfer (FRET), photon-induced electron transfer (PET) and rarely surface energy transfer (SET), dexter energy transfer (DET) mechanisms [26]. A simplified reaction that occurs during dynamic quenching is presented in Eq. 1.

$$F^* + Q \rightarrow F + Q \tag{1}$$

where * represents the excited state of QDs while Q denotes the quencher. This interaction obeys the Stern-Volmer relationship as indicated in Eq. 2 [24].

$$\frac{F_0}{F} = 1 + K_{SV}[Q] = \frac{\tau_0}{\tau} = 1 + kq\tau_0[Q] \tag{2}$$

where F and F_0 are the relative emission intensity of the donor (QDs) with and without the acceptor (quencher), respectively, K_{sv} is the Stern–Volmer quenching constant, [Q] is the concentration of the quencher, τ_0, and τ are the average lifetimes of the donor with and without the quencher, respectively, and k_q is the quenching rate constant.

The characteristics features of dynamic quenching are: (i) the fluorescence lifetime of the QDs changes in the presence or absence of the metal (ii) since this mechanism only affects the excited states of QDs the absorbance spectra do not display any changes (ii) raising the temperature increases the influence of dynamic quenching.

FRET

FRET occurs when non-radiative energy is transferred from excited QDs to a metal ion that is in the ground state through a dipole–dipole interaction [21, 34]. The application of QDs in FRET is easier when they function as donors because they can be excited at any wavelength as provided that excitation wavelength is shorter than the emission position. Also, QD emission can be tuned to overlap the absorbance spectra of the quencher. It is quite challenging to use QDs as FRET acceptors because of their broad absorption which makes it hard to find a donor with a PL spectrum that overlaps with QDs but is not concurrently excited with the QDs [34]. To overcome the QD-FRET predicament that is experienced when QDs are acceptors, fluorescent lanthanides are often employed as donors [35]. Alternatively, the up conversion of nanoparticles (NPs) which are excited in the near-infrared (NIR) region enabling their excitation but not directly exciting QDs are used as donors [36]. The main indicators of the FRET mechanism include, (i) overlapping of the QDs PL spectrum with that of the quencher, (ii) reduction of the fluorescent lifetime of the QDs upon interaction with metal ions, (iii) the distance of the QDs and quencher should be in the range 10–100 Å [37]. Recently, a novel QDs nanocomposite fluorescent probe for Fe^{3+} detection in aqueous solution was constructed using three components, namely the CdTe QDs as the energy donor, Rhodamine B derivative (RBD) was the energy acceptor while gamma-cyclodextrin (γ-CD), functioned as the linker between the QDs and RDB. The CdTe@γ-CD@RBD nanohybrid exhibited a strong fluorescence emission at 600 under 290 nm excitation. The RBD did not show any absorbance in the absence of Fe^{3+} but there was an apparent peak at 559 nm when Fe^{3+} was added, indicating a strong interaction between RBD and Fe^{3+}. The absorption peak of RBD-Fe^{3+} at 559 nm overlapped with the PL peak of the QDs signalling a FRET mechanism. The consistent decrease of the fluorescent intensity of the nanohybrid probe with the increase in the Fe^{3+} concentration between 10 and 60 μM confirmed an adjustable FRET efficiency between the probe and the metal ion. Moreover, the nanocomposite recorded a fluorescence lifetime of 22.7 ns without the incorporation of Fe^{3+} which was significantly reduced to 4.4 ns after Fe^{3+} was added further demonstrating a FRET process from QDs to RBD-Fe^{3+} [38].

PET

PET is a reversible process that can either be reductive or oxidative. Reductive PET occurs when the QD accepts electrons from the quencher (metal ion) while oxidative PET is when the QDs donate electrons to the quencher [21]. So, in this quenching mechanism, QDs can function as donors or acceptors of electrons to form the cation and anion radicals respectively which is the ground state complex (QDs-acceptor). Oxidative PET is the commonly observed fluorescent response between QDs and transition metal ions [37]. The coordination complexes are formed between the QDs' surface functional groups and the metal ions which have unfilled orbitals. The transfer of electrons from the excited QDs to the unfilled orbitals of the metal ion leads to fluorescence quenching [42]. The PET quenching mechanism is confirmed when there is (i) a reduction of fluorescent life time of the QD composite-quencher (metal ion) complex which indicates that the process is dynamic, (ii) the kq of QD composite-metal ion is larger than $10^7\ M^{-1}\ s^{-1}$ which is attributed either to electron or to energy transfer process, (iii) an overlap between the PL spectrum of QDs/QDs composite and the absorption band of the metal ion is less pronounced, (iv) the PL intensity decreases linearly with the increase of the metal ion while the absorption spectra remains

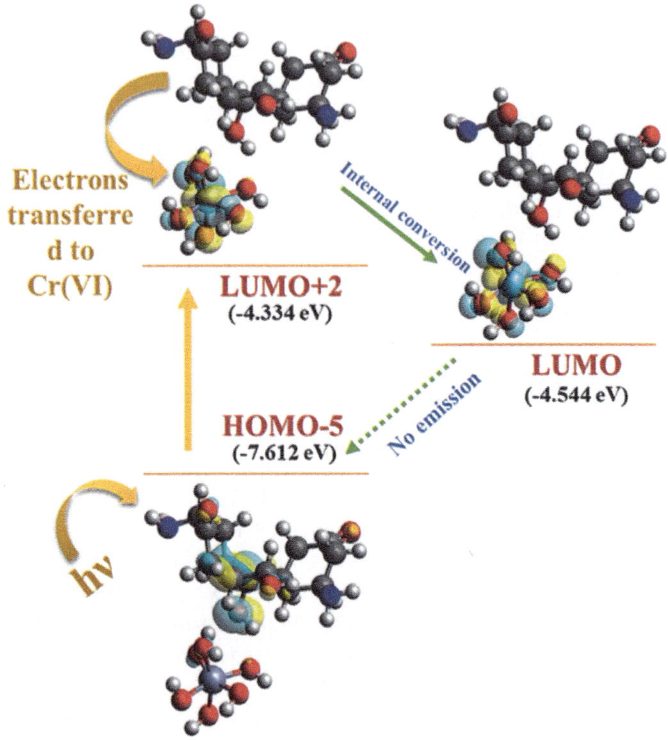

Fig. 1 Depiction of PET-based fluorescence quenching mechanism as per the TDDFT. Reproduced with permission from reference [22]

unchanged and finally (v) the PL intensity is restored with the addition of a chelating agent [39]. To enhance the stability and prevent the leaching of CQDs, Roy et al. [22] fabricated boehmite-carbon dot nanocomposite (BH@CQD) that was further encapsulated with a matrix of poly (vinylidene fluoride-co-hexafluoropropylene) (PVDF-HFP) polymer. The obtained nanocomposite served as a fluorometric sensor device to detect hexavalent chromium (Cr^{6+}) in water. A theoretical tool, the Time-Dependent Density functional (TDDFT) study showed that the quenching mechanism involved in the quenching detection of Cr^{6+} is PET. According to the TDDFT, upon excitation, the molecular orbitals at the Highest Energy Molecular Orbital (HOMO) −5 state go to the Least Unoccupied Molecular Orbital (LUMO) +2 state when QD composite-Cr^{6+} after a non-radiative internal conversion jump to the LUMO state of the complex system. During these transitions the molecular orbital shifts to the Cr^{6+} moiety and depicts electron transfer between QDs in the composite and the metal ion. Lastly, the PL quenching occurs because of a radiative transition that takes place from LUMO to HOMO-5 due to the orbital shifting towards the Cr^{6+} moiety and hence PET (Fig. 1) [40]. In simpler terms, excess Cr^{6+} ions block the fluorescence path of the QDs causing quenching [41].

2.1.4 Aggregation-Induced Quenching (ACQ)

ACQ is observed when the QDs display strong fluorescence in a dilute solution but the PL properties weaken and disappear at higher concentrations of the QDs [42]. The ACQ could result in (i) the loss of energy between the QDs, and the metal ion when they collide (ii) the generation of a non-radiated complex; and (iii) self-extinction due to excessive concentrations of the QDs [43]. For instance, Wang et al. [44] synthesized a very sensitive poly (β-cyclodextrin/chitosan (CTSCD)-sulphur QDs (SQDs) (SQDs/CTSD) nanocomposite via the self-assembly method for the selective detection of Ag^+ (Fig. 2). The SQD/CTSCD sensor detected Ag^+ in a good linear relationship in the range of 1.0×10^{-5} to 5.5×10^{-5} mol/L ($R^2 = 0.9992$) and a detection limit of ~85 nM. The negatively charged SQDs on the CTSCD chain coordinated with the positively charged silver ions to form S– Ag^+ –S bonds. The formed bonds are associated with non-covalent bonds resulting in Ag^+-SQD/CTSCD nanocomposite aggregates that lead to ACQ [44].

2.2 *Fluorescent Enhancement Mechanisms*

Fluorescence enhancement refers to the improvement or increment of the PL intensity when QDs interact with a molecule [23]. Fluorescent enhancement mechanisms can

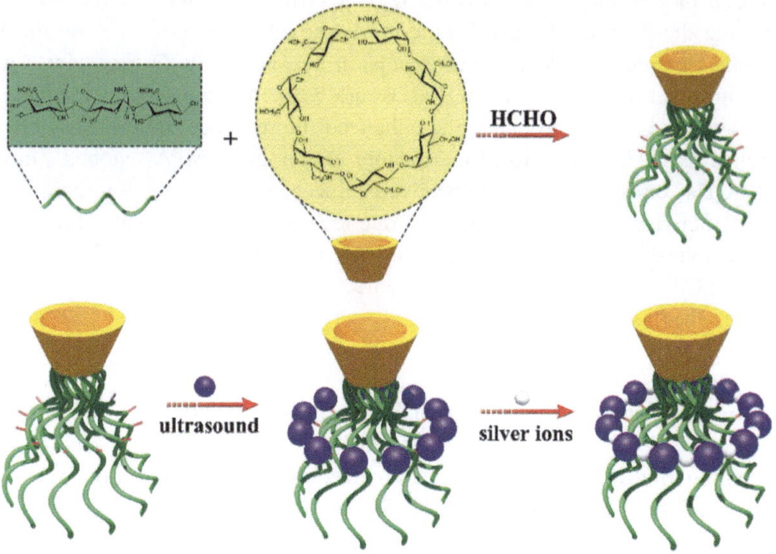

Fig. 2 Schematic diagram of the synthesis of SQD/CTSCD nanocomposite and the proposed mechanism for the detection of Ag^+ ions. Reproduced with permission from reference [44]

be broadly categorized into surface passivation, cation exchange, and aggregation-induced emission enhancement (AIEE).

2.2.1 Surface Passivation

QDs have surface defects such as interstitials, vacancies, and dangling bonds which act as trap sites for charge carriers, leading to the reduction of photoluminescence efficiency [45]. When the surface defects are passivated, the fluorescence becomes enhanced as the trap sites for the charge carriers are minimized on the surface of the QDs. Passivation in QDs has been realized when the QDs are used to detect biological molecules such as ascorbic acid [46]. Passivation of surface defects is also realized inorganically using ions such as defects by either Zn^{2+} or Cd^{2+} ions, which can also increase in the fluorescence lifetimes [47]. The size of the QDs determines the PL enhancement in that smaller QDs bear more surface defects than larger ones due to the large surface area-to-volume ratio. As such smaller QDs can accommodate larger concentrations of the metal ion to passivate the surface while larger ones may require lower concentration due to smaller surface area resulting in varied detection ranges [48]. Liu et al. [49] demonstrated that SnO_2 QDs exhibit an enhanced fluorescence intensity accompanied by high selectivity to Hg^{2+}. The fluorescent enhancement was observed because of the existence of dangling bonds due to the Sn vacancies on the QD surface which promote non-radiative recombination, resulting in low fluorescence efficiency that was passivated by Hg^{2+} which adsorbed onto the surface of SnO_2 QD to form extra bonds with the O atoms, and consequently eradicating the dangling bonds. Furthermore, the adsorption of Hg^{2+} on the SnO_2 QD surface created an electrostatic repulsion between the QDs and Hg^{2+}, which led to improved dispersion of SnO_2 QDs and weakened the self-quenching effect caused by aggregation and hence the enhanced fluorescence. In another study, Liu and colleagues [50] synthesized mercaptopropionic acid (MPA) capped Zn-doped $AgInS_2$ QDs composite that also displayed a linear increase in the fluorescent intensity with increased Cd^{2+} concentrations due to surface defect passivation (Fig. 3a), which was verified by the prolonged fluorescence lifetimes from 262.55 ns (in the absence of Cd^{2+}) to 292.33 ns (in the presence of Cd^{2+}) after the addition of 290 μM Cd^{2+}(Fig. 3b) [50].

2.2.2 Aggregation-Induced Emission Enhancement (AIEE)

The Aggregation-Induced Emission Enhancement (AIEE) mechanism was first reported by Tang's group in 2001 [51]. The group found that fluorophores in high concentrations exhibit strong fluorescence, mainly because in dilute solutions, the molecules rotate internally, while in aggregated form internal rotation is restricted due to space constraints thereby reducing the likelihood of non-radiative energy. Therefore, the excited QDs can only return to the ground state in the form of a radiative leap, resulting in enhanced fluorescence [51]. For example, Wang et al. [52]

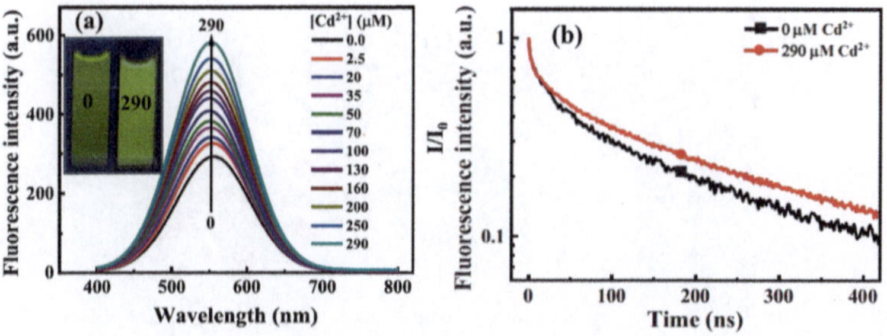

Fig. 3 PL **a** and time-resolved fluorescence spectra **b** of MPA-capped Zn-doped $AgInS_2$ QDs in the presence of Cd^{2+} ions at increased concentrations (0–290 μM). Reproduced with permission from reference [50]

synthesized glutathione (GSH) functionalized C-dots (g-CDs) to selectively detect Fe^{3+} in cells. The fluorescence intensity gradually increased with the increasing Fe^{3+} concentration. The absorbance peak at 433 nm showed an enhancement when Fe was 100 vol%, compared to the pure g-CDs in aqueous solution. The TEM micrographs of the g-CDs in the ethanol-water medium and in (Fe, 50 vol%) confirmed the well-dispersed QDs in the absence of Fe while an aggregation of the QDs was observed in the presence of 50% Fe, thus confirming the AIEE due to the agglomerated g-CDs.

This mechanism is also reported by Wei and colleagues [53] who displayed the detection of Cd^{2+} ions using L-cysteine-capped Zn-doped Ag-In-S (ZAIS) QDs composites. The L-cysteine ligands on the ZAIS composite surface coordinated with the Cd^{2+} via the thiol group which weakened the electrostatic repulsion between the composites encouraging aggregation. The reduction in the negativity charge of the ZAIS composite from −38.8 to −27.4 mV in the presence of Cd^{2+} confirmed the reduction of the repulsion efficiency in the system. In addition, the Cd-SR system passivated the surface defects of the ZAIS composite (ZAIS@Cd-R) thus decreasing the non-radiative recombination, which enhanced the PL intensity [53].

2.3 Cation-Exchange

Cation-exchange is the substitution of cations in an ionic crystal with guest cations while maintaining the original anionic structure and rarely changing the size [54]. Cation-exchange has been reported to result in either PL quenching or PL enhancement. The quenching or enhancement of the PL depends on the effect of the guest cation on the QD. A turn-on mechanism for the detection of Hg^{2+} was realized with Mn-doped ZnSe QDs composite. The Hg^{2+} ions replaced the Mn^{2+} in the Mn-doped ZnSe QDs composite, which eliminated the fluorescence quenching effect of Mn^{2+} (Fig. 4a, b) [55]. The ICP-AES revealed that before the addition of Hg^{2+}, the amount of Mn^{2+} in the doped QDs was 1.8% but was partially replaced by Hg^{2+} and its content decreased to 1.1%, while the content of Hg^{2+} increased from 0.0 to 1.4%. In

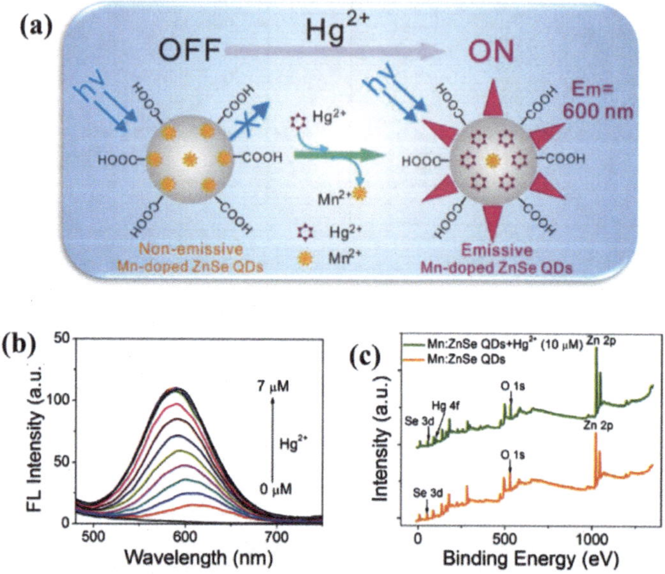

Fig. 4 **a** Schematic illustration of Mn-doped ZnSe sensor for Hg^{2+} detection. **b** Emission spectra of Mn-doped ZnSe QDs (40 μg/mL) upon the addition of different concentrations of Hg^{2+}, at an excitation wavelength of 400 nm. **c** XPS spectra of Mn-doped ZnSe QDs in the absence of Hg^{2+} and presence of 10 μM Hg^{2+}. Reproduced with permission from reference [55]

the absence of Hg^{2+}, the X-ray photoelectron spectroscopy (XPS) analysis showed the high-intensity characteristic peaks of Se, Zn, and O however, in the presence of Hg^{2+}, the intensity of the characteristic peak of Mn decreased (Fig. 4c), indicating the decrease of Mn^{2+} concentration on the surface of QDs, which is ascribed to the substitution of interfacial Mn^{2+} by Hg^{2+} [55].

In another development [56], a fluorescence quenching mechanism was attributed to the detection of Hg^{2+} ions using amino poly-(styrene-co-maleic anhydride) (NH_2-PSMA)-capped CdSe/ZnS core/shell QD nanocomposites due to the cation-exchange between Zn^{2+} on the ZnS shell and Hg^{2+} ions. This resulted in the formation of HgS particles on the CdSe/ZnS QD surface which destabilized the QD hence the quenching.

3 QD-Based Nanocomposites as Metal Sensors

QD-based nanocomposites are multi-functional composite materials that consist of QDs and one or more functional materials. The fabrication of QD-based nanocomposites is mainly aimed at enhancing the existing QD properties (i.e. optical, electrical, solubility, photostability), introducing other properties to the QDs (i.e. magnetic) and often achieved via core/shell structured and multi-material co-encapsulation approaches. In addition, the incorporation of QDs into other functional materials creates opportunities to introduce new sites for anchoring guest molecules such as metal ions. This has resulted in numerous reports of the detection of heavy metal ions using QD-based nanocomposites. The following sections briefly discuss the use of various types of QD-based nanocomposites in heavy metal ion detection including metal-doped QD, QD-based polymeric, QD-based magnetic, QD-based MOFs, and QD-based bioconjugate nanocomposites.

3.1 Metal-Doped QDs Nanocomposites as Metal Sensors

QD doping is the intentional introduction of atoms to QDs to either improve or introduce additional properties to the existing QD material. Consecutively, QDs have been doped with metals to color-tune their emission spectra, improve fluorescence, electrical properties, and stability as well as introduce magnetic properties for enhanced efficiency in biomedical, environmental monitoring, and energy applications [57, 58]. In some cases, metal doping has been reported to occur unintentionally during shell growth on core QDs. Here, the metal dopant tends to replace some of the existing cations within the QD lattice (cation-exchange). For instance, many authors have reported the replacement of cations (i.e. Cu^+, Ag^+, In^{3+}) in ternary I–III–VI-type core QDs with Zn^{2+} ions during the passivation of the core QDs with ZnS shell, which resulted in an unexpected blue-shift in the fluorescence spectra [47, 59]. As a result, Guo et al. [53] presented a blue-shift emission inhibition approach during ZnS shell growth on $CuInS_2$ QDs by intentionally introducing Zn ions into the $CuInS_2$ core QDs during their synthesis (CuInSZn composite) to inhibit the cation-exchange process that occurs between Zn^{2+} and the cations (Cu^{2+}/In^{3+}) during ZnS shell formation [53].

Several authors have reported metal doping as an effective approach for improving the metal sensing ability of QD materials [31, 54, 55]. For example, Chen et al. [60] discovered a remarkable enhancement in the fluorescence peak of Eu^{3+} ions when glutathione (GSH)-capped Eu-doped CdS nanocomposites were produced, indicating a fluorescence energy transfer from the CdS QDs to the Eu^{3+} ions (Fig. 5a). The fluorescence peak of Eu^{3+} continued to enhance at increased concentrations of the Eu-doped CdS nanocomposites, promoting wider detection ranges and detection limits for Hg^{2+} via fluorescence quenching (Fig. 5b, c). The strong interaction between the amide ($-NH_2$) groups on the GSH-Eu-doped CdS nanocomposite and the Hg^{2+} could be responsible for the electron transfer from the QD nanocomposite to Hg^{2+} which resulted in the quenching. Furthermore, the fluorescence lifetime of the nanocomposite gradually decreased at increased Hg^{2+} ions from 512.19 to 488.54 and 437.56 ns (500 nmol/L and 1200 nmol/L) respectively, which confirmed the enhanced electron transfer process at increased Hg^{2+} concentrations [60].

GSH-capped Zn-doped $CuInS_2$/ZnS (CuInZnS/ZnS) core/shell nanocomposites displayed fluorescence quenching due to mild cation-exchange during the detection of Cu^{2+} ions in aqueous solutions [61]. The cation-exchange was attributed to the

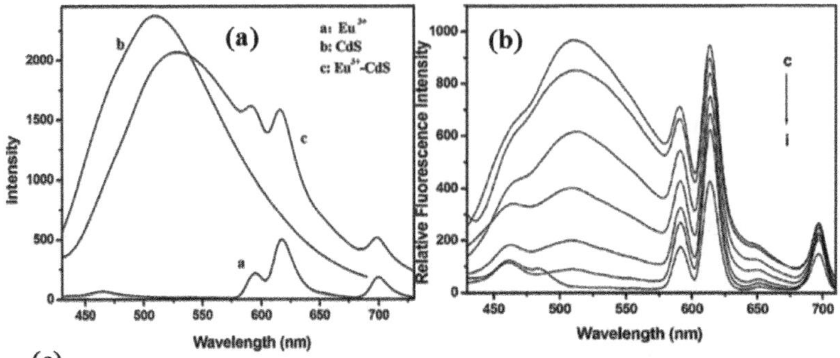

(c)

Concentration of Eu^{3+}-doped CdS with Cd^{2+} as a standard (mol L^{-1})	Linear range (nmol L^{-1})
0.1×10^{-4}	3–250
0.5×10^{-4}	5–650
2.0×10^{-4}	10–1500
3.5×10^{-4}	90–2000
5.0×10^{-4}	500–4500

Fig. 5 PL spectra of Eu^{3+}, CdS, Eu-doped CdS (**a**), and Eu-doped CdS with different concentrations of Hg^{2+} (c–i is 0, 60, 300, 500, 700, 900, and 1500 nmol/L, respectively) (**b**). Produced linear detection range for Hg^{2+} at increased concentrations of the Eu-doped CdS QD nanocomposite (**c**). Reproduced with permission from reference [60]

replacement of Zn^{2+} with Cu^{2+} ions which was also reflected in the red-shifted fluorescence and absorbance spectra probably due to some size changes in the wider band gap of ZnS after the cation-exchange process. The cation-exchange was confirmed by the decline in the In and Zn concentrations of the CuInZnS/ZnS nanocomposite from 55.0% to 54.3% and 37.55 to 33.7% at increased Cu^{2+} concentrations, respectively. The time-resolved PL spectra revealed a decrease in the fluorescence lifetimes of the nanocomposite in the presence of Cu^{2+} ions from 413 to 379 ns and 289 ns at 10 μM and 100 μM, respectively suggesting a gradual change in the electron–hole recombination process during emission at increased Cu^{2+} concentrations. The CuInZnS/ZnS nanocomposite produced a linear range and DL of 0.020–20 μM and 6.7 nM for the detection of Cu^{2+} ions. Good performance was achieved towards real water samples compared to the accuracy of the ICP-atomic emission spectroscopy (AES) method [61].

3.2 QD-Based Polymeric Nanocomposites as Metal Sensors

The QDs-based polymeric nanocomposites are fabricated by combining a QD/nanofiller with a polymer matrix [62]. The incorporation of QDs in a polymer is attempted in various ways, such as (i) ligand-exchange between polymer and QDs;

(ii) grafting polymer to QDs; (ii) grafting polymer from QDs; (iv) capping polymer onto QDs; and (iv) growing QDs within the polymeric template. The polymer encapsulation of QDs not only offers protection to the QD and sometimes enhances the optical properties and photo stability of the QD but also provides functional groups that target molecules such as heavy metal ions for their detection [63, 64].

For example, Zhou et al. [56] reported aminolyzed poly-(styrene-co-maleic anhydride) (NH_2-PSMA)-capped CdSe/ZnS QD nanocomposites for the detection of Hg^{2+} and Cu^{2+} ions via fluorescence quenching mechanism [56]. The fluorescence quenching of the NH_2-PSMA-capped CdSe/ZnS QD upon the addition of Hg^{2+} was attributed to the formation of HgS particles on the CdSe/ZnS QD surface due to the cation-exchange between Zn^{2+} and Hg^{2+} ions. The formation of the HgS particles generated surface defects which destabilized the QD hence the quenching. The formation of HgS particles was evidenced by the disappearance of the NH_2-PSMA-capped CdSe/ZnS QD absorbance peak, the red-shifted fluorescence peak, and the color change from colorless to yellow in the presence of Hg^{2+} ions. On the contrary, the fluorescence quenching of the NH_2-PSMA-capped CdSe/ZnS QD upon the addition of Cu^{2+} was attributed to the electron transfer from the NH_2-PSMA-capped QD to Cu^{2+} via the co-ordination of Cu^{2+} with the amide, hydroxyl, and carbonyl groups on the PSMA, which formed non-radiative surface channels. No changes were observed in the absorbance, fluorescence spectra, and solution color of the NH_2-PSMA-capped CdSe/ZnS QD in the presence of Cu^{2+}, indicating that no new molecules were formed. Both Hg^{2+} and Cu^{2+} ions could be detected in co-existence in the presence of a masking agent. Linear detection ranges of 0.02–0.7 μM (LOD = 6.94 nM) and 0.1–1.4 μM (LOD = 20.58 nM) were achieved for Cu^{2+} and Hg^{2+} ions, respectively [56].

In another development, He et al. [65] reported the simultaneous detection of Ag^+, Hg^{2+}, Cu^{2+}, and Fe^{3+} ions using a QD-polymeric nanocomposite based on the encapsulation of two hydrophobic green and red fluorescing QDs (Zn–CdSe/ZnS QDs_{523} and QDs_{600}) with a blue fluorescent amphiphilic block co-polymer, based on phenolphthalin grafted polyarylene ether nitrile (P-PEN). The detection of all the metal ions occurred via the fluorescence quenching of the P-PEN@Zn–CdSe/ZnS QDs_{523} and QDs_{600} nanocomposites. The fluorescence quenching that occurred in the detection of Ag^+ and Hg^{2+} was attributed to the generation of surface defects due to the cation-exchange reaction between Zn^{2+} on the ZnS shell and Ag^+/Hg^{2+} ions. The fluorescence quenching effect of Cu^{2+} was related to the combination of surface defect formation due to the cation-exchange between Zn^{2+} and Cu^{2+} as well as the coordination of Cu^{2+} with carbonyl groups on the P-PEN polymer. Similarly, to Cu^{2+}, Fe^{3+} ions coordinated with P-PEN polymer on the QD surface via carbonyl groups, and no evidence of cation-exchange between Zn^{2+} and Fe^{3+} was seen due to the lower Kpa of FeS (17.2) compared to that of ZnS (23.8) as observed in the detection of Hg^{2+}, Ag^+, and Cu^{2+} which exhibited Kpa's of 51.8, 49.2 and 35.2, respectively. The selective detection of metal ions was based on selective fluorescence quenching of the different-sized QDs within the nanocomposite, such that: Ag^+ and Hg^{2+} ions mainly quenched the green and red fluorescent QDs whereas high concentrations of Cu^{2+} and Fe^{3+} ions quenched the green, red fluorescent QDs and the blue fluorescent band of the P-PEN polymer. The selectivity of Ag^{+} and Hg^{2+} ions could be further observed in that the red fluorescent QD was quenched at a higher degree by the Ag^{+} than the green fluorescent QDs while Hg^{2+} quenched the red and green QDs at a similar degree. In the case of Cu^{2+}, a higher quenching degree was evident in red fluorescent QD compared to blue and green. Additionally, Fe^{3+} ions revealed similar quenching degrees for all three fluorescent peaks, however, a higher quenching degree on the blue fluorescent band was observed compared to other metal ions which were related to the detection mechanism based on the coordination with the P-PEN via carbonyl groups (Fig. 6). Linear detection ranges of 0.1–1, 1–5 and 0.05–3, 3–5 μM were achieved for Ag^{+} and Hg^{2+} for the green and red quenched fluorescence while 1–10, 10–50, and 0.5–10, 10–100 μM was achieved for Cu^{2+} and Fe^{3+} for the quenched fluorescence of all three emission peaks (Fig. 7). The P-PEN@Zn–CdSe/ZnS QDs_{523} and QDs_{600} nanocomposites showed potential in the simultaneous detection of Ag^{+}, Hg^{2+}, Cu^{2+}, and Fe^{3+} in environmental wastewater samples via spiked recoveries. Naked–eye detection of Ag^{+}, Hg^{2+}, Cu^{2+}, and Fe^{3+} using test strips was also demonstrated [65].

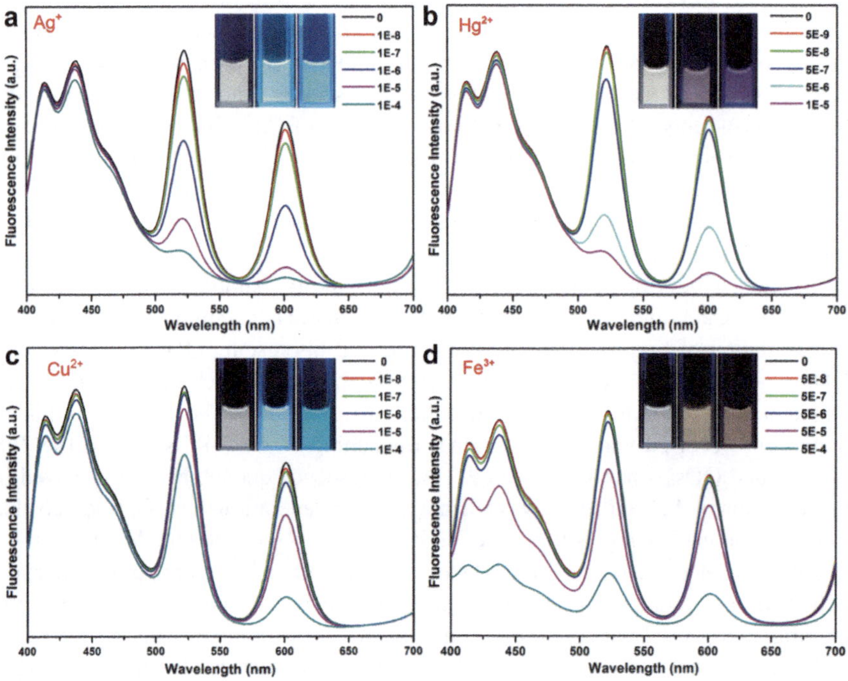

Fig. 6 PL spectra of P-PEN@Zn–CdSe/ZnS QDs_{523} and QDs_{600} NC in the presence of Ag^{+} (**a**), Hg^{2+} (**b**), Cu^{2+} (**c**), and Fe^{3+} (**d**) (mol/L). Reproduced with permission from reference [65]

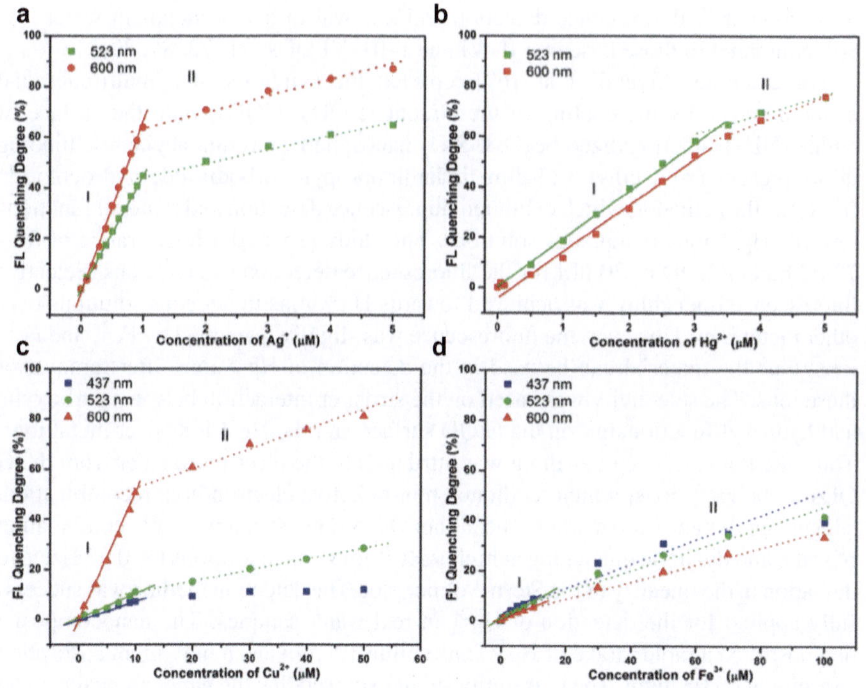

Fig. 7 Linear plot of fluorescence quenching degree of P-PEN@Zn–CdSe/ZnS QDs_{523} and QDs_{600} NC in the presence of Ag^{+} (**a**), Hg^{2+} (**b**), Cu^{2+} (**c**), and Fe^{3+} (**d**) (mol/L). Reproduced with permission from reference [65]

3.3 QD-Based Magnetic Nanocomposites

Magnetic QD nanocomposites are bi-functional (magneto-fluorescent) materials that are a result of combining magnetic nanoparticles (MNPs) such as iron oxide (Fe_3O_4) NMs and QDs. Magnetic QD nanocomposite has been mainly achieved by (i) co-encapsulation of individually synthesized MNPs and QDs in amphiphilic polymers or silica shell and (ii) core/shell formation where the QDs are deposited around the MNP to form a shell or vice versa [66].

Magnetic QD nanocomposites have mainly benefited biological applications due to their excellent performance in biological imaging, drug delivery, and medical diagnosis [67, 68]. Until recently, researchers explored the bi-functional ability of magneto-fluorescent Fe_3O_4 NM-QDs nanocomposites (NC) in the fluorescence detection and removal of heavy metals in water [69–72]. The use of magnetic Fe_3O_4 NM as a magnetic component in the nanocomposites was motivated by their popularity in the removal of heavy metals in water due to their ease of separation and high adsorption capabilities [73–75], especially those with layered adsorptive structures such as functionalized meso-porous materials (e.g. SiO_2 and $CaCO_3$) [76–78]. Indeed, magneto-fluorescent Fe_3O_4 NM-QDs NCs based on carbon NM remain the most popular in fluorescence detection and removal of heavy metals in water [79, 80], compared to those based on II–VI and I–III–VI QDs [71, 72, 81, 82].

For example, Alvand et al. [69] reported the synthesis of a multifunctional nanocomposite by the coating of the Graphene QDs (GQDs) onto the surface of amine (NH_2)-functionalized Fe_3O_4@SiO_2 nanospheres through acylamide binding in the presence of 1-ethyl-3-(3-dimethylaminopropyl) carbodiimide hydrochloride (EDC) as the activator, which exhibited fluorescence detection and removal functions towards Hg^{2+} ions in aqueous solutions. The study reported a linear range of 0.1–70 μM and a LOD of 30 nM for the fluorescence detection of Hg^{2+} ions. Selective fluorescence quenching was achieved towards Hg^{2+} ions in aqueous solutions over other metal ions. However, the fluorescence was slightly quenched by Fe^{2+} and Fe^{3+} ions, thus the material can be used in the detection of Hg^{2+} ions after removal of these ions. The selectivity was based on the stronger interaction between carboxylic and hydroxyl functionality on the G QD surface and the Hg^{2+} ions over metal ions. Thus, the fluorescence quenching was attributed to the electron transfer from the G QDs to the Hg^{2+} ions, which facilitated non-radiative electron/hole recombination, suggesting dynamic quenching. The authors also suggested a probable combination of static and dynamic quenching at higher Hg^{2+} ions concentrations (>70 μM) due to deviation in the linearity of the Stern-Volmer plot. The detection method was successfully applied for the detection of Hg^{2+} in real water samples. The nanocomposite demonstrated a rapid intake of Hg^{2+} ions within 1.5 min and a maximum adsorption capacity of 68.03 mg/g. The reusability studies showed that the nanocomposite could be used for four cycles with negligible loss of detection [69].

A novel magnetic-fluorescent nanocomposite based on carboxymethyl chitosan (CMCN)-functionalized Fe_3O_4 NPs conjugated to GSH-capped CdTe/ZnS core/shell QDs (CMCN-Fe_3O_4@CdTe/ZnS) was created for the simultaneous determination and removal of Hg^{2+} in water (Fig. 8a) [83]. Under ideal circumstances, the fluorescence of CMCN-Fe_3O_4@CdTe/ZnS nanocomposite was quenched at increased concentrations of Hg^{2+} ions. Thus, a linear range and detection limit of 0.3–5 μmol/L and 9.1 nmol/L was achieved for the detection of Hg^{2+} (Fig. 8b, c). The fluorescence quenching was attributed to the formation of a non-fluorescent ground-state CMCN-Fe_3O_4@CdTe/ZnS-Hg complex, which occurred via coordination of Hg^{2+} with COO−, $-NH_2$, and OH^- groups on the CMCN nanocomposite surface, suggesting static quenching. Additionally, the Hg^{2+} ions compete with the GSH-capped CdTe/ZnS QD for GSH functionalization and win (Hg^{2+} striping), leaving an imperfect QD surface hence the fluorescence quenching (Fig. 8a). Static quenching was confirmed by the unchanged fluorescence lifetime in the presence and absence of Hg^{2+} ions (33 ns) (Fig. 8d, f). Furthermore, the red-shifted and enhanced absorbance band (Fig. 8e) observed after the addition of Hg^{2+} to CMCN-Fe_3O_4@CdTe/ZnS could be due to the formation of a non-fluorescent ground-state CMCN-Fe_3O_4@CdTe/ZnS–Hg complex. Co-existing ions (Na^+, K^+, Ca^{2+}, Mg^{2+}, Al^{3+}, Ba^{2+}, Zn^{2+}, Fe^{2+}, and Fe^{3+}) showed low interference effect in detecting Hg^{2+} ions however, Ag^+, Pb^{2+}, and Cu^{2+} ions showed a higher quenching effect, which was attributed to their affinity for COO^-, $-NH_2$ and OH^- groups on exposed nanocomposite surface [83].

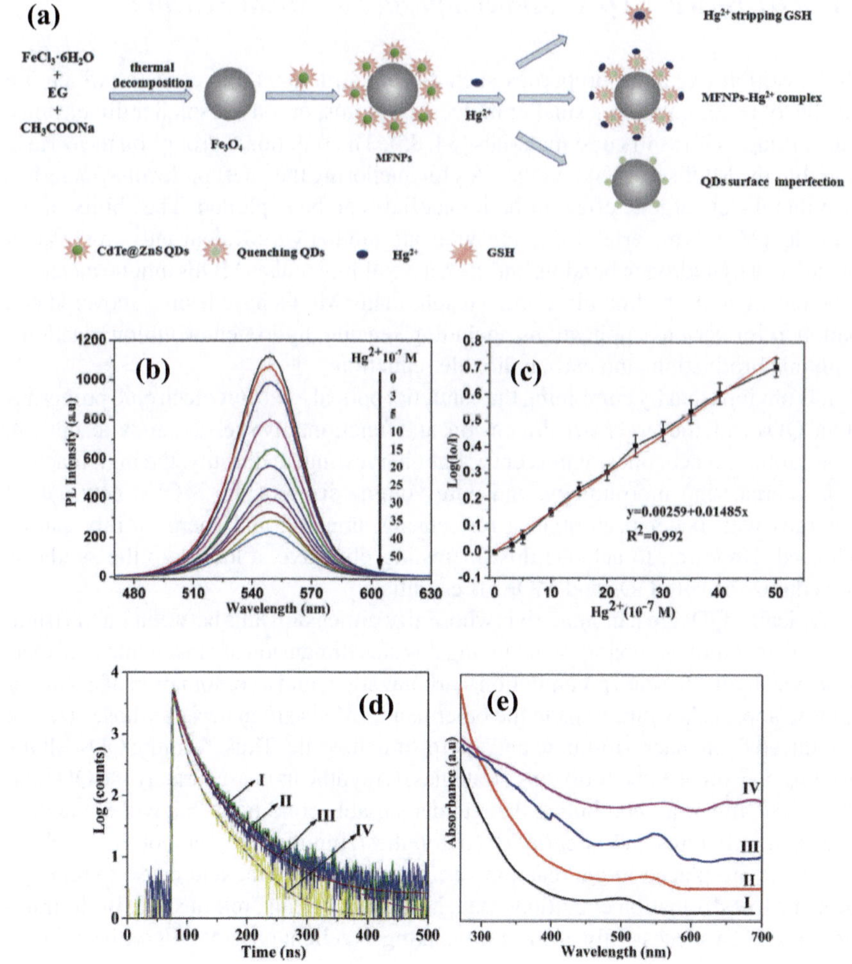

(f)

Sample	B_1(%)	τ_1(ns)	B_2(%)	τ_2(ns)	B_3(%)	τ_3(ns)	τ(ns)	F_0/F	τ_0/τ
I	26.17	3.99	56.97	13.22	16.86	53.84	33.56	1	1
II	23.74	6.62	62.33	18.86	13.93	58.95	33.25	1.36	1.01
III	23.35	6.14	61.65	18.09	16.50	55.49	33.05	2.0	1.01
IV	26.77	6.81	54.12	20.73	16.11	53.36	32.52	2.88	1.03

The concentration of Hg^{2+} (10^{-7} mol/L)(I) 0, (II) 10, (III) 20 and (IV) 30.

Fig. 8 Schematic diagram of the synthesis CMCN-Fe_3O_4@CdTe/ZnS nanocomposite for the detection of Hg^{2+} (**a**). PL spectra (**b**), linear fluorescence intensity plot (**c**), time-resolved fluorescence spectra (**d**), absorbance spectra (**e**), and fluorescence lifetimes (**f**) of CMCN-Fe_3O_4@CdTe/ZnS nanocomposites in the presence of Hg^{2+} (1×10^{-7} M–50×10^{-7} M). Reproduced with permission from reference [83]

3.4 QD-Based MOFs Nanocomposites as Metal Sensors

MOFs exhibit excellent properties such as ring structures, the cores of which are capable of accommodating smaller molecules, atoms, or ions of smaller dimensions, thus serving as hosts to guest materials [84, 85]. The functional groups on the surface or within the MOFs can serve as the sites for anchoring the guest molecules/materials such that the synergetic effect of both materials can be exploited. The ability of the individual MOF's materials to interact with one another via covalent and non-covalent means such as hydrogen bonding and $\pi-\pi$ assembling makes MOFs unique materials to be reckoned in the first place. As a result, many MOFs have been discovered and exploited for various applications, including sensing, hydrogen evolution reactions, ammonia production, and carbon dioxide reduction.

It is obvious that by combining the catalytic, optical, and optoelectronic properties of the QDs with the larger size dimension, and functionality such as catalytic, optical, photoluminescence, phosphorescence, light harvesting, and utility, the high reaction surface area, ring morphology, and pore volume sizes of the MOFs, new hybrid materials with better performance for sensing, imaging and energy utility can be achieved. However, to achieve this interesting objective, a look into the synthesis procedures for both QDs and MOFs is essential.

Typically, QDs are nanomaterials whose size dimension falls between 1 and 10 nm. Thus, all techniques capable of achieving this size dimension are taken into consideration, viz: hydrothermal, precipitation, carbonization, and a combination of a number of these approaches [86, 87]. On the other hand, MOFs are materials whose size can be reduced from macro-to-micro and micro-to-nanoscale. Thus, forming QDs-MOFs systems may involve the following strategies: (i) synthesizing separately the QDs and MOFs and afterward combining them under suitable conditions that will not tamper with their structures and integrity or (ii) synthesizing through one pot, in which all precursors are placed in the reaction vessel at the same time and cause to undergo the same transformation conditions, e.g., heating, refluxing, mixing, etc. Both strategies have been successfully utilized for making excellent QDs-MOFs with profound performance [6].

Due to the synergetic effects of their individual material properties, the effectiveness of using QDs-MOFs as efficient sensing agents for heavy metal ions has been explored. For example, Wang et al. [88] produced QD@MOF based on zeolitic imidazolate framework (ZIF) and CdTe QDs. The CdTe@ZIF-365 was utilized as a fluorescent ratiometric sensor for Cu^{2+} with high sensitivity and selectivity in the mixed cation ions' solution (Fig. 9a, b). Here, the fluorescence detection of the Cu^{2+} was mainly based on the quenching of the CdTe QDs fluorescence peak since the ZIF-365 fluorescence was only slightly quenched. A good linear relationship was established between the fluorescence intensity of the CdTe@ZIF-365 nanocomposite and Cu^{2+} ion concentration (0 0.0075 μmol/L) with a detection limit of 0.25 nM (Fig. 9c, d) [88].

Furthermore, Guo et al. [89] reported the use of QD@MOF based on the co-encapsulation of carbon (C) QDs and Au nanoclusters (NCs) with ZIF-8 (CQD/

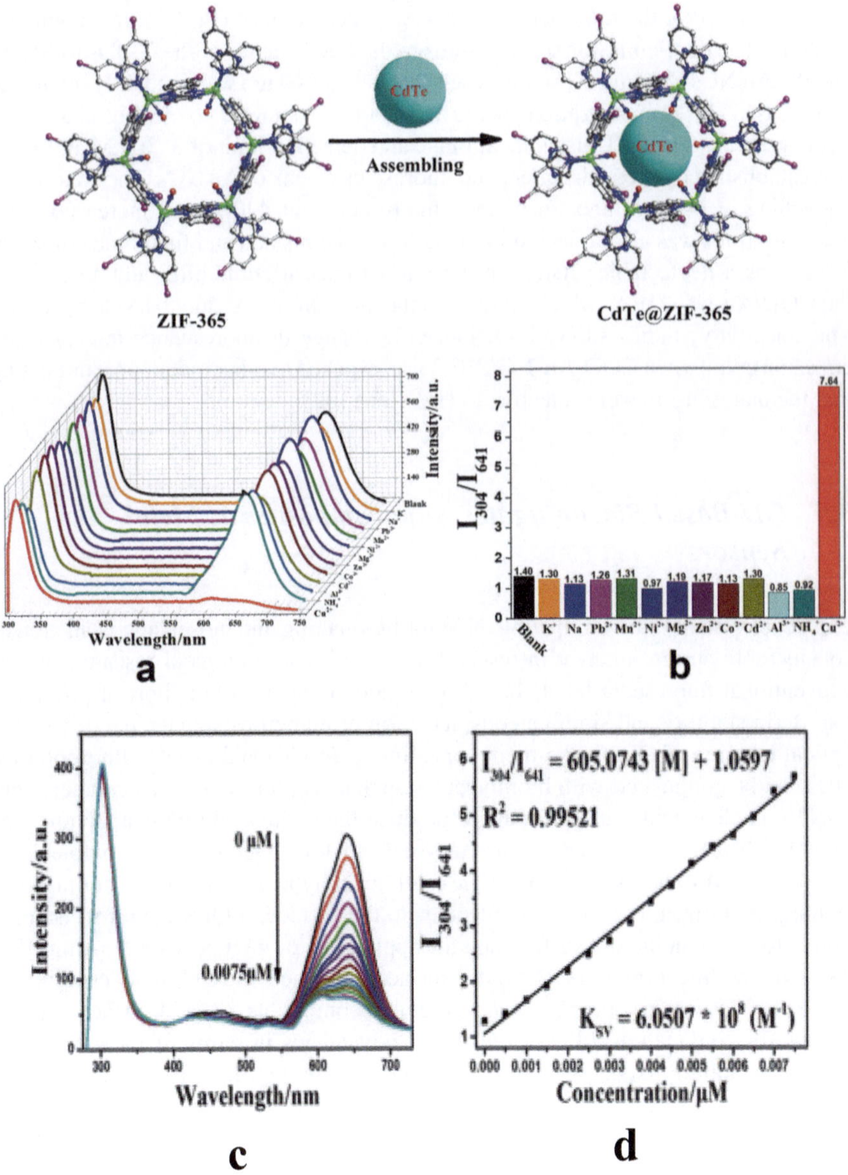

Fig. 9 Schematic diagram of the fabrication of CdTe@ZIF-365 NC (top). PL spectra and fluorescence intensity ratio (304 nm/641 nm) of CdTe@ZIF-365 NC in the presence of other metals (**a**, **b**). PL spectra and ratiometric plots (**c**, **d**) of CdTe@ZIF-365 NC in the presence of Cu^{2+} ions. Reproduced with permission from reference [88]

Au-NC/ZIF-8) for the ratiometric fluorescence detection of Hg^{2+} ion in an aqueous medium. The dual-emission originated from the fluorescence of the C QDs (440 nm) and the Au NCs (640 nm). The fluorescence peak at 640 nm was gradually quenched at increased Hg^{2+} concentrations while that at 440 nm showed no obvious changes in fluorescence intensity. Thus, a linear range and detection limit of 3–30 nM and 1 nM was established based on the quenched fluorescence peak of Au NCs. The continuous quenching at 640 nm and unchanged fluorescence at 440 nm at increased Hg^{2+} concentrations was also reflected in the fading red emission and final blue emission. There was a neglectable change in the fluorescence lifetime after adding Hg^{2+} to the CQD/Au-NC/ZIF-8, which confirmed the quenching was due to a static process. The selectivity studies showed a neglectable change in fluorescence intensity and emission color when CQD/Au-NC/ZIF-8 was exposed to other metal ions, indicating that the nanosensors were selective to Hg^{2+} ions [89].

3.5 QD-Based Bioconjugates Nanocomposites as Metal Sensors

The development of biocompatible NPs for bioimaging and biosensing is an area of considerable current interest in research [90]. In broad biological research, where conventional fluorescent labels based on organic molecules fall short of providing long-term stability and simultaneous detection of numerous signals, the distinctive optical features of QDs make them appealing as *in-vivo* and in-vitro fluorophores. QDs can be conjugated with biomolecules such as peptides, deoxyribonucleic acid (DNA), nucleic acids, and proteins to produce bioconjugated QD nanocomposites [91, 92]. Thus, the QD acts as a nano-scaffold for attachment to biomolecules, creating a multifunctional nanoparticle–biological hybrid. Some of the conjugation techniques utilized to attach biomolecules to the surface of QDs are direct adsorption onto the quantum dot surface and the application of inert polymer coatings [93, 94]. Any modification to a QD's outer surface must meet the following criteria: (1) make the QDs water-soluble; (2) make them biocompatible; (3) add surface reactive groups for conjugation; (4) keep all QDs' advantages in terms of photochemical long-term stability (spectral range, brightness, fluorescence, narrow emission); (5) produce monodispersed particles or particles with a narrow size distribution; and (6) add a thin shell and capping layer to prevent the engineered particle from getting bigger [95, 96].

QD-bioconjugate nanocomposites have been used to label, image, and detect a variety of materials, from large proteins and nucleic acids to metal ions and tiny molecules [5, 97, 98]. Both fluorescent turn-on and turn-off sensing techniques have been used for both biomolecule analysis and extracellular/ intracellular metal ion detection [5]. The sensitive detection of intracellular metal ions is important to understand biological functions and for various medical diagnoses. For instance, Zn is the second most abundant metal in the human body, essential for several biological processes but can be toxic when present excessively. Unusual Zn^{2+} patterns have been associated with Prostate cancer. Alzheimer's disease, diabetes, and epilepsy. Several QD-based nanocomposites have been explored for intracellular metal ion sensors [5, 99], but only a few reports are based on QD-bioconjugates. For example, QD-based Bi-FRET biopolymeric nanocomposite containing two colored QDs (517 nm and 560 nm), were fabricated for ratiometric sensing of Zn^{2+} in different media [4]. The QD scaffold was coated with poly(dA) homopolymer/double-stranded DNA, followed by the loading of meso-tetra(4-sulfonatophenyl)porphine dihydrochloride (TSPP). The BiFRET process is based on FRET-1, where an efficient overlap between QD 517 nm emission and TSSP absorption band at 515 nm occurs and in FRET-2 the overlap is inefficient due to the weak absorption peak at 555 nm. During Zn^{2+} sensing, Zn^{2+} chelates with TSPP, resulting in the reduction of the 515 nm absorption peak and enhancement of the 555 nm absorption peak. This causes a reduction in FRET-1 and an enhancement in FRET-2 (Fig. 10a, b). Consecutively, the QD-based Bi-FRET biopolymeric nanocomposite produced a linear detection range and detection limit of 0.05–4 μM and 1 nM (Fig. 10c). The nanocomposite was applied for sensing Zn^{2+} ions in solution or paper substrate and to visualize Zn^{2+} in living cells [4].

In another development, Wu et al. [99] demonstrated the use of DNAzymes coupled SiO_2@CdSe/ZnS QDs as selective sensors for Pb^{2+} and Cu^{2+} ions in aqueous solutions. The DNAzymes were made from two quenchers and one of the quenchers was designed with binding pockets that are selective to Zn^{2+} and Cu^{2+} ions. The sensing mechanism was based on the fluorescence turn-on of the QD upon the binding of metal ions to the DNAzymes due to the cleavage of DNAzymes, which separates from QD (Fig. 10d). Multiplexed detection of Cu^{2+} and Zn^{2+} was demonstrated by using two QDs of different emission colors within a range of 0.2–4 μM with detection limits of 0.2 nM and 0.5 nM, respectively [99].

4 Challenges, Conclusion, and Future Perspectives

For years, new functional materials based on QDs have been created and improved for analytical and bioanalytical applications. One of the realized shortcomings associated with QD-based nanocomposites is their fabrication. For instance, QDs-polymer composites suffer from problems associated with the compatibility of the polymer and the QD as well as obtaining proper homogeneity of QDs in the polymer host. In some cases, the QD nanocomposite loses its optical properties during the incorporation of QDs with magnetic materials and paramagnetic ions. Even though a lot of research on synthesis methods has been published in the previous decade, it is still challenging to discover a scalable, effective, but straightforward manufacturing pathway to produce QD nanocomposites. Another shortcoming is achieving selective detection of metals in the midst of other molecules and multi-analyte detection. Although many attempts have been successful with varied functionality, vast surface area, up-conversion photoluminescence, and the electron reservoir features of QDs,

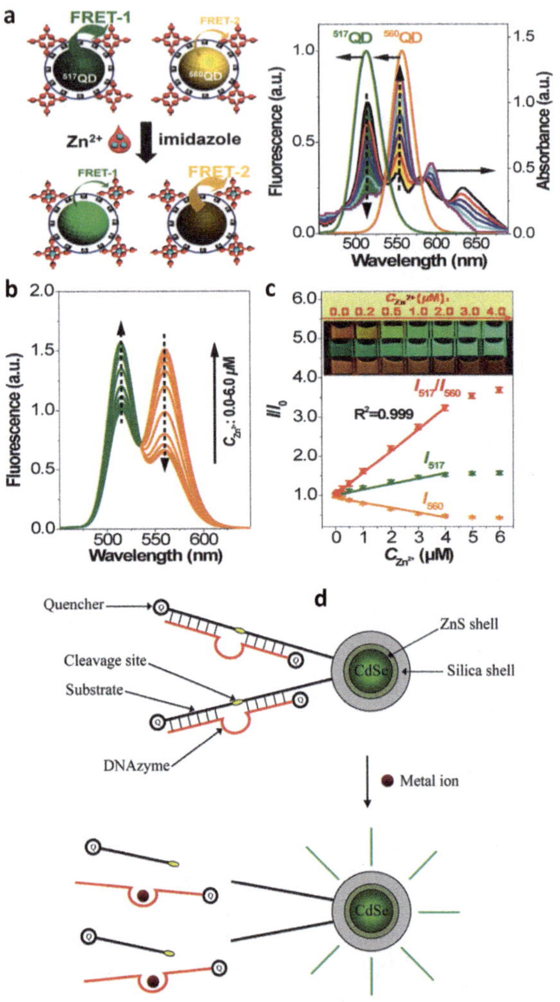

Fig. 10 Schematic diagram demonstrating the detection of Zn^{2+} ions using the QD-based Bi-FRET biopolymeric nanocomposite (left) and the corresponding spectral modeled response (right) (**a**). PL spectra **b** and linear plot of fluorescence intensity ratios (QD_{517}/QD_{560}) **c** of QD-based Bi-FRET biopolymeric in the presence of Zn^{2+} ions (0.2–4 μM). [99] Reproduced with permission from reference [4]. **d** The schematic diagram for the sensing mechanism of Pb^{2+} and Cu^{2+} ions using DNAzymes@SiO_2@CdSe/ZnS QDs. Reproduced with permission from reference

this field of research could benefit from tapping into molecular imprinting strategies and the incorporation of more than one QD type to target different analytes at once.

Metal detection is mostly accomplished by the interaction of QDs, however, the introduction of other functional materials to the QDs has demonstrated great potential in pollutant removal especially those based on carbon and magnetic nanomaterials. Due to their unmatched flexibility and integration for multi-analyte detection, these quantum dots nanocomposites have created a wide range of intriguing possibilities in environmental and biological sensing applications.

With the use of targeted ligands such as nucleic acids, peptides, antibodies, and small molecules, QD synthesis, and functionalization continues to improve. Studies on QD-based intracellular sensors are advancing their potential use due to their great biocompatibility and performance in super-targeted administration. However, QD technology still has considerable difficulties achieving the intended broad bioanalytical goals. The difficulties include some restrictions, such as the size compatibility, which restricts their tissue penetration; low stability (low quantum efficiency and high decay rates); as well as the toxicity consequences. We think that a major factor in their extensive use is the development of more efficient, reliable, and repeatable conjugation techniques. Looking ahead to QD technology, there is tremendous bioanalytical potential, and as soon as these limiting issues have been addressed, the field will make a huge advancement.

References

1. Xie, W., Peng, C., Wang, H., Chen, W.: Health risk assessment of trace metals in various environmental media, crops and human hair from a mining affected area. Int. J. Environ. Res. Public Health **14** (2017). https://doi.org/10.3390/ijerph14121595
2. Stankovic, S., Jovic, M., Stankovic, A.R., Katsikas, L.: Environmental Chemistry for a Sustainable World (2014). https://doi.org/10.1007/978-94-007-2442-6
3. Jaishankar, M., Tseten, T., Anbalagan, N., Mathew, B.B., Beeregowda, K.N.: Toxicity, mechanism and health effects of some heavy metals. Interdiscip. Toxicol. **7**, 60–72 (2014). https://doi.org/10.2478/intox-2014-0009
4. Liu, Y., Qu, X., Guo, Q., Sun, Q., Huang, X.: QD-biopolymer-TSPP assembly as efficient BiFRET sensor for ratiometric and visual detection of Zinc Ion. ACS Appl. Mater. Interfaces **9**, 4725–4732 (2017). https://doi.org/10.1021/acsami.6b14972
5. Ren, H.-B., Wu, B.-Y., Chen, J.-T., Yan, X.-P.: Silica-coated S2-enriched manganese-doped ZnS quantum dots as a photoluminescence probe for imaging intracellular Zn^{2+} ions. Anal. Chem. **83**, 8239–8244 (2011)
6. Bader, N.R.: Sample preparation for flame atomic absorption spectroscopy: an overview. Rasayan J. Chem. **4**, 49–55 (2011)
7. Inui, T., Kosuge, A., Ohbuchi, A., Fujita, K., Koike, Y., Kitano, M., Nakamura, T.: Determination of heavy metals at sub-ppb levels in water by graphite furnace atomic absorption spectrometry using a direct introduction technique after preconcentration with an iminodiacetate extraction disk. Am. J. Anal. Chem. **03**, 683–692 (2012). https://doi.org/10.4236/ajac.2012.310090
8. De Acha, N., Elosúa, C., Corres, J.M., Arregui, F.J.: Fluorescent sensors for the detection of heavy metal ions in aqueous media. Sensors (Switzerland) **19** (2019). https://doi.org/10.3390/s19030599
9. Zhang, D., Xu, Y., Liu, Q., Xia, Z.: Encapsulation of CH3NH3PbBr 3 perovskite quantum dots in MOF-5 microcrystals as a stable platform for temperature and aqueous heavy metal ion detection. Inorg. Chem. **57**, 4613–4619 (2018). https://doi.org/10.1021/acs.inorgchem.8b00355
10. Yang, W., Guo, W., Gong, X., Zhang, B., Wang, S., Chen, N., Yang, W., Tu, Y., Fang, X., Chang, J.: Facile synthesis of Gd–Cu–In–S/ZnS bimodal quantum dots with optimized properties for tumor targeted fluorescence/MR in vivo imaging. ACS Appl. Mater. Interfaces **7**, 18759–18768 (2015). https://doi.org/10.1021/acsami.5b05372
11. Wu, P., Zhao, T., Wang, S., Hou, X.: Semiconductor quantum dots-based metal ion probes. Nanoscale **6**, 43–64 (2014). https://doi.org/10.1039/c3nr04628a
12. Schiffman, J.D., Balakrishna, R.G.: Quantum dots as fluorescent probes: synthesis, surface chemistry, energy transfer mechanisms, and applications. Sens. Actuators B Chem. **258**, 1191–1214 (2018). https://doi.org/10.1016/j.snb.2017.11.189

13. May, B.M., Fakayode, O.J., Mashale, K.N., Bambo, M.F., Mishra, A.K., Nxumalo, E.N.: Detection and separation of silver ions from industrial wastewaters using fluorescent D-glucose carbon nanosheets and quaternary silver indium zinc sulphide quantum dots. J. Water Process Eng. **49**, 102944 (2022). https://doi.org/10.1016/j.jwpe.2022.102944
14. May, B.M., Bambo, M.F., Hosseini, S.S., Sidwaba, U., Nxumalo, E.N., Mishra, A.K.: A review on I–III–VI ternary quantum dots for fluorescence detection of heavy metals ions in water: optical properties, synthesis and application. RSC Adv. **12**, 11216–11232 (2022). https://doi.org/10.1039/d1ra08660j
15. Elfeky, S.A.: Facile sensor for heavy metals based on thiol-capped CdTe quantum dot. J. Environ. Anal. Chem. **05**, 1–5 (2018). https://doi.org/10.4172/2380-2391.1000232
16. Das, P., Maruthapandi, M., Saravanan, A., Natan, M., Jacobi, G., Banin, E., Gedanken, A.: Carbon dots for heavy-metal sensing, pH-sensitive cargo delivery, and antibacterial applications. ACS Appl. Nano Mater. **3**, 11777–11790 (2020). https://doi.org/10.1021/acsanm.0c02305
17. Guo, Y., Zhang, L., Zhang, S., Yang, Y., Chen, X., Zhang, M.: Fluorescent carbon nanoparticles for the fluorescent detection of metal ions. Biosens. Bioelectron. **63**, 61–71 (2015). https://doi.org/10.1016/j.bios.2014.07.018
18. Fakayode, O.J., Tsolekile, N., Songca, S.P., Oluwafemi, O.S.: Applications of functionalized nanomaterials in photodynamic therapy. Biophys. Rev. **10**, 49–67 (2018). https://doi.org/10.1007/s12551-017-0383-2
19. Frasco, M.F., Chaniotakis, N.: Semiconductor quantum dots in chemical sensors and biosensors. Sensors **9**, 7266–7286 (2009). https://doi.org/10.3390/s90907266
20. Zikalala, N., Parani, S., Oluwafemi, O.S.: Facile aqueous synthesis of ZnInS quantum dots and its application for selective detection of Co2+ ions. Nanotechnology **32** (2021). https://doi.org/10.1088/1361-6528/abf193
21. Muñoz, R., Santos, E.M., Galan-Vidal, C.A., Miranda, J.M., Lopez-Santamarina, A., Rodriguez, J.A.: Ternary quantum dots in chemical analysis, synthesis and detection mechanisms. Molecules **26**, 1–16 (2021). https://doi.org/10.3390/molecules26092764
22. Roy, S., Bardhan, S., Mondal, D., Saha, I., Roy, J., Das, S., Chanda, D.K., Karmakar, P., Das, S.: Polymeric carbon dot/boehmite nanocomposite made portable sensing device (Kavach) for non-invasive and selective detection of Cr(VI) in wastewater and living cells. Sens. Actuators B Chem. **348**, 130662 (2021). https://doi.org/10.1016/j.snb.2021.130662
23. Revesz, I.A., Hickey, S.M., Sweetman, M.J.: Metal ion sensing with graphene quantum dots: detection of harmful contaminants and biorelevant species. J. Mater. Chem. B. **10**, 4346–4362 (2022). https://doi.org/10.1039/d2tb00408a
24. Liu, J., Li, R., Yang, B.: Carbon dots: a new type of carbon-based nanomaterial with wide applications. ACS Cent. Sci. **6**, 2179–2195 (2020). https://doi.org/10.1021/acscentsci.0c01306
25. Zu, F., Yan, F., Bai, Z., Xu, J., Wang, Y., Yicun, H., Xuguang, Z.: The quenching of the fluorescence of carbon dots: a review on mechanisms and applications. Microchim. Acta **184**, 1899–1914 (2017). https://doi.org/10.1007/s00604-017-2318-9
26. Liu, W., Diao, H., Chang, H., Wang, H., Li, T., Wei, W.: Chemical Green synthesis of carbon dots from rose-heart radish and application for Fe^{3+} detection and cell imaging. Sens. Actuators B Chem. **241**, 190–198 (2017). https://doi.org/10.1016/j.snb.2016.10.068
27. Gauthier, T.D., Shane, E.C., Guerin, W.F., Seitz, W.R., Grant, C.L.: Fluorescence quenching method for determining equilibrium constants for polycyclic aromatic hydrocarbons binding to dissolved humic materials. Environ. Sci. Technol. **20**, 1162–1166 (1986). https://doi.org/10.1021/es00153a012
28. Molaei, M.J.: Principles, mechanisms, and application of carbon quantum dots in sensors: a review. Anal. Methods **12**, 1266–1287 (2020). https://doi.org/10.1039/C9AY02696G
29. Lu, S., Li, G., Lv, Z., Qiu, N., Kong, W., Gong, P., Chen, G., Xia, L., Guo, X., You, J., Wu, Y.: Facile and ultrasensitive fluorescence sensor platform for tumor invasive biomaker β-glucuronidase detection and inhibitor evaluation with carbon quantum dots based on inner-filter effect. Biosens. Bioelectron. **85**, 358–362 (2016). https://doi.org/10.1016/j.bios.2016.05.021

30. Huynh, T.V., Anh, N.T.N., Darmanto, W., Doong, R.A.: Erbium-doped graphene quantum dots with up- and down-conversion luminescence for effective detection of ferric ions in water and human serum. Sens. Actuators B Chem. **328**, 129056 (2021). https://doi.org/10.1016/j.snb.2020.129056
31. Wang, Y., He, J., Zheng, M., Qin, M., Wei, W.: Dual-emission of Eu-based metal-organic frameworks hybrids with carbon dots for ratiometric fluorescent detection of Cr(VI). Talanta **191**, 519–525 (2019). https://doi.org/10.1016/j.talanta.2018.08.078
32. Zheng, M., Xie, Z., Qu, D., Li, D., Du, P., Jing, X., Sun, Z.: On-off-on fluorescent carbon dot nanosensor for recognition of chromium(VI) and ascorbic acid based on the inner filter effect. ACS Appl. Mater. Interfaces **5**, 13242–13247 (2013). https://doi.org/10.1021/am4042355
33. Hildebrandt, N., Mihye, L., Namjum, K., Da Yeon, C., Jwam-Min, N.: Plasmonic quenching and enhancement: metal–quantum dot nanohybrids for fluorescence biosensing. Chem. Commun. **59**, 2352–2380 (2023). https://doi.org/10.1039/d2cc06178c
34. Hildebrandt, N., Charbonnière, L.J., Beck, M., Ziessel, R.F., Löhmannsröben, H.G.: Quantum dots as efficient energy acceptors in a time-resolved fluoroimmunoassay. Angew. Chem. Int. Ed. **44**, 7612–7615 (2005). https://doi.org/10.1002/anie.200501552
35. Mattsson, L., Wegner, K.D., Hildebrandt, N., Soukka, T.: Upconverting nanoparticle to quantum dot FRET for homogeneous double-nano biosensors. RSC Adv. **5**, 13270–13277 (2015). https://doi.org/10.1039/c5ra00397k
36. Wang, J., Teng, X., Wang, Y., Si, S., Ju, J., Pan, W., Wang, J., Sun, X., Wang, W.: Carbon dots based fluorescence methods for the detections of pesticides and veterinary drugs: response mechanism, selectivity improvement and application. Trends Anal. Chem. **144**, 116430 (2021). https://doi.org/10.1016/j.trac.2021.116430
37. Wu, W., Hu, Z., Shi, C., Xu, R., Zhao, Y., Ding, Y.: Construction of CdTe@γ-CD@RBD nanoprobe for Fe3+-sensing based on FRET mechanism in human serum. Spectrochim. Acta—Part A Mol. Biomol. Spectrosc. **296**, 122645 (2023). https://doi.org/10.1016/j.saa.2023.122645
38. Liu, C., Tang, B., Zhang, S., Zhou, M., Yang, M., Liu, Y., Zhang, Z., Zhang, B., Pang, D.: Photoinduced electron transfer mediated by coordination between carboxyl on carbon nanodots and Cu2+ quenching photoluminescence. J. Phys. Chem. C **122**, 3662–3668 (2018). https://doi.org/10.1021/acs.jpcc.7b12681
39. Lin, Z.J., Zheng, H.Q., Zheng, H.Y., Lin, L.P., Xin, Q., Cao, R.: Efficient capture and effective sensing of Cr2O72-from water using a zirconium metal-organic framework. Inorg. Chem. **56**, 14178–14188 (2017). https://doi.org/10.1021/acs.inorgchem.7b02327
40. Gabor, G., Walt, D.R.: Sensitivity enhancement of fluorescent pH indicators by inner filter effects. Anal. Chem. **63**, 793–796 (1991). https://doi.org/10.1021/ac00008a011
41. Li, X., Zhao, S., Li, B., Yang, K., Lan, M., Zeng, L.: Advances and perspectives in carbon dot-based fluorescent probes: mechanism, and application. Coord. Chem. Rev. **431**, 213686 (2021). https://doi.org/10.1016/j.ccr.2020.213686
42. Ma, P., Zuo, J., Li, Z., Xiao, D., Wu, H., Zhang, Y., Dong, A.: Application progress of green carbon dots in analysis and detection. Part. Particle Syst. Charact. **39**, 1–21 (2022). https://doi.org/10.1002/ppsc.202200104
43. Wang, S., Wang, H., Yue, J., Lu, Y., Jia, T., Chen, J.: Highly sensitive and selective fluorescent sensor for Ag+ detection using β-cyclodextrin/chitosan polymer-coated S QDs based on an aggregation-induced quenching mechanism. Cellulose **29**, 4083–4095 (2022). https://doi.org/10.1007/s10570-022-04524-7
44. Cho, E., Kim, T., Choi, S., Jang, H., Min, K., Jang, E.: Optical characteristics of the surface defects in InP colloidal quantum dots for highly efficient light-emitting applications. ACS Appl. Nano Mater. 17106–7114 (2018). https://doi.org/10.1021/acsanm.8b01947
45. May, B.M.M., Parani, S., Oluwafemi, O.S.: Detection of ascorbic acid using green synthesized AgInS2 quantum dots. Mater. Lett. **236**, 432–435 (2019). https://doi.org/10.1016/j.matlet.2018.10.155
46. Li, L., Pandey, A., Werder, D.J., Khanal, B.P., Pietryga, J.M., Klimov, V.I.: Efficient synthesis of highly luminescent copper indium sulfide-based core/shell nanocrystals with surprisingly long-lived emission. J. Am. Chem. Soc. **133**, 1176–1179 (2011). https://doi.org/10.1021/ja108261h

47. Xia, Y.S., Cao, C., Zhu, C.Q.: Two distinct photoluminescence responses of CdTe quantum dots to Ag (I). J. Lumin. **128**, 166–172 (2008). https://doi.org/10.1016/j.jlumin.2007.07.007
48. Jianqiao, L., Yu, B., Jingcheng, S., Qing, Y., Jifei, L., Jiayu, Y., Ce, F., Qianru, Z.: Selective detection of mercury ions based on tin oxide quantum dots: performance and fluorescence enhancement model. J. Mater. Chem. C **8**, 8274–8284 (2021). https://doi.org/10.1039/d1tc00824b
49. Liu, Y., Tang, X., Deng, M., Zhu, T., Edman, L., Wang, J.: Hydrophilic AgInZnS quantum dots as a fluorescent turn-on probe for Cd2+ detection. J. Alloys Compd. **864**, 158109 (2021). https://doi.org/10.1016/j.jallcom.2020.158109
50. Luo, J., Xie, Z., Lam, J.W.Y., Cheng, L., Chen, H., Qiu, C., Kwok, H.S., Zhan, X., Liu, Y., Zhu, D., Tang, B.Z.: Aggregation-induced emission of 1-methyl-1,2,3,4,5-pentaphenylsilole. Chem. Commun. 1740–1741 (2001). https://doi.org/10.1039/b105159h
51. Wang, C., Jiang, K., Xu, Z., Lin, H., Zhang, C.: Glutathione modified carbon-dots: from aggregation-induced emission enhancement properties to a "turn-on" sensing of temperature/Fe3+ ions in cells. Inorg. Chem. Front. **3**, 514–522 (2016). https://doi.org/10.1039/c5qi00273g
52. Wei, C., Wei, X., Hu, Z., Yang, D., Mei, S., Zhang, G., Su, D.: A fluorescent probe for Cd2+ detection based on the aggregation-induced emission enhancement of aqueous Zn–Ag–In–S quantum dots. Anal. Methods **11**, 2559–2564 (2019). https://doi.org/10.1039/c9ay00716d
53. Liu, Y., Deng, M., Zhu, T., Tang, X., Han, S., Huang, W., Shi, Y., Liu, A.: The synthesis of water-dispersible zinc doped AgInS2 quantum dots and their application in Cu2+ detection. J. Lumin. **192**, 547–554 (2017). https://doi.org/10.1016/j.jlumin.2017.07.018
54. Zhou, Z., Yan, R., Zhao, J., Yang, L., Chen, J., Hu, Y.: Highly selective and sensitive detection of Hg $2+$ based on fluorescence enhancement of Mn-doped ZnSe QDs by Hg2 $+ -$Mn2 $+$ replacement. Sens. Actuators B Chem. **254**, 8–15 (2018). https://doi.org/10.1016/j.snb.2017.07.033
55. Zhou, Q., Li, Z., Wang, Q., Peng, L., Luo, S., Gu, F.L.: Polymer-capped CdSe/ZnS quantum dots for the sensitive detection of Cu2+ and Hg2+ and the quenching mechanism. Anal. Methods (2021). https://doi.org/10.1039/d1ay00432h
56. Manna, G., Jana, S., Bose, R., Pradhan, N.: Mn-doped multinary CIZS and AIZS nanocrystals. J. Phys. Chem. Lett. **3**, 2528–2534 (2012). https://doi.org/10.1021/jz300978r
57. Gharghani, S., Zare, H., Shahedi, Z., Fazaeli, Y., Rahighi, R.: Synthesis of magnetic ions-doped QDs synthesized via a facial aqueous solution method for optical/MR dual-modality imaging applications. J. Fluoresc. **31**, 897–906 (2021). https://doi.org/10.1007/s10895-021-02720-5
58. Pons, T., Pic, E., Lequeux, N., Cassette, E., Bezdetnaya, L., Guillemin, F., Marchal, F., Dubertret, B.: Cadmium-free CuInS2/ZnS quantum dots for sentinel lymph node imaging with reduced toxicity. ACS Nano **4**, 2531–2538 (2010). https://doi.org/10.1021/nn901421v
59. Chen, H.Q., Fu, J., Wang, L., Ling, B., Bin Qian, B., Chen, J.G., Zhou, C.L.: Ultrasensitive mercury(II) ion detection by europium(III)-doped cadmium sulfide composite nanoparticles. Talanta**83**, 139–144 (2010). https://doi.org/10.1016/j.talanta.2010.08.052
60. Jiao, M., Li, Y., Jia, Y., Yang, Z., Luo, X.: Aqueously synthesized color-tunable quaternary Cu-In-Zn-S quantum dots for Cu(II) detection via mild and rapid cation exchange. Sens. Actuators B Chem. **294**, 32–39 (2019). https://doi.org/10.1016/j.snb.2019.05.027
61. Keskin, İÇ., Türemiş, M., Kati, M.İ, Kibar, R., Çetin, A.: Effects of CdS quantum dot in polymer nanocomposites: in terms of luminescence, optic, and thermal results. Radiat. Phys. Chem. **156**, 137–143 (2019). https://doi.org/10.1016/j.radphyschem.2018.11.006
62. Weaver, J., Zakeri, R., Aouadi, S., Kohli, P.: Synthesis and characterization of quantum dot-polymer composites. J. Mater. Chem. **19**, 3198–3206 (2009). https://doi.org/10.1039/b820204d
63. Matvienko, O.O., Savin, Y.N., Kryzhanovska, A.S., Vovk, O.M., Dobrotvorska, M.V., Pogorelova, N.V., Vashchenko, V.V.: Dispersion and aggregation of quantum dots in polymer-inorganic hybrid films. Thin Solid Films **537**, 226–230 (2013). https://doi.org/10.1016/j.tsf.2013.03.046
64. He, X., Jia, K., Bai, Y., Chen, Z., Liu, Y., Huang, Y., Liu, X.: Quantum dots encoded white-emitting polymeric superparticles for simultaneous detection of multiple heavy metal ions. J. Hazard. Mater. **405**, 124263 (2021). https://doi.org/10.1016/j.jhazmat.2020.124263

65. Thakur, D.: Fluorescent and magnetic nanocomposites for multimodal imaging. The Ohio State University (2010)
66. García, R.S., Stafford, S., Gun'ko, Y.K.: Recent progress in synthesis and functionalization of multimodal fluorescent-magnetic nanoparticles for biological applications. Appl. Sci. **8**, 12–16 (2018). https://doi.org/10.3390/app8020172
67. Tufani, A., Qureshi, A., Niazi, J.H.: Iron oxide nanoparticles based magnetic luminescent quantum dots (MQDs) synthesis and biomedical/biological applications: a review. Mater. Sci. Eng. C **118**, 111545 (2021). https://doi.org/10.1016/j.msec.2020.111545
68. Alvand, M., Shemirani, F.: A Fe3O4@SiO2@graphene quantum dot core-shell structured nanomaterial as a fluorescent probe and for magnetic removal of mercury(II) ion. Microchim. Acta **184**, 1621–1629 (2017). https://doi.org/10.1007/s00604-017-2134-2
69. Lingamdinne, L.P., Koduru, J.R., Karri, R.R.: A comprehensive review of applications of magnetic graphene oxide based nanocomposites for sustainable water purification. J. Environ. Manag. **231**, 622–634 (2019). https://doi.org/10.1016/j.jenvman.2018.10.063
70. Liu, Z., Li, G., Xia, T., Su, X.: Ultrasensitive fluorescent nanosensor for arsenate assay and removal using oligonucleotide-functionalized Cu-In-S2 quantum dot@magnetic Fe3O4 nanoparticles composite. Sens. Actuators B Chem. **220**, 1205–1211 (2015). https://doi.org/10.1016/j.snb.2015.06.111
71. Kurshanov, D.A., Khavlyuk, P.D., Baranov, M.A., Dubavik, A., Rybin, A.V., Fedorov, A.V., Baranov, A.V.: Magneto-fluorescent hybrid sensor CaCo3–Fe3o4–AgInS2/ZnS for the detection of heavy metal ions in aqueous media. Materials (Basel) **13**, 1–11 (2020). https://doi.org/10.3390/ma13194373
72. Warner, C.L., Addleman, R.S., Cinson, A.D., Droubay, T.C., Engelhard, M.H., Nash, M.A., Yantasee, W., Warner, M.G.: High-performance, superparamagnetic, nanoparticle-based heavy metal sorbents for removal of contaminants from natural waters. Chemsuschem **3**, 749–757 (2010). https://doi.org/10.1002/cssc.201000027
73. Carlos, L., Einschlag, F.S.G., González, M.C., Mártire, D.O.: Applications of magnetite nanoparticles for heavy metal removal from wastewater. Waste Water—Treat. Technol. Recent Anal. Dev. (2013). https://doi.org/10.5772/54608
74. Matei, E., Predescu, A.M., Râpă, M., Tarcea, C., Pantilimon, C.M., Favier, L., Berbecaru, A.C., Sohaciu, M., Predescu, C.: Removal of chromium(VI) from aqueous solution using a novel green magnetic nanoparticle-chitosan adsorbent. Anal. Lett. **52**, 2416–2438 (2019). https://doi.org/10.1080/00032719.2019.1601734
75. Ren, C., Ding, X., Fu, H., Li, W., Wu, H., Yang, H.: Core-shell superparamagnetic monodisperse nanospheres based on amino-functionalized CoFe2O4@SiO2 for removal of heavy metals from aqueous solutions. RSC Adv. **7**, 6911–6921 (2017). https://doi.org/10.1039/c6ra27728d
76. Tahergorabi, M., Esrafili, A., Kermani, M., Shirzad-Siboni, M.: Application of thiol-functionalized mesoporous silica-coated magnetite nanoparticles for the adsorption of heavy metals. Desalin. Water Treat. **57**, 19834–19845 (2016). https://doi.org/10.1080/19443994.2015.1106351
77. Mokadem, Z., Mekki, S., Saïdi-Besbes, S., Agusti, G., Elaissari, A., Derdour, A.: Triazole containing magnetic core-silica shell nanoparticles for Pb2+, Cu2+ and Zn2+ removal. Arab. J. Chem. **10**, 1039–1051 (2017). https://doi.org/10.1016/j.arabjc.2016.12.008
78. Li, L., Wang, F., Lv, Y., Liu, J., Bian, H., Wang, W., Li, Y., Shao, Z.: Cqds-doped magnetic electrospun nanofibers: fluorescence self-display and adsorption removal of mercury(II). ACS Omega **3**, 4220–4230 (2018). https://doi.org/10.1021/acsomega.7b01969
79. Li, L., Jia, C., Wang, F., Fan, H., Jiao, W., Shao, Z.: Facile synthesis of magnetic fluorescent nanoparticles: Adsorption and selective detection of Hg(II) in water. J. Mater. Chem. C. **6**, 2360–2369 (2018). https://doi.org/10.1039/c7tc05564a
80. Wang, Y., Tang, M., Shen, H., Che, G., Qiao, Y., Liu, B., Wang, L.: Recyclable multifunctional magnetic mesoporous silica nanocomposite for ratiometric detection, rapid adsorption, and efficient removal of Hg(II). ACS Sustain. Chem. Eng. **6**, 1744–1752 (2018). https://doi.org/10.1021/acssuschemeng.7b03040

81. Guo, Y., Liu, H., Chen, D., Qu, J., Yang, J.: High recycling Fe3O4–CdTe nanocomposites for the detection of organophosphorothioate pesticide chlorpyrifos. Green Energy Environ. **7**, 229–235 (2022). https://doi.org/10.1016/j.gee.2020.09.001
82. Yang, C.H., Ding, Y.L., Qian, J.: Design of magnetic-fluorescent based nanosensor for highly sensitive determination and removal of HG2+. Ceram. Int. **44**, 9746–9752 (2018). https://doi.org/10.1016/j.ceramint.2018.02.209
83. Gu, H., Huang, X., Chen, Q., Sun, Y., Ping Tan, C.: The influence of different metal atoms on the performance of metalloporphyrin-based sensor reaction with propanol. Mater. Technol. **36**, 823–830 (2021). https://doi.org/10.1080/10667857.2020.1800295
84. Xin, Z., Wang, Y.R., Chen, Y., Li, W.L., Dong, L.Z., Lan, Y.Q.: Metallocene implanted metalloporphyrin organic framework for highly selective CO2 electroreduction. Nano Energy **67**, 104233 (2020). https://doi.org/10.1016/j.nanoen.2019.104233
85. Wu, T., Liu, X., Liu, Y., Cheng, M., Liu, Z., Zeng, G., Shao, B., Liang, Q., Zhang, W., He, Q.: Application of QD-MOF composites for photocatalysis: energy production and environmental remediation. Coord. Chem. Rev. **403**, 213097 (2020). https://doi.org/10.1016/j.ccr.2019.213097
86. Wu, Y., Wang, D., Li, Y.: Understanding of the major reactions in solution synthesis of functional nanomaterials. Sci. China Mater. **59**, 938–996 (2016). https://doi.org/10.1007/s40843-016-5112-0
87. Wang, X.Z., Zhang, Z.Q., Guo, R., Zhang, Y.Y., Zhu, N.J., Wang, K., Sun, P.P., Mao, X.Y., Liu, J.J., Huo, J.Z., Wang, X.R., Ding, B.: Dual-emission CdTe quantum dot@ZIF-365 ratiometric fluorescent sensor and application for highly sensitive detection of L-histidine and Cu2+. Talanta **217**, 121010 (2020). https://doi.org/10.1016/j.talanta.2020.121010
88. Guo, M., Chi, J., Li, Y., Waterhouse, G.I.N., Ai, S., Hou, J., Li, X.: Fluorometric determination of mercury(II) based on dual-emission metal-organic frameworks incorporating carbon dots and gold nanoclusters. Microchim. Acta. **187** (2020). https://doi.org/10.1007/s00604-020-04508-z
89. Xing, Y., Chaudry, Q., Shen, C., Kong, K.Y., Zhau, H.E., Chung, L.W., Petros, J.A., O'regan, R.M., Yezhelyev, M.V., Simons, J.W.: Bioconjugated quantum dots for multiplexed and quantitative immunohistochemistry. Nat. Protoc. **2**, 1152–1165 (2007)
90. Medintz, I.L., Uyeda, H.T., Goldman, E.R., Mattoussi, H.: Quantum dot bioconjugates for imaging, labeling and sensing. Nat. Mater. **4**, 435–446 (2005)
91. Pathak, S., Davidson, M.C., Silva, G.A.: Characterization of the functional binding properties of antibody conjugated quantum dots. Nano Lett. **7**, 1839–1845 (2007)
92. Mahtab, R., Harden, H.H., Murphy, C.J.: Temperature-and salt-dependent binding of long DNA to protein-sized quantum dots: thermodynamics of "inorganic protein"—DNA interactions. J. Am. Chem. Soc. **122**, 14–17 (2000)
93. Pinaud, F., King, D., Moore, H.-P., Weiss, S.: Bioactivation and cell targeting of semiconductor CdSe/ZnS nanocrystals with phytochelatin-related peptides. J. Am. Chem. Soc. **126**, 6115–6123 (2004)
94. Mansur, H.S.: Quantum dots and nanocomposites. Wiley Interdiscip. Rev. Nanomed. Nanobiotechnol. **2**, 113–129 (2010)
95. Smith, A.M., Duan, H., Mohs, A.M., Nie, S.: Bioconjugated quantum dots for in vivo molecular and cellular imaging. Adv. Drug Deliv. Rev. **60**, 1226–1240 (2008)
96. Zhu, X., Chen, L., Shen, P., Jia, J., Zhang, D., Yang, L.: High sensitive detection of Cry1Ab protein using a quantum dot-based fluorescence-linked immunosorbent assay. J. Agric. Food Chem. **59**, 2184–2189 (2011)
97. Han, M., Gao, X., Su, J.Z., Nie, S.: Quantum-dot-tagged microbeads for multiplexed optical coding of biomolecules. Nat. Biotechnol. **19**, 631–635 (2001)
98. Ma, Q., Su, X.: Advances in the application of QD-based intracellular sensing systems. Appl. Spectrosc. Rev. **51**, 162–181 (2016). https://doi.org/10.1080/05704928.2015.1096276
99. Wu, C.S., Khaing Oo, M.K., Fan, X.: Highly sensitive multiplexed heavy metal detection using quantum-dot-labeledDNAzymes. ACS Nano **4**, 5897–5904 (2010). https://doi.org/10.1021/nn1021988

Quantum Dots Composites in Catalysis Applications

Krishna Hari Sharma, Nallin Sharma, and Chandan Srivastava

Abstract The quantum dot (QD) has displayed numerous outstanding roles in scientific advancements recently. Another aspect of QD development is their role in photocatalytic activity. The optical tunnability of QD-based composites has immensely attracted the scientific community to explore their capabilities. The various studied compositions are in accordance with carbon nanomaterials, transition metal oxide nanomaterials, metal chalcogenides nanomaterials, perovskites nanomaterials. These three are broader categories covered in the chapter with an interest towards various photocatalytic approaches.

Keywords Quantum dots · Composites · Catalysis · Charge-transfer

List of Abbreviations

CQDs	Carbon QDs
MX	Chalcogenides
ChQDs	Chalcogenides QDs
CB	conduction band and valence band (VB)
COFs	covalent organic frameworks
CTF-1	covalent triazine-based frameworks
EHP	electron-hole pair
g-C3N4	graphitic carbon nitride
MPA	mercaptopropionic acid
MOFs	metal-organic frameworks

K. H. Sharma
Department of Electrical Engineering, National Taiwan University, Taipei, Taiwan

N. Sharma (✉) · C. Srivastava
Department of Materials Science, Indian Institute of Sciences, Bangalore, India
e-mail: nallin_25@hotmail.com

C. Srivastava
e-mail: csrivastava@iisc.ac.in

S. Thomas et al. (eds.), *Quantum Dots Based Nanocomposites*, Engineering Materials,
https://doi.org/10.1007/978-3-031-54779-9_14

N-CQDs	nitrogen-doped CQDs
PQDs	Perovskite QDs
QD	Quantum Dots
rGO	reduced graphene oxide
SQDs	Semiconductor QDs
HNTs	silica/halloysite nanotubes
TC-HCl	tetracycline chloride
TGA	thioglycolic acid

1 Introduction

The Industrial Revolution expedited the growth of economic development which hauled the prosperity of humankind. However, the backbone of this opulence is fossil fuels for energy production. It is predicted that the energy consumption globally will reach around 25–27 TW, by 2050 [1–3]. Moreover, such heavy reliance on fossil fuels leads to severe problems such as the exhaustion of greenhouse gases (causing air pollution), water pollution (freshwater streams, seawater), and soil pollution (because of landfills). To foresee a better future for the next generation, a better alternative is required for the production of clean energy, and to revive the natural resources. To deal with this, the scientific community seeks immense hope in the catalysis process. The catalysis process facilitates the production of clean fuel (such as H_2), wastewater treatment (using photo-and electro-catalysis), and VOC abatement [4–7]. The applications of QD in photocatalytic reaction producing Hydrogen and oxygen are of keen interest to the scientific community. Hydrogen evolution reaction (HER) and Oxygen evolution reaction (OER) are being studied with prime emphasis in several atmospheric conditions. An efficient catalyst can be engineered by fabricating a hybrid composite containing semiconductors, metal nanoparticles, quantum dots etc. [8]. Among all the materials, quantum dots (QDs) are the focus of the interest in this chapter which play pivotal roles in designing efficient composites for catalytic applications. Being quasi-zero-dimensional materials, QDs possess the large surface area which makes them capable of rapid electron transfer, tunable chemical activity, renewability, nontoxicity, large two-photon absorption cross-sections, up-conversion ability, and excellent selectivity [9, 10]. Therefore, the performance of QDs-based composites can be modulated by modifying their surface and bulk features, and energy gap enlargement [11]. These properties make QDs an excellent candidate for catalytic applications. This chapter features the role of carbon QDs-, semiconductor QDs-, and perovskite QDs-based composites for their applications in catalytic based energy conversion processes.

2 Carbon QDs (CQDs)-Based Composites

Carbon-based materials, as of its abundance in nature, have always attracted researchers and scientists for their applications in catalysis. In addition, these materials are bio-compatible, less toxic, photostable, and have the ability of chemiluminescence, multiphoton excitation, and up-conversion photoluminescence [12–14]. Moreover, carbon QDs (CQDs), being a member of carbon family, inherits all the capabilities. Besides, CQDs also possess larger surface area, high conductivity, and supernal charge-transfer capabilities. The electronic properties of the CQDs can further be modified using the doped heteroatom (B, N, P, S, etc.), this strategy enhances the catalytic properties via intramolecular charge transfer [15, 16]. Simultaneously, the intermolecular charge transfer can also boost by the CQDs-based composites. Based on the synthesis precursors, different functional groups ($-NH_2$, $-OH$, $-COOH$, etc.) present on the surface of CQDs could provide favorable sites for highly active catalysis processes [17].

There are several approaches which should be in considerations while fabricating an efficient CQDs-based composite. Here, some of them will be discussed. To begin with, the use of CQDs as an active center to promote defect sites for catalytic reactions-CQDs/graphene composite was fabricated to achieve the overall water splitting [18]. In this work, the combination of CQDs and graphene facilitated the catalytic reaction in synergy-CQDs offer abundant active edges and defect sites for electrocatalytic reactions. Whereas, the graphene served as a conductive substrate which prevent the aggregation of CQDs, and ease the charge transfer process (Fig. 1).

Using another approach, the composites of CQDs and metal compounds can be used as an efficient catalyst. Generally, the abundance presence of functional groups on the surface of CQDs encourages the coordination between CQDs and metal ions with empty d-orbitals. This leads to the formation of stable CQDs-metal compounds composites [19]. Moreover, the employed CQDs boast to form ultrafine crystals with stable nanostructures [20]. In another report, it was reported that the nitrogen-doped CQDs (N-CQDs) helps to improve the surface area of MnO_2 nanoflowers

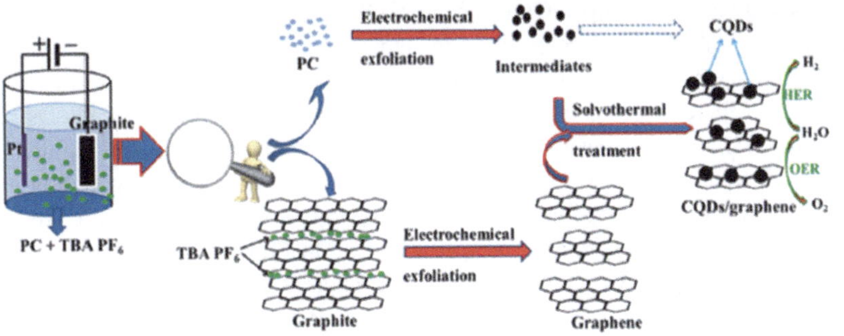

Fig. 1 Probable formation route of the CQDs/graphene heterostructure [18]. Copyright 2019, Royal Society of Chemistry

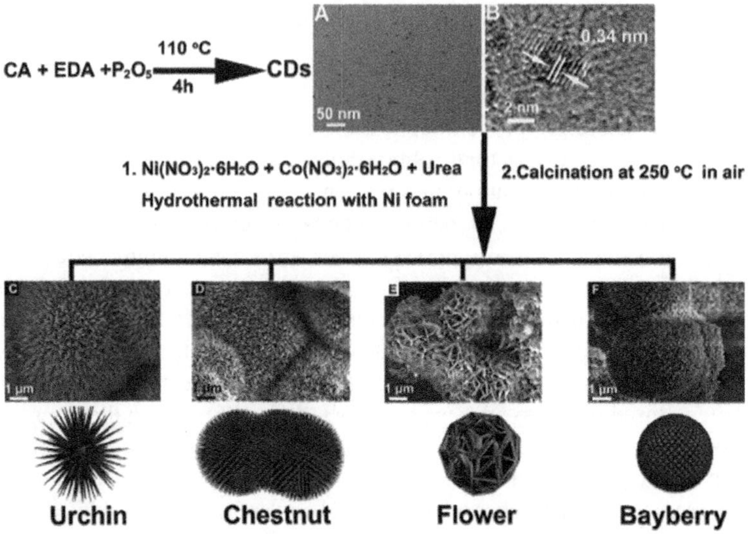

Fig. 2 **a, b** the hydrothermally synthesized CQDs and **c–f** CQDs/$NiCo_2O_4$ composite with different morphologies [22]. Copyright 2016, Wiley

in a N-CQDs/MnO_2 composite. In addition, it roughens the surface with enhanced conductivity and improved OER activity [21]. The same strategy was used by Xiong et al. where the morphology of $NiCo_2O_4$ was controlled by adjusting inventory rating of CQDs-from sea urchin to flower, and bayberry, as displays in figure [22] (Fig. 2).

Moreover, few reports claimed to enhance the electrochemical activity by employing the CQDs for enhancing the conductivity of overall composite. The use of CQDs effectively eliminates the Schottky barrier from the catalyst-electrolyte interface, and catalyst-electrode interface [23]. For instance, Kang et al. has reported CQDs/SnO_2–Co_3O_4 composite for electrocatalytic water oxidation where CQDs enhanced the conductivity of composite by facilitating the enhanced electron transfer between the surface and active sites [24]. In another report, Yang et al. reported the synthesis of Co, Fe co-doped CQDs where CQDs enhanced the electrocatalytic OER activity and stability by accelerating the electron transfer [25].

The remarkable performances of CQDs-based composites manifested themselves as a perfect fit for various electrocatalytic applications. Nevertheless, CQDs-based composites also show enhanced photocatalytic activities for various applications such as H_2 production, CO_2 reduction, and organic contamination degradation. For instance, Sargin et al. reported CQDs/TiO_2/Pt catalyst for H_2 production, where, up on the absorption of the photon, the sp^2-conjugated CQDs produce excited electrons in the valence-conduction-band bridge [26]. Thereafter, the electrons get transfer to the conduction band of TiO_2 which results in the production of H_2—the report claimed to produce 1458 μ mol g^{-1} of H_2 after 2 h (Fig. 3).

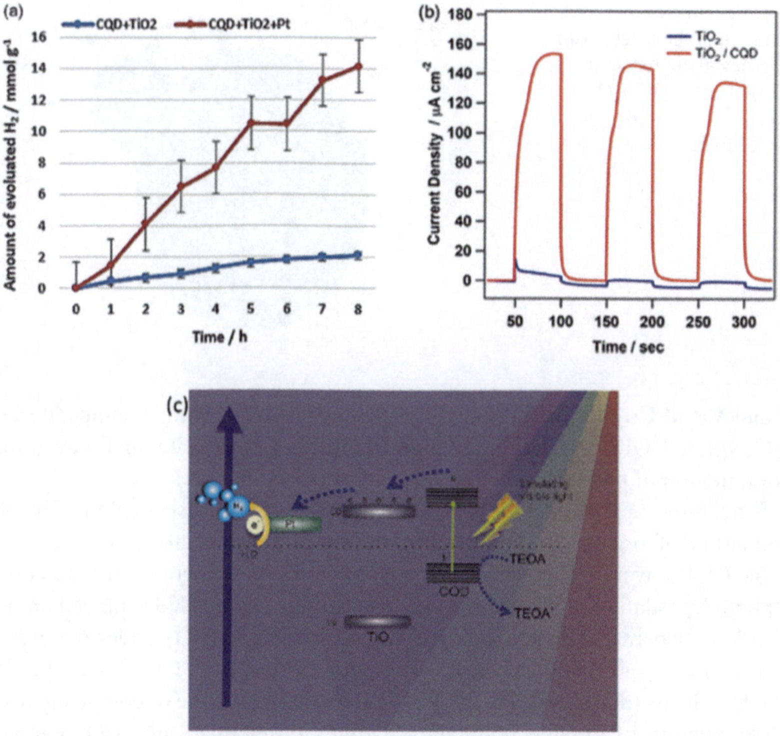

Fig. 3 **a** The photocatalytic hydrogen evolution obtained by gas chromatography, **b** the photocurrent density (using chronoamperometry method) of TiO_2, and CQD/TiO_2, and **c** Schematic representation of photocatalytic hydrogen evolution mechanism of CQD/TiO_2/Pt [26]. Copyright 2019, Elsevier

In addition, Gogoi et al. developed CQDs/CdS composite for UV-Vis driven photocatalytic water-splitting reaction-the use of CQDs enhanced the performance of catalyst by 50% as compared to bare CdS [27]. Moreover, Kulandaivalu et al. synthesized CQDs/Cu_2O nanocomposite for the photoreduction of CO_2. It is reported that CQDs improved the overall process by 54% as compared to bare Cu_2O. The mechanism of photoreduction of CO_2–C_2H_6 is reported as follows-Upon visible light-driven photoexcitation of Cu_2O, the electron-hole pairs generate in the conduction band (CB) and valence band (VB) of Cu_2O respectively. The redox potential of pure Cu_2O was reported -1.0 V and $+1.1$ V for VB and CB respectively which makes it competent to reduce CO_2–C_2H_6 ($E^o = -0.27$ V). The process began as the electrons in the CB of Cu_2O interacts with CO_2 to produce CO_2^- radicals, simultaneously, the holes in the VB of Cu_2O reacted with H_2O to form OH and H_2 which further oxidized into O_2 and H^+. The H radicals formed as the electrons reacted with H^+ ions which then reduce the CO_2^- radicals to form CH_3 which further dimerized to generate C_2H_6. However, in this process, the bare Cu_2O displayed poor efficiency because of rapid recombination rate of photogenerated electron and hole pairs in the

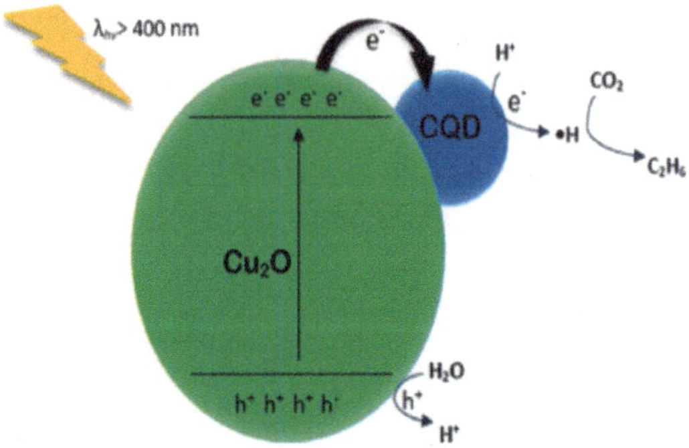

Fig. 4 Schematic illustration of the mechanism of photocatalytic reduction of CO_2 over CQD/Cu_2O composite [28]. Copyright 2018, Elsevier

CB and VB of Cu_2O. This obstacle was cleared out by making a composite CQDs/Cu_2O, where CQDs assisted in the photostability of composite, and preventing the recombination of photogenerated charge carriers [28] (Fig. 4).

Using same approach, CQDs-based composites were also exploited in photodegradation of organic pollutants. For example, Tyagi et al. immobilized water soluble CQDs over electro-spun TiO_2 nanofibers [29]. This composite was used for the photodegradation of methylene blue dye. In this report, CQDs played an important role to separate the photo-excited charge carriers in TiO_2 under the exposure. The production of O_2 (due to the electrons transferred from CB of TiO_2 to CQDs) and OH radicals (due to holes in the VB of TiO_2) at the surface of composite resulted the degradation of organic pollutant 2.5 times faster than only TiO_2 nanofibers. In another report, Ramar et al. compared the performance of undoped-CQDs and nitrogen-doped CQDs for the degradation of MB dye under the exposure of solar light [30].

3 Semiconductor QDs (SQDs)-Based Composites

Having the size comparable to three-dimensional exciton Bohr radius, the movement of internal electrons get restricted in QDs [31]. Here, in case of the semiconductors, if the size of nanomaterials gets smaller or equal to its exciton Bohr radius, then, the internal energy of that material converts from nearly continuous energy band (in bulk materials) to discrete energy levels. Therefore, the bandgap of semiconductor QDs (SQDs) can be tuned by modulating its size. This makes SQDs a highly efficient class of materials for the applications in sensing, imaging, medicines, and displays. Among them, SQDs-based catalysis, prominently in the field of photocatalysis- for hydrogen production, CO_2 reduction, and organic pollutants degradation, will be discussed later in this section.

In general, when the light energy (hν) equal or greater than the bandgap energy (E_g) of the SQDs incident on the SQDs, compel electrons of the VB (of SQDs) transit to the CB (of SQDs). Upon excitation of electrons to the CB, the holes remain in the

VB of SQDs, this process is called electron-hole pair (EHP) generation. These photogenerated EHPs later contribute in the catalytic reactions-electrons are reductive in nature which are necessary in hydrogen production and CO_2 reduction, whereas, the holes are oxidative in nature which required in organic pollutants degradations [32]. Typically, SQDs-driven photocatalytic reactions face several challenges, including the surface traps of EHPs, the active sites blockage by reaction intermediates or agglomerates, and rapid recombination of EHPs. Therefore, to make this process more efficient, a co-catalyst generally introduced with SQDs. This SQDs-based composite work in synergy to separate EHPs, and avail them for respective catalytic reactions [33].

Chalcogenides (MX) QDs (ChQDs) are the most exploited class of QDs for diverse range of applications, where M represent the transition metals, and X signify the chalcogens (S, Se, and Te). The VB and CB of ChQDs are mainly formed because of the chalcogen, and transition metal element respectively. In general, ChQDs have the characteristics of narrow and direct bandgap which gives them advantage of rapid EHP generation. But, the maximum of generated EHPs quickly recombine. In addition, ChQDs also have the tendency to agglomerate during use, and due to the long exposure of light, the photogenerated holes get accumulate on the surface, resulting in the photo-corrosion. However, there are various strategies which can solve this issue with the advantage of overall enhancement in the catalytic efficiency. For instance, Liang's group fabricated the composite of CdS QDs with graphitic carbon nitride (g-C_3N_4) [34]. In this report, phosphorus modified g-C_3N_4 tubes were used as a porous matrix with CdS QDs which could protect QDs from photo-corrosion and agglomeration. These strategies promote the EHPs separation, suppress radiative recombination, extend the lifetime of carriers, and promote the reusability of catalyst (Fig. 5).

In another report, as templates of CdS QDs, mesoporous silica/halloysite nanotubes (HNTs) was investigated [35]. This composite was employed for the production of H_2 gas, the production of H_2 could reach at 2600 μ mol h^{-1} g^{-1}, with the reusability of five cycles. Moreover, to achieve the H_2 production rate of 13129 μ mol h^{-1} g^{-1}, Zhuge et al. fabricated MoS_2 coated CdS QDs [36]. The encapsulated MoS_2 on the surface of CdS QDs promote internal charge separation, and prevent photo-corrosion (Fig. 6).

In other reports, it's been studied that ZnS shell could prevent the QDs from photo-corrosion. ZnS shell inhibited the formation of trapping state on the surface of QDs, and facilitated the transfer of photogenerated carriers from QD core to the surface [37, 38]. Furthermore, the photocatalytic activity of ChQDs-based cocatalysts can be improved by surface modification including surface stoichiometry manipulation, construction of core/shell structure, doping with inorganic ions, ligand removal and exchange [39]. The surface ligands not only protect QDs from aggregation but they also regulate the growth rate of ChQDs, depending upon its chain lengths [40]. In addition, Bajorowicz et al. reported that the thioglycolic acid (TGA) caped CdTe QDs showed better photocatalytic performance than the mercaptopropionic acid

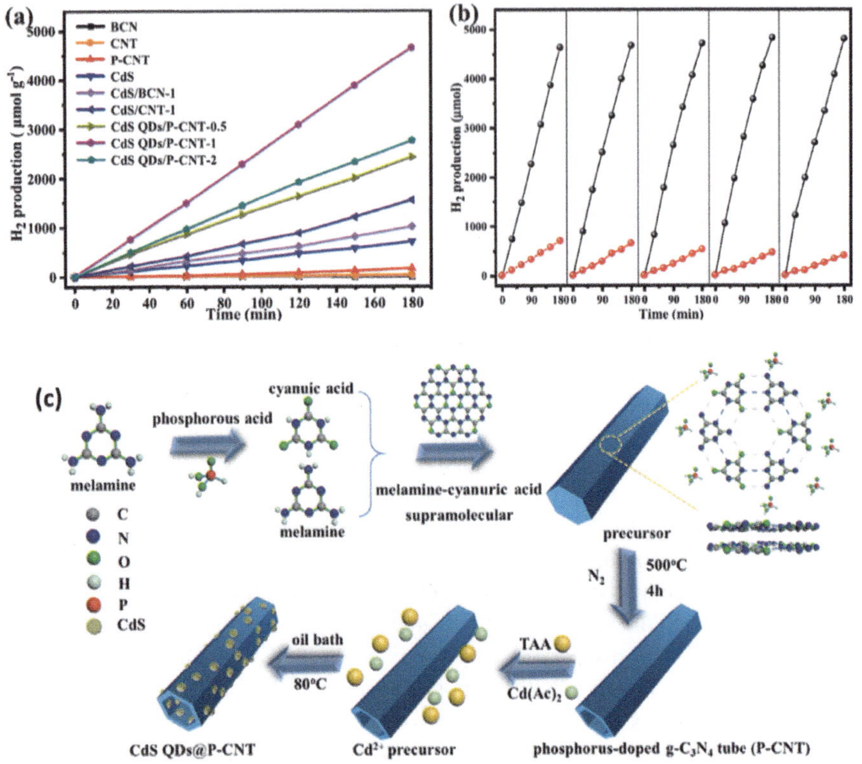

Fig. 5 **a** Photocatalytic H_2 production rates of the BCN, CNT, P-CNT, CdS, CdS/CNT, and CdS QDs/P-CNT, **b** stability tests for photocatalytic H_2 evolution, and **c** schematic illustration of CQD/g-C_3N_4 composite [34]. Copyright 2020, Elsevier

(MPA) capped CdTe QDs. TGA, being a shorter ligand as compared to MPA, minimizes the transfer distance of photogenerated electrons [41]. To improve the separation of EHPs, a suitable ligand induces intimate contact between the QDs and the semiconductor substrates. In this regard, Sakizadeh et al. found that the L-cysteine capped CdS QDs and reduced graphene oxide (rGO) don't integrate with each other because both have negative charge on the surface. Whereas, by exchanging the L-cysteine to cysteamine, the QDs became positively charged which interacted well with rGO through electrostatic attraction [42]. The use of organic ligands doesn't always improve the photocatalytic performance of ChQDs-based composites. Sometimes, it creates the physical or electronic barriers for photogenerated charge carriers from QDs to cocatalysts [43] (Fig. 7).

Along with, the surface stoichiometric ratio of M and X (in MX ChQDs) also manipulates the photocatalytic performance of ChQDs. For instance, Huang et al. synthesized Se rich CdSe QDs, these extra Se sites acted as hole traps which prohibit the interfacial charge transfer from CdSe QDs to the sacrificial agents, resulting

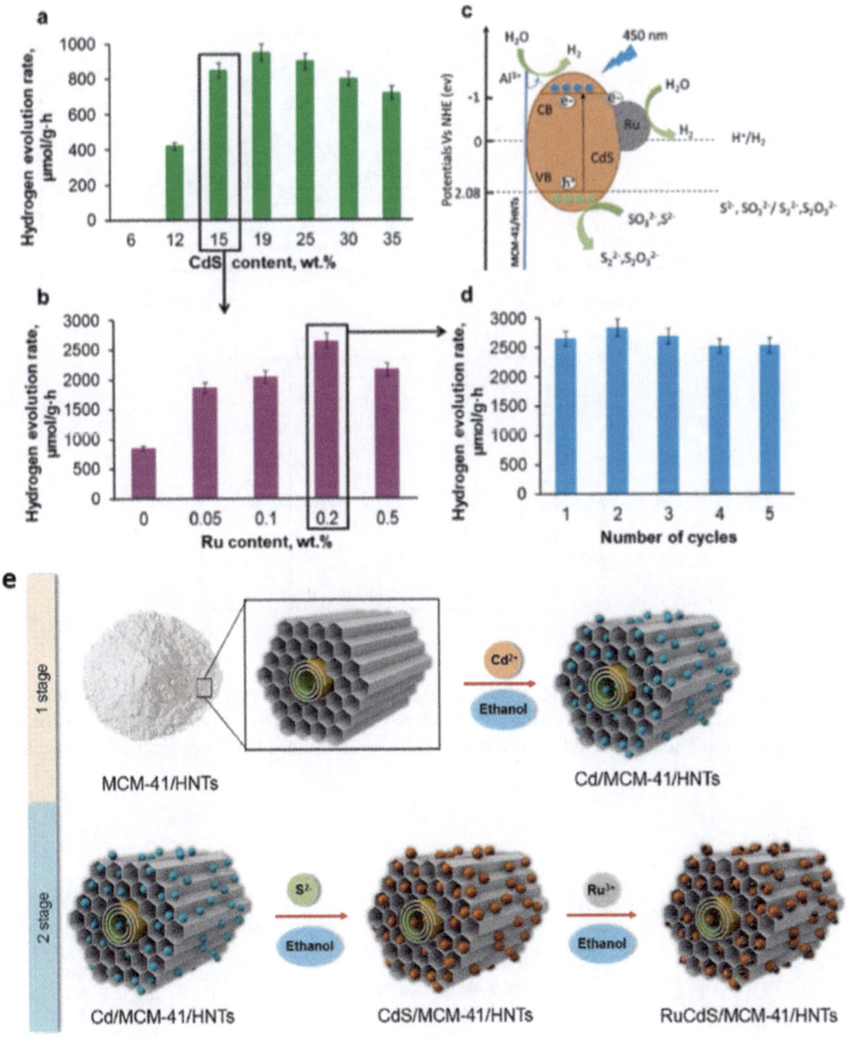

Fig. 6 Photocatalytic activity of **a** CdS and **b** $Ru_{15}CdS$ QDs stabilized on the surface of MCM-41/HNT (depending on CdS and Ru content), **c** schematic representation of the photocatalytic mechanism under visible light illumination, **d** stability of $0.2Ru_{15}CdS$/MCM-41/HNT composite for five cycles of H_2 evolution, and **e** schematic of fabrication steps of RuCdS/MCM-41/HNT composite [35]. Copyright 2021, American Chemical Society

the enhancement in EHPs recombination which leads to poor photocatalytic performance, in the contrary, the Cd-rich CdSe QDs, surface Cd atoms acted as active sites, resulting in the enhancement in photocatalytic performance [44] (Fig. 8).

The ionic doping/modification (cationic or anionic) in ChQDs also an effective approach for enhancing the photocatalytic performance. For instance, to prevent the

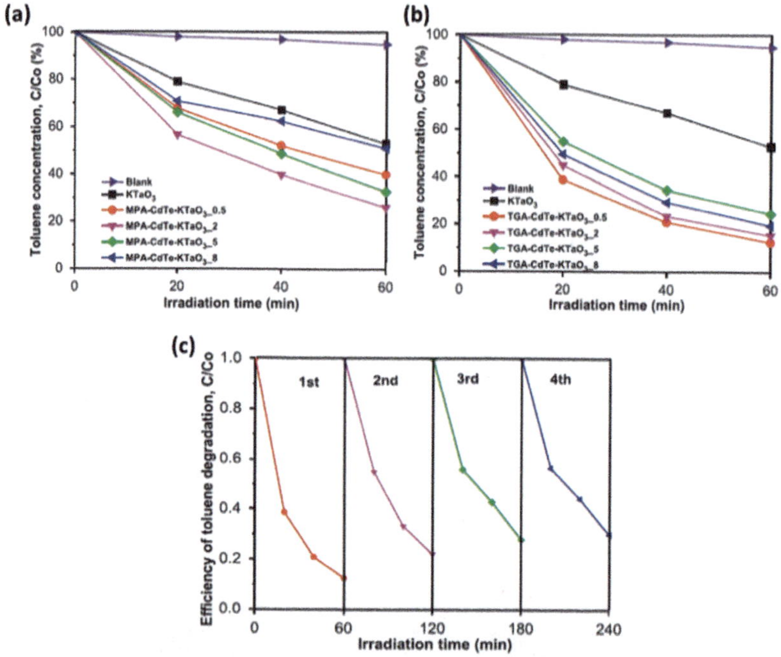

Fig. 7 Efficiency of **a** MPA-CdTe-KTaO3, and **b** TGA-CdTe-KTaO3 for the photodegradation of toluene, and **c** recycling performance of TGA-CdTe-KTaO3-0.5 composite toward toluene photodegradation under LED light illumination ($\lambda_{max} = 415$ nm) [41]. Copyright 2016, Elsevier

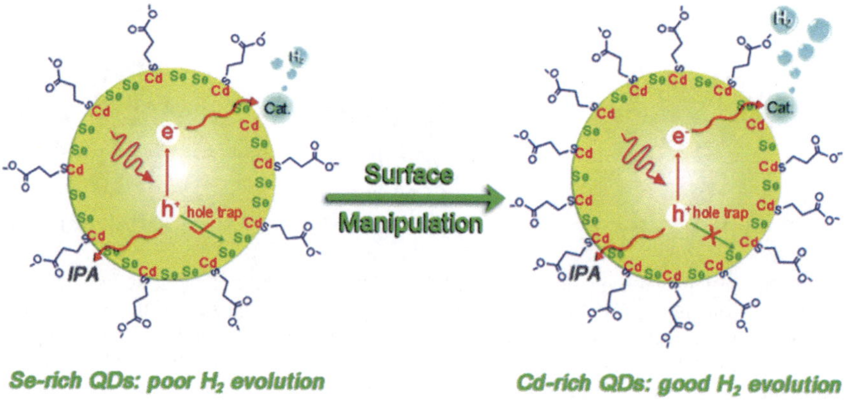

Fig. 8 Hydrogen evolution in Se-rich and Cd-rich CdSe QDs [44]. Copyright 2018, Royal Society of Chemistry

EHPs recombination, Pu et al. synthesized Mn^{2+} doped ZnSe QDs [45]. This report explored that the photoluminescence (PL) decay (from 50 to 1000 μs) of the crystal can be adjusted by varying the doping concentration. In another report, S^{2-} modified QDs show low dielectric barrier with its cocatalyst, and facilitate the transfer of photogenerated charge carriers.

In general, the structure of the cocatalyst of QDs also affect the photocatalytic performance of overall composite. For example, Ma et al. fabricated a nanocomposite of 0D CdS QDs and 2D ZnO nanosheets [46]. The direct Z-scheme heterojunction promoted the separation of photogenerated EHPs. Recently, the researchers have also explored the photocatalytic performance of ChQDs coupled with metal-organic frameworks (MOFs), and covalent organic frameworks (COFs) [47, 48]. The MOFs and their derivatives offer high surface area, and high porosity with adjustable pore size which is in favor for effective photocatalytic reaction. However, the wide bandgap of MOFs limits its use in visible-light-driven photocatalysis [49–51]. Thus, the coupling of MOFs with ChQDs facilitate their use in photocatalytic reactions. For example, Mao et al. used SILAR approach for the integration of CdS QDs into thiol-laced UiO-66 (UiOS) for photocatalytic H_2 evolution [52]. Similarly, COFs also have advantages of tunable structure, good thermal, and chemical stability, large surface area, and high charge carrier mobility [53]. However, COFs are organic porous crystalline materials, composed of covalent-bonded nonmetallic elements (such as C, N, O, B etc.) [54]. Huang et al. designed a composite composed of CdS QDs and covalent triazine-based frameworks (CTF-1) [48]. The prepared composite showed the enhanced photocatalytic H_2 evolution by the factor of 4 and 55 times as compared to bulk CdS and pristine CTFs, respectively (Fig. 9).

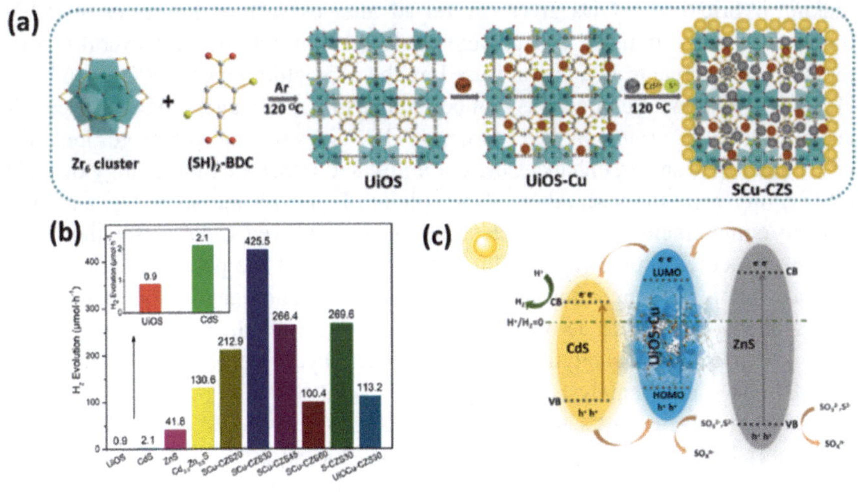

Fig. 9 **a** Schematic illustration of synthesis procedure of SCu-CZS composite, **b** the photocatalytic H_2 evolution performance, and **c** schematic illustration of interfacial charge transfer process in SCu-CZS catalyst [47]. Copyright 2020, Elsevier

4 Perovskite QDs (PQDs)-Based Composites

Perovskites QDs (PQDs), a unique class of photocatalysts, are promising in the field because of its structural simplicity and pliability, photo-absorption tunability, long diffusion lengths, prolonged lifetime and high mobility of charge carriers [55]. Having the chemical formula of ABX_3, these perovskites are most exploited in the field of photocatalysis, where A and B are cations, and X is an anion. The cation A can be an organic (methylammonium, formamidinium) or inorganic (Cs^+), the cation B represents a divalent metal ion (such as Pb, Sn, Bi, etc.), and X is the halide anion (Br, Cl, or I). By modulating the composition of cation/anion, the photo-absorption of PQDs can be enabled from UV to NIR region [56]. Furthermore, the size of PQDs play a significant role in dictating its photocatalytic performance. For instance, Hou et al. employed four different sizes (3.8, 6.1, 8.5, and 11.6 nm) of $CsPbBr_3$ PQDs (dispersed in ethyl acetate/water) for the CO_2 reduction application. To measure their comparative performance towards CO_2 reduction, as-mentioned PQDs were illuminated under solar exposure for 8 h individually. It was found that the PQDs of size 8.5 nm show better performance [57] (Fig. 10).

In addition, the crystalline phase of PQDs also elucidate its performance- Guo et al. reported that the PQDs of crystalline phase are more active than the orthorhombic phase $CsPbBr_3$ PQDs for their application in CO_2 reduction [58]. Moreover, there are several other strategies which enhances the performance of PQDs-based hybrids for their application in CO_2 reduction. For instance, Mu et al. reported cobalt-doped $CsPbBR_3/Cs_4PbBr_6$ QDs/hexafluorobutyl methacrylate [59]. To elaborate on this, Wu et al. investigated the visible light-driven CO_2 conversion performance by using the $MAPbI_3$/Fe-porphyrin-based MOFs composite [52]. Choosing MOFs for making hybrid with PQDs gives several advantages such as longevity of PQDs, improving charge transfer, and prevent agglomeration [60–62]. In several reports, the silica and organic layers were used as the protecting layer of PQDs [63–67]. However, the flow of charge gets hampered by using these layers. Alternatively, the use of conductive protecting layer to wrap PQDs gives several advantages including better electron transfer, enhances the durability and intact the morphology of PQDs [68]. Along with, the researchers also exploited PQDs for the photodegradation of organic contaminants. For example, Gao et al. employed $CsPbX_3$ for the degradation of methyl orange (MO). This report shows that the photocatalytic activity increases with the concentration of catalyst up to certain quantity, but beyond that limit, the photocatalytic activity tends to decrease [69]. In another report, Qian et al. studied the photocatalytic degradation of tetracycline chloride (TC-HCl) using $CsPbBr_3$ as a photocatalyst [70]. Recently, PQDs has also been employed as a visible light-driven photocatalyst for the stereoselective C–C oxidative coupling of α- aryl ketonitriles [71]. In another report, the surface tuning of PQDs was achieved using zwitter ions- $CsPbBr_3$ PQDs were capped by zwitterionic ligand (3-(N,N-dimethyloctadecylammonio)propanesulfonate, DMOA-PS) [72]. Being a diverse photocatalyst, PQDs-based hybrids also found their application in NO gas

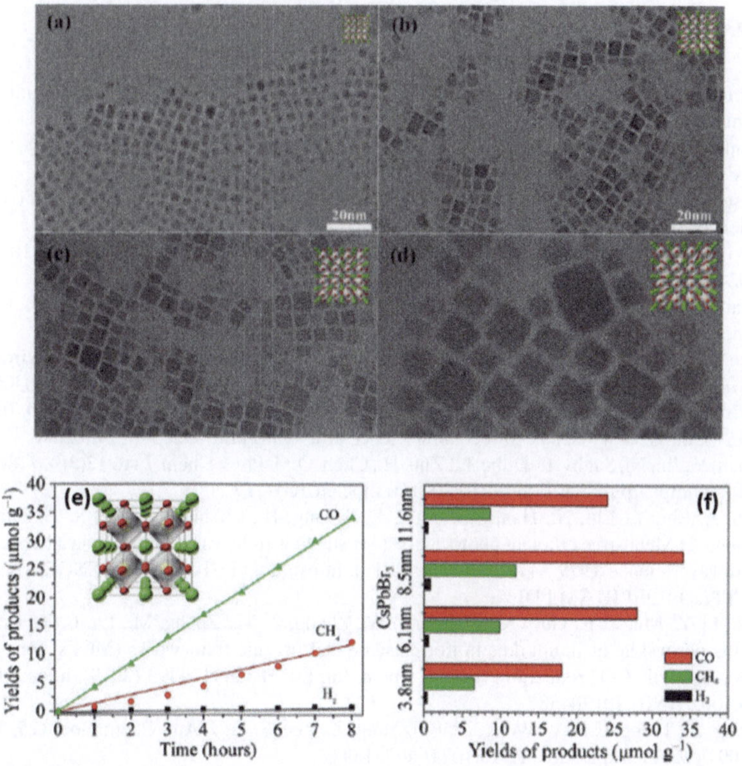

Fig. 10 TEM images of **a** 3.8, **b** 6.1, **c** 8.5, and **d** 11.6 nm $CsPbBr_3$ QDs, solar light-driven (300 W Xe lamp) CO_2 reduction, **e** 8.5 nm $CsPbBr_3$ QDs, and **f** for different sizes of $CsPbBr_3$ QDs [57]. Copyright 2017, Wiley

oxidation. For example, Huo et al. used $FAPbBr_3$/A-SiO_2 composite for NO gas degradation [72].

5 Conclusions

This chapter illustrated the role of CQDs-, SQDs-, and PQDs-based composites for their application in catalysis. As-mentioned researches promise the great potential of QDs based composites for the catalysis related applications. The range of electron hole pair (EHP) generation is critical in developing an efficient photocatalytic application. The studies promoted applications such as dye degradation, organic pollutant degradation, CO_2 reduction, hydrogen evolution reaction (HER), oxygen evolution reaction (OER) activity. However, rigorous research is still required to exploit the full potential of these composites.

References

1. Johnsson, F., Kjarstad, J., Rootzen, J.: Climate Policy **19**, 258 (2019). https://www.tandfonline.com/doi/full/10.1080/14693062.2018.1483885
2. Fontecave, M.: Angew. Chem. Int. Ed. **50**, 6704 (2011). https://onlinelibrary.wiley.com/doi/abs/10.1002/anie.201102819
3. Pasten, C., Santamarina, J.C.: Energy Policy **49**, 468 (2012). https://www.sciencedirect.com/science/article/abs/pii/S0301421512005617
4. Ren, J.T., Chen, L., Wang, H.Y., Tian, W.W., Yuan, Z.Y.: (2023). https://doi.org/10.1039/D3EE02467A
5. Islam, S.K.E., Hang, D.R., Chen, C.H., Chou, M.M.C., Liang, C.T., Sharma, K.H.: Chemosphere **266**, 129148 (2021). https://pubs.acs.org/doi/10.1021/acs.chemrev.8b00264
6. Lee, J.F., Ok, Y.S., Tsang, D.C.W., Song, J., Jung, S.C., Park, Y.K.: Sci. Total Environ. **719**, 137405 (2020). https://www.sciencedirect.com/science/article/abs/pii/S0048969720309153
7. Krishnamurty, A., Adebayo, B., Gelles, T., Rownaghi, A., Rezaei, F.: Catal. Today **350**, 100–119 (2020). https://www.sciencedirect.com/science/article/abs/pii/S0920586119302676
8. Yuan, Y., Jin, N., Saghy, P., Dube, L., Zhu, H., Chen, O.: J. Phys. Chem. Lett. **12**(30), 7180–7193 (2021). https://pubs.acs.org/doi/10.1021/acs.jpclett.1c01717
9. Liu, J., Liu, Y., Liu, N., Han, Y., Zhang, X., Huang, H., Lifshitz, Y., Lee, S.-T., Zhong, J., Kang, Z.: Metal-free efficient photocatalyst for stable visible water splitting via a two-electron pathway. Science **1979**(347), 970–974 (2015). https://doi.org/10.1126/SCIENCE.AAA3145/SUPPL_FILE/LIU.SM.PDF
10. Wu, L.-Y., Mu, Y.-F., Guo, X.-X., Zhang, W., Zhang, Z.-M., Zhang, M., Lu, B.-T.: Encapsulating perovskite quantum dots in iron-based metal-organic frameworks (MOFs) for efficient photocatalytic CO2 reduction. Angew. Chem. Int. Ed. **58**, 9491–9495 (2019). https://doi.org/10.1002/ANIE.201904537
11. Kang, Z., Tsang, C.H.A., Wong, N.-B., Zhang, Z., Lee, S.-T.: J. Am. Chem. Soc. **129**, 12090–12091 (2007). https://doi.org/10.1021/ja075184x
12. Hu, C., Li, M., Qiu, J., Sun, Y.-P.: Design and fabrication of carbon dots for energy conversion and storage. Chem. Soc. Rev. **48**, 2315–2337 (2019). https://doi.org/10.1039/C8CS00750K
13. Han, Y., Tang, D., Yang, Y., Li, C., Kong, W., Huang, H., Liu, Y., Kang, Z.: Non-metal single/dual doped carbon quantum dots: a general flame synthetic method and electro-catalytic properties. Nanoscale **7**, 5955–5962 (2015). https://doi.org/10.1039/C4NR07116F
14. Zou, G., Hou, H., Hu, J., Ji, X.: General synthesis of heteroatom-doped hierarchical carbon toward excellent electrochemical energy storage. Batter Supercaps **2**, 712–722 (2019). https://doi.org/10.1002/BATT.201900030
15. Li, Y., Zhao, Y., Cheng, H., et al.: 15–18 downloaded via NATL TAIWAN UNIV on March 30. J. Am. Chem. Soc. **134**, 22 (2012). https://doi.org/10.1021/ja206030c
16. Zhu, C., Zhai, J., Dong, S.: Bifunctional fluorescent carbon nanodots: green synthesis via soy milk and application as metal-free electrocatalysts for oxygen reduction. Chem. Commun. **48**, 9367–9369 (2012). https://doi.org/10.1039/C2CC33844K
17. Jouyandeh, M., Mousavi Khadem, S.S., Habibzadeh, S., et al.: Quantum dots for photocatalysis: synthesis and environmental applications. Green Chem. **23**, 4931–4954 (2021)
18. Zhao, M., Zhang, J., Xiao, H., et al.: Facile: In situ synthesis of a carbon quantum dot/graphene heterostructure as an efficient metal-free electrocatalyst for overall water splitting. Chem. Commun. **55**, 1635–1638 (2019). https://doi.org/10.1039/c8cc09368g
19. Li, W., Wei, Z., Wang, B., et al.: Carbon quantum dots enhanced the activity for the hydrogen evolution reaction in ruthenium-based electrocatalysts †. Mater. Chem. Front **4**, 277 (2020). https://doi.org/10.1039/c9qm00618d
20. Zhang, L., Yang, Y., Ziaee, M.A., et al.: Nanohybrid of carbon quantum dots/molybdenum phosphide nanoparticle for efficient electrochemical hydrogen evolution in alkaline medium. ACS Appl. Mater. Interfaces **10**, 9460–9467 (2018). https://doi.org/10.1021/ACSAMI.8B00211

21. Wei, J.S., Ding, H., Zhang, P., et al.: Carbon dots/NiCo2O4 nanocomposites with various morphologies for high performance supercapacitors. Small **12**, 5927–5934 (2016). https://doi.org/10.1002/smll.201602164
22. Chen, H., Ouyang, S., Zhao, M., et al.: Synergistic activity of Co and Fe in amorphous Cox−Fe−B catalyst for efficient oxygen evolution reaction. ACS Appl. Mater. Interfaces **9**, 2023 (2017). https://doi.org/10.1021/acsami.7b13939
23. Zhao, S., Li, C., Liu, J., et al.: Carbon quantum dots/SnO2–Co3O4 composite for highly efficient electrochemical water oxidation. Carbon N Y **92**, 64–73 (2015). https://doi.org/10.1016/J.CARBON.2015.03.002
24. Yang, M., Feng, T., Chen, Y., et al.: Synchronously integration of Co, Fe dual-metal doping in Ru@C and CDs for boosted water splitting performances in alkaline media. Appl. Catal. B **267**, 118657 (2020). https://doi.org/10.1016/J.APCATB.2020.118657
25. Sargin, I., Yanalak, G., Arslan, G., Patir, I.H.: Green synthesized carbon quantum dots as TiO2 sensitizers for photocatalytic hydrogen evolution. Int. J. Hydrog. Energy **44**, 21781–21789 (2019). https://doi.org/10.1016/J.IJHYDENE.2019.06.168
26. Gogoi, D., Koyani, R., Golder, A.K., Peela, N.R.: Enhanced photocatalytic hydrogen evolution using green carbon quantum dots modified 1-D CdS nanowires under visible light irradiation. Sol. Energy **208**, 966–977 (2020). https://doi.org/10.1016/J.SOLENER.2020.08.061
27. Kulandaivalu, T., Abdul Rashid, S., Sabli, N., Tan, T.L.: Visible light assisted photocatalytic reduction of CO2 to ethane using CQDs/Cu2O nanocomposite photocatalyst. Diam. Relat. Mater. **91**, 64–73 (2019). https://doi.org/10.1016/J.DIAMOND.2018.11.002
28. Tyagi, A., Tripathi, K.M., Singh, N., et al.: Green synthesis of carbon quantum dots from lemon peel waste: applications in sensing and photocatalysis. RSC Adv. **6**, 72423–72432 (2016). https://doi.org/10.1039/C6RA10488F
29. Ramar, V., Moothattu, S., Balasubramanian, K.: Metal free, sunlight and white light based photocatalysis using carbon quantum dots from Citrus grandis: a green way to remove pollution. Sol. Energy **169**, 120–127 (2018). https://doi.org/10.1016/J.SOLENER.2018.04.040
30. Xue, J., Wang, X., Jeong, J.H., Yan, X.: Fabrication, photoluminescence and applications of quantum dots embedded glass ceramics. Chem. Eng. J. **383**, 123082 (2020). https://doi.org/10.1016/J.CEJ.2019.123082
31. Li, K., Peng, B., Peng, T.: Recent advances in heterogeneous photocatalytic CO2 conversion to solar fuels. ACS Catal. **6**, 7485–7527 (2016). https://doi.org/10.1021/ACSCATAL.6B02089
32. Wu, H.L., Li, X.B., Tung, C.H., Wu, L.Z.: Semiconductor quantum dots: an emerging candidate for CO2 photoreduction. Adv. Mater. **31** (2019)
33. Liang, Q., Zhang, C., Xu, S., et al.: In situ growth of CdS quantum dots on phosphorus-doped carbon nitride hollow tubes as active 0D/1D heterostructures for photocatalytic hydrogen evolution. J. Colloid Interface Sci. **577**, 1–11 (2020). https://doi.org/10.1016/J.JCIS.2020.05.053
34. Stavitskaya, A., Glotov, A., Pouresmaeil, F., et al.: CdS quantum dots in hierarchical mesoporous silica templated on clay nanotubes: implications for photocatalytic hydrogen production. ACS Appl. Nano Mater. **5**, 605–614 (2022). https://doi.org/10.1021/ACSANM.1C03393
35. Zhuge, K., Chen, Z., Yang, Y., et al.: In-suit photodeposition of MoS2 onto CdS quantum dots for efficient photocatalytic H2 evolution. Appl. Surf. Sci. **539**, 148234 (2021). https://doi.org/10.1016/J.APSUSC.2020.148234
36. Guijarro, N., Campiña, J.M., Shen, Q., et al.: Uncovering the role of the ZnS treatment in the performance of quantum dot sensitized solar cells. Phys. Chem. Chem. Phys. **13**, 12024–12032 (2011). https://doi.org/10.1039/C1CP20290A
37. Yang, S.M., Huang, C.H., Zhai, J., et al.: High photostability and quantum yield of nanoporous TiO2 thin film electrodes co-sensitized with capped sulfides. J. Mater. Chem. **12**, 1459–1464 (2002). https://doi.org/10.1039/B105796K
38. Xiang, X., Wang, L., Zhang, J., et al.: Cadmium chalcogenide (CdS, CdSe, CdTe) quantum dots for solar-to-fuel conversion. Adv Photon. Res. **3**, 2200065 (2022). https://doi.org/10.1002/adpr.202200065

39. Pradhan, N., Reifsnyder, D., Xie, R., et al.: Surface ligand dynamics in growth of nanocrystals. J. Am. Chem. Soc. **129**, 9500–9509 (2007). https://doi.org/10.1021/JA0725089
40. Bajorowicz, B., Nadolna, J., Lisowski, W., et al.: The effects of bifunctional linker and reflux time on the surface properties and photocatalytic activity of CdTe quantum dots decorated KTaO3 composite photocatalysts. Appl. Catal. B **203**, 452–464 (2017). https://doi.org/10.1016/J.APCATB.2016.10.027
41. Sakizadeh, J., Cline, J.P., Snyder, M.A., et al.: Tailored coupling of biomineralized CdS quantum dots to rGO to realize ambient aqueous synthesis of a high-performance hydrogen evolution photocatalyst. ACS Appl. Mater. Interfaces **12**, 42773–42780 (2020). https://doi.org/10.1021/ACSAMI.0C11063
42. Chang, C.M., Orchard, K.L., Martindale, B.C.M., Reisner, E.: Ligand removal from CdS quantum dots for enhanced photocatalytic H2 generation in pH neutral water. J. Mater. Chem. A Mater. **4**, 2856–2862 (2016). https://doi.org/10.1039/C5TA04136H
43. Huang, M.Y., Li, X.B., Gao, Y.J., et al.: Surface stoichiometry manipulation enhances solar hydrogen evolution of CdSe quantum dots. J. Mater. Chem. A Mater. **6**, 6015–6021 (2018). https://doi.org/10.1039/C8TA00385H
44. Pu, C., Ma, J., Qin, H., et al.: Doped semiconductor-nanocrystal emitters with optimal photoluminescence decay dynamics in microsecond to millisecond range: synthesis and applications. ACS Cent. Sci. **2**, 32–39 (2016). https://doi.org/10.1021/ACSCENTSCI.5B00327
45. Ma, D., Shi, J.W., Zou, Y., et al.: Highly efficient photocatalyst based on a CdS quantum dots/ZnO nanosheets 0D/2D heterojunction for hydrogen evolution from water splitting. ACS Appl. Mater. Interfaces **9**, 25377–25386 (2017). https://doi.org/10.1021/ACSAMI.7B08407
46. Mao, S., Shi, J.W., Sun, G., et al.: Cu (II) decorated thiol-functionalized MOF as an efficient transfer medium of charge carriers promoting photocatalytic hydrogen evolution. Chem. Eng. J. **404**, 126533 (2021). https://doi.org/10.1016/J.CEJ.2020.126533
47. Huang, H., Xu, B., Tan, Z., et al.: A facile in situ growth of CdS quantum dots on covalent triazine-based frameworks for photocatalytic H2 production. J. Alloys Compd. **833**, 155057 (2020). https://doi.org/10.1016/J.JALLCOM.2020.155057
48. Wei, J., Chen, Y., Zhang, H., et al.: Hierarchically porous S-scheme CdS/UiO-66 photocatalyst for efficient 4-nitroaniline reduction. Chin. J. Catal. **42**, 78–86 (2021). https://doi.org/10.1016/S1872-2067(20)63661-0
49. Liu, H., Xu, C., Li, D., Jiang, H.L.: Photocatalytic hydrogen production coupled with selective benzylamine oxidation over MOF composites. Angew. Chem. **130**, 5477–5481 (2018). https://doi.org/10.1002/ANGE.201800320
50. Wang, H., Zhu, Q.L., Zou, R., Xu, Q.: Metal-organic frameworks for energy applications. Chem 2:52–80 (2017). https://doi.org/10.1016/J.CHEMPR.2016.12.002
51. Mao, S., Zou, Y., Sun, G., et al.: Thio linkage between CdS quantum dots and UiO-66-type MOFs as an effective transfer bridge of charge carriers boosting visible-light-driven photocatalytic hydrogen production. J. Colloid Interface Sci. **581**, 1–10 (2021). https://doi.org/10.1016/J.JCIS.2020.07.121
52. Wang, H., Wang, H., Wang, Z., et al.: Covalent organic framework photocatalysts: structures and applications. Chem. Soc. Rev. **49**, 4135–4165 (2020). https://doi.org/10.1039/D0CS00278J
53. Côté, A.P., Benin, A.I., Ockwig, N.W., et al.: (2005) Chemistry: porous, crystalline, covalent organic frameworks. Science **310**, 1166–1170 (1979). https://doi.org/10.1126/SCIENCE.1120411
54. Peighambardoust, N.S., Sadeghi, E., Aydemir, U.: Lead Halide perovskite quantum dots for photovoltaics and photocatalysis: a review. ACS Appl. Nano Mater. **5**, 14092–14132 (2022)
55. Shyamal, S., Pradhan, N.: Halide perovskite nanocrystal photocatalysts for CO2 reduction: successes and challenges. J. Phys. Chem. Lett. **11**, 6921–6934 (2020). https://doi.org/10.1021/ACS.JPCLETT.0C00191
56. Hou, J., Cao, S., Wu, Y., et al.: Inorganic colloidal perovskite quantum dots for robust solar CO2 reduction. Chem. Euro. J. **23**, 9481–9485 (2017). https://doi.org/10.1002/CHEM.201702237
57. Guo, S.H., Zhou, J., Zhao, X., et al.: Enhanced CO2 photoreduction via tuning halides in perovskites. J. Catal. **369**, 201–208 (2019). https://doi.org/10.1016/J.JCAT.2018.11.004

58. Mu, Y.-F., Zhang, W., Guo, X.-X., et al.: Water-tolerant lead halide perovskite nanocrystals as efficient photocatalysts for visible-light-driven CO2 reduction in pure water. ChemSusChem **12**, 4769–4774 (2019). https://doi.org/10.1002/CSSC.201902192
59. Wu, L.-Y., Mu, Y.A.-F., Guo, X.-X., et al.: CO2 reduction very important paper encapsulating perovskite quantum dots in iron-based metal-organic frameworks (MOFs) for efficient photocatalytic CO2 reduction. https://doi.org/10.1002/ange.201904537
60. Chen, Z., Gu, Z.G., Fu, W.Q., et al.: A confined fabrication of perovskite quantum dots in oriented MOF thin film. ACS Appl. Mater. Interfaces **8**, 28737–28742 (2016). https://doi.org/10.1021/ACSAMI.6B11712
61. He, H., Cui, Y., Li, B., et al.: Confinement of perovskite-QDs within a single MOF crystal for significantly enhanced multiphoton excited luminescence. Adv. Mater. **31**, 1806897 (2019). https://doi.org/10.1002/ADMA.201806897
62. Raja, S.N., Bekenstein, Y., Koc, M.A., et al.: Encapsulation of perovskite nanocrystals into macroscale polymer matrices: enhanced stability and polarization. ACS Appl. Mater. Interfaces **8**, 35523–35533 (2016). https://doi.org/10.1021/ACSAMI.6B09443
63. Sun, C., Zhang, Y., Ruan, C., et al.: Efficient and stable white LEDs with silica-coated inorganic perovskite quantum dots. Adv. Mater. **28**, 10088–10094 (2016). https://doi.org/10.1002/ADMA.201603081
64. Huang, S., Li, Z., Kong, L., et al.: Enhancing the stability of CH3NH3PbBr 3 quantum dots by embedding in silica spheres derived from tetramethyl orthosilicate in "waterless" toluene. J. Am. Chem. Soc. **138**, 5749–5752 (2016). https://doi.org/10.1021/JACS.5B13101
65. Huang, H., Chen, B., Wang, Z., et al.: Water resistant CsPbX3 nanocrystals coated with polyhedral oligomeric silsesquioxane and their use as solid state luminophores in all-perovskite white light-emitting devices. Chem. Sci. **7**, 5699–5703 (2016). https://doi.org/10.1039/C6SC01758D
66. Palazon, F., Akkerman, Q.A., Prato, M., Manna, L.: X-ray lithography on perovskite nanocrystals films: from patterning with anion-exchange reactions to enhanced stability in air and water. ACS Nano **10**, 1224–1230 (2016). https://doi.org/10.1021/ACSNANO.5B06536
67. Chen, K., Deng, X., Dodekatos, G., Tüysüz, H.: Photocatalytic polymerization of 3,4-ethylenedioxythiophene over Cesium lead iodide perovskite quantum dots. J. Am. Chem. Soc. **139**, 12267–12273 (2017). https://doi.org/10.1021/JACS.7B06413
68. Gao, G., Xi, Q., Zhou, H., et al.: Novel inorganic perovskite quantum dots for photocatalysis. Nanoscale **9**, 12032–12038 (2017). https://doi.org/10.1039/c7nr04421f
69. Qian, X., Chen, Z., Yang, X., et al.: Perovskite cesium lead bromide quantum dots: a new efficient photocatalyst for degrading antibiotic residues in organic system. J. Clean. Prod. **249**, 119335 (2020). https://doi.org/10.1016/J.JCLEPRO.2019.119335
70. Yuan, Y., Zhu, H., Hills-Kimball, K., et al.: Stereoselective C−C oxidative coupling reactions photocatalyzed by zwitterionic ligand capped CsPbBr 3 perovskite quantum dots. Angew. Chem. Int. Ed. **59**, 22563–22569 (2020). https://doi.org/10.1002/ANIE.202007520
71. Krieg, F., Ochsenbein, S.T., Yakunin, S., et al.: Colloidal CsPbX3 (X = Cl, Br, I) nanocrystals 2.0: zwitterionic capping ligands for improved durability and stability. ACS Energy Lett. **3**, 641–646 (2018). https://doi.org/10.1021/ACSENERGYLETT.8B00035
72. Pu, Y.C., Fan, H.C., Liu, T.W., Chen, J.W.: Methylamine lead bromide perovskite/protonated graphitic carbon nitride nanocomposites: interfacial charge carrier dynamics and photocatalysis. J. Mater. Chem. A Mater. **5**, 25438–25449 (2017). https://doi.org/10.1039/C7TA08190A

Quantum Dots Nanocomposites Bioimaging Probes

Seyed Ahmad Dehdast and Omid Pourdakan

Abstract Quantum dots nanocomposites (QDNCs) are highly fascinating nanomaterials for bioimaging application due to their distinctive optical properties. By using the unique optical properties of quantum dots based nanocomposites, the performance of biomedical imaging can be significantly improved in terms of their contrast, resolution, sensitivity, specificity, and penetration depth. For this reason, recently, the use of QDNCs in bioimaging and medical imaging studies has received much attention and is increasingly being used. Recent QDNCs as a bioimaging probes were used in fluorescent imaging, positron emission tomography (PET imaging) and magnetic resonance imaging (MRI) investigations. In this chapter different types of QDNCs, their properties, advantages, and emerging applications of these nanomaterials in ex-vivo and preclinical study summarized and presented.

Keywords Quantum dots nanocomposites (QDNCs) · Bioimaging probe · Fluorescence imaging · PET imaging · Magnetic resonance imaging (MRI)

List of Abbreviations

A-GQD	Adenine-modified Graphene Quantum Dots
Ag_2Se	Silver Selenide
AKI	Acute Kidney Injury
AOM	Azoxymethane
BP	Black Phosphorus
BPQD	Black Phosphorous quantum dots
C18-PMH	Poly(maleic anhydride-alt-1-octadecene)
Cd	Cadmium
CdS	Cadmium Sulfide
CdSe	Cadmium Selenide

S. A. Dehdast (✉) · O. Pourdakan
School of Medicine, Iran University of Medical Sciences, Tehran, Iran
e-mail: arashdehdast@gmail.com

S. Thomas et al. (eds.), *Quantum Dots Based Nanocomposites*, Engineering Materials,
https://doi.org/10.1007/978-3-031-54779-9_15

CIS (CuInS)	Copper Indium Sulfide
CKD	Chronic Kidney Disease
CQDNCs	Carbon Quantum Dots based nanocomposites
CQD-WS	Carbon Quantum Dot-tungsten sulfide
CRET	Cerenkov resonance energy transfer
CT	Computed Tomography
DAPI	4′,6-Diamidino-2-phenylindole
DMF	Dimethylformamide
DSS	Dextran Sodium Sulfate
EDTA	Ethylenediaminetetraacetic acid
FA	Folic Acid
FL	Fluorescence
FLIM	Fluorescence Lifetime Imaging
FR	Folate Receptor
FRET	Förster Resonance Energy Transfer
Gd-DTPA	Gadolinium-diethylenetriaminepentaacetic acid
Gd-DOTA	Gadolinium –2,2′,2″,2‴-(1,4,7,10-tetraazacyclododecane-1,4,7,10-tetrayl)tetraacetic acid
GQDNCs	Graphene quantum dots based nanocomposites
GSH	Glutathione
ICG	Indocyanine Green
In	Indium
InP	Indium Phosphate
InS	Indium Sulfide
LED	Light-emitting diode
MB	Methylene Blue
MMP	Matrix metalloprotease
MoS_2	Molybdenum Disulfide
MR	Magnetic Resonance
MRI	Magnetic Resonance Imaging
MTT	2,5-Diphenyl-2H-tetrazolium bromide
N-GQD	Nitrogen doped-Graphene Quantum Dots
NIR-I	Near-Infrared I
NIR-II	Near-Infrared II
OCT	Optical Coherence Tomography
OPFI	One-photon Fluorescence Imaging
PA	PhotoAcoustic
PbS	Lead Sulfide
PEG	Polyethylene glycol
PET	Positron Emission Tomography
PET	Photoexcited Electron Transfer
QCE	Quantum Confinement Effect
QD	Quantum Dots
QDNC	Quantum dots nanocomposites
QY	Quantum Yield

RGD	Arginine-Glycine-Aspartic Acid
RP	Red-Phosphorus
RQDNCs	Radioactive Quantum Dots nanocomposites
TMDs	Transition metal dichalcogenides
TPFI	Two-photon Fluorescence Imaging
WS_2	Tungsten Disulfide
WP	White Phosphorus
ZnS	Zinc Sulfide
ZnSe	Zinc Selenide

1 Introduction

Quantum dots nanocomposites (QDNCs) are a multipurpose material that can be employed in a broad range of applications. This is due to special physical and chemical traits of QDNCs. Their surface chemistry can be easily manipulated and changed. Because of these unique properties, quantum dots are one of the attractive candidate used nanomaterial in research and medicine with high potential for use in many medical devices and industrial products. Due to the small size of quantum dots (less than 20 nm), they are ideal for imaging applications and drug delivery to various tissues. Also, QDNCs have proved to be efficient catalysis material. These properties make that's applicable in medicine [1]. Moreover, QDNCs have the unique optical attributes, practical and fantastic physical capacity to specific wavelengths of light for use in bioimaging. Recently, have been discovered quantum dots can be useful in various scientific and industrial disciplines such as medical engineering, imaging, electronic devices and solar cells. These nanostructures have also been employed to develop light-emitting diodes' quantum and electroluminescence efficiency. Providing a wide range of colors and optical coding in many colors for biological experiments is one the most important their characteristics [2]. Anyway, according to the mentioned features, quantum dots possess a variety of features that make them great nanobiotechnology instruments. These nanomaterials are useful for imaging because of high capacity for chemical functionalization and the wide variety of surface functions and bioconjugation that may be attached to the quantum dots and their exceptional chemo-physical characteristics. Modification of QDNCs makes that's useful highly sensitive and selective for diagnostic and theranostic systems. QDNCs have high therapeutic potential, for the reason of high surface loading of medication, genes and their controllable release of the loaded materials. Collectively, quantum dots are very fantastic and adaptable nanomaterial for future medical engineering, bioengineering and nanomedicine uses [3]. In this chapter, the potential and recent research applications of quantum dots nanocomposites for imaging in medicine and biology are explored.

2 Bioimaging

Bioimaging is a scientific method that enables the real-time perception of biological processes in body, with the use of non-invasive tools. Its purpose is to cause negligible interference to live processes and provide information on the 3D structure of the observed construction. Bioimaging technologies are used in various fields such as fundamental and medical sciences for anatomical and physiological examining, and in gathering research information [4].

Bioimaging has numerous applications in various fields, including medicine, biology, chemistry, materials science, and environmental science. In medicine, bioimaging is used for diagnosis, treatment planning, and monitoring of diseases. For example, X-ray imaging is commonly used to detect fractures and tumors, while magnetic resonance imaging (MRI) is used to visualize soft tissues and organs. In biology, bioimaging is used to study cellular processes, such as mitosis and apoptosis, and to investigate the structure and function of biomolecules, such as proteins and nucleic acids. In chemistry, bioimaging is used to study chemical reactions and to develop new materials, while in environmental science, bioimaging is used to study the distribution and transport of pollutants in the environment [5].

Complex bioimaging applications require multi-modal and multi-scale imaging techniques that can involve ultrasound and light imaging to comprehend complicated structures and interactions profound inside the body [6]. Many clinical imaging techniques such as ultrasound imaging (sonography), X-ray computed tomography (X-ray CT), optical coherence tomography (OCT), and magnetic resonance imaging (MRI) utilize the entire energy spectrum for imaging. Furthermore, research methods include electron microscopy, mass spectrometry imaging, fluorescence tomography, biochemical luminescence, various forms of OCT, and optoacoustic imaging. Light microscopy methods include confocal, multi-photon, total internal reflection, and super-resolution fluorescence microscopy employ the whole energy spectrum for analysis [7].

3 Quantum Dots Nanocomposites as Fluorescence Imaging Probe

Biocompatible QDNCs are typically composed of a nanoscale core made of semiconductor materials such as CdSe or InP, which emit light when excited by an external energy source such as a laser. The core is then coated by a semiconductor substance for example ZnS or ZnSe, which serves to passivate the surface of the core and improve its photostability.

In addition, the outer layer of biocompatible QDNCs is often modified with a coating of silica, which can provide a high surface area for further modifications with biomolecules including nucleic acids, antibodies, peptides, and small-molecule ligands. These modifications can enable the QDs to target specific molecules on the cancer cells surface, allowing for highly selective imaging and even therapy [8].

Dong et al. successfully prepared Ag_2Se quantum dots (QDs) using a solvothermal method, which emitted light at a wavelength of 1300 nm in the second near-infrared (NIR-II) window. The Ag_2Se QDs were then modified to enhance their properties such as water solubility, colloidal stability, photostability, and biocompatibility. These modified Ag_2Se QDNCs were used for deep imaging of arteries and organs in vivo, providing higher spatial resolution than indocyanine green (ICG), as a near-infrared (NIR-I) fluorescent substance that using in medical diagnosis. Investigation of in vitro cell toxicity were conducted using PEGylated poly (maleic anhydride-alt-1-octadecene) Ag_2Se (C18-PMH-PEG-Ag_2Se) QDNCs on L929 cell line, and the results showed that the cells remained viable with different concentrations of the QDNCs, without any significant changes in their viability. In vivo experiments involved administering C18-PMH-PEG-Ag2Se QDNCs via tail intravenous injection, resulting in high spatial resolution images of vascular structures and deep organs within the body, with a resolution of approximately 100 μm. The minimal autofluorescence, low scattering, and absorption of the NIR-II emission at 1300 nm enabled highest penetration depth for deep organ imaging with high resolution, making it possible to image organs like the liver and spleen [9].

The utilization of QDNCs in clinical applications has been constrained due to concerns regarding their potential toxicity, particularly with regards to their core material, cadmium. Yaghini et al. have developed a novel type of QDNCs that are free of heavy metals and cadmium (Cd), and are biocompatible, relying on indium (In) as the core material. These QDNCs have good photoluminescence quantum yield (QY) and have been evaluated for lymph node mapping using fluorescence lifetime imaging (FLIM) (Fig. 1). In vitro toxicity studies were conducted on MCF-7 cells using MTT assays, and No notable cell viability reduction was observed in QDNCs treated cells. Subsequent ex vivo imaging of regional lymph nodes following subcutaneous injection through the paw of rats demonstrated the potential of these QDNCs in the context of lymph node mapping. Upon injection, the QDNCs exhibited a preferential accumulation in two specific lymph nodes that drain the thoracic mammary fat pads in rats. The investigation on biodistribution, conducted through chemical extraction, revealed that the retention of the QDNCs is highly limited to these regional lymph nodes, with only negligible amounts detected in other organs. Remarkably, the QDNCs promptly localized within minutes of injection, and their photoluminescent properties remained stable for a prolonged duration. In contrast, blue dyes tend to swiftly drain from lymph nodes. This protracted retention of the QDNCs could prove advantageous in a clinical context, as it has the potential to prevent false positive outcomes associated with the rapid migration of dyes through the lymphatic channel, all while being logistically more manageable [10].

Jeong et al. have developed a protease-activatable near-infrared II (NIR-II) quantum dot nanocomposites (QDNCs) probe that was sensitive to matrix metalloprotease (MMP) activity. The high performance luminescent PbS/CdS/ZnS multishell NIR QDNCs with protease-activatable (PA) modulators (AcM) linked to probe

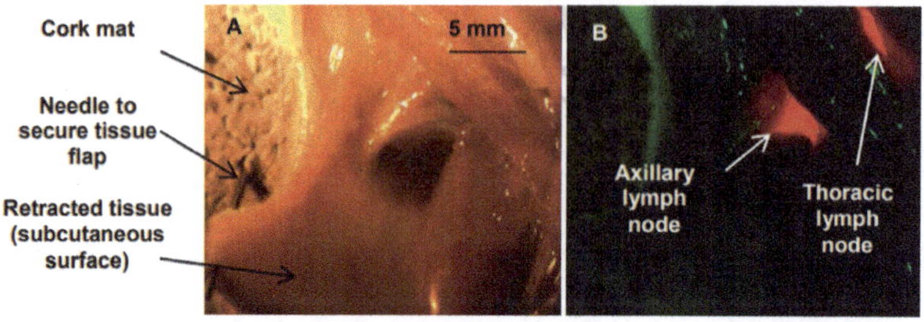

Fig. 1 **a** The image on the left side displays color, while **b** the right part exhibits photoluminescence. This photoluminescence image showcases the armpit area after the QDNCs administration. In order to illuminate the tissues, a collimated LED lamp with a peak emission at 455 nm was employed. The green emission observed in image, **b** is indicative of tissue autofluorescence, while the red emission corresponds to the photoluminescence of the QDNCs in the axillary and thoracic lymph nodes. Reproduced with permission [10]. Copyright 2016, Elsevier

surface. The photoexcited electron transfer (PET) depends on QDNCs and the methylene blue (MB) quencher distance and has an effect on fluorescence (FL) intensity which can be modulated by AcM. The modulation mechanism can be applied to NIR-II wavelength and longer and is different from FRET mechanism however it is not fully discovered yet. Signal amplification was observed at colorectal tumor site by ex-vivo fluorescent imaging and deep imaging potential was approved by phantom experiments. PET-mediated NIR-II QDNCs can be extended to other protease probes, enabling high-resolution, real-time, noninvasive FL imaging (Fig. 2) [11].

Carbon quantum dots based nanocomposites (CQDNCs) are a novel class of carbon nanomaterials that have gained growing interest in recent times owing to their distinct physical and chemical characteristics. These nanomaterials are generally characterized by their sub-10 nm dimensions, exhibiting a spherical morphology accompanied by a crystal lattice configuration and surface-bound chemical moieties that play a pivotal role in their luminescent features [12].

One of the most important features of CQDNCs is the quantum confinement effect (QCE), which occurs when the size of the CQDNCs is smaller than its exciton Bohr radius. This effect leads to discrete energy levels within the CQDNCs, which in turn results in tunable emission wavelengths and high quantum yields. Furthermore, CQDNCs are less cytotoxic and environmentally hazardous than traditional semiconductor quantum dots, making them a more attractive option for biomedical and environmental applications [13, 14].

In addition, CQDNCs have suitable water solubility, chemical stability, and photobleaching resistance, as well as ease of surface functionalization and large-scale preparation. These properties make them an attractive candidate for various applications, such as bioimaging, drug delivery, and optoelectronics [14].

Graphene quantum dots based nanocomposites (GQDNCs), on the other hand, are derived from graphene, a two-dimensional carbon material that has unique electronic and mechanical properties. GQDNCs are typically smaller than 10 nm in size and have a tunable bandgap, making them attractive for electronic and optoelectronic

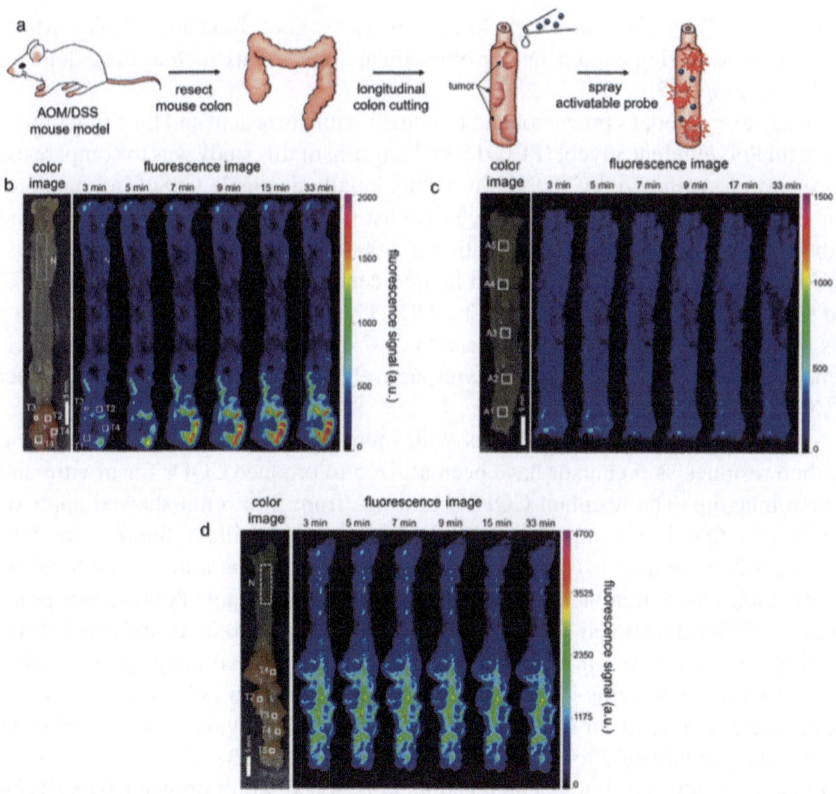

Fig. 2 **a** The diagram illustrates the process of ex vivo imaging of colon cancer in mice using near-infrared II fluorescent (NIR-II FL) technology by applying an activatable probe. At first, colon tissue is extracted from a mouse with a tumor treated with Azoxymethane/dextran sodium sulfate (AOM/DSS). The tissue is then cut longitudinally, and an even distribution of the PA-NIR QDNCs probe is sprayed onto the surface of the tissue. **b** Observe the results of the ex vivo FL imaging of colon cancer by applying the protease-activatable probe, PA-NIR QDNCs. The color image showcases the intact model of mouse colon cancer, while the time-dependent FL image demonstrates the changes that occur after the spraying of the probe. The designated tumor regions, T1–T4, and the normal colon tissue, N, are identified. **c** Similarly, examination of the ex vivo FL imaging of a healthy colon by employing the same spraying technique with PA-NIR QDNCs. The color image illustrates the intact model of the mouse colon, and the time-dependent FL image reveals the alterations that transpire following the application of the probe. Once again, the tumor regions, T1–T4, and the normal colon tissue, N, are explicitly labeled. **d** Investigation of colon cancer's ex vivo FL imaging using NIRQDs without the activatable probe. The color image presents the model of mouse colon cancer in its entirety, whereas the time-dependent FL image depicts the changes that arise after the spraying of the probe. Reproduced with permission [11]. Copyright 2017, American Chemical Society

applications. They also have high surface area and good biocompatibility, which make them promising candidates for biomedical applications such as drug delivery and bioimaging [15].

Using carbon soot as precursor and treating it with nitric acid and then functionalized with Polyethylene glycol (PEG)1500, Yang aim of this study was to compare the properties of synthesized CQDNCs to commercially available CdSe/ZnS quantum dots (QD) in terms of imaging and cytotoxicity. CQDs-PEG and QDs were excited with femtosecond pulsed laser excitation at 800–880 nm and revealed that CQDs-PEG had higher biocompatibility and luminescence, with a quantum yield of 20%, and smaller size compared to CdSe/ZnS QDs. Cytotoxicity in vitro and in vivo test was performed on human breast cancer MCF-7 and human colorectal adenocarcinoma HT-29 cell line and CD-1 mice respectively and showed no significant effect of CQDNCs compare to QD [16].

A one-pot hydrothermal synthesis with monosaccharides in Wheat straw and bamboo residues as precursor have been utilized to produce CQDs for in vitro and in vivo imaging. The resultant CQDs that range from 2 to 6 nm showed approximately 13% QY. In a comparison to the commercially available fluorophore 4′,6-diamidino-2-phenylindole (DAPI), in a different range temperature from 4 to 60 °C and photobleaching treatment, Huang et al. CQDs exhibit stable fluorescence properties. L929 fibroblasts cell line was used to investigate cytotoxicity and showed cell viability more than 90% in highest concentration. For in vivo imaging application a nude mouse with Smmc-7721 tumor was injected through tail's vein and mouse organs were imaged from 0 to 24 h. CQDs showed low cytotoxicity, antioxidant activity and good feature for imaging application [17] (Fig. 3).

Liu et al. synthesized graphene quantum dots doped with nitrogen (N-GQD) by solvothermal method and using Dimethylformamide (DMF) as solvent and nitrogen source, which had a size of 3 nm and excitation in the region of 390 nm, and were used for two-photon and deep tissue imaging. To investigate the ability of two-photon imaging, the nanoparticles were incubated with HeLa cells and excited by a femtosecond laser with a wavelength of 800 nm, which showed high fluorescent contrast with green color around the nuclei of the cells and the ability to label the membrane and cytoplasm of HeLa cells. They showed no penetration into the nucleus. Cell viability studies in different doses showed that these nanoparticles are not toxic. In order to evaluate the deep tissue imaging of GQD, a tissue phantom was used and the obtained results show that the nanoparticles penetrate deeply. The large imaging penetration depth of 1800 μm achieved by N-GQD in tissue phantoms significantly extends the fundamental imaging depth limit of two-photon microscopy (Fig. 4) [18].

Lou et al. synthesized adenine-modified graphene quantum dots (A-GQDs) with microwave-assisted method. The A-GQDs showed 3–5 nm lateral size, single-few-layer thickness, 21.63% quantum yield at 350 nm excitement wavelength and 4.47 ns average lifetime. Duo it's biocompatibility and two photon green fluorescence production is suitable for two-photon fluorescent cell imaging. The novel white light-activated antibacterial properties of the A-GQDs was observed for the first time. With their exceptional photoluminescent performance and facile biological conjugation,

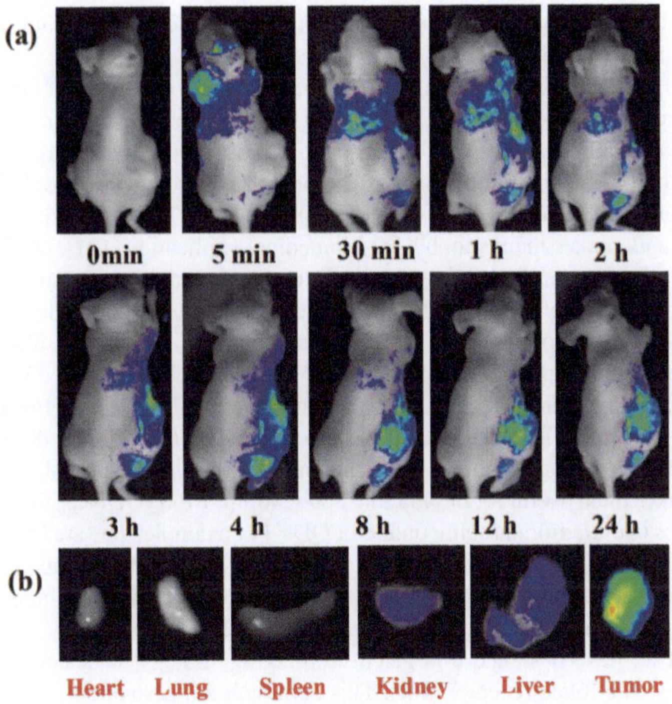

Fig. 3 **a** In vivo fluorescence imaging of nude mice was conducted following the intravenous administration of a solution containing colloidal quantum dot-tungsten sulfide (CQD-WS). **b** Illustrative fluorescence images of dissected organs from a mouse were obtained 24 h after the intravenous injection of a CQD-WS solution. reproduced with permission [17]. Copyright 2019, MDPI

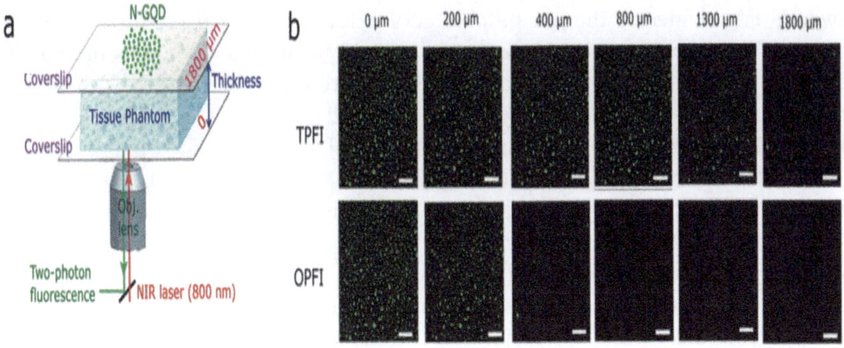

Fig. 4 **a** The diagram illustrates the arrangement utilized for conducting Two-photon fluorescence imaging (TPFI) of N-GQDs in a tissue phantom with varying thickness. **b** The depth of penetration of N-GQDs is presented for TPFI (on the right panel) and One-photon fluorescence imaging (OPFI) (on the left panel) within the tissue phantom. Reproduced with permission [18]. Copyright 2013, American Chemical Society

A-GQDs hold great promise as an emerging tool for a wide array of multifunctional biomedical applications [19].

Transition metal dichalcogenides (TMDs) are a family of layered compounds that Their three-atom thick unit cell is formed by a layer of transition metal atoms (Mo, W, Ta, etc.) sandwiched between two layers of chalcogen atoms (S, Se, Te) [20]. Among TMDs, molybdenum disulfide (MoS_2) has emerged as a promising candidate for these applications. MoS_2 can be synthesized as quantum dots (QDs) with a size range of 1–10 nm, which makes them suitable for biomedical applications [21]. MoS_2 QDNCs have several properties that make them attractive for drug delivery and imaging. First, they are water-soluble, which allows for easy dispersion in biological fluids and reduces toxicity concerns. They are also biocompatible and exhibit high stability, which makes them suitable for use in vivo. MoS_2 QDNCs also have a high surface area, which allows for high loading of drug molecules, and they can be functionalized with targeting moieties such as antibodies or peptides for targeted drug delivery. In addition, MoS_2 QDNCs have been shown to exhibit strong photoluminescence, which makes them useful for bioimaging applications. MoS_2 QDNCs also have other advantages over traditional semiconductor QDs. For example, they are less toxic than cadmium-based QDs, which are commonly used in bioimaging applications. MoS_2 QDNCs also exhibit a higher quantum yield and a broader excitation range than other TMDs such as tungsten disulfide (WS_2) QDNCs [22].

Roy et al. have designed a targeted bioimaging strategy to detect cancer cells that express the folate receptor (FR). This approach utilizes nanoprobes composed of free folic acid (FA) and MoS_2 quantum dots nanocomposites (QDNCs). The MoS_2 QDNCs were synthesized through a hydrothermal method, resulting in water-soluble particles measuring 4–5 nm with cysteine functionalization. Upon excitation at 370 nm, these MoS_2 QDs exhibit a blue emission with the highest intensity occurring at 444 nm. The MoS_2 QDNCs display a considerable level of sensitivity towards FA, rendering them a suitable choice for FA-targeted "turn-off" imaging probes. These probes can be effectively utilized in in vivo studies involving FR-overexpressed cancer cells that have undergone pretreatment with FA. Through the utilization of confocal microscopy, it was observed that B16F10 cancer cells, which were pretreated with FA, exhibited a higher population of dimmed fluorescence in comparison to both untreated cancer cells and HEK-293 normal cells. Additionally, flow cytometry analyses revealed a noteworthy disparity in the geometric mean of fluorescence between FA-pretreated B16F10 cancer cells and their untreated counterparts. These nanoprobes, based on MoS_2 QDNCs, have the potential to be utilized as nanoprobes for the pre-diagnosis of cancer via targeted bioimaging. The benefits of employing this method encompass the simple preparation of the nanoprobes, the exceptional selectivity and sensitivity of MoS_2 QDNCs towards FA, and the capability to achieve targeted bioimaging without the requirement of bioconjugating FA with nanoparticles [23].

Similar to MoS_2 QDNCs, WS_2 QDNCs have a size range of 1–10 nm, which makes them suitable for biomedical applications. They also exhibit unique electronic and optical properties due to their nanoscale size and quantum confinement effects. WS_2 QDNCs have several advantages over other types of QDs. For example, they are less toxic than traditional cadmium-based QDs, which makes them more suitable for biomedical applications. They are also more stable under biological conditions, which makes them more suitable for in vivo imaging. In addition, WS_2 QDNCs exhibit strong photoluminescence, which makes them useful for bioimaging applications. They also have a large surface area, which allows for high loading of drug molecules, and can be functionalized with targeting moieties such as antibodies or peptides for targeted drug delivery [24].

Phosphorus is an essential element for maintaining good health in the human body, and the final degradation products of Black Phosphorus (BP) nanomaterials are harmless, making them suitable for use in biomedical applications [25]. There are three major allotropes of phosphorus, including white phosphorus (WP), red phosphorus (RP), and BP. WP, which has a tetrahedral structure, is chemically unstable, while RP possesses an amorphous structure and better chemical stability and BP has the most stable allotrope [26].

BP structures can be found in various forms, including 2D layered structures, 1D phosphorene nanoribbons, and zero-dimensional quantum dots. Unlike graphene, which has a zero band gap, and transition metal dichalcogenides (TMDs), which have a very large band gap, BP exhibits a tunable, direct, and thickness-dependent band gap. This property makes BP a promising candidate for a range of electronic and optoelectronic applications, including photovoltaics, transistors, and sensors. BP also has a sufficiently high carrier mobility, making it useful for high-speed electronics [27].

As a narrow band gap semiconductor, they are expected to show strong second near-infrared (NIR-II) fluorescence. Cholesterol-modified PEGylated BP nanoparticles (BP@lipid-PEG nanospheres) for in vitro on chicken slices and in vivo NIR-II imaging in normal nude mice were used by Xu et al. These nanospheres emit broad emissions from 900 to 1650 nm under an 808 nm laser, with a quantum yield of 8% of the standard dye IR-26. The NIR-II fluorescence image provides sharp contrast and can be used for blood vessel diameter measurements. Biocompatibility, biodegradability, and stability of nanospheres alongside gradually metabolizing by the liver in 48 h, making them promising for biomedical applications like imaging, drug delivery, and photothermal therapy (PTT) (Fig. 5) [28].

Post-imaging contrast agent's renal excretion poses a considerable challenge and may give rise to long-term toxicity. Specifically, contrast agents that are not readily eliminated from the body, such as certain agents employed in magnetic resonance imaging (MRI) and computed tomography (CT) scans, present a difficulty. Zhang et al. have synthesized black phosphorous quantum dots (BPQDs) possessing an ultra-small size (1.74 ± 0.23 nm) and modified with polyethylene glycol (PEG), thereby demonstrating suitable short-term stability (12 h). These BPQDs exhibit near-infrared (NIR) absorption increase and can produce robust photoacoustic (PA) signals when excited in the near-infrared. PEGylated BPQDs have the capacity to identify kidney dysfunctions, such as acute kidney injury (AKI) and chronic kidney disease (CKD), in living mice with a remarkable level of sensitivity and accuracy. This can be used in real-time observation for treatment response in kidney injury.

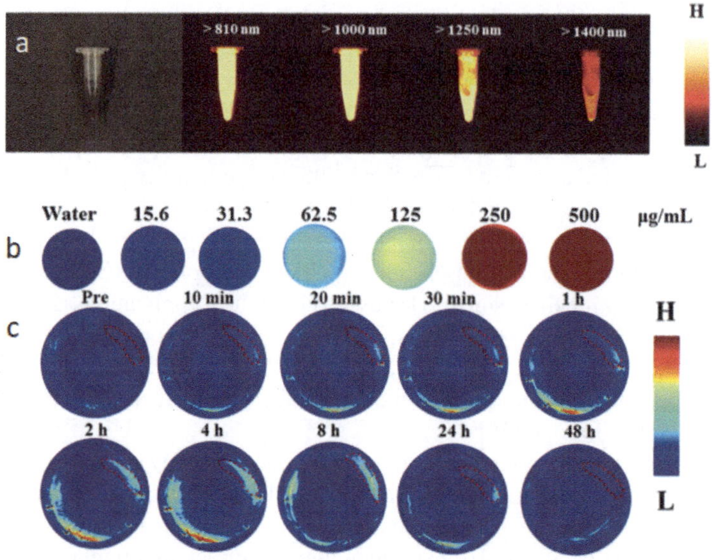

Fig. 5 **a** Optical images of a solution containing BP@lipid-PEG nanosphere, with a concentration of 1.3 mg ml^{-1}, were captured both in the absence and presence of irradiation by an 808 nm laser. **b** In vitro, photoacoustic (PA) images were obtained for aqueous solutions of BP@lipid-PEG nanosphere. c) Furthermore, in vivo PA imaging of the liver was conducted at various time intervals following intravenous administration of aqueous solutions of BP@lipid-PEG nanosphere to mice. Reproduced with permission [28]. Copyright 2019, American Chemical Society

Alongside of being biocompatibility and biodegradability, good renal-clearance of PEG-BPQDs improved depth of PA imaging [29].

4 Quantum Dots Nanocomposites Imaging Probe and PET Imaging

Fluorescence imaging is a useful technique for preclinical qualitative applications due to its high sensitivity, low cost, and short acquisition time. It has been used to provide intraoperative guidance with various fluorophore conjugated probes. However, fluorescence imaging suffers from limitations such as lack of quantitation, shallow depth of penetration, and tissue autofluorescence, which can limit its use in some applications. To overcome such limitation, another powerful imaging technique can be used alongside of fluorescence imaging. Positron emission tomography (PET) is widely used for diagnostic applications in clinical oncology. PET imaging utilizes radiolabeled compounds that emit positrons, which are detected by the PET scanner. This allows for highly sensitive and quantitative imaging, as well as the ability to image deep tissues within the body [30, 31]. When a positron (β^+) undergoes decay, it produces high-energy positron particles that can travel through tissue at velocities greater than the speed of light in the same medium. This results in the emission of Cerenkov radiation, which is detectable by optical imaging techniques. However, Cerenkov radiation is highly absorbed in living organisms, meaning that only a small fraction of the emitted radiation can be detected. This limits the use of Cerenkov radiation for imaging applications in vivo. To overcome this limitation, researchers are developing new approaches to enhance the detection of Cerenkov radiation. One such approach is to use radioluminescent nanoparticles that can be excited by the Cerenkov radiation and emit light in the visible or near-infrared range. This allows for the detection of Cerenkov radiation with higher sensitivity and specificity [32].

Guo et al. used 64CuCl2 as precursor to synthesis [64Cu]CuInS/ZnS QDNCs. Positron-emitting radioisotopes attachment to QDNCs helps PET functionality and results in self-illuminating QDNCs, with no need for an external light source due to Cerenkov resonance energy transfer (CRET). [64Cu]CuInS/ZnS QDNCs showed great stability with less than 3% free 64Cu when exposed to serum containing Ethylenediaminetetraacetic acid (EDTA) for 24 h. To improve biocompatibility, radioactive QDNCs (RQDNCs) were PEGylated and showed higher tumor uptake in U87MG mouse xenograft model to a maximum of 10.8%ID/g. [64Cu]CuInS/ZnS QDNCs were successfully applied as an efficient PET/self-illuminating luminescence in vivo imaging agents [33] (Fig. 6).

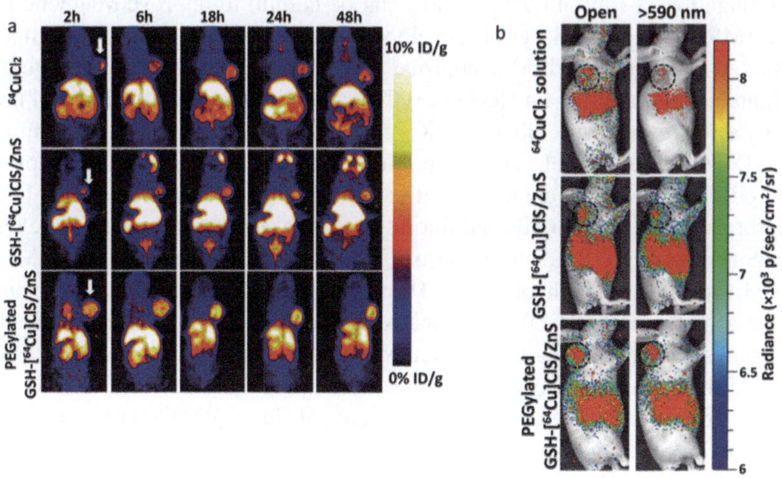

Fig. 6 A whole-body coronal PET images of U87MG tumor-bearing mice from 2 to 48 h were captured after intravenous administration of 100 μL (50 μg, 300 μCi) of 64CuCl2, Glutathione (GSH)-[64Cu]CIS/ZnS, and PEGylated GSH-[64Cu]CIS/ZnS RQDNCs (each group contained three mice's). The tumor location is indicated by an arrow. b) CRET images of U87MG tumor-bearing mice were obtained at 6 h after injection of 100 μL (300 μCi) of 64CuCl2, GSH-[64Cu] CIS/ZnS, and PEGylated GSH-[64Cu] CIS/ZnS RQDNCs. The tumor area was circled in these luminescence images (with 3 mice in each group). Notably, these images were captured without the use of excitation light, utilizing an open and red filter (>590 nm). Reproduced with permission [33]. Copyright 2015, American Chemical Society

Tu et al. synthesis dextran-coated silicon quantum dots (QDNCs) ($d = 15.1 \pm 7.6$ nm, $\lambda_{emi} = 440$ nm) and labled with a macrocyclic ligand-64Cu2 + with 78% radiolabeling yield. Four mice were injected by radiolabeled QDNCs to investigate bio distribution in PET imaging and ex vivo gamma counting. Results showed that in a short period after injection QDNCs were accumulated in bladder and liver but after hours only liver showed strong signal. Ex vivo gamma counting approved in vivo imaging results [34].

5 Quantum Dots Nanocomposites Application in Magnetic Resonance Imaging

Magnetic resonance imaging (MRI) is a noninvasive imaging technique that give high spatial resolution in clinical diagnoses. To obtain images with better and clearer resolution contrast agents can be helpful. Most of contrast agents that are used for MRI, are gadolinium (Gd3+) chelates such as gadolinium-diethylenetriaminepentaacetic acid (Gd–DTPA) and gadolinium –2,2′,2″,2‴-(1,4,7,10-tetraazacyclododecane-1,4,7,10-tetrayl)tetraacetic acid (Gd-DOTA). These complex have high Gd^{+3} content and relatively low proton's longitudinal relaxivity (r1) which lead to high administration of them [35].

Conjugation of carbon quantum dots can be helpful to improve Magnetic resonance (MR) efficiency. Xu et al. developed carbon quantum dots nanocomposite -stabilized gadolinium hybrid nanoprobe (Gd–CQDNCs) that exhibit low toxicity and high biocompatibility on HepG2 cell line and mice model as in vitro and invivo toxicity assay respectively. Gd–CQDNCs exhibited a greater MR response compared to Gd–DTPA due to their elevated gadolinium concentration and hydrophilic properties. Additionally, there were no indications of the CQDs' fluorescence signal being compromised. The in vivo MR and fluorescence dual-modality imaging were tested on zebrafish embryo. Biodistribution of Gd-QDNCs was investigated in mice and showed no change in weight of mice's. The Gd–CQDNCs modification with arginine-glycine-aspartic acid (RGD) was done to enhance its binding affinity with U87 cancer cells, thereby facilitating targeted imaging. While the MR response exhibited a notable ability to penetrate depths and visualize spatial regions, fluorescence-based analysis provided a detailed distribution imaging of Gd–CQDNCs within tissues due to its superior resolution and sensitivity [36].

In another study Ding et al. anchord a single Gd atom on graphene quantum dots (SAGd-GQDNCs), which result in a dendrite-like structure and compared it to the commercial Gd-DTPA. MTT Test on HepG2 cells survive more than 90% and in vivo test on 4T1 tumor bearing mice showed that SAGd-GQDNCs had 25 times higher longitudinal relaxivity at low Gd^{+3} concentration as 2 μ mol/kg. The obtained result offer Gd-CQDNCs different complex form a potential carrier to be used as MRI contrast agent (Fig. 7) [37].

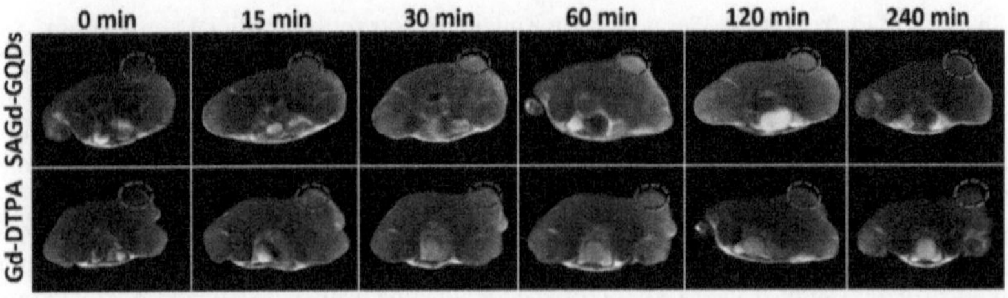

Fig. 7 In vivo magnetic resonance imaging (MRI) was performed on mice with tumors after an intravenous administration of two different samples, SAGd-GQDNCs and Gd-DTPA, at varying time intervals. The changes in the MRI signal of the tumor were indicated by a red circle. Reproduced with permission [37]. Copyright 2021, American Chemical Society

6 Conclusion and Future Perspectives

Quantum dots based nanocomposite (QDNCs) are amazing nanomaterials for diagnosis and bioimaging applications. The quantum dots specific chemical, physical and optical properties enable high functionalization capacity, excellent loading efficiency for imaging, sensing, diagnostic and theranostic application. Multimodal imaging or multiplexed imaging such as Photoacoustic/Fluorescence, MR/Fluorescence, PET/Fluorescence imaging and theranostic imaging application are the future of quantum dots based bioimaging. However, more animal and preclinical study should be done to understand the optimum dose of quantum dots and quantum dots nanocomposites without toxicity for clinical bioimaging application. However, quantum dots and quantum dots base nanocomposites offer a brilliant applicant for developing new bioimaging strategies and theranostic medicine as an imaging probe.

References

1. Bentolila, L.A., Ebenstein, Y., Weiss, S.: Quantum dots for in vivo small-animal imaging. J. Nucl. Med. **50**(4), 493–496 (2009)
2. Vannoy, C.H., Tavares, A.J., Noor, M.O., Uddayasankar, U., Krull, U.J.: Biosensing with quantum dots: a microfluidic approach. Sensors **11**(10), 9732–9763 (2011)
3. Volkov, Y.: Quantum dots in nanomedicine: recent trends, advances and unresolved issues. Biochem. Biophys. Res. Commun. **468**(3), 419–427 (2015)
4. Lahoti, H.S., Jogdand, S.D.: Bioimaging: evolution, significance, and deficit. Cureus. **14**(9), e28923 (2022). https://doi.org/10.7759/cureus.28923.PMID:36225412;PMCID:PMC9541884
5. James, M.L., Gambhir, S.S.: A molecular imaging primer: modalities, imaging agents, and applications. Physiol. Rev. **92**(2), 897–965 (2012). https://doi.org/10.1152/physrev.00049.2010. PMID: 22535898
6. Moseley, M., Donnan, G.: Multimodality imaging: introduction. Stroke **35**, 2632–2634 (2004)
7. Weissleder, R., Nahrendorf, M.: Advancing biomedical imaging. Proc. Natl. Acad. Sci. U S A **112**(47), 14424–14428 (2015). https://doi.org/10.1073/pnas.1508524112. PMID: 26598657, PMCID: PMC4664297

8. Shao, L., Gao, Y., Yan, F.: Semiconductor quantum dots for biomedicial applications. Sensors (Basel) **11**(12), 11736–11751 (2011). https://doi.org/10.3390/s111211736. PMID: 22247690. PMCID: PMC3252007
9. Dong, B., Li, C., Chen, G., Zhang, Y., Zhang, Y., Deng, M., et al.: Facile synthesis of highly photoluminescent Ag2Se quantum dots as a new fluorescent probe in the second near-infrared window for in vivo imaging. Chem. Mater. **25**(12), 2503–2509 (2013). https://doi.org/10.1021/cm400812v
10. Yaghini, E., Turner, H.D., Le Marois, A.M., Suhling, K., Naasani, I., MacRobert, A.J.: In vivo biodistribution studies and ex vivo lymph node imaging using heavy metal-free quantum dots. Biomaterials **104**, 182–191 (2016). https://doi.org/10.1016/j.biomaterials.2016.07.014. PMID: 27454064. PMCID: PMC4993815
11. Jeong, S., Song, J., Lee, W., Ryu, Y.M., Jung, Y., Kim, S.Y., et al.: Cancer-microenvironment-sensitive activatable quantum dot probe in the second near-infrared window. Nano Lett. **17**(3), 1378 (2017). https://doi.org/10.1021/acs.nanolett.6b04261
12. Das, S., Mondal, S., Ghosh, D.: Carbon quantum dots in bioimaging and biomedicines. Front. Bioeng. Biotechnol. **11**, 1333752 (2024). https://doi.org/10.3389/fbioe.2023.1333752
13. Zaini, M.S., et al.: Quantum confinement effect and photoenhancement of photoluminescence of PbS and PbS/MnS quantum dots. Appl. Sci. **10**(18), 6282 (2020)
14. Li, H., et al.: Water-soluble fluorescent carbon quantum dots and photocatalyst design. Angew. Chem. Int. Ed. **49**(26), 4430–4434 (2010)
15. Tajik, S., et al.: Carbon and graphene quantum dots: a review on syntheses, characterization, biological and sensing applications for neurotransmitter determination. RSC Adv. **10**(26), 15406–15429 (2020)
16. Yang, S.-T., Wang, X., Wang, H., Lu, F., Luo, P.G., Cao, L., et al.: Carbon dots as nontoxic and high-performance fluorescence imaging agents. J. Phys. Chem. C **113**(42), 18110–18114 (2009). https://doi.org/10.1021/jp9085969
17. Huang, C., Dong, H., Su, Y., Wu, Y., Narron, R., Yong, Q.: Synthesis of carbon quantum dot nanoparticles derived from byproducts in bio-refinery process for cell imaging and in vivo bioimaging. Nanomaterials **9**(3), 387 (2019). https://doi.org/10.3390/nano9030387
18. Liu, Q., Guo, B., Rao, Z., Zhang, B., Gong, J.R.: Strong two-photon-induced fluorescence from photostable, biocompatible nitrogen-doped graphene quantum dots for cellular and deep-tissue imaging. Nano Lett. **13**(6), 2436–2441 (2013). https://doi.org/10.1021/nl400368v. PMID: 23675758
19. Luo, Z., Yang, D., Yang, C., Wu, X., Hu, Y., Zhang, Y., et al.: Graphene quantum dots modified with adenine for efficient two-photon bioimaging and white light-activated antibacteria. Appl. Surface Sci. **434**, 155–162 (2018). https://doi.org/10.1016/j.apsusc.2017.10.121
20. Khaledian, S., et al.: Two-dimensional nanostructure colloids in novel nano drug delivery systems. Colloids Surf. A **585**, 124077 (2020)
21. Zhang, W., et al.: Tumor acidity and near-infrared light responsive drug delivery MoS2-based nanoparticles for chemo-photothermal therapy. Photodiagn. Photodyn. Ther. **38**, 102716 (2022)
22. Li, L., et al.: Facile synthesis of MoS 2 quantum dots as fluorescent probes for sensing of hydroquinone and bioimaging. Anal. Methods **11**(26), 3307–3313 (2019)
23. Roy, S., Bobde, Y., Ghosh, B., Chakraborty, C.: Targeted bioimaging of cancer cells using free folic acid-sensitive molybdenum disulfide quantum dots through fluorescence "Turn-Off." ACS Appl. Bio Mater. **4**(3), 2839–2849 (2021). https://doi.org/10.1021/acsabm.1c00090. PMID: 35014323
24. Niknam, S., Dehdast, S.A., Pourdakan, O., Shabani, M., Koohi, M.K.: Tungsten disulfide nanomaterials (WS2 NM) application in biosensors and nanomedicine: a review. Nanomed. Res. J. **7**(3), 214–226 (2022). https://doi.org/10.22034/nmrj.2022.03.001
25. Shao, J., Xie, H., Huang, H., et al.: Biodegradable black phosphorus-based nanospheres for in vivo photothermal cancer therapy. Nat. Commun. **7**, 12967 (2016). https://doi.org/10.1038/ncomms12967
26. Khandelwal, A., Mani, K., Karigerasi, M.H., Lahiri, I.: Phosphorene—The two-dimensional black phosphorous: properties, synthesis and applications. Mater. Sci. Eng.: B **221**, 17–34 (2017). ISSN 0921-5107. https://doi.org/10.1016/j.mseb.2017.03.011

27. Doganov, R.A., O'Farrell, E.C., Koenig, S.P., Yeo, Y., Ziletti, A., Carvalho, A., Campbell, D.K., Coker, D.F., Watanabe, K., Taniguchi, T., Castro Neto, A.H., Özyilmaz, B.: Transport properties of pristine few-layer black phosphorus by van der Waals passivation in an inert atmosphere. Nat. Commun. **6**, 6647 (2015)
28. Xu, Y., Ren, F., Liu, H., Zhang, H., Han, Y., Liu, Z., et al.: Cholesterol-modified black phosphorus nanospheres for the first NIR-II fluorescence bioimaging. ACS Appl. Mater. Interfaces (2019). https://doi.org/10.1021/acsami.9b05825
29. Zhang, W., Shen, Z., Wu, Y., Zhang, W., Zhang, T., Yu, B.Y., Zheng, X., Tian, J.: Renal-clearable and biodegradable black phosphorus quantum dots for photoacoustic imaging of kidney dysfunction. Anal. Chim. Acta **1204**, 339737 (2022). ISSN 0003-2670. https://doi.org/10.1016/j.aca.2022.339737
30. So, M.-K., Xu, C., Loening, A.M., Gambhir, S.S., Rao, J.: Self-illuminating quantum dot conjugates for in vivo imaging. Nat. Biotechnol. **24**, 339–343 (2006)
31. Ametamey, S.M., Honer, M., Schubiger, P.A.: Molecular imaging with PET. Chem. Rev. **108**, 1501–1516 (2008)
32. Robertson, R., Germanos, M., Li, C., Mitchell, G., Cherry, S., et al.: Optical imaging of Cerenkov light generation from positron-emitting radiotracers. Phys. Med. Biol. **54**, N355–N365 (2009)
33. Guo, W., Sun, X., Jacobson, O., Yan, X., Min, K., Srivatsan, A., Chen, X.: Intrinsically radioactive [64Cu]CuInS/ZnS quantum dots for PET and optical imaging: improved radiochemical stability and controllable Cerenkov luminescence. ACS Nano **9**(1), 488–495 (2015). https://doi.org/10.1021/nn505660r
34. Tu, C., Ma, X., House, A., Kauzlarich, S.M., Louie, A.Y.: PET imaging and biodistribution of silicon quantum dots in mice. ACS Med. Chem. Lett. **2**(4), 285–288 (2011). https://doi.org/10.1021/ml1002844.PMID:21546997;PMCID:PMC3086380
35. Gulani, V., Calamante, F., Shellock, F.G., Kanal, E., Reeder, S.B.: Gadolinium deposition in the brain: summary of evidence and recommendations. Lancet Neurol. **16**, 564–570 (2017)
36. Xu, Y., Jia, X.-H., Yin, X.-B., He, X.-W., Zhang, Y.-K.: Carbon quantum dot stabilized gadolinium nanoprobe prepared via a one-pot hydrothermal approach for magnetic resonance and fluorescence dual-modality bioimaging. Anal. Chem. **86**(24), 12122–12129 (2014). https://doi.org/10.1021/ac503002c
37. Ding, H., Wang, D., Sadat, A., Li, Z., Hu, X., Xu, M., de Morais, P.C., Ge, B., Sun, S., Ge, J., Chen, Y., Qian, Y., Shen, C., Shi, X., Huang, X., Zhang, R.Q., Bi, H.: Single-atom gadolinium anchored on graphene quantum dots as a magnetic resonance signal amplifier. ACS Appl. Bio Mater. **4**(3), 2798–2809 (2021). https://doi.org/10.1021/acsabm.1c00030. PMID: 35014319

Quantum Dots Nanocomposites as Drug Delivery Vehicle

Seyed Ahmad Dehdast and Omid Pourdakan

Abstract Quantum dots nanocomposites (QDNCs) has triggered enormous interest in, its applications in various fields of medicine and pharmaceutical science due to its unique properties. While QDNCs has demonstrated suitable potential as an imaging probe in medical imaging, its application as a nano-vehicle in drug delivery system for drugs and biomolecules in biomedicine, has only begun to be consideration. Advanced nano-carriers with slow, controllable release and excellent carrier efficiency are now considered the ideal vehicle for developing drug delivery systems. Remarkable properties of QDNCs include chemical stability, extremely large surface area, and low toxicity making it a promising material to design of novel drug delivery systems. Recent research has demonstrated significant success in the use of QDNCs for medical applications, including its utilization as a vehicle for controlled drug delivery, as well as in photothermal therapy, photodynamic therapy, and theranostic application. This chapter, presents a comprehensive overview of the current studies of drug delivery systems based on QDNCs vehicle. Moreover, various strategies for nano-vehicle design, functionalization methods for effective drug delivery, and reduction of quantum dots toxicity and the essential features for design of new Quantum dots nanocomposites vehicle in photothermal therapy, photodynamic therapy, and controlled drug delivery presented.

Keywords Quantum dots nanocomposites · Drug delivery system · Controlled release · Vehicle · Cancer · Photothermal therapy

List of Abbreviations

Ag	Silver
BBB	Blood-Brain Barrier
BPQD	Black Phosphorous quantum dots
BPQDs@Lipo	Black Phosphorous quantum dots liposome composite

S. A. Dehdast (✉) · O. Pourdakan
School of Medicine, Iran University of Medical Sciences, Tehran, Iran
e-mail: arashdehdast@gmail.com

S. Thomas et al. (eds.), *Quantum Dots Based Nanocomposites*, Engineering Materials,
https://doi.org/10.1007/978-3-031-54779-9_16

CdSe	Cadmium Selenide
CSCs	Cancer Stem Cells
CLSM	Confocal Laser Scanning Microscope
CMOS	Complementary Metal-Oxide-Semiconductor
CQD	Carbon Quantum Dots
DAPI	4′,6-Diamidino-2-phenylindole
DOX	Doxorubicin
EPR	Enhanced Permeability and Retention
FA	Folic Acid
FAR	Folate Receptors
FRET	Förster Resonance Energy Transfer
GQD	Graphene quantum dots
GOQD	Graphene oxide quantum dots
HP	HematoPorphyrin
IC50	Inhibitory concentration
In	Indium
InP	Indium Phosphate
MAP	Maximum Amplitude Projection
MDDS	Multi Drug Delivery Systems
MoS2	Molybdenum Disulfide
MUA	Mercaptoundecanoic
NIR	Near-Infrared
PA	Photoacoustic
PDT	Photodynamic Therapy
PEG	Polyethylene glycol
PLGA	Poly (lactic-co-glycolic acid)
PTT	Photothermal Therapy
QD	Quantum Dots
QDNC	Quantum dots nanocomposites
R&D	Research and Development
ROS	Reactive Oxygen Species
S	Sulfur
SOSG	Singlet Oxygen Sensor Green
TDD	Targeted Drug Delivery
WPU	Waterborne polyurethane
Zn	Zinc
ZnS	Zinc Sulfide
ZnSe	Zinc Selenide

1 Introduction

The major problems that threaten the health of people in the world today have created challenges and motivation for research and development (R&D) of various drugs and drug delivery systems using new methods. On the other hand, conventional medicine has some disadvantages such as poor bioavailability, weak targeting specificity, organ and systemic toxicity [1]. For this reason, R&D as to new drug delivery systems and therapies is growing and proceeding at an accelerated rate. The use of nanoscience and nanotechnology is one of the keys strategies of researchers to improve conventional treatment methods.

The use of nanoproducts requires detailed investigation due to the large chemical diversity and insufficient knowledge of these products. These studies should be designed in such a way that they provide detailed information about toxicity, biocompatibility, physiological effects, and the effects on other organs of the body in addition to the target organ. In case of appropriate response, the nanobiotechnology products can be used in clinical trials studies and possibly in treatment [2]. Quantum dots nanocomposites (QDNCs) are among the most popular semiconductor nanomaterials that have uniqueness physical and chemical, and optical properties such as High surface-to-volume ratio, high potential bio-chemical functionalization, photoluminescent properties, high surface energy, thermal, electrical, and optical features. These outstanding properties of quantum dots enable an extensive range of applications, such as optics, electronics, environmental science, and medicine. Quantum Dots and Carbon Quantum Dots have biomedical potential application, including diagnostic, bioimaging, antimicrobial, anticancer, drug delivery and, theranostic fields [3, 4]. The evaluation of benefits and challenges of utilizing quantum dots such as cytotoxicity, safety, inhibition concentration and drug release platforms in biomedical applications is important. Nanomaterials, including quantum dots, are taken up by various cellular methods, such as active and passive targeting, phagocytosis, and non-phagocytosis routes. Functionalization capacity of QDNCs and diversity in the entry routes of nanomaterials into the cell makes it possible to design different drug carriers with different strategies for drug delivery and therapy [5]. Recently, several studies have been conducted based on QDNCs drug delivery systems for various diseases treatment such as malaria, cancer, etc. This chapter offers an exciting overview of recent researches of this emerging area for researchers in the fields of medicine, biomedical engineering, pharmaceutical science and material science, as well as for nanomedicine groups. In this chapter, we discuss about Quantum dots nanocomposites based drug delivery systems that have been studied in vitro and preclinical studies.

2 Controlled Drug Delivery

Controlled drug release is a strategy used to regulate the rate and timing of drug delivery to a specific target site in the body. This approach is particularly useful in cases where sustained drug exposure is required, or when drugs have a narrow therapeutic window and must be delivered in precise doses.

There are several ways to achieve controlled drug release, including the use of drug delivery systems such as nanoparticles, graphene quantum dots, liposomes, and hydrogels. These systems can be designed for drugs release in response to specific stimuli, such as light, pH changes, temperature, and the presence of enzymes or other biomolecules [6, 7].

Near-infrared (NIR) light has acquired popularity in biomedical applications because of its non-invasive nature and its capability to penetrate tissues with minimal absorption and scattering. NIR light has a wavelength range of 700–2500 nm, which allows it to pass through biological tissues with minimal damage, making it an attractive option for non-invasive diagnostic and therapeutic applications.

In addition to its tissue-penetrating ability, NIR light can be adjusted using a range of parameters, including wavelength, duration, and intensity, enabling precise control over the amount of energy delivered to a specific target site. This makes it possible to tailor the light to the specific needs of the application, whether it's for imaging, therapy, or a combination of both [8, 9].

Geng et al. put black phosphorus quantum dot in the hydrophobic bilayer and Doxorubicin (DOX) in liposome aqueous core to controlled release of drug by NIR light irradiation. Biocompatibility on two normal and cancer cell line L929 (mouse fibroblast cells), QSG-7701 (normal human liver cells), A549 (human lung carcinoma cells) and MCF-7 showed good biosafety. The released rate can be controlled by laser power, means with increase in power release rate increase also. The in vitro therapeutic assessment showed that group which incubated with DOX-loaded black phosphorous quantum dots liposome composite (BPQDs@Lipo) and irradiate with 808 nm laser for 10 min achieved maximum cell death among other groups result in promising efficiency of this chemo-photothermal therapy (10).

Liu created a versatile nanocarrier, Mo_S2-PEG, by modifying highly fluorescent MoS_2 QDs with biocompatible Polyethylene glycol (PEG). This modification resulted in improved physiological stability and decreased toxicity. MoS_2-PEG-DOX, the resulting drug carrier, demonstrated desirable pH-responsive drug release properties. Results from a confocal laser scanning microscope (CLSM) study showed that DOX detached from MoS_2-PEG inside the cytoplasm and entered the nucleolus to inhibit cellular growth. The cellular uptake of MoS_2-PEG and drug release can be monitored simultaneously using two separate fluorescence channels, which reveal their localization and relative internalization kinetics (Fig. 1). These findings suggest that pH-responsive drug delivery systems based on MoS_2 QDs may be a promising option for cancer therapy [11].

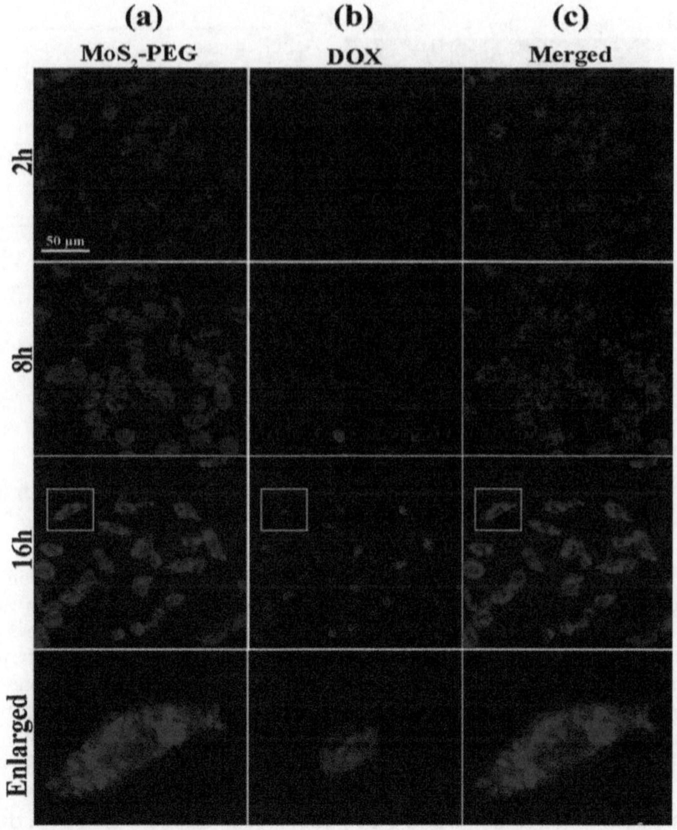

Fig. 1 CLSM images were acquired of U251 cells that were subjected to incubation with MoS_2-PEG-DOX for varying periods of time. The obtained images encompass three distinct categories: **a** MoS_2-PEG, **b** DOX, and **c** merged images of (**a**) and (**b**). Reproduced with permission [11]. Copyright 2020, Elsevier

To increase the effectiveness of chemotherapy in killing cancer cells, it is beneficial for large doses of anticancer drugs to enter the nuclei of cancer cells. Su et al. cancer-stem-cell-nucleus-penetrable red-emissive carbon quantum dots (CSCNP-R-CQDs) that are capable of entering the nuclei of both cancer cells and cancer stem cells due to nitrogen surface group. When doxorubicin was loaded onto the surfaces of the carbon quantum dots at a concentration of 30 μg/mL, the average viability of HeLa cells was reduced to 21%, while free doxorubicin only decreased cell viability to 50% (Fig. 2). The doxorubicin-loaded carbon quantum dots also showed promise in eliminating cancer stem cells. This study suggests that carbon quantum-dot-based drug carriers have potential for effectively eradicating cancer and represents a potential strategy for developing new anticancer drugs [12].

Pancreatic cancer has a poor prognosis and is resistant to traditional therapies, making it the most deadly of all cancers. Nigam et al. investigate the potential of

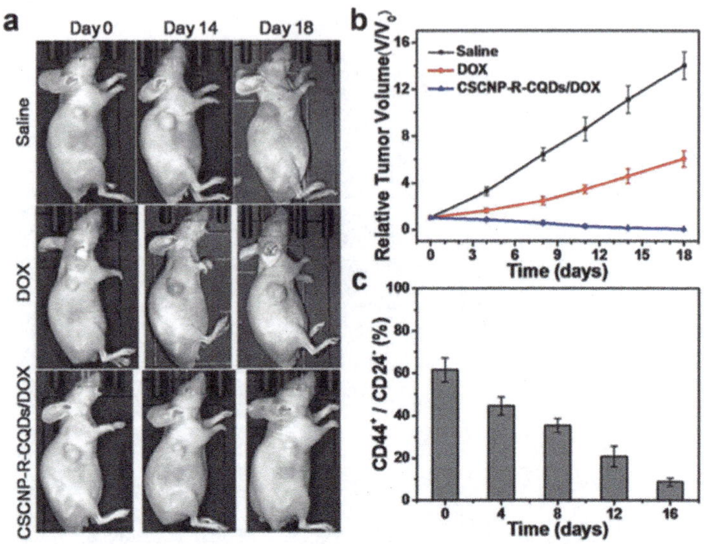

Fig. 2 **a** Exhibit a tumor-bearing mice images after treatment with saline solution, DOX, and CSCNP-R-CQDs/DOX on various days. **b** The relative volume of tumors in the mice was measured following the administration of diverse treatments. **c** Furthermore, the percentage of Cancer Stem Cells (CSCs), specifically CD44+/CD24− levels, present in the cells obtained from the tumors were also measured after 0, 4, 8, 12, and 16 days of treatment. All the provided data is presented as the mean value ± standard deviation, with a sample size of n = 5. Reproduced with permission [12]. Copyright 2020, American Chemistry Society

hyaluronic acid functionalized and green fluorescent graphene quantum dot (GQD)-labeled human serum albumin nanoparticles for pancreatic cancer-specific drug delivery and bioimaging. GQDs have become more popular than their metal semiconductor counterparts due to their tunable fluorescence properties and biocompatibility. The synthesized GQDs had excellent fluorescence and a good quantum yield of approximately 14%. Gemcitabine, the preferred drug for pancreatic cancer treatment, was encapsulated in albumin nanoparticles. The presence of hyaluronic acid as a targeting moiety on the human serum albumin nanoparticles improved the efficiency of gemcitabine on resistant pancreatic cancer cells. The nanoformulation significantly enhanced the bioavailability and sustained release property of the drug to pancreatic cancer cells in vitro [13].

Lung cancer is a common and deadly form of cancer worldwide, and conventional chemotherapy is the primary treatment option for small cell lung cancer. However, it lacks specificity and often results in severe side effects [14]. To address this issue, Ruzycka-Ayoush developed a new targeted drug delivery (TDD) system using Ag–In–Zn–S quantum dots (QDs) modified with various molecules, including folic acid (FA), to target doxorubicin (DOX) to folate receptors (FARs) on adenocarcinoma human alveolar basal epithelial cells (A549). The QD–FA–DOX nanoconjugates showed higher cytotoxicity than other synthesized QD samples. In contrast,

QD–MUA–FA–DOX nanoconjugates were the most cytotoxic against A549 cells, confirming FARs as molecular targets. The QD–MUA–FA–DOX nanoconjugates also significantly inhibited A549 cell migration in an in vitro scratch assay. Furthermore, all nanoconjugates induced significantly more DNA breaks than non-treated cells at the IC_{50} dose (Fig. 3). Overall, the QD–MUA–FA–DOX nanoconjugates showed the greatest cytotoxicity and genotoxicity and significantly inhibited the migratory potential of A549 cells [15].

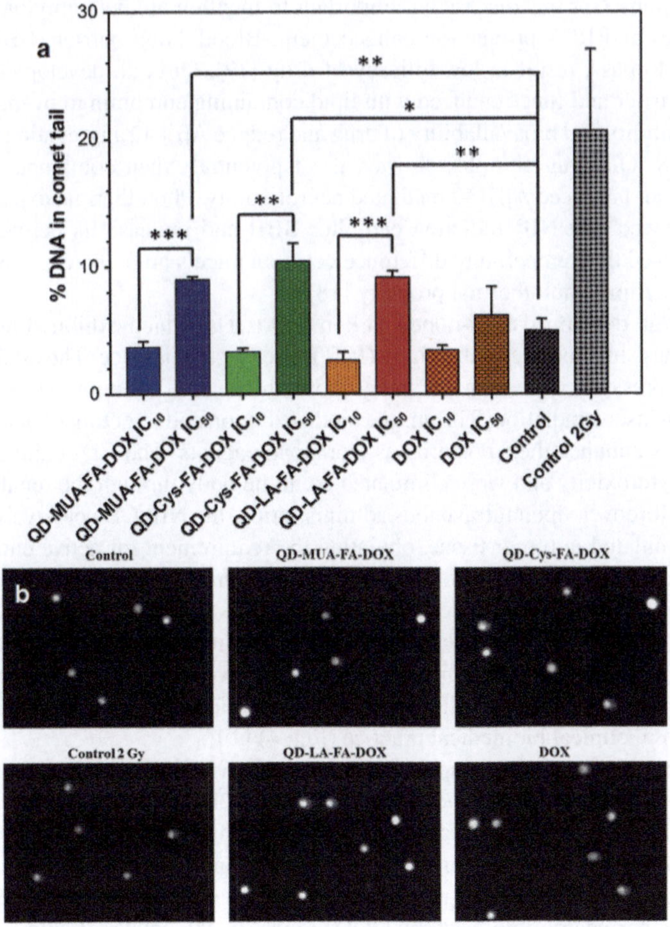

Fig. 3 **a** Represent the DNA damage percentage in A549 cells after treatment with QD–MUA–FA–DOX, QD–Cys–FA–DOX, QD–LA–FA–DOX, and DOX at IC10 and IC50. $*P < 0.05$, $**P < 0.01$, $***P < 0.001$, and $****P < 0.0001$, between the investigated concentrations (unpaired t-test). **b** Fluorescent images were obtained of the DNA damage of A549 cells subsequent to incubation with the compounds at IC50 after 24 h. Damage detected by the comet assay. Ag–In–Zn–S QD modified with: 11-mercaptoundecanoic acid (MUA), Cys L-cysteine, LA lipoic acid, FA folic acid, DOX doxorubicin. Reproduced with permission [15]. Copyright 2021, Springer Nature

3 Photothermal Therapy

Alzheimer's disease is a progressive neurodegenerative disorder that affects the brain, leading to memory loss, cognitive decline, and behavioral changes. It is the most common cause of dementia and typically affects people over the age of 65, although early-onset forms of the disease can occur in younger individuals. The exact cause of Alzheimer's disease is not yet fully understood, but it is thought to involve a combination of genetic, environmental, and lifestyle factors [16]. Multiple factors are in response for this disease but important to mention are beta-amyloid plaques, tau tangles and ROS production enhancement. Blood–brain barrier (BBB) hardly let drugs to pass, result in low efficacy of drug [17]. Qi et al. developed a MoS_2-QDs as carrier and functionalized with lipid containing curcumin to overcome poor water solubility and bioavailability of drug and reduce Aβ1-42 aggregation and eliminate ROS. QD-drug complex showed great potential when combined with NIR treatment and reduced Aβ1-42 mediated neurotoxicity. Photothermal experiment on mices showed that NIR radiation can alter BBB and increase its permeability. It was observed that temperature difference between mice's brain increased and reflect MoS_2 QDs/lipid photothermal property [18].

Bao et al. developed S, N-doped carbon dots (CDs) which exhibited remarkable effectiveness in photothermal therapy (PTT) when tested on mice. These CDs exhibited an impressive conversion efficiency of 59% in aqueous solutions when subjected to 655-nm laser irradiation. The unique biodistribution and targeting of near-infrared (NIR) CDs enhance their potential as theranostic agents. The CDs exhibited a low level of cytotoxicity and were eliminated from the body through the renal filtration system. Moreover, upon intravenous administration, the NIR CDs passively targeted and accumulated in tumor tissue, obviating the requirement for active entities such as antibodies. This resulted in robust NIR fluorescence and photoacoustic signals. The tumors were eradicated after intravenous administration of the CDs, followed by exposure to a 655 nm laser. The CDs' accumulation in tumor tissue without the need for intratumor injection, their high photothermal conversion, exceptional optical and photoacoustic imaging capabilities, and renal excretion render them well-suited for utilization in clinical biomedical practice (Fig. 4) [19].

Perini et al. found that graphene quantum dots (GQDs) with surface carboxyl groups can increase membrane permeability in a 2D and 3D model of glioblastoma multiform, a type of brain cancer. This enhances the effectiveness of two chemotherapeutic drugs, doxorubicin and temozolomide, even when administered at sub-therapeutic doses. The proposed mechanism of action is that the increased membrane permeability induced by GQDs improves the uptake of antitumor drugs inside glioblastoma cells. The study suggests that biocompatible GQDs can increase membrane permeability through photothermal conversion in a reliable tumor model, allowing the use of sub-therapeutic doses of antitumor drugs, enhancing their efficacy specifically in cancer regions while reducing side effects, potentially improving the quality of life for patients. Additionally, the increased membrane permeability due to GQDs-mediated PTT enhances the release of reactive oxygen species, resulting in

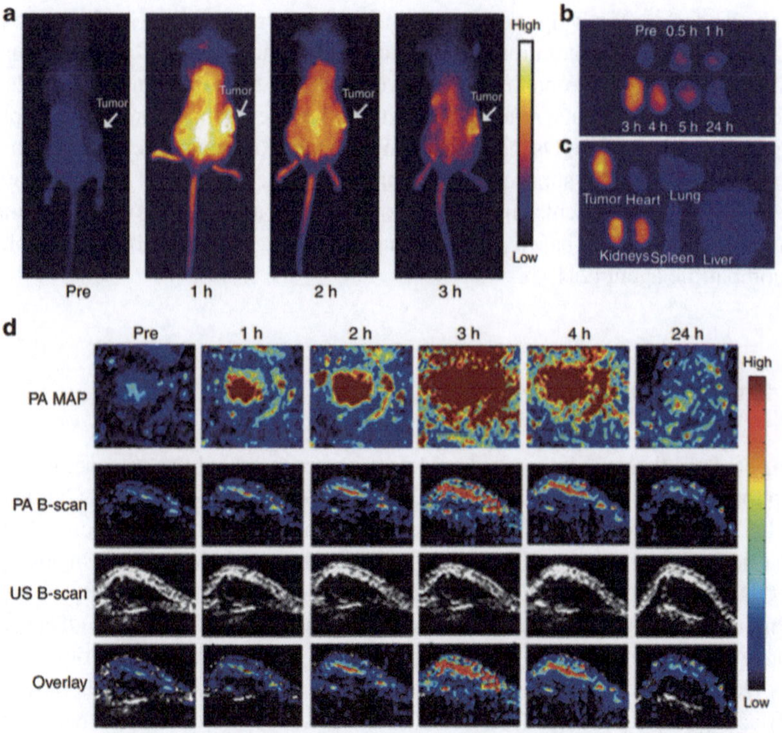

Fig. 4 Exhibit in vivo CDs that passively targeted. **a** Mouse bodies NIR fluorescence images were captured after the intravenous administration of CDs at different time intervals. **b** The near-infrared fluorescence of H22 tumors that were removed from mice at different time intervals after injection were analyzed. **c** The near-infrared fluorescence of major organs and H22 tumors that were dissected from mice three hours after injection is measured. The near-infrared fluorescence images of the mouse bodies, major organs, and tumors were captured using a 655 nm laser with an intensity of 10.6 mW cm^{-2}. The emitted light was filtered using a 700 nm long pass filter and recorded using a CMOS camera for near-infrared imaging. **d** 4T1 tumors Photoacoustic (PA) maximum amplitude projection (MAP) and B-scan PA images in mice were obtained after the intravenous injection of CDs at different time points. Reproduced with permission [19]. Copyright 2018, Nature Publication Group

improved efficacy of immunotherapy with strong migration of the immune system towards irradiated cancer spheroids [20].

Shao et al. developed biodegradable nanospheres by loading black phosphorus quantum dots (BPQDs) into Poly (lactic-co-glycolic acid) (PLGA). BPQD rapid degradation was prevented due to PLGA hydrophobicity, improving their photothermal performance in a physiological environment. The nanospheres showed no significant toxicity and good biocompatibility in hematological, blood biochemical, and histological analyses. The nanospheres exhibit prolonged blood circulation and efficacious tumor accumulation upon intravenous administration through the enhanced permeability and retention (EPR) effect. Remarkably, even in deep tissues, the nanospheres effectively induce apoptosis in cancer cells when subjected to near-infrared (NIR) laser irradiation, leading to highly efficient tumor ablation. Furthermore, the BPQDs/PLGA nanospheres have exhibited efficacy in animal models with different immune systems. After fulfilling their therapeutic function, the nanospheres degrade and release BPQDs that transform into nontoxic phosphate and phosphonate compounds, which are subsequently cleared from the body. This gradual degradation of BPQDs circumvents any local phosphorus imbalance invivo, highlighting the clinical potential of BP-based photothermal therapy as a uniquely biodegradable and biocompatible agent [21].

4 Photodynamic Therapy

Dong et al. synthesize MoS_2 quantum dots to compare their ability to generate oxygen for photodynamic therapy with commercial photodynamic agent PpIX. Cytotoxicity on HeLa cells showed no obvious result on cell which indicate good biocompatibility of particles. In comparison to MoS_2 QDs, PpIX showed 22% decrease in absorption but QDs showed photostability after 1h irradiation. To evaluate the $^1O_2$ production within cells, a singlet oxygen sensor green (SOSG) reagent was used. QDs showed stronger fluorescence signal which was a sign of more oxygen production compare to PpIX which provide great promising application in photodynamic therapy [22].

Black phosphorus quantum dots (BPQDs) are of interest in cancer therapy due to their excellent biocompatibility and photothermal and photodynamic efficacies. However, their low environmental stability and poor dispersity have limited their practical applications. Lu et al. synthesized, biocompatible anionic waterborne polyurethane (WPU) nanoparticles from castor oil to encapsulate BPQDs, resulting in much-improved dispersion and environmental stability without affecting the photothermal efficiency of BPQDs. Interestingly, the encapsulation of BPQDs in WPU nanoparticles led to a significant enhancement in reactive oxygen species (ROS) generation, indicating an improved photodynamic efficacy compared to bare BPQDs. The pH regulation effect of WPU-BPQDs was investigated, and it was found that WPU-BPQDs boosted ROS production in acidic tumor environments compared to bare BPQDs. In vitro and in vivo experiments demonstrated that WPU-BPQDs effectively improved both photodynamic and photothermal therapy for cancer treatment. Overall, WPU-BPQDs have excellent dispersity, biocompatibility, and biodegradability and are a promising agent for photodynamic and photothermal cancer treatments [23].

Yan et al. created a biocompatible near-infrared (NIR) photosensitizer using a self-assembled CdSe quantum dot (QD)-aza-a@PEG conjugate for targeting tumors with concurrent imaging and photodynamic therapy. The aza-boron-dipyrromethene (aza-BODIPY) moiety in the conjugate emits strong NIR light at 750 nm upon excitation of the QD donor via efficient FRET, making it useful for bioimaging. Additionally, the conjugate can generate singlet oxygen for photodynamic therapy using laser light at 635 nm. The water-solubility of this conjugate and its ability to target specific tumor cells is promising for various applications, including cellular labeling, deep-tissue imaging, assay labeling, diagnosis, and photodynamic therapy [24].

Ahirwar et al. demonstrated the efficacy of graphene quantum dots (GQDs) and graphene oxide quantum dots (GOQDs) in photodynamic therapy (PDT) for the eradication of cancer cells. GQDs and GOQDs were employed as photosensitizers for the PDT process, while 365 nm UV tube light was the irradiation source. The GQDs and GOQDs possessed a size range spanning from 1.5 to 5.5 nm and demonstrated photoluminescence behavior dependent on excitation, encompassing the UV to visible regions and yielding singlet oxygen. Furthermore, these QDs exhibited minimal toxicity towards skin cancer (B16F10) and breast cancer (MCF-7) cells, along with facile cellular uptake and distribution within the cells, thus positioning them as promising candidates for PDT application. Photodynamic therapy (PDT) was conducted on B16F10 and MCF-7 cells utilizing GQDs/GOQDs, resulting in a remarkable mortality rate of over 90% within a brief exposure period of 5 min. Conversely, cells that were not subjected to GQDs/GOQDs and were instead exposed to irradiation for the same duration exhibited a survival rate of more than 80%. GQD/GOQD PDT offers numerous advantages, such as its potent efficacy in killing cells through micro-watt power UV irradiation, reduced exposure time, and the ability to provide uniform and extensive coverage. Nonetheless, it is essential to acknowledge that the scope of this therapy may be confined to skin-related cancers due to the limited ability of UV irradiation to penetrate deeply into tissues [25].

The low solubility, poor photostability, and tendency to aggregate of conventional photosensitizers have posed obstacles to their application in photodynamic therapy (PDT) for cancer cells. This has led to the development of complex synthetic procedures, such as using nanomaterials to load photosensitizers, to effectively use photosensitizers in PDT. However, Murali et al. have developed a simple and efficient method for generating hematoporphyrin (HP) photosensitizer-encapsulated carbon quantum dots (CQDs) (HP-CQDs) through a one-step microwave reaction involving HP monomer as one of the precursor materials. The resultant HP-CQDs exhibited the same optical and chemical characteristics as HP, while also demonstrating exceptional solubility in water. Moreover, HP-CQDs displayed remarkable capability for producing reactive oxygen species under deep red light, which rendered them highly effective in the photodynamic therapy (PDT)-assisted eradication of human breast cancer cells (MCF-7), exhibiting significant phototoxicity and minimal dark toxicity in comparison to HP alone (Fig. 5). Overall it was demonstrates that that the integration of photosensitizer-implanted CQDs can be accomplished by carefully selecting appropriate precursors and synthesis conditions to yield CQDs, thereby offering a promising approach for PDT-assisted cancer treatment [26].

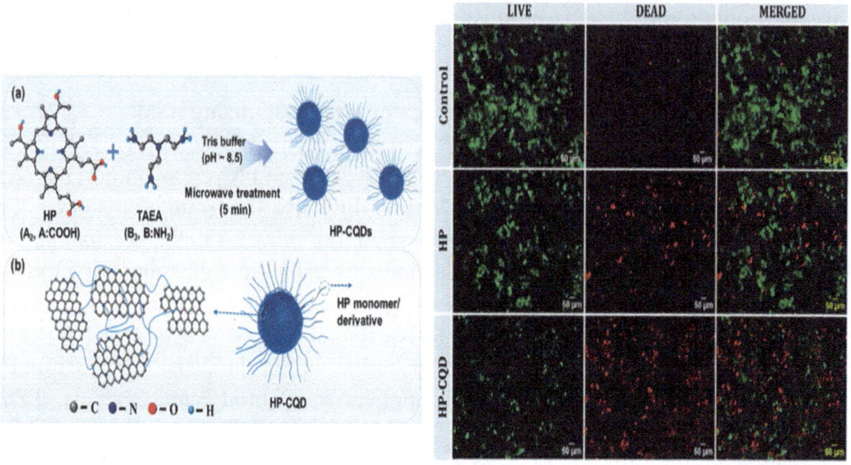

Fig. 5 Display schematic picture of **a** synthesis steps of HP-CQD and **b** structures that HP-CQD possibly can achieve through cross-linking in the left panel. A 635 nm laser at 50 mW cm^{-2} was used for 1 min to irradiate MCF-7 cells that were incubated with HP and HP-CQDs. After irradiation, cells were stained with green dye calcein-AM and red dye EthD-1 (red) to exhibit live and dead cells in confocal microscopy images, respectively, in the right panel. Reproduced with permission [26] Copyright 2022, American Chemical Society

5 Conclusion and Future Perspectives

Quantum dots nanocomposites are outstanding vehicle materials for drug delivery applications. The quantum dots specific chemophysical associated properties enable high functionalization capacity, excellent loading efficiency for imaging, diagnostic, controllable drug release, and drug targeting. Controlled drug delivery, multi drug delivery systems (MDDS) and theranostic application are the future of quantum dots and carbon dots based drug delivery.

However, more in vitro and in vivo and preclinical investigation must be performed to understand the biocompatibility, biosafety, toxicity and the long-term health effects of the associated with these nanocomposites. Finally, quantum dots based nanocomposites offer an excellent candidate for developing new nanodrugs and suitable vehicle for drug delivery systems in medicine.

References

1. de Bono, J.S., Ashworth, A.: Translating cancer research into targeted therapeutics. Nature **467**(7315), 543–549 (2010)
2. Peer, D., Karp, J.M., Hong, S., Farokhzad, O.C., Margalit, R., Langer, R.: Nanocarriers as an emerging platform for cancer therapy. Nat. Nanotechnol. **2**(12), 751–760 (2007)

3. Loretz, B., Jain, R., Dandekar, P., Thiele, C., Yamada, H., Mostaghaci, B., et al.: Chances of nanomaterials for pharmaceutical applications. Safety Aspects Eng. Nanomater. 279–317 (2013)
4. Das, P., Sherazee, M., Marvi, P.K., Ahmed, S.R., Gedanken, A., Srinivasan, S., Rajabzadeh, A.R.: Waste-derived sustainable fluorescent nanocarbon-coated breathable functional fabric for antioxidant and antimicrobial applications. ACS Appl. Mater. Interfaces **15**(24), 29425–29439 (2023)
5. Kelf, T.A., Sreenivasan, V.K.A., Sun, J., Kim, E.J., Goldys, E.M., Zvyagin, A.V.: Non-specific cellular uptake of surface-functionalized quantum dots. Nanotechnology **21**(28), 285105 (2010). https://doi.org/10.1088/0957-4484/21/28/285105
6. Torkashvand, H., Dehdast, S.A., Nateghpour, M., Haghi, A.M., Fard, G.C., Elmi, T., Shabani, M., Tabatabaie, F.: Antimalarial nano-drug delivery system based on graphene quantum dot on plasmodium falciparum: preparation, characterization, toxicological evaluation. Diamond Related Mater. **132**, 109670 (2023)
7. Qiu, M., Wang, D., Liang, W., Liu, L., Zhang, Y., Chen, X., Sang, D.K., Xing, C., Li, Z., Dong, B., Xing, F., Fan, D., Bao, S., Zhang, H., Cao, Y.: Novel concept of the smart NIR-light-controlled drug release of black phosphorus nanostructure for cancer therapy. Proc. Natl. Acad. Sci. U S A **115**(3), 501–506 (2018). https://doi.org/10.1073/pnas.1714421115. PMID: 29295927, PMCID: PMC5776980
8. Shum, P., Kim, J.M., Thompson, D.H.: Phototriggering of liposomal drug delivery systems. Adv. Drug Deliv. Rev. **53**(3), 273–284 (2001). https://doi.org/10.1016/s0169-409x(01)00232-0. PMID: 11744172
9. Hirsch, L.R., Stafford, R.J., Bankson, J.A., Sershen, S.R., Rivera, B., Price, RE., Hazle, J.D., Halas, N.J., West, J.L.: Nanoshell-mediated near-infrared thermal therapy of tumors under magnetic resonance guidance. Proc. Natl. Acad. Sci. U S A **100**(23), 13549–54 (2003). https://doi.org/10.1073/pnas.2232479100. PMID: 14597719, PMCID: PMC263851
10. Geng, S., Wu, L., Cui, H., Tan, W., Chen, T., Chu, P.K., Yu, X.-F.: Synthesis of lipid–black phosphorus quantum dot bilayer vesicles for near-infrared-controlled drug release. Chem. Commun. **54**(47), 6060–6063 (2018). https://doi.org/10.1039/c8cc03423k
11. Liu, L., Jiang, H., Dong, J., Zhang, W., Dang, G., Yang, M., Li, Y., Chen, H., Ji, H., Dong, L.: PEGylated MoS2 quantum dots for traceable and pH-responsive chemotherapeutic drug delivery. Colloids Surf. B: Biointerfaces**185**, 110590. ISSN 0927-7765. https://doi.org/10.1016/j.colsurfb.2019.110590
12. Su, W., Guo, R., Yuan, F., Li, Y., Li, X., Zhang, Y., Zhou, S., Fan, L.: Red-emissive carbon quantum dots for nuclear drug delivery in cancer stem cells. J. Phys. Chem. Lett. **11**(4), 1357–1363 (2020). https://doi.org/10.1021/acs.jpclett.9b03891
13. Nigam, P., Waghmode, S., Louis, M., Wangnoo, S., Chavan, P., Sarkar, D.: Graphene quantum dots conjugated albumin nanoparticles for targeted drug delivery and imaging of pancreatic cancer. J. Mater. Chem. B **2**(21), 3190–3195 (2014). https://doi.org/10.1039/c4tb00015c
14. Yuan, M., Huang, L.L., Chen, J.H., Wu, J., Xu, Q.: The emerging treatment landscape of targeted therapy in non-small-cell lung cancer. Signal Transduct. Target Ther. **4**, 61 (2019). https://doi.org/10.1038/s41392-019-0099-9. PMID: 31871778, PMCID: PMC6914774
15. Ruzycka-Ayoush, M., Kowalik, P., Kowalczyk, A., et al.: Quantum dots as targeted doxorubicin drug delivery nanosystems in human lung cancer cells. Cancer Nano **12**, 8 (2021). https://doi.org/10.1186/s12645-021-00077-9
16. Zeng, H., Qi, Y., Zhang, Z., Liu, C., Peng, W., Zhang, Y.: Nanomaterials toward the treatment of Alzheimer's disease: recent advances and future trends. Chin. Chem. Lett. **32**, 1857–1868 (2021)
17. Cheignon, C., Tomas, M., Bonnefont-Rousselot, D., Faller, P., Hureau, C., Collin, F.: Oxidative stress and the amyloid beta peptide in Alzheimer's disease. Redox Biol. **14**, 450–464 (2018)
18. Qi, X., Ye, P., Xie, M.: MoS2 quantum dots based on lipid drug delivery system for combined therapy against Alzheimer's disease. J. Drug Deliv. Sci. Technol. **82**, 104324 (2023). ISSN 1773-2247. https://doi.org/10.1016/j.jddst.2023.104324

19. Bao, X., Yuan, Y., Chen, J., et al.: In vivo theranostics with near-infrared-emitting carbon dots—Highly efficient photothermal therapy based on passive targeting after intravenous administration. Light Sci. Appl. **7**, 91 (2018). https://doi.org/10.1038/s41377-018-0090-1
20. Perini, G., Palmieri, V., Friggeri, G., et al.: Carboxylated graphene quantum dots-mediated photothermal therapy enhances drug-membrane permeability, ROS production, and the immune system recruitment on 3D glioblastoma models. Cancer Nano **14**, 13 (2023). https://doi.org/10.1186/s12645-023-00168-9
21. Shao, J., Xie, H., Huang, H., Li, Z., Sun, Z., Xu, Y., Xiao, Q., Yu, X.F., Zhao, Y., Zhang, H., Wang, H., Chu, P.K.: Biodegradable black phosphorus-based nanospheres for in vivo photothermal cancer therapy. Nat. Commun. **30**(7), 12967 (2016). https://doi.org/10.1038/ncomms12967
22. Dong, H., Tang, S., Hao, Y., Yu, H., Dai, W., Zhao, G., et al.: (2016). Fluorescent MoS2 quantum dots: ultrasonic preparation, up-conversion and down-conversion bioimaging, and photodynamic therapy. ACS Appl. Mater. Interfaces **8**(5), 3107–3114 (2016). https://doi.org/10.1021/acsami.5b10459
23. Lu, F., Li, Z., Kang, Y., Su, Z., Yu, R., Zhang, S.: Black phosphorus quantum dots encapsulated in anionic waterborne polyurethane nanoparticles for enhancing stability and reactive oxygen species generation for cancer PDT/PTT therapy. J. Mater. Chem. B (2020). https://doi.org/10.1039/d0tb02101f
24. Yan, Y., Tian, J., Hu, F., Wang, X., Shen, Z.: A near IR photosensitizer based on self-assembled CdSe quantum dot-aza-BODIPY conjugate coated with poly(ethylene glycol) and folic acid for concurrent fluorescence imaging and photodynamic therapy. RSC Adv. **6**(115), 113991–113996 (2016). https://doi.org/10.1039/c6ra23113f
25. Ahirwar, S., Mallick, S., Bahadur, D.: Photodynamic therapy using graphene quantum dot derivatives. J. Solid State Chem. **282**, 121107 (2020). ISSN 0022-4596. https://doi.org/10.1016/j.jssc.2019.121107
26. Murali, G., Kwon, B., Kang, H., Modigunta, J.K.R., Park, S., Lee, S., Lee, H., Park, Y.H., Kim, J., Park, S.Y., Kim, Y.J.: Hematoporphyrin photosensitizer-linked carbon quantum dots for photodynamic therapy of cancer cells. ACS Appl. Nano Mater. **5**(3), 4376–4385 (2022). https://doi.org/10.1021/acsanm.2c00443

Quantum Dots Based Composites for Diagnostics

Ravikiran Maheshrao Suryawanshi, Shailesh S. Chalikwar, Pravin Tajane, Shahajan Baig, A. A. Kazi, M. Akiful Haque, Vasudha Bakshi, Falak A. Siddiqui, and Sharuk L. Khan

Abstract In recent years, quantum dots (QDs) have emerged as a promising nanomaterial with the potential to revolutionize diagnostic applications due to their unique optical and electrical capabilities. In this setting, there has been a lot of interest in incorporating QDs into composite structures to create cutting-edge diagnostic tools and methods. This summary summarizes the present status of diagnostic composites based on Quantum Dots. It describes the fundamentals of QDs and how their fluorescence can be tuned and their properties vary depending on their size, making them promising candidates for sensitive and multiplexed detection. The synthesis, functionalization, and applications of QDs-based composites across multiple diagnostic platforms are all covered in this article. Their importance in improving the efficiency of in vitro and in vivo diagnostic techniques, such as biosensing, immunoassays, cellular imaging, and targeted drug delivery, is highlighted in particular. The difficulties and factors related to QDs-based composites' biocompatibility, toxicity, and clinical translation are also discussed in the abstract. Overall, this chapter highlights the revolutionary potential of Quantum Dots based composites in diagnostics, allowing for more precise, efficient, and individual monitoring of disease.

R. M. Suryawanshi · S. S. Chalikwar
Industrial Pharmacy & Quality Assurance Department, R. C. Patel Institute of Pharmaceutical Education & Research, Karwand Naka, Shirpur, Maharashtra 425 405, India

P. Tajane
Department of Pharmaceutical Chemistry, S.V.N.H.T's College of B. Pharmacy, Rahuri Factory, Rahuri, Ahmednagar, Maharashtra 413706, India

S. Baig
Clinical Research Associate, Clinnex, Ahmedabad, Gujarat 380054, India

A. A. Kazi · F. A. Siddiqui (✉) · S. L. Khan
Department of Pharmaceutical Chemistry, N.B.S. Institute of Pharmacy, Ausa 413520, India
e-mail: falakarjumand26@gmail.com

M. A. Haque · V. Bakshi · F. A. Siddiqui · S. L. Khan
Department of Pharmaceutical Chemistry, School of Pharmacy, Anurag University, Hyderabad 500088, India

S. Thomas et al. (eds.), *Quantum Dots Based Nanocomposites*, Engineering Materials,
https://doi.org/10.1007/978-3-031-54779-9_17

1 Introduction

The remarkable and versatile optical features of quantum dots (QDs) make them an intriguing family of nanoscale semiconductor materials [1]. The rules of quantum mechanics control the behaviour of these minuscule crystals, which typically range in size from 1 to 10 nm [2]. Due to quantum confinement effects, their properties deviate greatly from those of bulk materials, allowing for amazing applications in domains as diverse as electronics, photonics, and biomedicine. The electrical and optical properties of quantum dots vary with their size, making this one of their defining characteristics. Similarly to atoms and molecules, quantum dots have discrete energy levels rather than a continuous electronic state, as is the case with bulk materials [3]. The optical characteristics of a QD alter depending on its size because the bandgap (the difference in energy between its most occupied and least occupied energy levels) is sensitive to this difference. As a result, quantum dots exhibit size-tunable emission, a property that allows them to absorb and emit light at varying wavelengths. The high photoluminescence efficiency of quantum dots is also crucial. QDs undergo photoluminescence, in which they emit light at a larger wavelength after being excited by photons [4]. The nanoscale confinement of charge carriers leads to decreased non-radiative decay routes and enhanced radiative recombination rates, resulting in efficient light emission. Quantum dots are a promising technology due to their high photoluminescence efficiency, which has potential uses in a wide range of fields. Quantum dots may also absorb numerous photons of varying energy thanks to their broad absorption spectra [5]. Multiphoton absorption is a remarkable phenomenon that has paved the way for breakthroughs in fields like quantum computing and high-tech imaging. Exciting quantum dots with lower-energy photons via multiphoton absorption enables for deep-tissue imaging in biological systems with reduced risk of tissue injury [6]. Due to their extreme stability and suitability for long-term imaging applications, quantum dots are also a wonderful choice. Repeated stimulation causes fluorophores to gradually lose their fluorescence, a process known as photobleaching. However, quantum dots are resistant to photobleaching, meaning their fluorescence will be stable during your experiment [7]. The electrical characteristics of quantum dots are also unique. Applications in transistors, solar cells, and LEDs are possible due to their ability to function as efficient charge carriers in electrical devices [8]. Quantum dots are a possible contender for future electrical and optoelectronic devices because their electronic properties may be tailored for specific purposes by controlling their size, shape, and composition [9]. The possible toxicity of quantum dots is one of the problems that must be solved. There are worries regarding the effects of using some quantum dots made with heavy metals, including cadmium-based ones, on ecosystems and living organisms [10]. Quantum dots based on indium or carbon, which are less poisonous and more biocompatible, are among the alternatives being developed by scientists.

More than two thousand years ago, a method was established for synthesising PbS Qdots from inexpensive natural components including PbO, $Ca(OH)_2$, and water [11]. Hair dyes made from these ingredients were popular among Romans and Greeks. One of the oldest and most often used methods for altering glass's hue is by manipulating the size of Qdots inside silicate glasses. CdS and CdSe were added to silicate glasses to produce red-yellow hues at the turn of the twentieth century. Using x-ray diffraction (XRD), Rocksby was able to determine in 1932 that precipitates of CdS and CdSe were responsible for the colours [12]. Doped glasses made from semi-conductor particles were previously employed as filters in optics. In 1981, Ekinov and Onushchenko reported that nanometer-sized CuCl in silicate glass caused a blue shift in the optical spectrum [13]. In 1982, Efros and Efros proposed that the colour of glass might be modified by adjusting the size or stoichiometry of CdSxSe1-x due to quantum size effects (the change of optical and optoelectronic properties with size) [14]. In 1991, Rosetti et al. discussed the topic of colour shifts in semiconductor colloidal solutions [15]. There were a number of advances in synthesis over this time span [16]. Many basic features [17] of Qdots have drawn commercialization efforts [18], leading to a surge in experimental and theoretical study on these nanoparticles over the past two decades [2].

2 Optical Properties of Quantum Dots

QDs' unusual optical features are the result of their nanocrystal grain boundary and electron hole pairs (ex-citons). Among the many pho-tophysical properties of QDs are their resistance to photobleach and decreased fluorescence intermit-tency, as well as their high yields of fluorescent quan-tum and broad absorption and size-tunable spectra [19]. Due to their unique photo-physical properties, QDs have numerous potential applications in the biomedical industry, including but not limited to: diagnostics, medication delivery, gene therapy, etc.

Light emission properties that can be tuned by changing QD size are shown in Fig. 2. As was previously stated, the diameter of QDs is between 2 and 10 nm. By tuning their sizes, QDs emit light over a broad range of wavelengths, from 400 to 4000 nm, allowing them to be detected and identified in the visible, ultraviolet (UV), and infrared regions of the electromagnetic spectrum. As the diameter of the QDs is reduced, the degree of confinement increases, leading to a larger bandgap and, ultimately, an exciton with a higher energy [20]. As a result of the high energy in the bandgap, the emission shifts to the low WL of blue spectra. Light WL is inversely proportional to energy. Quantum confinement causes electrons in nanomaterials to exist in discrete energy levels, as opposed to the continuous levels seen in bulk semiconductor material. The energy gap has of the QDs are changing with their dimension which is depicted in Fig. 1.

Absorption spectra are those produced when an atom absorbs enough photon energy (h) to jump from a lower to a higher energy level. When a molecule or atom emits a photon of energy, it is transitioning from a higher energy level to a lower one, and this process is known as the emission spec-tra. Due to the wide absorption band, numerous quantum dots can be stimulated by a single light source with WL less than emission WL. The aforementioned quality makes it easier to create fast and cheap

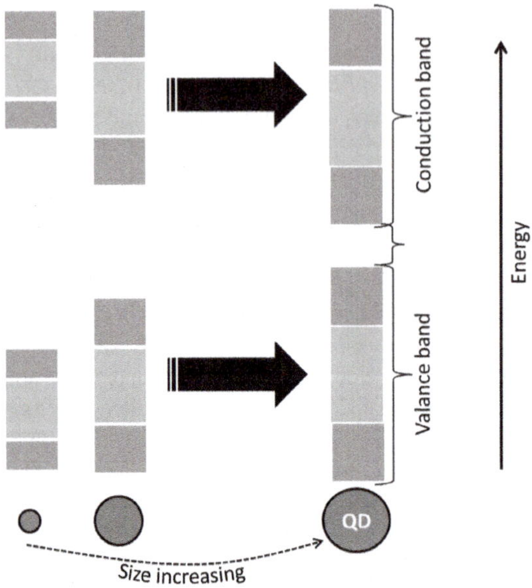

Fig. 1 The size-tunable optical characterization of QDs

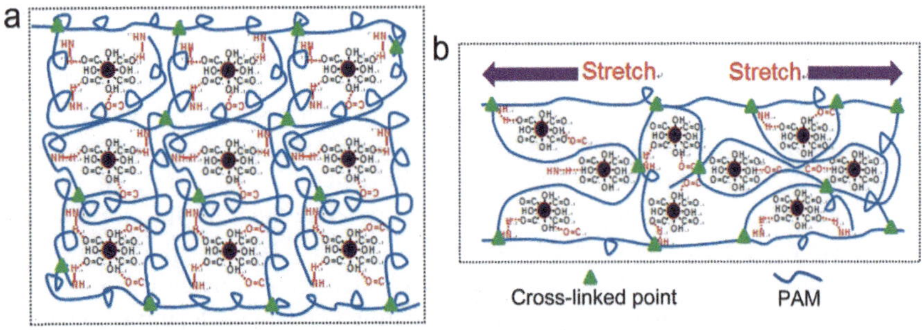

Fig. 2 Carbon dots reinforced superstretchable hydrogel microstructure in ambient condition and in stretched condition. Reproduced with the permission form Ref. [36]. © 2016 American chemical society

setups. By narrowing the QDs' emission spectra to 20 nm in the visible spectrum, we can identify individual signals without interference [21].

Photobleaching is the irreversible degradation of luminescent materials caused by light-responding processes or optical excitation, and it leads to a decrease in fluorescence intensity [29]. QDs, in general, exhibit a blinking characteristic under continuous stimulation of the molecule, which is an appealing phenom. This is because the QDs alternate between non-emitting and emitting states as a result of the trapping and detraining of photo-excited carriers. In contrast to fluoro-phores, the 'quantum jumps' in intensity of fluctuations typically only last for brief intervals of time [22]. The crystalline shells of the quantum dots prevented them from blinking, though.

3 Properties and Advantages of Quantum Dots-Based Composites

The novel features of quantum dots (QDs) have attracted a lot of attention, and when QDs are embedded in composites, they open up exciting new avenues of research and application. These QD-based composites show a synergy of properties, combining the benefits of QDs and the host matrix, which leads to improved performance and multiple uses. Let's learn more about the benefits and features of composites made with Quantum Dots. QD-based composites have improved optical characteristics, which is one of their main advantages. The emission spectra of quantum dots can be tuned to produce a wide range of colors with great photoluminescence efficiency. These characteristics, when combined with a composite material, pave the way for the creation of cutting-edge optical devices including LEDs, displays, and sensors. Industries like electronics and photonics stand to benefit greatly from the brighter, more efficient gadgets made possible by the QDs' improved optical characteristics. Customized electrical conductivity can be achieved by incorporating Quantum Dots into conductive matrices and altering the composite's electrical properties. In terms of electronic activity, QDs can function either as charge carriers or charge traps. This property has applications in energy storage, flexible electronics, and conductive inks. Quantum confinement effects of quantum dots combined with conductivity of the host material can create composites with novel electronic characteristics. QD-based composites can also improve the mechanical properties of materials, such as their strength and flexibility. It is possible to increase the composite's mechanical strength, durability, and flexibility by tailoring the host matrix and QD size. This benefit can be put to use in a variety of settings, including structural materials, coatings, and even wearable technology, where improved mechanical durability and functionality are valued. Improved Sensing and Imaging: The photostability and brightness of quantum dots can be used to create very sensitive and precise detection systems when combined with sensor matrices or contrast agents. Bioimaging, medical diagnostics, and environmental sensing can all benefit from the use of QD-based composites [23–27]. QDs' adjustable emission wavelengths improve diagnostic platform throughput by enabling the simultaneous detection of several objects at once. Composites with specific thermal characteristics can be created by incorporating Quantum Dots into thermally conductive matrix. Applications like thermal management in electronics and energy-efficient materials benefit greatly from this property. The thermal conductivity of the composite can be precisely adjusted by altering the size and concentration of QDs. QD-based composites can be functionalized with a wide range of molecules including polymers, biomolecules, and nanoparticles, greatly increasing their potential applications. This adaptability paves the way for precise targeting, monitored medication release, and a wide range of environmental interactions. Drug delivery and catalysis are just two of the many uses for composites based on QDs, which can be tailored to display specific chemical, biological, or catalytic functions. Quantum dots can be used as a greener alternative to more conventional materials in a number of different contexts [18, 28–32]. For instance, QDs can be used to lessen the environmental impact of paints and coatings by replacing pigments made with harmful heavy metals. In addition, QD-based solar cells and lighting systems show potential for eco-friendly and productive power production and use.

4 Types of QDs Nanocomposites

The optical or electronic properties of a quantum dot (QD) are determined by its inorganic semiconductor core, while the core is protected and functional groups for conjugation of biological probes are provided by the QD's organic coating layer. Cores can be formed from compounds like cadmium telluride (CdTe), cadmium selenide (CdSe), indium arsenate (InAs), cadmium sulfide (CdS), or lead sulfide (PbS), from which heavy metal ions can seep out and cause severe toxicity [20]. The introduction of QD into a biosystem raises serious concerns about the potential for core material-induced nanotoxicity. Despite recent advances in developing QDs that do not rely on heavy metals, their lackluster quantum yields (QY) and limited size-tunable emission range severely limit their usefulness in biomedicine [33].

4.1 QD-Polymer Nanocomposite

QDs are a potential new family of carbon nanomaterials due to their simple synthesis, unusual characteristics, and structural and compositional variety. QDs offer extensive surficial polar capabilities that, when combined with their other desirable properties, make them ideal candidates for further synthesis into QD/polymer nanocomposites, which can bring about significant performance increases and a host of new purposes.

Several polymers have been used to encapsulate QDs in an effort to lessen their cytotoxicity and make them more biocompatible. Diffusion inhibition of released Cd^{2+} ions from the QDs was shown to be the mechanism by which an amphiphilic polymer coating layer on CdSe QD greatly reduced QD cytotoxicity in an earlier study [34]. After encapsulating QDs in polyethylene glycol (PEG)-graftedpolyethylenimine (PEI), Nie and Duan discovered that the cytotoxicity decreased with increasing PEG grafts on each PEI molecule [38]. This finding indicates that polymer chains may wrap around QDs, creating a tight barrier that prevents the release of Cd^{2+} ions. Since PEI is a cytotoxic polymer, having PEG grafts present may reduce the likelihood of PEI molecules coming into contact with cells [35].

Fabricating hydrogels requires the use of both hydrophilic polymers and QDs with hydrophilic surface character. QDs have strong water dispersibility and minimal cytotoxicity because of their hydrophilic functionalities, which are typically hydroxyl, carboxyl, or epoxide functionalities. Hydrogels are in high demand as a soft material for biomedical applications, but their mechanical qualities are often insufficient for the purpose. Mechanical benefits (Fig. 2) were achieved through the use of CDs to alter the chemically crosslinked hydrogel network of polyacrylamide (PAM) [36].

Since CDs are known to be able to hold medicines with the help of intermolecular interactions and give rise to PL under UV/visible light, hydrogels incorporating CDs are promise for thermo-responsive drug administration with simultaneous bioimaging. Using the thermoresponsive polymer PNIPAM, CDs were revealed in a PAM network to create smart hydrogels that fluoresced blue at room temperature when exposed to UV light [37].

Polymer can improve QD biocompatibility, but it also has a number of limitations that reduce cell penetration capacity, water solubility, and imaging efficacy, such as increasing hydrodynamic size. Polymer coating layer may potentially significantly alter the physiochemical characteristics of the QD surface.

4.2 Silica-QD Nanocomposites

Since silica can inhibit flocculation of particles and preserve photoluminescence while also accepting addition of several functional groups like thiol, amine, phosphate, and carboxyl groups [38], it has been used to coat QDs. The Ströber method, a sol–gel procedure in which a silica shell grows with QDs as seeds in a water–ethanol mixture, has been used to create silica-encapsulated QD composites [39]. Nanocomposites made up of quantum dot (QD)/silica hybrid (Fig. 3) particles were studied for their resistance to high temperature and humidity in terms of their photoluminescence [40]. The maintenance of QDs' PL stability in the presence of heat and moisture is a challenge in their application as a color-converting material. Nanocomposites of QD/ silica hybrid particles of varying morphologies were produced, and their PL stabilities in conditions of high temperature and humidity were studied. Additionally, they discovered that while photoactivation of QDs in the hybrid particles did improve the PL property at first, the PL decreased further due to subsequent photocorrosion.

In recent times, there has been a significant surge in research interest surrounding inorganic halide perovskite quantum dots (QDs) with the chemical formula $CsPbX_3$, where X represents the halide elements chlorine (Cl), bromine (Br), and iodine (I). This heightened attention can be attributed to the considerable potential these QDs hold for utilization in cost-effective light sources and displays. Nevertheless, the inadequate structural and chemical stabilities of these materials pose significant challenges that impede the reliable operation of devices. The present study achieved the successful synthesis of $CsPbBr_3$/silica quantum dot (QD) composites using a straightforward sol–gel reaction with tetramethoxysilane as the sole molecule precursor

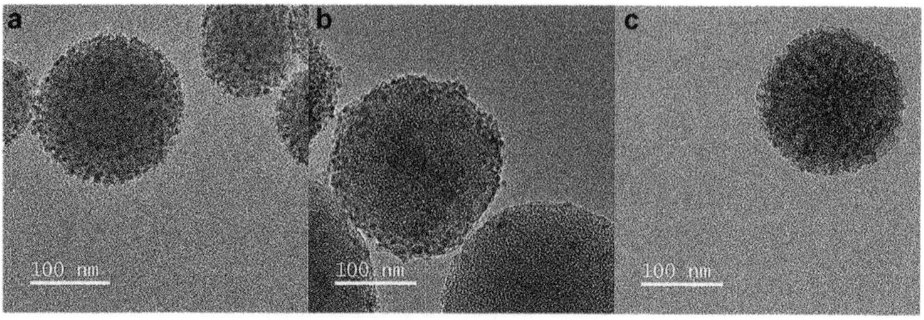

Fig. 3 TEM images of surface engineered QDs nanoparticles based on silica. Reproduced with the permission from Ref. [40] © 2020 American chemical society

[41]. The composites of $CsPbBr_3$/silica quantum dots, as prepared, exhibited significantly enhanced stability when exposed to heat, light, and environmental factors such as oxygen and moisture. In addition, the utilization of $CsPbBr_3$ QDs resulted in a comparatively reduced photoluminescence linewidth and an increased quantum yield when compared to the original $CsPbBr_3$ QDs. In their study, Jeon et al. employed silica particles as a template for the synthesis of silica–quantum dot–silica (SQS) core–shell–shell particles. They successfully regulated the refractive index of these hybrid particles by modifying their surface, aiming to reduce the refractive index disparity between the particles and the surrounding solvent and polymer matrix. The choice of silica particles as the template was motivated by their amenability to surface modification with a phenyl group, which is recognized for its substantial refractive index. The hybrid particles were employed for the production of a nanocomposite film, wherein poly(methyl methacrylate) was utilized as the matrix polymer. Subsequently, an analysis of their optical properties was conducted. It was observed that through the implementation of surface modification, enhancements were achieved not only in the transmittance but also in the mitigation of light scattering characteristics of the nanocomposite film [42]. Consequently, this resulted in an increase in the quantum efficiency (QE). In conclusion, nanocomposites incorporating hybrid particles and unmodified quantum dots (QDs) were utilized in the fabrication of a light-emitting diode (LED), and a comparative analysis was conducted to evaluate their respective efficiencies and stabilities.

The utilization of both silica and polymers in the modification of quantum dot (QD) nanocomposites has been employed in order to achieve the most desirable properties, as the surface chemistry of QDs significantly impacts their solubility, dispersity, optical properties, and probe conjugation. Polymer or silica matrices have been utilized to create a range of multifunctional nanocomposites incorporating quantum dots (QDs). This approach enables the encapsulation of multiple components within a single nanocomposite, thereby significantly broadening the potential applications of QDs in various fields such as bioimaging, biosensing, drug delivery, and cancer therapy.

5 Applications of Quantum Dots-Based Composites in Diagnostics

5.1 Drug Delivery

QDs have great potential as a medicine delivery and targeting technology. There has been a lot of excitement about the potential of QDs in the field of targeted drug delivery research. Their one-of-a-kind qualities have led to intensive research of them [43]. This evaluation took into account their significant function as medication delivery and targeting agents for anticancer medicines. This article provides a review of recent efforts to use QDs as a drug delivery and targeting system, with a special emphasis on their application in cancer therapy. Numerous research focus on the use of QDs in cancer therapy, despite the fact that multiple additional methods have also been lately presented for drug delivery. QDs have potential as useful instruments for microenvironment-specific medication delivery. Wei et al. assessed graphene QDs for this application [44]. One of the biggest issues in clinical practice is hypoxia-mediated chemoresistance in oral squamous cell cancer, thus they've produced Pt-loaded and PEG-modified graphene QDs (GPt) to combat this. In tumor biopsies, Pt biodistribution was observed, and more Pt was reported to accumulate in tumors after treatment with GPt than after treatment with free cisplatin. When considering the systemic effects, the Pt distribution in the kidneys and liver was also examined. In the GPt group, significantly less Pt was collected in the kidneys and livers compared to the free cisplatin group. The data demonstrated that oral squamous cell carcinoma cells were made more sensitive to therapy with GPt under both normoxic and hypoxic conditions, with the latter resulting in greater Pt accumulation in tumor cells. Furthermore, GPt was found to strongly limit tumor development while exhibiting far less systemic toxicity than free cisplatin. Microenvironment-specific cancer treatment with GPt has been proposed as an innovative approach. Researchers are looking into the synthesis of magnetic Fe_3O_4/SiO_2/graphene–CdTe QDs/chitosan nanocomposites (FGQCs) to see if they can improve the drug loading content over magnetic/fluorescent bifunctional nanocomposites [45]. The first observation of the in vitro interaction between the FGQCs and the heptoma cell line SMMC-7721 cells was conducted using TEM ultrathin section imaging. The graphene-quantum dots (QDs) demonstrate a significant luminescent response in aqueous surroundings when excited at a wavelength of 365 nm. The loading capacity and encapsulation efficiency of the fabricated graphene quantum dots conjugates (FGQCs) were determined to be 70% and 50%, respectively. Ina nother work the Fe_3O_4 nanoparticles were encased within the hyperbranched poly L-lysine citramid (HBPLC) [46]. Various techniques were employed to thoroughly characterize the prepared magnetic nanocarrier. The evaluative analysis focused on the diverse potential of the magnetic nanocarrier. The results of the in-vitro drug release studies indicate that the nanocomposite that was prepared demonstrates pH-responsive behavior. The study on antioxidants demonstrated the favorable antioxidant characteristics of the nanocarrier. Additionally, the nanocomposite exhibited a high level of photoluminescence, as evidenced by its quantum yield of 48.5%. The results of cellular uptake investigations demonstrated that Fe_3O_4-HBPLC-Arg/QD exhibited a substantial level of cellular internalization in MCF-7 cells, indicating its potential utility in bioimaging applications. The nanocarrier that was prepared underwent in-vitro cytotoxicity, colloidal stability, and enzymatic degradability studies. The results indicated that the nanocarrier exhibited non-toxic properties, with a cell viability rate of 94%. Additionally, it demonstrated stability in colloidal form and a biodegradability rate of approximately 37%. Javanbakht and Namazi presented a study on the fabrication of a novel hydrogel nanocomposite film incorporating graphene quantum dots (QDs) and doxorubicin (DOX) within carboxymethyl cellulose. Based on the findings, it was concluded that the nanocomposite hydrogel films possess potential as a viable platform for the delivery of anticancer drugs. This system has the capability to achieve pH-dependent and prolonged release of DOX [47]. Sesamol exhibits anticancer properties and demonstrates a notable reduction in tumor size and lipid peroxidation levels within cancerous cells. Nevertheless, due to its water solubility, sesamol encounters substantial challenges in terms of cellular penetration and internalization. A study was conducted to synthesize CdS-modified chitosan with the aim of enhancing the bioavailability of sesamol. The system under investigation was effectively loaded with sesamol, resulting in a notable enhancement of the cytotoxic activity associated with sesamol [48].

The incorporation of inorganic nanoparticles and anticancer agents within polymeric nanoparticles, resulting in a core–shell structure, offers several advantages. Firstly, it enhances localized contrast, enabling improved visualization through various diagnostic techniques, thereby facilitating precise diagnosis. Additionally, this encapsulation strategy ensures heightened therapeutic efficacy, leading to increased effectiveness in cancer treatment [49]. In the present context, quantum dots (QDs) have undergone thorough assessment for their potential in theranostic applications, specifically in the field of cancer diagnosis and therapy, as demonstrated by numerous researchers. The present study introduces a multifunctional drug delivery system that consists of two inorganic nanocrystals and an anticancer drug, all of which are enclosed within a biodegradable poly (lactic-co-glycolic acid) (PLGA) shell. In this study, inorganic nanocrystals, specifically superparamagnetic iron oxide nanoparticles (SPION) for in vivo magnetic resonance imaging (MRI) and cadmium-free manganese-doped zinc sulfide (Mn: ZnS) quantum dots (QDs) for fluorescence in vitro imaging, were employed. The selection of busulfan as a lipophilic anticancer agent was made.

In a separate investigation, a nanocomposite comprising Fe_3O_4/CQD was synthesized, exhibiting combined magnetic and fluorescent characteristics [50]. In the initial step, the Fe_3O_4 nanoparticles were synthesized and subsequently subjected to functionalization using glutaric acid. The surface of the nanoparticles was then modified to facilitate the attachment of CQDs, as depicted in Fig. 4a. The utilization of magnetic/fluorescent nanocomposites (Fe_3O_4/CQDs) has facilitated the incorporation and application of DOX in the field of medicine. The primary novelty of the current study lies in the multifunctional capabilities exhibited by Fe_3O_4/CQDs magnetic-fluorescent nanocomposites, which surpass those observed in prior research. These nanocomposites demonstrate significant potential for utilization in medical diagnostics and therapeutics. The utilization of the nanocomposite as a contrast agent in magnetic resonance imaging (MRI) for diagnostic purposes is facilitated by its possession of superparamagnetic characteristics. Furthermore, it possesses the ability to serve as a tool for the labeling of cancer cells and for optical imaging, owing to its inherent fluorescence characteristics. The utilization of Fe_3O_4/CQD nanocomposites for loading the DOX anti-cancer drug in therapeutic performance offers a controlled and targeted drug delivery system, as illustrated in Fig. 4b.

A conjugate, namely QD-MUC1-DOX, consisting of pH-responsive quantum dots (QDs), mucin1 (MUC1), and aptamer, has been developed and assessed for its potential in targeted drug delivery to ovarian cancer. The utilization of DNA aptamers, known for their specificity towards mutated mucin 1 that is overexpressed in ovarian cancer, involved their conjugation with QDs [51]. To ensure the stability of the system within the systemic circulation and facilitate drug release specifically within the acidic pH environment of cancer cells, the drug DOX was conjugated to QDs using a pH-sensitive hydrazone bond. This bond exhibits swift hydrolysis under acidic pH conditions. In a study conducted by Senel et al., N-doped graphene quantum dots were synthesized and subsequently coated with Eudragit RS 100 [52]. These QDs were further utilized as carriers for the loading of ephrin receptor-A2-siRNA. The system underwent testing to evaluate its performance in terms of DNA interactions, cell viability, antimicrobial efficacy, and antioxidant properties. The study unveiled that N-doped graphene quantum dots were rapidly taken up by the cells and exhibited

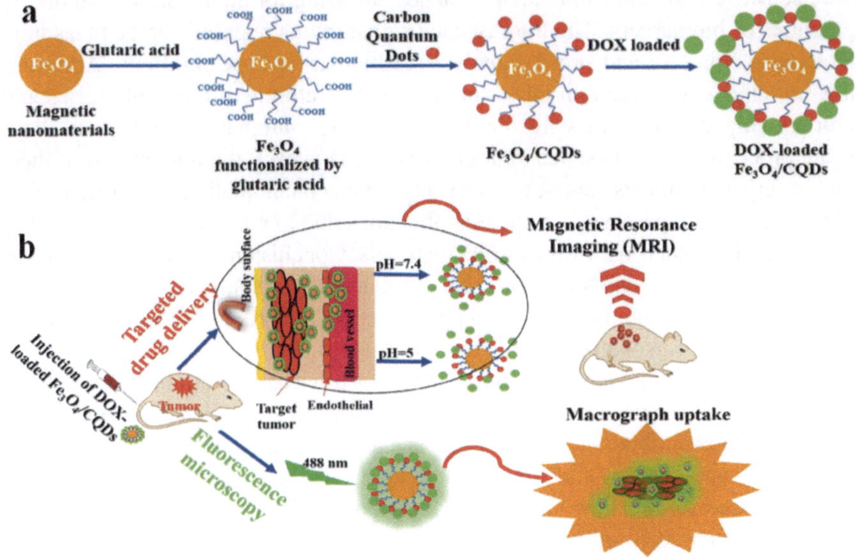

Fig. 4 **a** Preparation of DOX-Loaded Fe_3O_4/CQDs; **b** Applications of Nanocomposites in MRI, Targeted Drug Delivery, and Cellular Uptake. Reproduced with the permission from Ref. [50] © 2022 American chemical society

cell viability that varied with the dosage. Due to their inherent cell-impermeability, various strategies are required to enhance the translocation of quantum dots (QDs) across cellular membranes. Various techniques, including microinjection, electroporation, and lipofectamine-mediated transfection, have been examined as physical means to enhance the incorporation of quantum dots (QDs) into viable cells.

5.2 Immunoassays for Biomarker Detection

Due to their distinct optical features and adaptable functionalization options, quantum dots (QDs) have recently emerged as useful instruments in the field of biomarker detection using immunoassays. Immunoassays based on QDs, which utilize these tiny semiconductor particles, have the potential to radically alter the diagnostic and investigative processes. QDs are "sensitized" by having target-specific biomolecules, like antibodies or antigens, attached to their surfaces so that they can form a binding complex with the matching biomarker in a patient sample. Upon excitation, the QDs emit a unique fluorescence upon binding, providing a quantitative readout of biomarker presence and concentration. These immunoassays benefit greatly from QDs due to their remarkable fluorescence features, such as their size-tunable emission and high photostability. Multiplexing makes it possible to use a small number of samples to detect a large number of biomarkers in a single experiment by using QDs that produce light of different colors. Furthermore, QDs-based immunoassays are more sensitive than conventional approaches, allowing for the identification of low quantities of biomarkers. The fluorescence signal that is released can be measured, allowing for precise and precise analysis of biomarker concentrations. Additionally, immunoassays based on QDs can be included into microfluidic platforms and lab-on-a-chip devices, allowing for the downsizing, automation, and quick diagnostic procedures that these technologies enable. However, the importance of thorough safety assessments cannot be overstated when contemplating the translation of QDs-based immunoassays to clinical applications due to potential biocompatibility difficulties associated with various QD materials. Conclusion: Quantum dots-based immunoassays are a state-of-the-art tool for detecting biomarkers, heralding a sea change in clinical diagnostics, scientific inquiry, and the promise of game-changing developments in personalized medicine.

The use of bioconjugated QD nanocomposites to visualize cellular substructures in real time was reported by Kim et al. Here, we present a bioresponsive delivery system in which antibody-coated QDs are encapsulated in biodegradable polymeric nanospheres and undergo endolysosomal to cytosolic translocation (Fig. 5) via pH-dependent reversal of nanocomposite surface charge polarity [53]. The QD bioconjugates are released from the polymer nanospheres once they enter the cytoplasm and undergo hydrolysis. Without the need for cell fixation or membrane permeabilization, this method permits multiplexed tagging of subcellular components within living cells. This method permits high throughput cytoplasmic administration of QDs with negligible cellular toxicity, in contrast to traditional intracellular delivery methods. This discovery also exemplifies a key rational approach to designing a nanosystem with multiple functions for use in biology.

CdTe quantum dots (QDs) and polymer nanocomposites are studied for x-ray scintillation and imaging. 510–680 nm aqueous CdTe QDs were integrated into polyvinyl alcohol or polymethyl methacrylate polymer matrices. Nanocomposite films had 5 lines/mm x-ray luminescent resolution. CdTe QD nanocomposites are promising for X-ray imaging due to their short decay time, nonafterglow, and excellent spectral match to charge coupled devices [54]. Lim et al. developed multispectrally encoded nanoprobes, perfluorocarbon (PFC)/quantum dots (QDs) nanocomposite emulsions, for multispectral MR and multicolor optical imaging. Our technique used four PFC materials' multispectral MR capabilities and three colored CdSe/ZnS QDs' multicolor emission properties [55]. PFC/QDs nanocomposite emulsions were made by exchanging hydrophobic ligands coated onto CdSe/ZnS QDs with 1H,1H,2H,2H-perfluorooctanethiol to make the QDs dispersible in PFC liquids. Phospholipids emulsified QD-containing PFC liquids into aqueous solutions for biocompatibility. Selective excitation of PFC/QDs nanocomposite emulsions with magnetic resonance

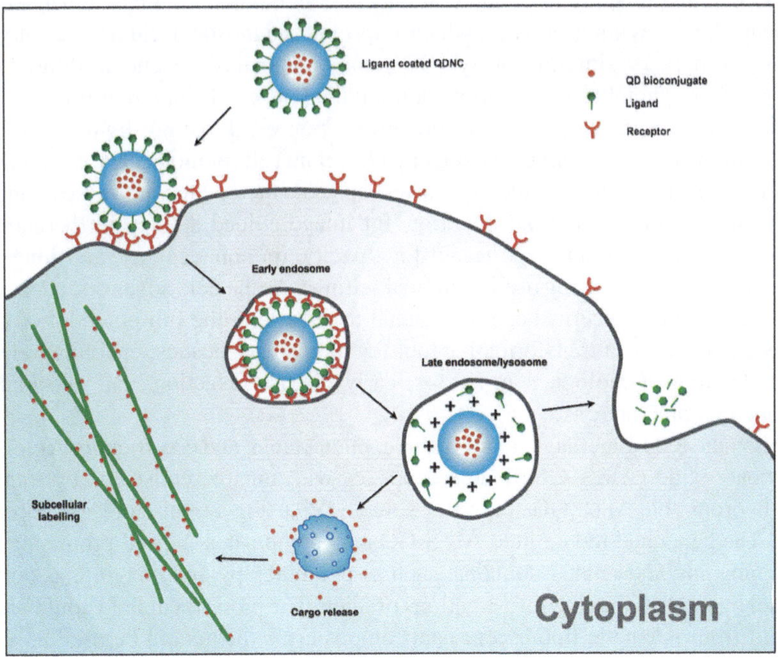

Fig. 5 QDNC cytosolic delivery and subcellular targeting. Schematic illustrating QDNC escape from the endolysosomal compartment and cytosolic release after cellular uptake. This method labels subcellular organelles and other molecular targets with antibody-conjugated QDs. *Note* QDNC charge distribution may be uneven inside acidic vesicles due to partial peptide bond breaking on the surface. Reproduced with permission from Ref. [53] © 2008 American Chemical Society

frequencies of each PFC yielded separate 19F-based MR pictures, while appropriate optical filters could pick a specific fluorescence image. PFC/QDs nanocomposite emulsions were taken up by macrophages (90.55%) and dendritic cells (85.34%), but not by T cells (33%). In vivo, nanocomposite emulsions differentiated immunotherapeutic cells (macrophages, dendritic cells, and T cells).

5.3 In-Vivo Diagnostics

Quantum dots-nanocomposites combine the unique features of quantum dots (QDs) with various nanomaterials to provide non-invasive, real-time biological process monitoring in live organisms. These hybrid structures combine QDs' size-tunable emission and long-lasting fluorescence with nanocomponents' multifunctionality. QDs-nanocomposites with biocompatible coatings and targeting ligands can be targeted to specific cells or tissues. These nanocomposites offer unprecedented imaging, disease detection, and treatment monitoring in vivo. QD fluorescence allows detailed imaging of cellular architecture and dynamics. QDs' tunable emission spectra allow multiplexing, which improves diagnostic accuracy by detecting many biomarkers simultaneously. QDs-nanocomposites' extended fluorescence and reduced photobleaching allow longitudinal research and real-time surveillance of disease development and therapy responses. Targeting ligands on QDs-nanocomposites ensure targeted contact with certain cells or molecules, reducing off-target effects. QDs-nanocomposites can encapsulate therapeutic molecules, function as diagnostic probes, and deliver drugs for image-guided therapy in theranostics. QDs-nanocomposites must be tested for toxicity, immunogenicity, and long-term consequences before being used in clinical settings. As the field advances, addressing biocompatibility concerns becomes crucial to unlocking the full potential of QDs-nanocomposites as a transformative tool for in vivo diagnostics, enabling a deeper understanding of biological processes, early disease detection, and personalized therapeutic interventions.

Another study generated biocompatible, photostable, and eco-friendly AgInZnS–graphene oxide (AIZS–GO) nanocomposites with tunable emissions by transferring hydrophobic AIZS QDs on water-soluble GO using a mini-emulsion approach [56]. The generated red emitted AIZS-GO nanocomposites showed promising uses in bioimaging and related domains, such as phototherapy and imaging, as demonstrated by in vivo imaging investigations of mice. The biomedical field could greatly benefit from magnetic-fluorescent nanocomposites, which could be used as a new dual-modality probe tool to improve the accuracy of medical detection. However, getting high-quality magnetic-fluorescent nanocomposites is difficult due to intricate and wasteful postprocesses. Therefore, it is still difficult to find a simple, efficient, and appropriate way for fabricating magnetic-fluorescent functional nanocomposites. The new synthesis methods are quickly becoming necessities, and they will likely allow us to solve the aforementioned issues [57]. Instead of using complex post-processing techniques, we describe a unique self-assembly synthesis method for building magnetic-fluorescent bimodal imaging nanocomposites. Rapid cross-linking between Fe_3O_4 and QDs was achieved by cerium(III) ion induced coordination bonds, with cerium(III) ions acting as the connecting node and carboxylate groups as the bridging ligands. Mice with tumors were used to verify the potential utility of dual-modality imaging (Fig. 6). This clever method worked wonderfully for constructing the magnetic-fluorescent Fe_3O_4-QDs nanocomposite. Our cerium(III) ion driven self-assembly approach may have broad relevance for nanoparticles and organic molecules with carboxyl groups; however, further investigation is required.

Carbon nanohorns/chitosan/quantum dots (CNHs/CS/QDs) nanocomposite has been synthesized in a unique way. Amino-CNHs were encapsulated using CS to create amino-around surfaces, and they coupled with carboxyl-QDs in an effective manner [58]. The produced nanocomposite was then used for in vivo imaging of C. elegans and for marking HeLa cells, both of which relied on its robust red fluorescence. The nanocomposite's advantageous space property also makes it a promising candidate as a medicine delivery vehicle. Emerging as a useful tool for directing minimally invasive surgical operations is dual-modality imaging, which makes use of complementing modalities like all-optical ultrasound and photoacoustic imaging. Because of their wavelength-selective optical absorption, quantum dots are an exciting material for application in these dual-modality imaging devices. This paper presents the first quantum dot nanocomposite designed for simultaneous co-registration of laser-generated ultrasound and photoacoustic imaging [59]. The created nanocomposites are applied to the tips of tiny optical fibers; they are made up of $CuInS_2$ quantum dots and medical-grade polydimethylsiloxane (CIS-PDMS). High optical absorption (>90%) at 532 nm for ultrasonic generation and low optical

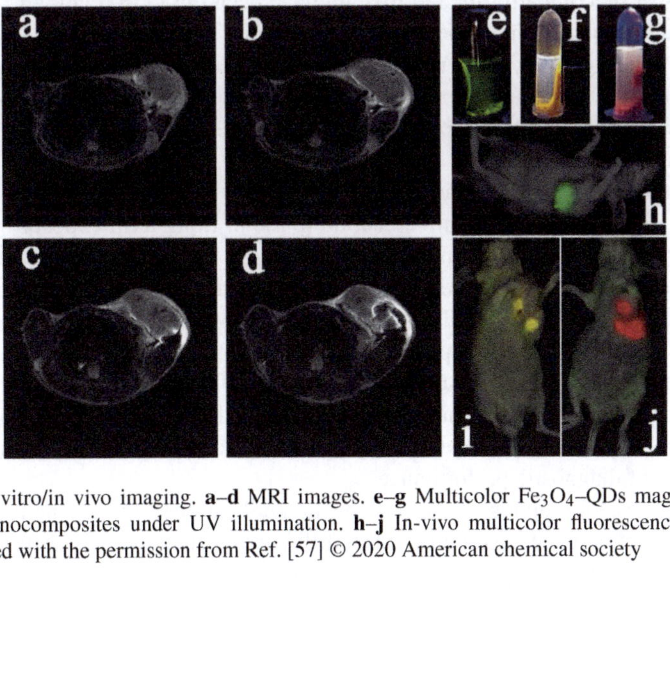

Fig. 6 In vitro/in vivo imaging. **a–d** MRI images. **e–g** Multicolor Fe_3O_4–QDs magnetic–fluorescent nanocomposites under UV illumination. **h–j** In-vivo multicolor fluorescence imaging. Reproduced with the permission from Ref. [57] © 2020 American chemical society

absorption (5%) at near-infrared wavelengths higher than 700 nm are just two examples of the films' wavelength-selective optical characteristics. Using the film's wavelength selectivity, an ink-filled tube phantom may be imaged using both all-optical ultrasound and photoacoustic techniques, which are then registered to one another. This study indicates the feasibility of using quantum dots as selective absorbers at specific wavelengths for producing all-optical ultrasound.

In order to act as both an imaging agent in the visible-light region and a photothermal cancer-therapy agent in the NIR region, a unique multifunctional, low-toxicity QD-rGO nanocomposite has been produced [60]. By using a surfactant layer to reduce QD toxicity and a carefully regulated spacer to minimize fluorescence quenching, this nanocomposite is able to solve several of the aforementioned challenges. Intriguingly, as the QD-rGO absorbs NIR irradiation, causing photothermal therapy and cell killing, the generated heat from the QD-rGO causes a temperature increase and a marked decrease in the QD brightness, providing a means for in situ heat/temperature sensing and an indicator of the progress of the photothermal therapy. In the meantime, coupling of the QD-rGO with targeting/internalization ligands has enabled active targeting and improved cell internalization.

6 Challenges and Limitations

While Quantum Dots (QDs) based composites show great promise for use in diagnostics, they still face a number of obstacles before they can be widely implemented in practice.

Issues of biocompatibility and toxicity have been raised in relation to the use of QDs in diagnostic settings. Heavy metals included in some forms of QDs may be harmful to living organisms. Avoiding negative effects on cells, tissues, and organs requires that QDs and the composite materials they are typically used with be biocompatible. Biocompatibility and possible buildup in live beings need extensive research to determine safety.

The stability, distribution, and clearance of QDs and their nanocomposites can be affected by the immune system's reaction to them. Long-term exposure to QDs in the environment may cause immunogenic reactions due to unfavorable interactions with the immune system. Preventing unwanted immune responses requires strategies to reduce immunogenicity and improve QDs' removal from the body.

Although QDs-based composites have shown promise in experimental experiments, there are still many obstacles to overcome before they may be used in clinical practice. Critical measures that must be performed to ensure safe and successful clinical deployment include regulatory approval, standardization of protocols, and validation of diagnostic performance.

For successful diagnostic and therapeutic applications, it is essential to achieve high levels of targeted specificity. It is a difficult task that involves accurate surface functionalization and careful design to ensure that QDs-nanocomposites reach the appropriate target cells or tissues while avoiding off-target effects.

Concerns have been raised about the long-term stability of QDs in biological settings. The accuracy of diagnostic tests may degrade over time due to aging-induced changes in optical characteristics, aggregation, or deterioration. Sustaining and reliable diagnostic results necessitate the implementation of strategies to improve the long-term durability of QDs-based composites.

Bright fluorescence is a hallmark of QDs, however the quantum yield and brightness of different QD formulations can vary. For effective and trustworthy diagnostics, it is essential to achieve uniform and optimum fluorescence emission across all batches of QDs.

Ethical and Regulatory Considerations The clinical application of nanomaterials such as QDs presents ethical and regulatory concerns. Potential effects on the environment, long-term safety, and patient consent are all important considerations.

Complexity and expense arise when manufacturing on a big scale, as in the synthesis and functionalization of QDs. One obstacle to wider implementation is the lack of established, low-cost methods for manufacturing QDs-based composites and incorporating them into diagnostic procedures.

The accuracy and specificity of QDs-based diagnostics can be impacted by interference and background signals, both of which can arise from complicated biological samples. Methods to reduce background noise and maximize signal-to-noise ratios are essential for producing accurate results.

Although QDs-based composites present an opportunity for quantitative analysis, it may be difficult to achieve precise and accurate quantification of biomarkers due to factors including differences in QD labeling efficiency and binding kinetics.

7 Conclusion

Quantum dots (QDs) integrated into composite structures have opened up novel diagnostic avenues and have the potential to radically alter the course of medical research and patient care. QDs' adaptability to functionalization and hybridization, along with their distinctive optical and electrical properties, have paved the way for a wide variety of ground-breaking diagnostic uses. The combination of nanotechnology and diagnostics has the potential to dramatically improve the efficiency, accuracy, and speed with which diseases can be identified, monitored, and treated.

QDs-based composites provide a flexible diagnostic toolkit for tackling complex problems in a wide range of fields. QDs have enabled the development of ultrasensitive biosensors and immunoassays, allowing for the quick and precise identification of biomarkers at trace levels in in vitro diagnostics. Multiplexed assays have been made possible by QDs due to their tunable fluorescence spectra and resilience to photobleaching, allowing for the simultaneous detection of many analytes in a single sample. The implications for improved diagnostic efficiency and thorough disease profiling are enormous.

Now, QDs-based composites are being used for in vivo diagnostics, allowing for the real-time viewing of cellular and molecular activities inside living creatures, expanding the field of study far beyond the limitations of the traditional laboratory. Incorporating QDs into nanocomposites allows for precise functionalization and targeting, allowing for delivery to targeted organs and tissues. This paves the way for tailored treatment and monitoring, as well as non-invasive imaging of disease progression. Because of their versatility, QDs-nanocomposites are gaining increasing attention in the developing field of theranostics, where diagnosis and treatment are combined.

While QDs-based composites hold revolutionary diagnostic potential, we must first overcome obstacles related to biocompatibility, toxicity, and clinical translation. To ensure their feasibility and efficacy in clinical applications, these nanocomposites require rigorous studies of their safety profile. Successful clinical adoption of QDs-based composites will depend on addressing these problems through the integration of interdisciplinary research spanning materials science, biology, and medicine.

In conclusion, Quantum Dots based composites have sparked a paradigm change in the diagnostics landscape, providing a wealth of resources that reimagine our potential in the areas of early diagnosis, continuous monitoring, and individualized care. Nanotechnology combined with diagnostics has the potential to revolutionize healthcare by providing more precise, rapid, and tailored treatment plans for each patient. As knowledge of QDs-based composites increases, so does the scope of potential diagnostic innovations, highlighting the enormous promise that these cutting-edge technologies represent for the future of healthcare.

References

1. Ma, Q., Su, X.: Recent advances and applications in QDs-based sensors. Analyst **136**(23), 4883–4893 (2011)
2. Ganguly, S., et al.: Advancement in science and technology of carbon dot-polymer hybrid composites: a review. Funct. Compos. Struct. **1**(2), 022001 (2019)
3. Kumar, D.S., B.J. Kumar, Mahesh, H.: Quantum nanostructures (QDs): an overview. Synthesis of inorganic nanomaterials pp. 59–88 (2018)
4. Aguilera-Sigalat, J., Bradshaw, D.: Synthesis and applications of metal-organic framework–quantum dot (QD@ MOF) composites. Coord. Chem. Rev. **307**, 267–291 (2016)
5. Wood, V., Bulović, V.: Colloidal quantum dot light-emitting devices. Nano Rev. **1**(1), 5202 (2010)
6. Cai, Y., et al.: Optical nano-agents in the second near-infrared window for biomedical applications. Chem. Soc. Rev. **48**(1), 22–37 (2019)
7. Orfield, N.J., et al.: Photophysics of thermally-assisted photobleaching in "giant" quantum dots revealed in single nanocrystals. ACS Nano **12**(5), 4206–4217 (2018)
8. Shirota, Y., Kageyama, H.: Charge carrier transporting molecular materials and their applications in devices. Chem. Rev. **107**(4), 953–1010 (2007)
9. Das, P., et al.: Optical Properties of Quantum Dots. In: Quantum Dots and Polymer Nanocomposites, pp. 69–85. CRC Press (2022)
10. Das, P., et al.: Graphene based emergent nanolights: A short review on the synthesis, properties and application. Res. Chem. Intermed. **45**, 3823–3853 (2019)
11. Walter, P., et al.: Early use of PbS nanotechnology for an ancient hair dyeing formula. Nano Lett. **6**(10), 2215–2219 (2006)
12. Rooksby, H.: The colour of selenium ruby glasses. J. Soc. Glas. Technol. **16**, 171–181 (1932)

13. Ekimov, A.I.: Quantum size effect in three-dimensional microscopic semiconductor crystals. Jetp Lett. **34**, 345 (1981)
14. Efros, A.L., Efros, A.L.: Interband absorption of light in a semiconductor sphere. Sov. Phys. Semicond. **16**(7), 772–775 (1982)
15. Rossetti, R., et al.: Size effects in the excited electronic states of small colloidal CdS crystallites. J. Chem. Phys. **80**(9), 4464–4469 (1984)
16. Spanhel, L., et al.: Photochemistry of colloidal semiconductors. 20. Surface modification and stability of strong luminescing CdS particles. J. Am. Chem. Soc. **109**(19), 5649–5655 (1987)
17. Klimov, V., et al.: Optical gain and stimulated emission in nanocrystal quantum dots. Science **290**(5490), 314–317 (2000)
18. Das, P., et al.: Tailor made magnetic nanolights: fabrication to cancer theranostics applications. Nanoscale Adv. **3**(24), 6762–6796 (2021)
19. Adam, G.O., Sharker, S.M., Ryu, J.H.: Emerging biomedical applications of carbon dot and polymer composite materials. Appl. Sci. **12**(20), 10565 (2022)
20. Devi, S., et al.: Quantum dots: an emerging approach for cancer therapy. Front. Mater. **8**, 798440 (2022)
21. Voura, E.B., et al.: Tracking metastatic tumor cell extravasation with quantum dot nanocrystals and fluorescence emission-scanning microscopy. Nat. Med. **10**(9), 993–998 (2004)
22. Mallick, T., Karmakar, A., Sultana, Z.: Quantum Dots: Potential Cell Imaging Agent. In: Application of Quantum Dots in Biology and Medicine: Recent Advances, pp. 191–207. Springer (2022)
23. Das, P., et al.: MXene/0D nanocomposite architectures: design, properties and emerging applications. Mater. Today Nano **24**, 100428 (2023)
24. Das, P., et al.: Waste-Derived Sustainable Fluorescent Nanocarbon-Coated Breathable Functional Fabric for Antioxidant and Antimicrobial Applications. ACS Applied Materials & Interfaces (2023)
25. Rajabzadeh, A.R., et al.: Magnetic Quantum Dots for Bioimaging. CRC Press (2023)
26. Das, P., et al.: Naturally derived carbon dots in situ confined self-healing and breathable hydrogel monolith for anomalous diffusion-driven phytomedicine release. ACS Appl. Bio Mater. **5**(12), 5617–5633 (2022)
27. Das, P., et al.: Carbon dot biopolymer-based flexible functional films for antioxidant and food monitoring applications. ACS Appl. Polym. Mater. **4**(12), 9323–9340 (2022)
28. Das, P., et al.: Carbon-dots-initiated photopolymerization: an in situ synthetic approach for MXene/poly (norepinephrine)/copper hybrid and its application for mitigating water pollution. ACS Appl. Mater. Interfaces **13**(26), 31038–31050 (2021)
29. Das, P., et al.: Immobilization of heteroatom-doped carbon dots onto nonpolar plastics for antifogging, antioxidant, and food monitoring applications. Langmuir **37**(11), 3508–3520 (2021)
30. Das, P., et al.: Carbon dots for heavy-metal sensing, pH-sensitive cargo delivery, and antibacterial applications. ACS Appl. Nano Mater. **3**(12), 11777–11790 (2020)
31. Ganguly, S., et al.: Microwave-synthesized polysaccharide-derived carbon dots as therapeutic cargoes and toughening agents for elastomeric gels. ACS Appl. Mater. Interfaces **12**(46), 51940–51951 (2020)
32. Das, P., et al.: Biocompatible carbon dots derived from κ-carrageenan and phenyl boronic acid for dual modality sensing platform of sugar and its anti-diabetic drug release behavior. Int. J. Biol. Macromol. **132**, 316–329 (2019)
33. Mohammadnia, M.S., et al.: Fluorescent cellulosic composites based on carbon dots: Recent advances, developments, and applications. Carbohyd. Polym. **294**, 119768 (2022)
34. Kirchner, C., et al.: Cytotoxicity of colloidal CdSe and CdSe/ZnS nanoparticles. Nano Lett. **5**(2), 331–338 (2005)
35. Duan, H., Nie, S.: Cell-penetrating quantum dots based on multivalent and endosome-disrupting surface coatings. J. Am. Chem. Soc. **129**(11), 3333–3338 (2007)
36. Hu, M., et al.: Low chemically cross-linked PAM/C-dot hydrogel with robustness and superstretchability in both as-prepared and swelling equilibrium states. Macromolecules **49**(8), 3174–3183 (2016)

37. Zhou, L., He, B., Huang, J.: Amphibious fluorescent carbon dots: one-step green synthesis and application for light-emitting polymer nanocomposites. Chem. Commun. **49**(73), 8078–8080 (2013)
38. Wolcott, A., et al.: Silica-coated CdTe quantum dots functionalized with thiols for bioconjugation to IgG proteins. J. Phys. Chem. B **110**(11), 5779–5789 (2006)
39. Rogach, A.L., et al.: "Raisin bun"-type composite spheres of silica and semiconductor nanocrystals. Chem. Mater. **12**(9), 2676–2685 (2000)
40. Jo, J.H., Heo, H.S., Lee, K.: Assessing stability of nanocomposites containing quantum dot/silica hybrid particles with different morphologies at high temperature and humidity. Chem. Mater. **32**(24), 10538–10544 (2020)
41. Zhang, F., et al.: Silica coating enhances the stability of inorganic perovskite nanocrystals for efficient and stable down-conversion in white light-emitting devices. Nanoscale **10**(43), 20131–20139 (2018)
42. Jeon, H., et al.: Improvement in efficiency and stability of quantum dot/polymer nanocomposite film for light-emitting diodes using refractive index-controlled quantum dot–silica hybrid particles. J. Mater. Chem. C **7**(38), 11764–11769 (2019)
43. Liu, H., et al.: Construction of biomass carbon dots based fluorescence sensors and their applications in chemical and biological analysis. TrAC Trends Anal. Chem. **118**, 315–337 (2019)
44. Wei, Z., et al.: Antitumor effect of a Pt-loaded nanocomposite based on graphene quantum dots combats hypoxia-induced chemoresistance of oral squamous cell carcinoma. Int. J. Nanomedicine 1505–1524 (2018)
45. Ou, J., et al.: Fabrication and cyto-compatibility of Fe_3O_4/SiO_2/graphene–CdTe QDs/CS nanocomposites for drug delivery. Colloids Surf., B **117**, 466–472 (2014)
46. Aslani, R., Namazi, H.: Fabrication of a new photoluminescent and pH-responsive nanocomposite based on a hyperbranched polymer prepared from amino acid for targeted drug delivery applications. Int. J. Pharm. **636**, 122804 (2023)
47. Javanbakht, S., Namazi, H.: Doxorubicin loaded carboxymethyl cellulose/graphene quantum dot nanocomposite hydrogel films as a potential anticancer drug delivery system. Mater. Sci. Eng. C **87**, 50–59 (2018)
48. Abdelhamid, H.N., et al.: Synthesis of CdS-modified chitosan quantum dots for the drug delivery of Sesamol. Carbohyd. Polym. **214**, 90–99 (2019)
49. Ye, F., et al.: Biodegradable polymeric vesicles containing magnetic nanoparticles, quantum dots and anticancer drugs for drug delivery and imaging. Biomaterials **35**(12), 3885–3894 (2014)
50. Fattahi Nafchi, R., et al.: In vitro study: synthesis and evaluation of Fe_3O_4/CQD magnetic/fluorescent nanocomposites for targeted drug delivery, MRI, and cancer cell labeling applications. Langmuir **38**(12), 3804–3816 (2022)
51. Savla, R., et al.: Tumor targeted quantum dot-mucin 1 aptamer-doxorubicin conjugate for imaging and treatment of cancer. J. Control. Release **153**(1), 16–22 (2011)
52. Şenel, B., et al.: Graphene quantum dots: synthesis, characterization, cell viability, genotoxicity for biomedical applications. Saudi Pharm. J. **27**(6), 846–858 (2019)
53. Kim, B.Y., et al.: Biodegradable quantum dot nanocomposites enable live cell labeling and imaging of cytoplasmic targets. Nano Lett. **8**(11), 3887–3892 (2008)
54. Kang, Z., et al.: CdTe quantum dots and polymer nanocomposites for x-ray scintillation and imaging. Appl. Phys. Lett. **98**(18) (2011)
55. Lim, Y.T., et al.: Multiplexed imaging of therapeutic cells with multispectrally encoded magnetofluorescent nanocomposite emulsions. J. Am. Chem. Soc. **131**(47), 17145–17154 (2009)
56. Zang, Z., et al.: Tunable photoluminescence of water-soluble AgInZnS–graphene oxide (GO) nanocomposites and their application in-vivo bioimaging. Sens. Actuators B Chem. **252**, 1179–1186 (2017)
57. Li, J., et al.: Self-assembly fabrication of honeycomb-like magnetic-fluorescent Fe_3O_4–QDs nanocomposites for bimodal imaging. Langmuir **36**(48), 14471–14477 (2020)

58. Li, J., et al.: Synthesis of carbon nanohorns/chitosan/quantum dots nanocomposite and its applications in cells labeling and in vivo imaging. J. Lumin. **145**, 74–80 (2014)
59. Bodian, S., et al.: $CuInS_2$ quantum dot and polydimethylsiloxane nanocomposites for all-optical ultrasound and photoacoustic imaging. Adv. Mater. Interfaces **8**(20), 2100518 (2021)
60. Hu, S.-H., et al.: Quantum-dot-tagged reduced graphene oxide nanocomposites for bright fluorescence bioimaging and photothermal therapy monitored in situ. Adv. Mater. **24**(13), 1748–1754 (2012)

Suitability of Hybrid Quantum Dots as Payload for Plants

Uma Sankar Gorla, Jeetendra Kumar Gupta, Prasanthi Samathoti, Alapati Sahithi, Mohana Vamsi Nuli, Vakkalagadda Ravi Kumar, and Rasapelly Ramesh Kumar

Abstract The use of hybrid quantum dots represents a unique opportunity to transform crop management and environmental monitoring through the integration of nanotechnology and agriculture. The combination of quantum dot characteristics with biological ingredients in these nanoparticles makes it possible to track plant health in real time, optimise nutrient uptake, and evaluate the surrounding environment. Hybrid quantum dots provide data-driven decision making, resource efficiency, and ecosystem preservation by delivering continual insights into plant conditions. However, issues like safety and cost need to be resolved before widespread use is possible. Nonetheless, hybrid quantum dots are a viable tool in the goal of smarter, more sustainable agriculture due to their potential benefits, which include enhanced crop yields and ecological sustainability.

U. S. Gorla
Department of Pharmacology, Viswanadha Institute of Pharmaceutical Sciences, Sontyam, Anandapuram, Visakhapatnam 530017, India

J. K. Gupta
Institute of Pharmaceutical Research, GLA University, 17 Km Stone, NH-2, Mathura-Delhi Road Mathura, Chaumuhan, Uttar Pradesh, India

P. Samathoti
Department of Pharmaceutics, MB School of Pharmaceutical Sciences, Mohan Babu University, Sree Sainathnagar, A.Rangampet, Tirupati, Andhra Pradesh 517102, India

A. Sahithi
Department of Pharmaceutical Analysis, School of Pharmacy, Nalla Narasimha Reddy Educational Society's Group of Institutions, Hyderabad, Telangana 500088, India

M. V. Nuli
Department of Pharmaceutical Analysis, Raghavendra Institute of Pharmaceutical Education and Research (RIPER)—Autonomous, Anantapuramu, Chiyyedu, Andhra Pradesh 515721, India

V. R. Kumar
School of Pharmacy, Guru Nanak Institutions Technical Campus, Ranga Reddy, Khanapur, HyderabadTelangana 501506, India

R. R. Kumar (✉)
Department of Pharmaceutical Chemistry, Marri Laxman Reddy Institute of Pharmacy Hyderabad, Dundigal, T.S 500043, India
e-mail: rameshkumarrasapelly@gmail.com

S. Thomas et al. (eds.), *Quantum Dots Based Nanocomposites*, Engineering Materials,
https://doi.org/10.1007/978-3-031-54779-9_18

1 Introduction

It is expected that nanomaterials will play a significant role in helping the agricultural sector deal with these problems in the future. The agricultural sector has benefited greatly from the proliferation of nanomaterials over the past few decades, with their use facilitating the monitoring of crop health, the promotion of crop development, the improvement of fertiliser and pesticide efficiency, the management of diseases, pests, and environmental challenges, and so on. Because their carbon backbone is non-toxic, carbon-based nanomaterials have substantially less environmental toxicity and greater biocompatibility than metal-based nanomaterials. Therefore, nanoparticles made of carbon have found extensive use in farming. Carbon dots (CDs) are a novel type of carbon nanomaterial with intrinsic photoluminescence (PL) capabilities and a size of less than 10 nm. Due to their superior and universal properties such as high photostability, small size, biocompatibility, water solubility, highly tunable PL properties, easy functionalization with biomolecules, and chemical inertness, CDs have garnered a lot of attention from researchers in recent years. Carbon quantum dots (CQDs) have been the topic of extensive optical research studies for applications including biomonitoring, sensing, photocatalysis, medicine and gene delivery, solar energy conversion, and light-emitting diodes (LEDs), despite the fact that CDs were only discovered in 2004 [1]. As a result of the toxicity and environmental concerns associated with regular semiconductor quantum dots, CDs have been proposed as a replacement [2].

A synthetic structure with dimensions of 100 nm or less, engineered nanoparticles (ENPs) has unique physicochemical features like conductivity and optical sensitivity. (i) Semiconductor quantum dots (QDs), one of the significant ENPs, typically consist of elements from either group II–VI or III–V. Due to the quantum size effect (iii), QDs have distinct optical (ii) and electronic features, such as narrow and symmetrical emission, high photostability, and a strong Stokes shift, that are not seen in bulk materials. In the biological sciences (iv), QDs can be utilised as a replacement for the more common fluorescent dyes. (v) Biomolecule-templated synthesis, live organism synthesis, polymer-templated synthesis, bottom-up colloidal synthesis, and top-down synthesis approaches are some of the sophisticated QD synthesis methods for different pathways. Sixth, when synthesising highly dispersed QDs for bioapplications, bottom-up colloidal synthesis is the method of choice. Surface modification, often through ligand exchange procedures, surface silanization, and coating with cross-linked shells [3], allows QDs to acquire additional optimised features such as solubility, electrophoretic mobility, and targeted activities. Quantum dots (QDs) are often affixed to functional biomolecules (proteins, enzymes, or nucleic acids) by adsorption or covalent coupling for bioapplications including bioimaging or drug delivery systems.

The domestication of plants for subsistence purposes is among the most important human activities ever. The number of hungry individuals has dropped by half in the last 20 years because to improvements in agricultural production techniques [4]. However, there are still about a billion people who aren't getting enough to eat. Agriculture requires an enormous quantity of resources to sustain the ever-growing population and accompanying demand from industry. An estimated 187 million metric tonnes of fertiliser, over 4 million tonnes of insecticides, and 2.7 trillion cubic metres of water (approximately 70% of all worldwide freshwater use) are used each year to produce the world's 3 billion metric tonnes of crops [5]. The widespread and inefficient use of pesticides and herbicides has led to significant resistance to these compounds, which poses a serious danger to biodiversity and the ecosystem. There will be additional pressure on food production and distribution because of climate change impacts like harsh weather and pest and disease epidemics. To meet the task of feeding a growing population, the agriculture sector must immediately adopt new eco-friendly technologies.

As a cutting-edge invention in nanotechnology, hybrid quantum dots are receiving attention for their potential to transform in different industry sectors [6–13]. The unique properties of quantum dots enable them to be used in promising applications [14–16]. We explore the exciting world of hybrid quantum dots as payloads for plants, including their applicability, benefits, and drawbacks, in this paper. The unique optical and electrical features of these nanoscale semiconductor particles represent a fresh strategy for boosting agricultural output and plant health. Precision farming, in which technology plays a central role in maximising resource utilisation and crop management, has become increasingly popular in recent years. Because of their distinctive properties, hybrid quantum dots are among the most intriguing options to aid in this transition. Let's dig into the reasons why they're so promising. Hybrid quantum dots, first and foremost, have extraordinary potential for nutrition monitoring in plants. These nanoparticles can be modified to bond with nutrients like nitrogen, phosphorus, and potassium by being equipped with the appropriate receptors. With this development, nutrient levels in plant tissues may be tracked in real time. With this level of specificity, farmers can adjust their fertilisation practises to be more efficient and gentler on the environment. The detection of plant diseases and pests is another intriguing use of hybrid quantum dots in agriculture. These nanoparticles may function as early warning systems if sensors capable of recognising particular infections or insect pests were attached to them. Quantum dots can alert farmers to the presence of a threat while a plant is under attack. By taking preventative measures, farmers can lessen the amount of chemical pesticides they use, which is good for the environment and the economy. Additionally, hybrid quantum dots can be used as supplemental artificial light sources for plant photosynthesis. Quantum dots augment daily "light period" for plants by generating light in the photosynthetically active radiation (PAR) spectrum, which is not found in natural sunlight. As a result, plant growth and productivity are increased even in areas with less sunshine or during seasons with shorter daylight hours. The benefits of hybrid quantum dots in the agricultural supply chain extend beyond their role in plant health and growth. It is now feasible to track the origin of agricultural goods with unprecedented accuracy thanks to the use of unique quantum dot identifiers applied during the growing cycle of a plant. In addition, the markers allow for constant quality check, guaranteeing that goods continue to be authentic and up to par as they make their way through the supply chain. Finally, the promising new field of contemporary agriculture is the use of hybrid quantum dots as payloads for plants. Crop yields, environmental impact, and the efficiency and sustainability of agricultural practises can all benefit from their abilities to monitor nutrients, identify pests and diseases, boost photosynthesis, and provide traceability. However, in order to fully utilise the potential of hybrid quantum dots in agriculture, it is crucial to overcome the obstacles connected with their deployment, such as potential toxicity and environmental effects.

2 Different Types of Nanocarriers for Plants

Improved agricultural productivity and pathogen resistance are only two examples of the many uses for nanomaterials (NCs) such metal NPs, carbon-based materials (CNTs, CNFs, graphene, and graphene oxide), polymers, and lipids. At larger concentrations, however, these NCs can have toxic effects on the plant. In most cases, NCs result in much greater seed germination, root/shoot length, floral output, and fruit yield.

The delivery of agrochemicals makes considerable use of numerous metal-based NCs (including Cu, Zn, and Ag) and metal oxides (including TiO_2, CuO, ZnO, and AgO). In addition, most metal-based NCs are incorporated into plant tissues as micronutrients, promoting healthier plant development [17]. To measure NP accumulation and plant growth, for instance, researchers utilise Cu-NPs in mung (Phaseolus radiatus) and wheat (Triticum aestivum). The results indicated that increased concentrations of Cu metal ions within the plants led to slower growth in young seedlings. The accumulation rate is proportional to the amount of Cu that plants are subjected to [18]. The effects of selenium (Se) and copper (Cu) nanoparticles (NPs) on tomato plants were studied by Hernández et al. It appeared that the antioxidant capacity of tomato fruits was enhanced by both NPs. The concentration of Se in the Se-NPs boosted the yield of tomatoes. Tomato yield was likewise boosted, albeit at a lower concentration, when Cu-NPs were used. Tomato yield is not affected by the concentration of Cu-NPs [19]. There have been a number of studies indicating that using metal-NPs in plant growth is beneficial. These metal NPs improve plant growth by boosting the metal NP concentration. However, phytotoxicity may be induced by metal NPs at larger concentrations. Metal NPs, in general, served as both micronutrients and NCs, boosting plant development.

Activated carbon, activated carbon nanofibers (ACFs), carbon beads, and carbon-based nanomaterials like graphene, graphene oxide, carbon nanotubes (CNTs), and carbon nanofibers (CNFs) are just some of the many carbon-based materials that find widespread use in fields as diverse as energy, environment, medicine, sensors, and agriculture [20]. Plants make effective use of these carbon-based NCs, which are employed to stimulate growth. Translocation within plants increases growth, and it is interesting to note that CNTs and CNFs can penetrate the seed coat [21]. Tomato seeds were used as a test subject for graphene. Graphene was found to improve a plant's germination rate, root length, and shoot length, according to the data [22]. Graphene quantum dots (G-QDs) were synthesised by Chakravarty et al. and tested on coriander and garlic plants. The results indicated that G-QDs acted as growth stimulants, aiding in the development of the plant's leaves, roots, shoots, flowers, and fruits [23]. Wheat (*Triticum aestivum L.*) has also been shown to respond to graphene in a separate study. The results indicated that graphene promoted root growth in wheat plants but had no effect on leaf development. Graphene exposure also impairs photosynthesis and biomass production by damaging roots and causing oxidative stress. In addition, the research indicated that the *Triticum aestivum L.* plant may have negative consequences from prolonged exposure [24]. Graphene oxide (GO) is a graphene derivative that can be easily functionalized with a surface functional group that has great biocompatibility, leading to its widespread use in a wide range of end applications. For instance, Hu et al. studied how GO interacted with plant cells. Inhibition of cell development, production of reactive oxygen species (ROS), and disruption of enzymatic function were all observed, suggesting that GO translocates easily within cells that are damaging to plant cells [25]. Researchers are continually interested in CNTs for a wide range of uses due to their exceptional physicochemical properties, including medication delivery, energy, environment, sensing, and agriculture. Cu/Zn-CNFs were also synthesised by the same set of researchers, who likewise integrated them into a polymeric (PVA-starch) matrix. The results indicated that the soil application of the developed Cu/Zn-CNF-PVA-starch-based formulation promoted greater chickpea plant development. In comparison to Cu/Zn-CNFs, the polymeric formulation's controlled-release of CNFs and Cu/Zn micronutrients is a benefit. The biodegradability of the Cu/Zn-CNF-PVA-starch formulation was also enhanced by the inclusion of starch. A related group also synthesised a bacterial homoserine lactone with Fe-CNF incorporation for use against the chickpea plant. Fe-CNFs were found to be effective in transporting bacterial homoserine lactone, which promotes plant growth and provides resistance to pathogens [26]. The same team also looked at the usage of Cu-CNFs to scavenge chromium while absorbing nutrients. In spite of Cr contamination, the data revealed that the Cu-CNFs improved water absorption capability [27]. Research has shown that CNFs can improve plant growth by acting as NCs in the delivery of micronutrients, increasing water intake, accelerating seed germination, and lengthening the plants' roots and shoots.

Applications ranging from medication administration to transdermal patches to wound healing to agriculture rely heavily on a wide variety of polymers such polyvinyl alcohol (PVA), starch, polylactic acid (PLA), poly (lactic-co-glycolic acid) (PLGA), hyaluronic acid (HA), gelatin, and chitosan [28]. These polymers were able to efficiently transport fertilisers and micronutrients inside plant tissues. These polymers are typically encapsulated within the micronutrients, fertilisers, or metal NPs, changing the surface charge of the materials and allowing for effective translocation within plants and controlled release of micronutrients. Deshpande et al., for instance, synthesised Zn-loaded chitosan NPs and evaluated their efficacy as a plant micronutrient. The results indicated that Zn-loaded chitosan NPs could be an effective vehicle for transporting micronutrients [29]. Synthesis of gibberellin phytohormone-containing nanoalginate-chitosan and nanochitosan-alginate composites was the topic of a separate study. The results indicated that polymeric NCs successfully release the growth-promoting phytohormones with no negative side effects on the plants [30].

As research into nanotechnology has progressed, it has become increasingly applicable in the industrial sector. The primary goal of a controlled-release delivery system in agricultural and other biological applications is to provide crops with the necessary fertilisers, agrochemicals, and micronutrients over an extended period of time while reducing their exposure to toxic levels. The NCs' readily functionalizable nature, large surface area, and quick mass transfer make them ideal for the efficient administration of fertilisers, agrochemicals, and micronutrients into plants. From this vantage point, various NCs (materials with dimensions of 100 nm or less) are introduced into fertilisers, agrochemicals, and micronutrients to facilitate their entry into plant cells. Several different approaches are taken to include NCs in fertilisers: Agrochemicals can be attached to NCs in four different ways: (1) encapsulation, in which fertilisers are encased on the surface of NCs; (2) absorption, in which NCs are absorbed onto fertilisers or fertilisers are absorbed onto NCs; (3) attachments through bonds; and (4) entrapment. The integration procedure contributes to the development of a reliable delivery system that allows for the measured dispersal of agrochemicals. In addition, NCs improve the stability of fertilisers, agrochemicals, and micronutrients, protecting them from degradation and leaching out into ecosystems and guaranteeing good efficiency even at low concentrations of agrochemicals and reduced toxicity. The size and surface charge of the NCs are the two most important elements in NC delivery within plants. Translocation of NCs (or micronutrients) within plants is influenced in part by their surface charge. While NCs can be either positively or negatively charged, plant cell walls are typically negative in charge. Higher buildup and reduced translocation of the positively charged NCs occurs due to positive–negative interactions. There may be negative consequences on plant growth from the increased accumulation. The repulsion forces between the negatively charged NCs and the positively charged plant components allow for efficient translocation of the NCs from root to shoot to leaf. Increased NC translocation ability benefits plant development [31]. The magnitude of NCs is, like surface charges, a crucial element. Translocation of NCs or nanomaterials up to 500 m is straightforward within plants. There are two main routes that NCs and other nanomaterials take: (1) apoplastic, which involves transport through the xylem, cell walls, and extracellular space, and (2) symplastic, which involves movement through the cytoplasm of neighbouring cells or plasmodesmata. About 50 nm-sized particles use symplastic pathways, while NCs larger than that use apoplastic pathways to move around inside plants.

3 Uptake, Translocation and Accumulation of CDs in Plants

QDs' distinctive fluorescence signal offers a useful method for tracking them in plants. Using fluorescence imaging, transmission electron microscopy (TEM) observations, or Raman spectroscopy, researchers have seen plants take in QDs [32] (Fig. 1). For instance, Loukanov et al. studied CD uptake in live tobacco bright yellow-2 (BY-2) plant cells using fluorescence and confocal imaging [33]. For our prior research into plant CD uptake and translocation, we used the mung bean as a model plant [34]. The CDs are made readily available for absorption and uptake by the roots of the mung bean plants thanks to hydroponics. The mung bean seeds germinated and grew rapidly after 5 days of cultivation at 25 °C. Increased absorption of CDs by the mung bean plant was demonstrated by the evident concentration-dependent reddish-orange fluorescence under 365 nm UV light (Fig. 2a). After 5 days of incubation, the mung bean seedlings were analysed for CDs by taking images of their roots, stems, cotyledons, and leaves with a confocal laser scanning microscope. In Fig. 2b and c, we notice that the fluorescence signals of the CDs are mostly localised in the vascular system of the root, stem, and leaves of the mung bean sprouts. The CDs were agglomerated and formed considerably bigger clusters in the intercellular space, as seen in TEM images of cross sections of control and CD-treated mung bean seedlings (Fig. 2d). In order to hasten the germination process, the CDs must first break through the seed coat via the intercellular space and then aggregate in the cotyledons. The CDs were absorbed by the root's surface when it emerged and then moved into the root's vascular bundles. The CDs were then transported via the vascular system from the roots to the stems and leaves, where they settled in the veins rather than the mesophyll system (Fig. 2e). Kang et al. observed a similar mechanism for the upward transport of CDs from the mung bean root to the stem and leaves [35]. It is important to note that the surface modification of plants determines the pattern of CD distribution in those plants [36].

Genetic engineering is crucial to plant research. Due to cell wall inhibition, current delivery methods like Agrobacterium-mediated gene transfer and gene gun rapid gene expression system are difficult or limited in host range. Using both plant and animal cells, as well as whole plants, Wang et al. proved the efficacy of a nanocomposite (CDP) developed through mixing polyethyleneimine (PEI) with carbon dot (CD) as a vector for gene transfection [37]. Wheat (Triticum aestivum), mung bean (Phaseolus radiatus), and rice (Oryza Sativa japonica) leaves and rice roots are all capable of efficient DNA delivery and robust gene expression without DNA integration (Fig. 2). Notably, the CDP induced hygromycin resistance in rice by successfully delivering the hydamycin resistance gene into the rice. Furthermore, CDP is capable of delivering and expressing the -glucuronidase gene in mature rice embryo generated callus. In addition to easing plasmid entry into cells, the CDP safeguards DNA from degradation by DNase. Our research yields a very effective DNA delivery system for rapid gene expression, which may prove to be a time-saving and beneficial gene engineering tool for studies in plant physiology.

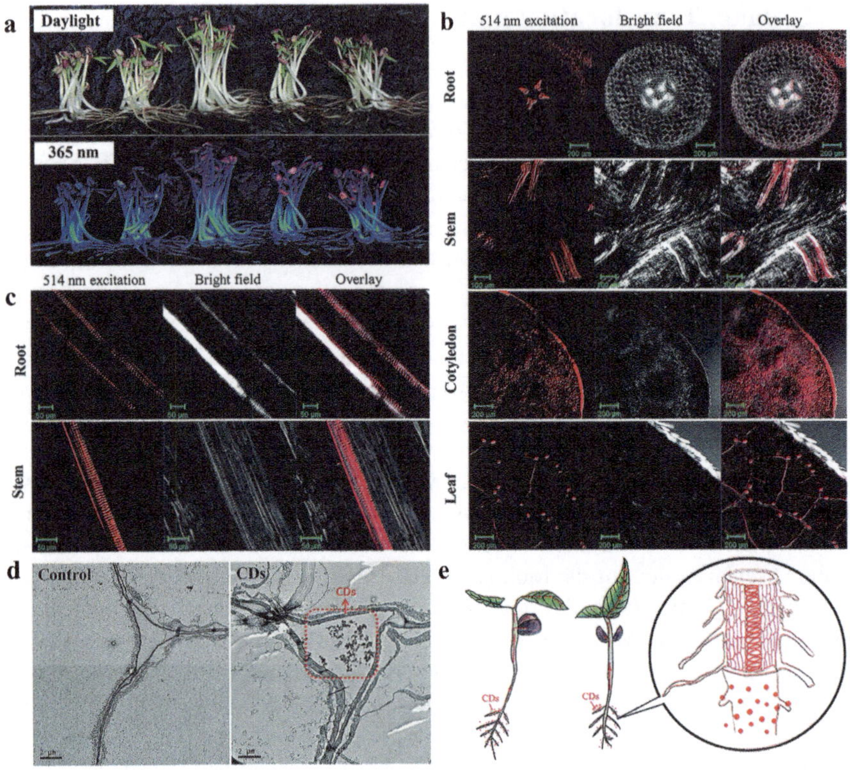

Fig. 1 CD uptake and translocation in mung beans. **a–e** Mung bean sprouts under daylight and 365 nm UV light; LSM images of root, stem, cotyledon, and leaf (**b**) and longitudinal sections (**c**) cultured with CDs (1.0 mg mL − 1) for 5 days; TEM image of roots cultured with pure water and CDs (1.0 mg mL − 1) after 5 days; and schematic illustration of CD uptake, transport, and accumulation. Reproduced with the permission from ref. [34] © 2016 American Chemical Society

4 QDs Delivery in Plants

The capacity to engineer plant function is hampered by the fact that current methods for nanomaterial delivery in plants lack the precision necessary to target specific subcellular compartments. One study developed a nanoscale platform that directs the delivery of nanomaterials containing biochemicals to the chloroplasts of plants, the cells' photosynthetic organelles [38]. Confocal microscopy imaging and quantitative detection by elemental analysis are both made possible by the fluorescence emission of quantum dots (QDs) in a low background window. Nanoparticle coating with a rationally chosen and conserved guiding peptide focuses their distribution to chloroplasts, while QD functionalization with -cyclodextrin molecular baskets enables loading and delivery of different drugs. In plant cells in vivo, peptide biorecognition allows for very efficient and precise delivery of QD with chemical cargos to chloroplasts (74.6 10.8%) and more finely tunable alterations in chloroplast redox activity

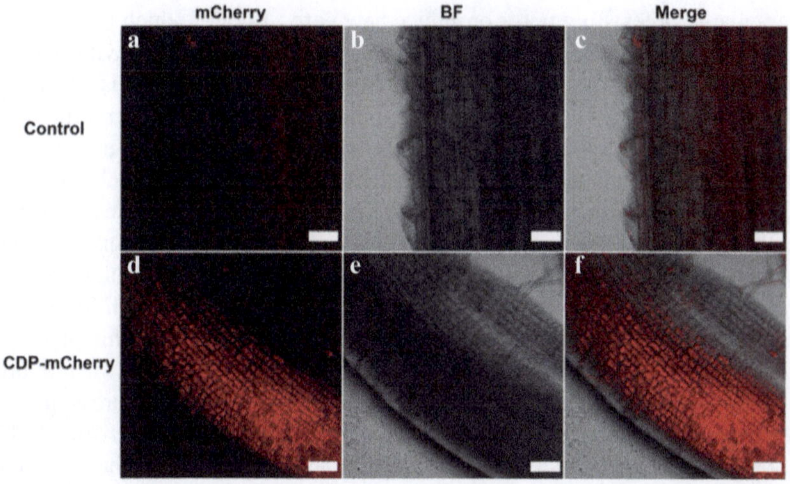

Fig. 2 CDP gene delivery and transient mCherry expression in rice root integrity. **a–c** Rice root was treated with 10 ng/μL pJIT163-mCherry plasmid without CDP. **d–f** Rice root treated with 10 ng/μL CDP-pJIT163-mCherry plasmid. The majority of cells showed red fluorescence (scale bar: 40 μm). Fluorescence confocal micrographs were taken after 12 h of soaking. Reproduced with the permission from ref. [37] © 2020 American Chemical Society

than can be achieved with chemicals alone. Applications of nanotechnology in plant biology, bioengineering, nanoparticle-plant interactions, and nano-enabled agriculture abound due to the precision with which chemical payloads can be delivered in the form of nanomaterials using biorecognition patterns. To research, augment, or impart unique plant functionalities, plant nanobiotechnology is an interdisciplinary area at the intersection of nanotechnology and plant biology. Nanoparticle interactions with plants and the effect of nanoparticles on plant function need the first step of delivering nanoparticles to plants in vivo. The uptake, transport, and dispersion of nanoparticles in plants can be studied with readily available confocal imaging equipment since quantum dots are smaller than plant cell wall pores, have variable surface chemistry, strong fluorescence, and do not photobleach. Another work revealed three approaches to transporting quantum dots into the leaves of living plants: lamina infiltration, entire shoot vacuum infiltration, and translocation from the roots to the leaves. Nanosensors and nanoparticles used in medicine administration are two examples of nanoparticles that could benefit from these techniques [39]. Plants are a key link in the bioaccumulation of nanoparticles into the food web and a possible route for their transfer to the environment. Currently, scientists are conducting research to determine how various nanoparticles affect the development and metabolism of plants. The plant's cell wall prevents outside substances, such as nanoparticles, from entering the cell easily. Pore diameter of cell wall varies from 5 to 20 nm, which establishes sieving properties [40]. This means that only nanoparticles or nanoparticle aggregates with a diameter smaller than the pore diameter of the cell wall can easily go through to the plasma membrane [41]. In addition, tailored nanoparticles can interact with cell walls to either enlarge preexisting pores or induce new pores, both of which boost nanoparticle uptake. During endocytosis, the plasma membrane forms a cavity-like shape around the nanoparticles, facilitating their further internalisation. It is also possible for them to pass across the membrane via in-membrane transport carrier proteins or ion channels. It is possible that the nanoparticles connect with various cytoplasmic organelles and disrupt metabolic activities there [42]. When nanoparticles build up on a plant's photosynthetic surface, they increase the temperature of the leaves, which in turn alters gas exchange via stomatal blockage and has knock-on effects on a number of other physiological and cellular processes [43]. The utilisation of plants as a source for nanoparticle synthesis has been the subject of additional research following the discovery of the mechanism of nanoparticle uptake and creation within plants.

An efficient aqueous method for preparing biocompatible QDs for monitoring nanoparticle-mediated gene transfer in conjunction with Jatropha curcas molecular breeding has been devised. Self-assembly was used to create water-soluble CdSe nanoparticles using L-Cysteine as a stabiliser, and an optimal synthesis technique was determined with the help of fluorescence spectroscopy. The QDs were utilised to electrostatically label chitosan-DNA nanoparticles, and the resulting QD-labeled chitosan-DNA complexes exhibited enhanced fluorescence qualities, as seen by a red shift in the emission and absorption spectra compared to those of the CdSe QDs alone. Instead of using Agrobacterium to induce genetic transformation in Jatropha curcas, this approach is being investigated as a potential improvement. Transfection of Jatropha curcas cells using CdSe/CS-DNA complexes does not appear to have resulted in DNA digestion, as evidenced by PCR amplification of the entire length of the carried reporter gene (GFP). In addition, fluorescence detection of the GFP gene indicates that the target DNA was integrated into the plant genome in the transfected callus cells [44]. In a separate investigation, researchers presented the new idea of producing quantum dots using plant materials. The aqueous extract of the root of the plant Rubia cardifolia was used to synthesise the copper nanoparticles in the form of quantum dots. SEM, TEM, AfM, and fluorescence microscopy were all used to characterise the produced quantum dots. The 22.68 nm quantum dots (QDs) synthesised were spherical particles with rough surfaces. The use of AfM demonstrated the green fluorescence characteristic of these quantum dots. The antimicrobial properties of Cu quantum dots were also examined. CuQDs exhibited strong antibacterial activity against *Klebsiella pneumoniae, Shigella spp., Escherichia coli, Plesiomonas shigelloides, Streptococcus aureus*, and *Pseudomonas aeruginosa*, as measured by colony forming capability testing and diameter of inhibitory zone measurement [45]. In order to determine how CdS-QDs compare to Cd^{2+} ions and bulk-CdS in terms of bioaccumulation and biological response, they were applied to soybean seedlings at concentrations ranging from 50 to 200 mg/L. In order to learn how different polarity and surface charge ligands affect QD stability, uptake, translocation, subcellular localization, and cellular response, bare CdS-QDs were coated with them [46]. QD-TOPO, QD-PVP, QD-MAA, and QD-GLY were all synthesised using four different ligands: trioctylphosphine oxide, polyvinylpyrrolidone, mercaptoacetic acid, and glycine. QD-MAA particles were the most stable, remaining 332 nm in size with minimal Cd^{2+} dissolution (9%) after 14 days in aqueous suspension, while QD-TOPO produced huge aggregates of 3861 nm in size with 27% Cd^{2+} dissolution. Soybean roots collected Cd at a rate of 568 (QD-MAA) to 1010 (QD-PVP) g/g tissue dry weight (DW) after being exposed to 100 mg/L bare and coated CdS-QDs in vermiculite for 14 days; this was statistically equal to the bulk-CdS treatment (639 g/g DW). QD-TOPO accumulated mostly in the membranes, while Cd from $CdCl_2$, bulk-CdS, QD-MAA, and QD-GLY accumulated primarily in the cell wall (40–55%) followed by organelles (28–40%), suggesting apoplastic route. The exception was QD-PVP, which translocated Cd from the roots to the shoots, sequestered it primarily in organelles (49%), and hence lowered leaf biomass. Peroxidases were found to have a crucial function in mitigating the oxidative stress that was induced by CdS-QD exposure. Root lignification permitted the plants to limit Cd accumulation even at the highest CdS-QD treatment (200 mg/L), with the exception of QD-PVP, where lignification was reduced by 21%, resulting in increased Cd content in the shoots. The plants that were subjected to 200 mg/L QD treatments exhibited stress resistance by increasing the amino acid content of their leaves. This research demonstrates how surface coating significantly affects the fate and impacts of QDs in a planted system.

Plants produce phytochemicals like melatonin and serotonin to help them adapt to adverse conditions. Although their synthesis pathways and enzymes have been extensively studied, and a phytomelatonin receptor was recently identified, the actual molecules themselves remain elusive. Understanding where indolamines are found in biological tissues can be a useful resource for studying their role in organisms of all kinds. Quantum dot nanoparticles were used in this investigation for the first time to enable direct visualisation of melatonin and serotonin in axenic roots, yielding exciting new insights. Epidermal cells took up melatonin, transported it laterally, and it eventually built up in endodermal and rapidly dividing pericycle cells [47]. There was a rapid polar migration of serotonin-absorbing cells from the crown region to the root tip. Melatonin and serotonin lost their localisation and were disseminated throughout cells as a result of thermal stress. These results show that melatonin and serotonin travel naturally in the root system to guide cell proliferation and imply that plants have a system to distribute the indolamines throughout tissues as antioxidants in response to environmental challenges. The field of biomedicine is looking to quantum dots (QDs) as a potentially useful tool [48]. They could be used in bio-labels and devices that aid in genetic manipulation and post-harvest management in the agricultural sector, as well as in the production of innovative pesticide formulations. We studied the cytotoxicity and genotoxicity of CdSe QDs in a model plant (Allium cepa) and identified connections between QDs' genotoxic activity and oxidative stress since interactions with higher plants are of significant environmental and ecological importance. Three different concentrations of CdSe QDs (12.5, 25, and 50 nM) were applied to Allium cepa bulbs with healthy roots. For cytotoxicity, we looked at cell viability and mitotic rates; for genotoxicity, we checked for DNA lesions, chromosome abnormalities, and micronuclei. We found that QDs caused oxidative stress and genotoxicity in a considerable way. The maximum uptake of CdSe QD occurred at 50 nM, which may be connected to the fact that Cd is retained in Allium roots in a dose-dependent manner. Both superoxide dismutase (SOD) and catalase (CAT) and glutathione peroxidase (GSH) were triggered in response to the oxidative stress caused by CdSe QD treatment. CdSe QDs were reported to be cytotoxic at 25 nM and genotoxic at 50 nM in a plant study. These results pave the way for figuring out what concentrations to utilise when putting these nanodevices to use on plants in the real world. CdTe quantum dots (CdTe-QDs) are a popular type of NP, and their influence on wheat seedlings when combined with ultraviolet B (UV-B) light was studied by Chen et al. The physical features of 5-day-old seedlings were used to test various concentrations of CdTe-QDs and UV-B radiation, both separately and in combination. Wheat antioxidant enzymes were activated when seedlings were treated with either CdTe-QDs (200 mg/L) or UV-B radiation (10 $kJ/m^2/d$). DNA laddering revealed that the buildup of CdTe-QDs in plant root cells triggered programmed cell death. Both CdTe-QDs and UV-B radiation were found to reduce root and shoot development [49]. Combination therapy appeared to have an additive inhibitory impact. The impact of UV-B and CdTe-QDs on plant development was detailed in this study. The discovery that CdTe-QDs accumulate in plants over the course of their lives emphasises the importance of continuing to monitor these connections.

The overspray of manmade nanoparticles into agricultural land is a major concern to food safety. Although there is great heterogeneity amongst rice cultivars in terms of cadmium accumulation, nothing is known about their reactions to Cd-based nanoparticles [50]. Two rice cultivars with varying Cd accumulation capability were studied for their ability to accumulate cadmium telluride quantum dots (CdTe QDs at 0.5, 1.0, 2.5, and 5.0 mg-Cd/L). After 7 days, the high-Cd-accumulating variety (T705) exposed to 5.0 mg-Cd/L of CdTe QDs showed minimal growth suppression in comparison to the low-Cd-accumulating variety (X24). Cd levels in the roots of both rice cultivars were similar, however T705 had a greater Cd level in its shoots than X24. The accumulation and transfer of CdTe QDs from roots to shoots in T705 has been seen using transmission electron and confocal laser scanning microscopy. T705 leaf activities and gene expression of antioxidative enzymes rose more than X24 leaf activities and gene expression. Our research is the first to confirm that various rice types accumulate Cd differently when exposed to Cd-based QDs, and that this variation has a genetic basis that needs to be explored further.

5 Challenges and Future Perspectives

QDs, or quantum dots, are nanoscale semiconductor materials that have garnered significant interest in many disciplines—from biology and agriculture to physics and chemistry. Scientists have recently investigated the use of quantum dots in plant studies, with the goal of better comprehending plant absorption mechanisms and developing more effective agricultural practises. Although there is much to be gained from incorporating nanoparticles into plant science, there are also several obstacles that must be overcome before they can be used safely and effectively. Quantum dots present a number of hurdles for plant researchers due to the complexity of the systems involved in their uptake and dispersion throughout plant tissues. Scientists have made great strides in following quantum dots as they go from the soil to the roots of plants and then to the various organs of those plants. But the fundamental mechanisms are still poorly understood. Root absorption and foliar application are two methods by which quantum dots can penetrate plants. How they get from one place to another inside the plant, like across cell membranes and through the vascular system, are still a mystery. Understanding these mechanisms is crucial for weighing the costs and benefits of using quantum dots in fields like agriculture and environmental monitoring. The potential toxicity of quantum dots to plant life and ecosystems is another major challenge linked with quantum dot research on plants. Heavy metals like those found in quantum dots, such as cadmium, lead, or selenium, are deadly to living things when ingested in sufficient quantities. The potential of environmental contamination cannot be completely ruled out, despite efforts to build less hazardous or non-toxic quantum dots. To ensure that the use of quantum dots in agriculture does not affect plant health or disturb ecosystems, researchers must conduct rigorous toxicity analyses. In order to safely use quantum dots, it is essential to understand the threshold concentrations at which they become damaging to plants. Quantum dots have the potential to revolutionise agriculture, but overcoming issues of scalability and cost-effectiveness are necessary for their broad use. Large-scale manufacturing and deployment of quantum dots is now cost prohibitive in many agricultural settings due to the high cost and resource requirements of current techniques of synthesis. There are current efforts to provide greener, more economical ways to produce quantum dots. If this works, it might cut down on production costs and open the door to widespread use in farming. The application of quantum dots in plant research and agriculture necessitates careful consideration of regulatory and ethical considerations, as is the case with any new technology. It is crucial to make sure these nanoparticles are employed in a way that doesn't endanger people or the environment. The novel difficulties presented by quantum dots in farming necessitate the creation or modification of regulatory structures. This includes creating disposal and waste management procedures, monitoring their use for safety, and setting standards for use.

The investigation of quantum dots in plants has the potential to significantly advance the fields of agriculture and environmental monitoring. Uptake mechanisms, potential toxicity issues, enhancing scalability, and establishing regulatory frameworks are just some of the obstacles that researchers and stakeholders must overcome. If these challenges can be surmounted, quantum dots have the potential to become useful instruments for improving crop management, gaining insight into plant physiology, and increasing sustainable agriculture. To ensure that agriculture and the environment benefit from and are not endangered by the use of this technology, it must be approached with caution.

6 Summary

At the crossroads of nanotechnology and agriculture, hybrid quantum dots are an exciting new development with the potential to significantly impact farming and environmental monitoring. These hybrid nanoparticles, which combine the biological components with the amazing optical and electrical capabilities of quantum dots, have the potential to revolutionise our understanding of and interaction with plant life. Hybrid quantum dots can be used as payloads for plants, allowing for highly precise monitoring and manipulation of many aspects of plant life. Hybrid quantum dots' capacity to deliver real-time insights on plant health and physiological processes is a major benefit of their use in agriculture. These tiny sensors may be programmed to detect a wide range of environmental and biological cues, enabling for real-time tracking of situations including nutrient availability, water stress, and disease risk. Such monitoring enables farmers and researchers to improve resource utilisation, react quickly to threats to crop health, and make data-driven decisions. Hybrid quantum dots may also play a pivotal role in the future of crop nutrition control. By modifying these nanoparticles to slowly release vital nutrients, we can give plants exactly what they need, when they need it. This specific method not only increases crop productivity but also reduces the environmental damage caused by runoff fertilisers, a major issue in today's farming. Hybrid quantum dots also have use in environmental monitoring. These nanoparticles can be used to monitor the movement of pollutants and heavy metals in water and soil, revealing important information about the state of ecosystems. Their ability to detect and mitigate environmental problems early on is a major factor in their value in protecting biodiversity and ensuring agriculture's long-term viability. Yet there are obstacles to overcome before hybrid quantum dots can be widely used in farming. First and foremost is making sure these nanoparticles don't hurt anyone or the planet too much. Guidelines for safe usage and reduction of potential toxicity should be prioritised. Scalability and cost-effectiveness are other important considerations for making this technology widely available to farmers and scientists. In conclusion, using hybrid quantum dots as payloads for plants is a forward-thinking strategy for dealing with the complex problems facing modern agriculture and environmental protection in this day and age. They are a game-changer in the agricultural sector because of their capacity to provide real-time monitoring, improve nutrient management, and contribute to environmental sustainability. While there are still obstacles to overcome, the potential gains in crop yields, resource efficiency, and environmental protection offered by this technology are too great to be disregarded. Hybrid quantum dots have the potential to usher in a new era of intelligent, environmentally friendly farming if current research and development efforts are successful.

References

1. Molaei, M.J.: A review on nanostructured carbon quantum dots and their applications in biotechnology, sensors, and chemiluminescence. Talanta **196**, 456–478 (2019)
2. Zuo, P., et al.: A review on syntheses, properties, characterization and bioanalytical applications of fluorescent carbon dots. Microchim. Acta **183**(2), 519–542 (2016)
3. Howes, P.D., Chandrawati, R., Stevens, M.M.: Colloidal nanoparticles as advanced biological sensors. Science **346**(6205), 1247390 (2014)
4. Pulizzi, F.: Nano in the future of crops. Nat. Nanotechnol. **14**(6), 507–507 (2019)
5. Beketov, M.A., et al.: Pesticides reduce regional biodiversity of stream invertebrates. Proc. Natl. Acad. Sci. **110**(27), 11039–11043 (2013)
6. Das, P., et al.: MXene/0D nanocomposite architectures: Design, properties and emerging applications. Mater. Today Nano **24**, 100428 (2023)
7. Das, P., et al.: Waste-derived sustainable fluorescent nanocarbon-coated breathable functional fabric for antioxidant and antimicrobial applications. ACS Appl. Mater. Interfaces (2023)
8. Rajabzadeh, A.R., et al.: Magnetic Quantum Dots for Bioimaging. CRC Press (2023)
9. Das, P., et al.: Naturally derived carbon dots in situ confined self-healing and breathable hydrogel monolith for anomalous diffusion-driven phytomedicine release. ACS Appl. Bio Mater. **5**(12), 5617–5633 (2022)
10. Das, P., et al.: Carbon dot biopolymer-based flexible functional films for antioxidant and food monitoring applications. ACS Appl. Polym. Mater. **4**(12), 9323–9340 (2022)
11. Das, P., et al.: Tailor made magnetic nanolights: fabrication to cancer theranostics applications. Nanoscale Adv. **3**(24), 6762–6796 (2021)
12. Das, P., et al.: Carbon-dots-initiated photopolymerization: An in situ synthetic approach for MXene/poly (norepinephrine)/copper hybrid and its application for mitigating water pollution. ACS Appl. Mater. Interfaces **13**(26), 31038–31050 (2021)
13. Das, P., et al.: Biocompatible carbon dots derived from κ-carrageenan and phenyl boronic acid for dual modality sensing platform of sugar and its anti-diabetic drug release behavior. Int. J. Biol. Macromol. **132**, 316–329 (2019)
14. Ganguly, S., et al.: Microwave-synthesized polysaccharide-derived carbon dots as therapeutic cargoes and toughening agents for elastomeric gels. ACS Appl. Mater. Interfaces. **12**(46), 51940–51951 (2020)
15. Das, P., et al.: Carbon dots for heavy-metal sensing, pH-sensitive cargo delivery, and antibacterial applications. ACS Appl. Nano Mater. **3**(12), 11777–11790 (2020)
16. Das, P., et al.: Immobilization of heteroatom-doped carbon dots onto nonpolar plastics for antifogging, antioxidant, and food monitoring applications. Langmuir **37**(11), 3508–3520 (2021)
17. Afreen, S., et al.: Polymer/metal/carbon-based hybrid materials for the detection of heavy metal ions. In: Multifunctional Hybrid Nanomaterials for Sustainable Agri-Food and Ecosystems, pp. 335–353. Elsevier (2020)
18. Lee, W.M., et al.: Toxicity and bioavailability of copper nanoparticles to the terrestrial plants mung bean (Phaseolus radiatus) and wheat (Triticum aestivum): plant agar test for water-insoluble nanoparticles. Environ. Toxicol. Chem.: Int. J. **27**(9), 1915–1921 (2008)
19. Hernández-Hernández, H., et al.: Impact of selenium and copper nanoparticles on yield, antioxidant system, and fruit quality of tomato plants. Plants **8**(10), 355 (2019)
20. Kumar, V., et al.: Development of bi-metal doped micro-and nano multi-functional polymeric adsorbents for the removal of fluoride and arsenic (V) from wastewater. Desalination **282**, 27–38 (2011)
21. Verma, S.K., et al.: Applications of carbon nanomaterials in the plant system: a perspective view on the pros and cons. Sci. Total Environ. **667**, 485–499 (2019)
22. Zhang, M., et al.: Effects of graphene on seed germination and seedling growth. J. Nanopart. Res. **17**, 1–8 (2015)
23. Chakravarty, D., Erande, M.B., Late, D.J.: Graphene quantum dots as enhanced plant growth regulators: effects on coriander and garlic plants. J. Sci. Food Agric. **95**(13), 2772–2778 (2015)

24. Zhang, P., et al.: Toxic effects of graphene on the growth and nutritional levels of wheat (Triticum aestivum L.): short-and long-term exposure studies. J. Hazard. Mater. **317**, 543–551 (2016)
25. Hu, X., et al.: Interactions between graphene oxide and plant cells: regulation of cell morphology, uptake, organelle damage, oxidative effects and metabolic disorders. Carbon **80**, 665–676 (2014)
26. Gupta, G.S., Kumar, A., Verma, N.: Bacterial homoserine lactones as a nanocomposite fertilizer and defense regulator for chickpeas. Environ. Sci. Nano **6**(4), 1246–1258 (2019)
27. Kumar, A., Gahoi, P., Verma, N.: Simultaneous scavenging of Cr (VI) from soil and facilitation of nutrient uptake in plant using a mixture of carbon microfibers and nanofibers. Chemosphere **239**, 124760 (2020)
28. Chen, B.Z., et al.: Self-implanted tiny needles as alternative to traditional parenteral administrations for controlled transdermal drug delivery. Int. J. Pharm. **556**, 338–348 (2019)
29. Deshpande, P., et al.: Zinc complexed chitosan/TPP nanoparticles: a promising micronutrient nanocarrier suited for foliar application. Carbohyd. Polym. **165**, 394–401 (2017)
30. Pereira, A.d.E.S., Oliveira, H.C., Fraceto, L.F: Polymeric nanoparticles as an alternative for application of gibberellic acid in sustainable agriculture: a field study. Sci. Rep. **9**(1), 7135 (2019)
31. Pérez-de-Luque, A.: Interaction of nanomaterials with plants: what do we need for real applications in agriculture? Front. Environ. Sci. **5**, 12 (2017)
32. Wang, H., et al.: Carbon dots promote the growth and photosynthesis of mung bean sprouts. Carbon **136**, 94–102 (2018)
33. Loukanov, A., et al.: Real time monitoring and quantification of uptake carbon nanodots in eukaryotic cells. Microsc. Res. Tech. **81**(12), 1541–1547 (2018)
34. Li, W., et al.: Phytotoxicity, uptake, and translocation of fluorescent carbon dots in mung bean plants. ACS Appl. Mater. Interfaces **8**(31), 19939–19945 (2016)
35. Zhang, M., et al.: One-step hydrothermal synthesis of chiral carbon dots and their effects on mung bean plant growth. Nanoscale **10**(26), 12734–12742 (2018)
36. Qian, K., et al.: Distribution of different surface modified carbon dots in pumpkin seedlings. Sci. Rep. **8**(1), 7991 (2018)
37. Wang, B., et al.: Carbon dots enable efficient delivery of functional DNA in plants. ACS Appl. Bio Mater. **3**(12), 8857–8864 (2020)
38. Santana, I., et al.: Targeted delivery of nanomaterials with chemical cargoes in plants enabled by a biorecognition motif. Nat. Commun. **11**(1), 2045 (2020)
39. Wu, H., et al.: In vivo delivery of nanoparticles into plant leaves. Curr. Protoc. Chem. Biol. **9**(4), 269–284 (2017)
40. Fleischer, A., O'Neill, M.A., Ehwald, R.: The pore size of non-graminaceous plant cell walls is rapidly decreased by borate ester cross-linking of the pectic polysaccharide rhamnogalacturonan II. Plant Physiol. **121**(3), 829–838 (1999)
41. Navarro, E., et al.: Environmental behavior and ecotoxicity of engineered nanoparticles to algae, plants, and fungi. Ecotoxicology **17**, 372–386 (2008)
42. Jia, G., et al.: Cytotoxicity of carbon nanomaterials: single-wall nanotube, multi-wall nanotube, and fullerene. Environ. Sci. Technol. **39**(5), 1378–1383 (2005)
43. Silva, L.C.D., et al.: Responses of restinga plant species to pollution from an iron pelletization factory. Water Air Soil Pollut. **175**, 241–256 (2006)
44. Wang, Q., et al.: Synthesis of water soluble quantum dots for monitoring carrier-DNA nanoparticles in plant cells. J. Nanosci. Nanotechnol. **11**(3), 2208–2214 (2011)
45. Mariselvam, R., et al.: Green synthesis of copper quantum dots using Rubia cardifolia plant root extracts and its antibacterial properties. J. Acad. Ind. Res. **3**(4), 191–194 (2014)
46. Majumdar, S., et al.: Surface coating determines the response of soybean plants to cadmium sulfide quantum dots. NanoImpact **14**, 100151 (2019)
47. Erland, L.A., et al.: Direct visualization of location and uptake of applied melatonin and serotonin in living tissues and their redistribution in plants in response to thermal stress. J. Pineal Res. **66**(1), e12527 (2019)

48. Banerjee, R., et al.: Cadmium selenide (CdSe) quantum dots cause genotoxicity and oxidative stress in Allium cepa plants. Mutat. Res./Genet. Toxicol. Environ. Mutagen. **865**, 503338 (2021)
49. Chen, H., Gong, Y., Han, R.: Cadmium telluride quantum dots (CdTe-QDs) and enhanced ultraviolet-B (UV-B) radiation trigger antioxidant enzyme metabolism and programmed cell death in wheat seedlings. PLoS ONE **9**(10), e110400 (2014)
50. Lian, F., et al.: Variety-dependent responses of rice plants with differential cadmium accumulating capacity to cadmium telluride quantum dots (CdTe QDs): Cadmium uptake, antioxidative enzyme activity, and gene expression. Sci. Total Environ. **697**, 134083 (2019)

Quantum Dots in Light Emitting Diode and Their Applications

Nallin Sharma and Chandan Srivastava

Abstract The development of confined dimensionality in nanoscopic materials has paved the path for many promising applications for human societies. One aspect is zero-dimensional quantum dots (QD) synthesized by various methods. These QDs have a significant role in technological advancements of various healthcare and environmental applications. In this current section, methodologies involved in fabricating quantum dot light emitting diode (QLED) are discussed. The phenomenon of emitting light from provided electrical energy is termed electroluminescence (EL). The mechanisms involved in obtaining electroluminescence are vital to build a working understanding. Three major mechanisms namely, energy transfer, field inducing, charge injection is prominently discussed in the coming sections. Furthermore, the advancements made in developing QLEDs are of huge interest, detailed of type I, II, and III structures are provided. Eventually the day-to-day modern applications of QLEDs in data communication technologies, sensor technologies, and data display technologies are covered with wider information range.

Keywords Quantum dots · Quantum light emitting diode · Color rendering index (CRI) · Luminous efficacy (LE) · Light fidelity (Li-Fi) · Augmented reality (AR) · Virtual reality (VR) · Temperature sensors

1 Introduction

In the ever-developing world scenario, a need for outperforming technologies is everlasting. Devoting a group of nanomaterials in Light emitting diodes have resulted in many outperforming applications. Quantum Dots (QD) as discussed in earlier sections are zero-dimensional confined nanostructures possessing atomically precise

N. Sharma (✉) · C. Srivastava
Department of Materials Engineering, Indian Institute of Sciences, CV Raman Road, Bengaluru 560012, India
e-mail: nallin_25@hotmail.com

C. Srivastava
e-mail: csrivastava@iisc.ac.in

S. Thomas et al. (eds.), *Quantum Dots Based Nanocomposites*, Engineering Materials,
https://doi.org/10.1007/978-3-031-54779-9_19

orientation capabilities. Utilizing QDs in fabricating efficient devices is one approach that evolved many research interests into futuristic technologies. One remarkable possibility is by developing optically tunable light emitting diodes. These Quantum dot light emitting diode (QLED) have replaced high power lasers in multiple arenas. QLEDs have grown to market capturing technology of our present age. The overwhelming performance of QLEDs has been a result of continuous and rigorous research work performed by the scientific community. The assembling of QLEDs is one prominent area of study that in general deals with enhancement of quantum yield (QY), lifetime, and luminance of overall devices. Incorporating QDs in light emitting diode (LED) was a breakthrough of its time that paved the way to zenith of modern technology.

Many synthesis approaches were reported as discussed in earlier sections, here the emphasis is more towards QLED designing, architecture, luminance mechanism, and applications. As the theory of semiconductor engineering begins with p–n junction diodes, similarly LEDs have assembling of many layers to promote electron transfer and recombination at the end of its journey. The desired product of the whole process is generation of visible light or photons having certain wavelengths. These generated photons have energy which is eventually either directly utilized or transferred in another form of operational entity. Owing to QDs size and orientation orders, marvelously engineered products can be seen in day-to-day life. Precisely small sized QDs brighten scenarios of complex device architecture with an ease of operations, functional components thus undergo comparatively lesser frictional, tribological, mechanical, and optical losses. QLEDs thus have been an astounding candidate of nanomaterial engineering to demonstrate miraculous scientific and technological achievements. The new age displays, data communication, sensor technology, and many others are vitally based on QLEDs.

Here in the following section, details of QLED functioning and evolving developments in various applied research are discussed with utmost interest to their fabrication developments till date. The photoluminescence quantum yield (PLQY) is one critical aspect to measure the efficiency of QLEDs and various other factors such as color rendering index (CRI), color quality scale (CQS) are discussed in detail. The later sections of the chapter encompass modern applications of QLEDs in data communication, sensors, and various data display technologies.

1.1 Color Scheme

Chromaticity is a fundamental unit used to define the visible wavelength ranges being emitted from any photon source. Scientifically, the range is defined by tristimulus coordinate system (x, y) that signifies color code as perceived by human vision. The coordinate system comprises of three axes namely x, y, and z but the dependence of $z = 1 - x - y$; delivers complete information with-in two axes i.e., x and y. Conventional chromaticity database is represented by standards of Commission Internationale d l'Eclairage, the international commission for illumination [1, 2]. The Commission

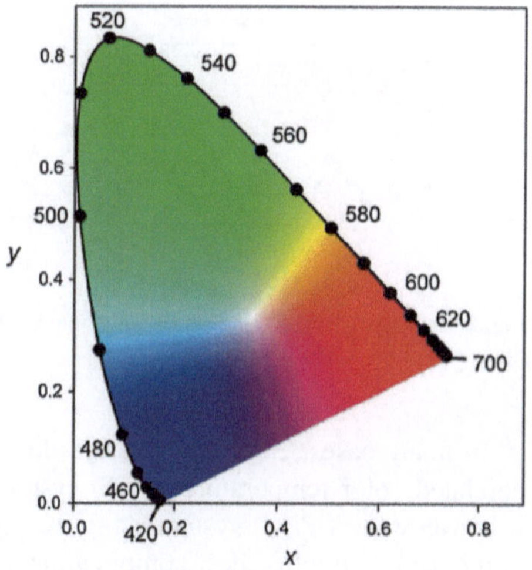

Fig. 1 Two-dimensional representation of standard CIE-1931 chromaticity diagram. (inspired from citation [3])

Internationale d l'Eclairage society has various charting schemes for representing 32 million deep focused colors, CIE-1931 is one standard widely acknowledged for defining visible chromaticity from LEDs (Fig. 1).

Every tristimulus value represents color codes in terms of basic three colors Red (R), Green (G), Blue (B) color, where $\overline{x}$ (λ), $\overline{y}$ (λ), and $\overline{z}$ (λ) are color match function and P(λ) is power density.

$$X = \int_{\lambda} \overline{x}(\lambda) P(\lambda) d\lambda$$

$$Y = \int_{\lambda} \overline{y}(\lambda) P(\lambda) d\lambda$$

$$Z = \int_{\lambda} \overline{z}(\lambda) P(\lambda) d\lambda$$

The equal energy locus located in the center of the chromaticity at (x, y, z) = $(\frac{1}{3}, \frac{1}{3}, \frac{1}{3})$, yet this ideal point is not the observed locus for each observation. Hence, the values are comprehended to (x, y) coordinates.

1.2 Color Temperature and Color Rendering Index

Black body radiation paves the basic understanding of color temperature, deduced from the Planckian locus plotted on standard CIE-1931 representation in x and y coordinate system. The color temperature of light source is determined with respect to black body radiation temperature in Kelvin (K) [1] (Fig. 2).

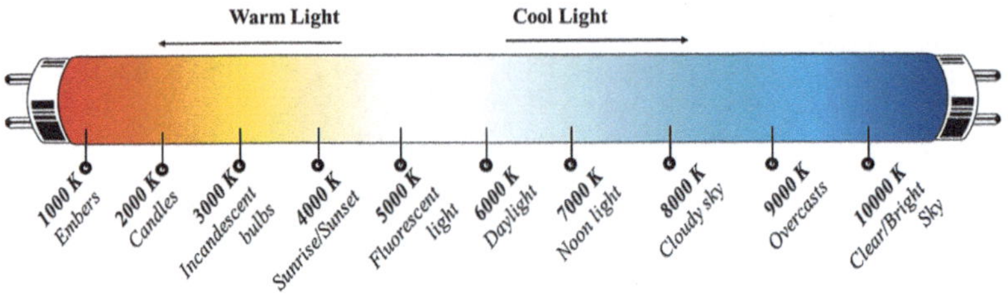

Fig. 2 Schematic representation of color temperatures with respect to observed natural illuminations. (Inspired from citation [3])

In many cases, color temperature does not fall in Planckian locus plot, and a corelated color temperature (CCT) system is used. The (x, y) coordinate system is converted to (*u'*, *v'*) system using the equation mentioned below. CIE-1976 was introduced to simplify these complications, and standard plot with (*u'*, *v'*) coordinates along the respective color temperatures could be defined. The coordinates system was derived using mathematical equation as stated below:

$$u' = \frac{4x}{-2x + 12y + 3}$$

$$v' = \frac{9y}{-2x + 12y + 3}$$

Capability of light source to reveal true color of illuminated object is termed as Color rendering index (CRI) [4]. The measure scale is between -100 to 100, where the best score is determined by 100 and vice versa. CRI measuring of perfect illumination source is relative to 100% blackbody radiation, maintained relative to measure color rendering index of usual illumination sources. Typical household illumination sources are above 80 CRI score, so that human vison is not tempered or harmed due to poor color rendering of objects illuminated. In cases particular of LEDs, color rendering of illuminated objects differs from blackbody standards and hence an enhanced scaling is observed for such case termed as Color Quality Scale (CQS). The new scale has gained tremendous impact over past few years to determine illumination quality of LEDs. CQS was developed by W. Davis and Y. Ohno in years 2004–05 [5, 6]. CQS implies 15 different reflective Munsell samples and records respective saturation factor of illuminated object that has gained intensified chroma, finally an enhanced information of colors is acquired depending upon illuminated saturation factor. This technique is effective against bright illuminating white light LEDs as compared to CRI; colors can be distinguished with percentage level of illumination under ambient conditions. CQS have amplified color rendering procedures in digital image processing techniques as well, by defining precise coloring score range under controlled illuminations [7]. In general, unlike CRI measuring the CQS technique scores between 0–100 that provides critical assessment by calculating root mean square (RMS) average of illuminated colors with respect to reference and sample light source illumination. Groups of CQS values are now referred as standard for procuring ideal illumination conditions for any given area/volume, providing significant functional pioneering in lighting (luminous) devices.

Luminous efficacy is another prime concern for calibrating optimal or desired illumination intensity. Luminous efficacy of optical radiation (LER) and luminous efficiency of device (LE) are two distinct measurements performed to obtain lumen counts. An optical emission comprises of visible and in-visible radiations, visible are helpful in observing illuminated objects whereas, invisible radiations are source of infra-red (heat) or deep ultra-violet rays in some cases. The ratio of visible to invisible radiation or emissions helpful to human vision with respect to total emitted radiation is known as luminous efficacy of optical radiation (LER), measured in lumen per optical watt. On contrary luminous efficacy (LE) of device defines same ratio in terms of electrical power i.e., lumen per electrical watts. These are represented using mathematical equations as shown below:

$$LER = \left(683\frac{lm}{W_{opt}}\right)\frac{\int P_{opt}(\lambda)v(\lambda)d\lambda}{\int P_{opt}(\lambda)d\lambda}$$

$$LE = \left(683\frac{lm}{W_{opt}}\right)\frac{\int P_{opt}(\lambda)v(\lambda)d\lambda}{I \times V}$$

where, $P(\lambda)$ is spectral density signifying light power emitted per unit wavelength, $v(\lambda)$ is sensitivity function of human vision, I is current injection level and V is applied voltage level at operating point of LED.

2 QLEDs Structure and Mechanisms

The structure of QLEDs comprises of three major components namely, emitter, electron transport, hole transport. For the very primitive designs of QLED reported by *Colvin* et al. in 1994 [8], had Indium based transparent conduction oxide, QD-polymer based emitter, and base metal electrode. The direct use of metal electrode caused severe losses in overall device performance, yet a new era for light emitting sources was born. Next improvisation was made by Coe et al. by using organic charge transport layer for sandwiched QD emission layer and this kind of QLED fabrication technique is termed as type-I design. Eventually, inorganic charge transport layers were implied to enhance efficiency, Au/p-GaN/CdSe/n-GaN/In was overall design reported by *Mueller* et al. in 2005 and termed as type-II [9]. With considerable enhancement factor due to inorganic transport layer, next proposed type-III [10, 11] design had both charge transport layer of inorganic material being considered of par efficient till date. Many more improvisations are explored by the scientific community to address the size and efficiency challenges (Fig. 3).

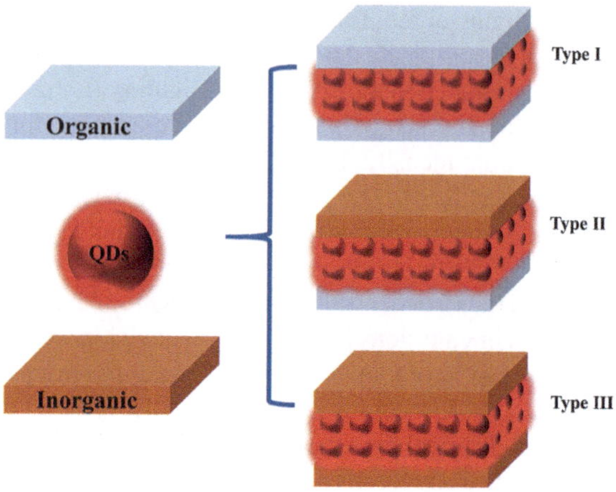

Fig. 3 General representation of QLED types. All inorganic as Type I, hybrid assembly as Type II, and all inorganic materials as Type III. (adapted from citation [12])

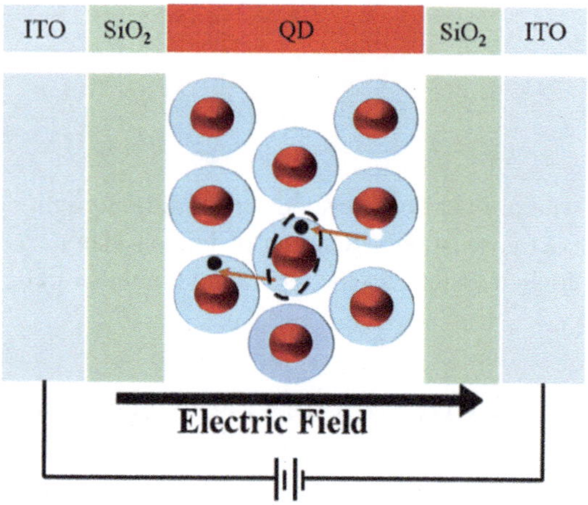

Fig. 4 Working mechanism of Field induced electroluminescence. (adapted from citation [12])

2.1 QLED-Electroluminescence Mechanism

Understanding the mechanism behind luminescence properties of QLEDs is crucial to build an efficient structure and fabrication process. Electroluminescence (EL) in general, is the process of releasing photons when excited electrons observe relaxation inhibition. For QLEDs, three major EL processes observed are namely, energy transfer, charge injection, and field induced.

2.1.1 Energy Transfer

The process where an excited donor generates electromagnetic fields in resonance to ground state acceptor, and later energy transfers to radiative localized state is termed as Förster resonance energy transfer (FRET). A sandwich QD layer present between the dipole acts as radiative emission layer and quantum confinement facilitates FRET in QD layer [13]. As QD are of 5×10^{-9} m scale and FRET is observed in a limited distance of $10–50 \times 10^{-9}$ m. This physical alignment has proved beneficial to enhance efficiency by controlling QD concentration to 7 × 10 [19] nanocrystals/L, the value further increases in solid-state assembling [14]. Apart from layer-to-layer energy transfer, inter-dot FRET or self-quenching phenomenon is also prominent in small packed QDs [15]. The surface-to-surface distance (D) in QD films is a beneficiary and limiting factor at same time, as it increases luminescence but on other hand degrades emission lifetime due to self-quenching.

2.1.2 Charge Injection

The process is considered crucial by many scientific researchers due to freedom of operations. Positioning of transport layers such that charge directly generates excitons in QD layer and radiative emission is observed, this phenomenon is termed as direct charge injection method [11, 16]. Sub-bandgap electroluminescence onset, formation of pseudo-band potential higher than threshold turn-on voltage capable of producing 0.1 cd/m [2] luminescence. It is essential to overcome various charge barriers or energy dissipation losses, termed as sub-bandgap turn-On phenomenon in charge injection QLEDs. The phenomenon was first suggested by Qian et al. and reported as Auger assisted energy up-conversion hole injection process in year 2010 [17]. Later an inversed transport layer structure was reported with exception direct charge injection due to electromagnetic coupling between QD layer and inorganic transport layer by Mashford et al. in 2013 [18]. This Auger assisted hole injection method achieved quantum efficiency of 90% by lowering driving voltage to sub-bandgap turn-On potential and gaining an overall 18% efficiency. Electrically pumped single-nanocrystal spectroscopy enlightens a conclusive remark to the long pursuing debate for charge injection phenomenon. Accumulating multiple scientific reports, a rational agreement is drawn that electron is injected first to charge neutral QD and thus forms a negatively charged intermediate (QD^-) resulting in formation of exciton state (QD^x). Hole injection is considered more demanding than electron injection to QD, supported by coulomb attracting effects. Once an electron injected to neutral QD state then hole injection potential barrier is lowered eventually, supporting QD to QD^- to QD^x to hole injection [19]. This strategy evidently limits multi carrier quenching effects due to presence of QD^- intermediate and inhibits over-hauling of charged electrons to QD layer causing Auger recombination. The feasibility of this method can be estimated by large scale production units agreeing the method as cost-effective.

2.1.3 Field Induced

As the name suggests, process of achieving radiative emission from QD layer sandwiched between two insulating or wide bandgap transparent conducting oxide is termed as field induced emission in QLEDs. The applied electric field across two oxide layer acts as charge transporter for sandwich QD layer, obeying coulomb attraction phenomenon [20]. In field induced emission, QDs are subjected to ionization due to induced potential from one charge transport layer, the ionization then leads to forming a hole and forces ionization to adjacent QDs. This process is continued until the charge recombination is directed through two wide bandgap transport layers. Eventually, excitons are formed in pairs of adjacent QDs resulting a radiative emission of narrow wavelength. Further, adding an insulating layer between QD layer and transport layer provided boost to quantum efficiency of QLEDs. These additional layers provided stability to QD charging and carrier imbalance, earlier prevailing due to electron and hole injection inconsistencies [21]. On the other hand, multiple layers as wide bandgap transporter and insulation gradually raised the need of higher electric field to operate and deteriorate the photoelectric property of QDs such as operational stability and emission efficiency (Fig. 4).

2.2 *QLED Design*

An ideal limitless performing QLED shall possess optimum balance of carrier injection capable of suppressing charge quenching effect and provide an inert interface from oscillating hindrances rising due to charge transport layers. Determining these two fundamental limitations of QLED designing an immediate development to irradicate quenching due to charge injection and charge transport layer materials is demanded. An efficient way to achieve charge balance is by minimizing overpopulation of electrons in transport layers. As per reports, using poly(methylmethacrylate) (PMMA) as an interface to inorganic ZnO/QD layer, the advantage was limiting excessive electron injection by reducing interaction sites between ZnO and QD layer. This optimization resulted in achieving EQE of 20.5% with luminance of 100 cd.m^{-2} for over 100,000 h. Apart from PMMA, other organic insulating layers namely CsCO3, Al2O3, Polyethylenimine (PEI) [22–25], and Polyethylenimine ethoxylated (PEIE) [26–29] have also been reported with significant improvisation in lifetime or luminance.

Charge transport layers have a crucial role in QLEDs and the interface interaction between ETL/HTL with QD layer is usually a retarding behavior in case of inorganic transport layers. To counter these hinderances an Al_2O_3 passivating layer is introduced to NiO layer so that NiOOH passivation could be achieved and quenching of QD emission is minimized. LiF was also introduced in similar understanding to control diminishing emissions, as a result an improvement of 6.52% with EQE of 21,600 cm.m^{-2} was observed [30]. Another critical aspect related to Hole transport layers is hole-induced quenching, this phenomenon is gradual and increases with lifetime of QLEDs. Inserting 1,3,5-tris(N-phenylbenimidazole-2-yl) benzene commonly known as TPBi as hole separation layer between QD/HTL was considered. This improvisation was closely studied for InP based QLEDs and surprisingly 100% quantum efficiency was achieved [31]. Inorganic separation layers are then being explored to overcome lifetime enhancement of QLEDs that can support unwanted leaching due to heating and recoil effects caused after elongated operations.

Formulating another vital aspect to enhance QLEDs emission is excessive exciton formation at interfacial layers, causing an unwanted charge field. The generation of unbiased fields causes randomization of charge transport mechanism or even pseudo-splitting of device architecture in several fragmented layers. This scenario is fatally alarming that can invoke inevitable failure to the overall device. In order to maintain proper functioning and avoid such circumstances, QD emissive layer shall be introduced away from such p-n junctions having potential to form strong charge fields. This strategy is advantageous in two significant manners i.e., firstly, the photo emission wavelength is unaltered by charge clouds and secondly, the electron path is evenly populated to support emissive properties. There are several other measures to optimize QLEDs performance depending upon their application interests, some of them are detailed in following sections.

2.2.1 III–V QLEDs

The elements from III and V group of periodic table have ample of exciton building combinations, among them InP is one configuration having ~1.35 eV bandgap [32] and excitonic Bohr radius of ~10 nm [33]. Synthesizing InP QD was a reckless task to researchers from several years, until coordination solvents could impart a role. Initially trioctyl-phosphine (C24H51P) was considered a strong coordinating solvent but due to its limitation in synthesis process as long reactivity time and uneven QD size distribution, more suitable methodology came up. Incorporating 'In-carboxylates' with a non-coordinating solvent significantly reduced the reaction time with fairly even size distribution. Further modification in reaction method led to formation of single shell InP/ZnS QD nanostructure with as high 70% QY in PL and ~45 nm in FWHM of green emitters. The efficiency for other emitters could be improvised with similar synthesis protocol but was limited due to finitely thin ZnS outer shell and the lattice mismatch between core and shell layer. To eliminate such hinderance an intermediate layer was proposed which could effectively thicken the overall structure and minimize lattice mismatch. The proposed layers were GaP, ZnSe, and composition-gradients of ZnSeS. These modifications crucially functionalized the PL QY to 85%, 82% for green and red QLEDs respectively.

Another aspect was to monitor losses due to charge transport layers; balancing charge injection and optical quenching is of critical interest. The competing interests to under-incline these losses were handled discreetly, imparting organic solvent and inorganic counterparts were implied. Using ZnO nanostructure as charge carrier in transport layer provided immediate push in emitter efficiency. This was later devised with multiple configurations in inverted QLEDs. Multi-shelled structure comprising InP/ZnSeS/ZnS QD that increased illumination to a record 70%. On similar notes using ZnMgO as carrier transport layer dopant, that raised the illumination to 10,000 cd/m [2] and current efficiency of 4.4 cd/A. with these many improvements projected in InP QD based LEDs the highest electroluminescence quantum efficiency ranges up to 3.4–4.4%, 2.5–3% for green and red QLEDs respectively. This provides considerable room for further improvisation and adaptations for III–V group based QLEDs.

2.2.2 II–VI QLEDs

It was difficult to get high-quality non-Cd blue QDs with the requisite emission wavelengths of 450–460 nm, which are appropriate for using QLEDs in displays and lighting. The new class of QLEDs are widely acknowledged for deep and ultra-blue emitting light sources. With the innovation of InP QDs, a new understanding was developed for such low bulk bandgap quantum confined nanostructure ranging below 1 nm. Many multi-cored quanta shelled nanostructures have been investigated since then for achieving higher quantum yield, yet the maximum efficiency achieved is 5–40%. Another prominent ZnSe QLED have attained immense attention in past decade with astounding QY. The PL emission ranges from 440 to 455 nm but with limited QY, whereas short wavelength emission within 420–430 nm have been utilized for many high-end applications due to its reliable QY of 83%. Despite reasonably high QY lesser reports is available on ZnSe QLEDs, but several different architectures have been implied among those reports. An inverted architecture using ZnSe QLEDs was reported to have QY of 60% with peak luminescence at 1170 cd/m [2] and current efficiency of 0.51 cd/A.

2.2.3 IV QLEDs

Carbon and Silicon are the top choices in single element based QLEDs owing to their quantum confined zero-dimensional structures. Interestingly, both these elements can emit from visible-to-far infrared region and the optical property is highly dependent on confinement dimensionality or size specific. The very primitive carbon-dot based QLED was reported to have luminance of 35 cd/m [2] and QY higher than 50%. The experimental understandings grew with graphene QDs with remarkable performances, as ETL materials, separation membranes, and emitter dopants as well. Eventually many reports came-up detailing the impact of Boron (B), Oxygen (O), Nitrogen (N), Phosphorous (P), Sulphur (S) doping in Carbon and Silicon QDs nanostructures. Boron is used to increase the LUMO levels, whereas N, P, and S are doped to enhance the HOMO levels. Introducing these dopant elements during synthesis results in developing p-type or n-type electronic vacancies to overall nanostructures. It was evidently supported that doping of N to carbon QDs significantly minimizes the LUMO–HOMO gap, resulting in a considerable red shift in photoluminescence.

Similar photo-emission phenomena under UV source are equally observed with uni-metallic nanosheets as well in earlier reports [34, 35]. Later higher functional groups as amine, phenylenediamine, aminophenazine, and others were explored as ligands to carbon-dot LEDs. Using o-phenylenediamine (O–PD) outperformed many deep red emitting QLEDs with two-photon red fluorescence with emission at 630 nm and 680 nm. In another report protonation of 2,3-diaminophenazine (2,3-DAPN) helped in achieving sharp and narrow red peak of ~20 nm width [36]. Similar reportable emissions were observed with Si-based QLEDs, their red sifted emission had much higher intensity but limited photon lifetime. This part of group IV QLEDs still require many more evidence to underlay fundamental phenomenon between ligands and metallic core that improves multiple excitons within visible electromagnetic spectrum [37].

3 QLED Applications

In recent years QLED has captivated limitless technological options filling the improvement gaps of data communication, quantum computing, displays, sensors, and many other applied research interests.

3.1 Sensors

The devices capable of measuring alteration in standard atmosphere that may occur due to changing temperature, pressure, gas concentration, healthcare scenarios and many other applications require high-end calibrated sustainable methodology promised through using QLEDs. Intriguing demands of high performing sensors in industrial and healthcare scenarios have brought the need to develop non-invasive methods to counter ambient changes caused due to physical factors. Here are several studies reported by various scientific groups that developed different sensors.

3.1.1 Temperature Sensors

These are the newly developed devices based on QLEDs to record the changing degree of heat in an environment. The fundamental working of such temperature sensors is based on temperature dependent photoluminescent property. Initially two different carbon dot based fluorescent nanomaterials were studied for this application, the PL wavelength differs ~20 nm from each other and eventually temperature of incident surface was recorded. The reversible change in fluorescence wavelength of nanosized carbon dots could visually detect the change in relative PL emission that is characteristic dependent to stimuli from temperature of subjected surface [38]. These PL dependent temperature sensing shortly gained immense interest due to highly non-invasive and contact-less method that became need of hour in several industry and healthcare procedures [39]. Later several studies were made to increase the sensitivity and temperature sensing range up to 300 K. In the newly surfaced scenario, carbon dots grafted using $CaAl_{12}O_{19}:Mn^{4+}$ hybrid phosphor could effortlessly read temperature variance from 80–300 K [40]. To further embark reliability in sensing two fluorescence methodology was studied, where a blue phosphor with yellow-green emissive layer was developed and the latter was extremely temperature sensitive. This technique improvised visual detection method, as the temperature of incident surface rises that induces gradual decrease in yellow-green emission intensity and finally blue color was seen contrasting bright enough.

3.1.2 Gas Sensor

This is another critical aspect of sensing technology worldwide; gaseous analyte monitoring devices require stringent measuring capabilities in ppb scales. This requirement is readily projected by multiple industries at a specific volume. The modeling of device should be carried out in a manner to facilitate analyte molecule or atom to surface at reactivity site of nanomaterial. Gas sensing is fundamentally a chemical reactivity monitored at sub-atomic levels. Firstly, a highly sensitive nanomaterial is synthesized having affinity towards the gaseous content of a specific element. This specificity is later quantified in terms of parts per million/billion (ppm/b) ratio, when present in an ambient environment. In every prerequisite of gas sensors require them to monitor a thermodynamically specific class of gaseous element in an environment composing many mixed or adulterer elemental vapors. Due to incompetency of chemical sensors in real-world scenarios to regulate gaseous composition, QLED based sensors prompted and demonstrated a remarkable efficiency in addition to longer sustainability. Photoluminescent QDs are considered for fabricating high-end gas sensors for heavy-duty applications. The methodology is based on Aggregation induced emission or quenching in most of QLEDs based gas sensors. As the analyte element interacts to nanomaterial surface, it induces certain change in optical properties that are recorded by using QLEDs. These QLEDs are immensely precise in collecting the change of intensity which is calibrated in terms of ppb concentrations for respective gaseous element. Many different sensing platforms are developed for gases derivatives of Oxygen, Phosphorous, Ammonium, Sulfur, Nitrogen, and others depending upon the quantification needs [41–43].

In reports CD/SBA-15, a hybrid nanomaterial synthesized by using mesoporous activity of ethyl orthosilicate in conjunction with carbon dots that possess fluorescent properties [44]. The sensitivity of CD/SBA-15 was concentration dependent that induces an inversely proportional relation with fluorescence intensity. Another interesting report using dual band emission i.e., blue, and red region wavelength were induced for quantifying Oxygen concentration. A hybrid of $CaSiO3:Eu3^{+}$ was synthesized and the overall composition was found to have high affinity towards oxygen [45]. The method reported a minimum response time of 4–7 s and recovery period of 18–29 s. Various other gaseous elements and mixtures such as NO_x,

Methane (CH_4), petroleum gases are also studied and quantified in close to pico-molar range using QLEDs.

3.2 Data Communications

Data communications limitations related to speed and capacity have generally been addressed with semiconductor lasers. For future applications in human-centric systems, short-distance communications, and indoor wireless data services, alternative solutions have been explored due to the comparatively higher production costs and the very complicated driver circuits of lasers, as well as eye safety concerns. A low-cost and high-power substitute are Quantum dot light-emitting diodes. Particularly for Internet of Things (IoT) and 6G technologies, as well as in moderate- to high-speed photonic interconnects, visible light communications (VLC), underwater communications, and precise indoor locating applications, LED-based networks are anticipated to be widely deployed [46–48].

After an electrical signal is received in a QLED communication connection, charge injection, transport, and recombination produce the equivalent optical signal [48, 49]. An LED's usual equivalent circuit may be reduced to a single diode resistor (R_d), capacitor (C), and series resistor (R_s) network. The series resistance is the equivalent resistance of the total resistance linked to the contacts, wire, and drive circuit, which is anticipated to be a constant value, whereas the diode resistance fluctuates nonlinearly with the applied electrical field. The diode capacitance results from charge buildup at the interfaces between several functional layers and the device's dielectric behavior. As a result, the RC effect, charge transit, and recombination are all delayed in time. The mathematical relation between injected current I(t) and applied voltage V(t) is expressed as:

$$\frac{\partial I(t)}{\partial t}+\left(\frac{1}{R_d}+\frac{1}{R_s}\right)\frac{I(t)}{C}=\frac{1}{R_s}\left(\frac{\partial V(t)}{\partial t}+\frac{V(t)}{R_d C}\right)$$

Here, RC time constant be expressed as

$$\tau_{RC}=\frac{C}{\frac{1}{R_d}+\frac{1}{R_s}}$$

Frequency response of a LED is denoted as

$$f_{-3\mathrm{dB}}=\frac{1}{2\pi\tau}$$

where, τ shall be corresponding value of either lifetime carrier τ_C or τ_{RC} and depending on whatever component is the primary limiting factor for modulation performance. Here, the modulation frequency at which the electrical power available from a detector is cut in half is described as the −3 dB modulation bandwidth $f_{-3\text{ dB}}$ of an LED. The frequency at which optical power is equal to half of its unmodulated value is known as the corresponding 3 dB optical bandwidth and is denoted by $\sqrt{3(2\pi\tau)^{-1}}$. However, compared to III–V semiconductors, which have substantially higher carrier mobilities (for example, 10^2–10^3 $cm^2\ V^{-1}\ s^{-1}$), fast LEDs based on developing materials, particularly those with organic functional layers, have much lower carrier mobilities. The time-period of injected carriers drift from the terminals to recombination region is charge transit time τ_{tr}—that is found crucial for optimizing high-speed data communicating QLEDs.

3.2.1 Light Fidelity (Li-Fi)

The communication technology expected to surpass Wi-Fi in terms of data transmission rates and users to uni-station will be Light Fidelity (Li-Fi). In general, the use of visible light radiation as data carriers in any given area of interest is known as Li-Fi. The connections to several different users in terms of internet of things (IoT), vehicle guided system, mobile users, and computing units are displayed with relation to Li-Fi. Initially the technique was deployed to a narrow remote purpose such as digital controller units of household appliances like air-conditioners, television controllers, etc., but considering operational power usage; this technology gained interest. Utilizing for mass communication, the head (Tx) access point is positioned at central light field, and other receivers (Rx) are dissipated over a large area of coverage. The transmitted data packs are passed using multiple QLEDs or wavelengths in visible light channel (VLC). Then Infra-red (IR) wavelength is used by receivers to communicate back with access point head in center. The multiple users in Li-Fi communication are marked respectively (Fig. 5).

There are numerous advantages to the QLEDs based data communication technology over conventional wireless data transmission methods such as WI-Fi and radio communication. The InGaN/GaN QLEDs have reportedly crossed benchmark of one gigahertz data transmission bandwidth in past few years. Currently the transmission rates have surpassed 4 Gbps rate using blue emitting Al/Mg doped GaN by *L Wang et. al.*, in year 2021 [50].

3.3 QLED Displays

These days technology has boosted multi-dimensional arena where information must be presented to audiences. The vastness of such information presenting surfaces commonly known as digital displays, ranges from deep sea to far-space applications. These displays require different viewing features due to the operational circumstances. for instance, the deep-sea display requires low luminescence with higher

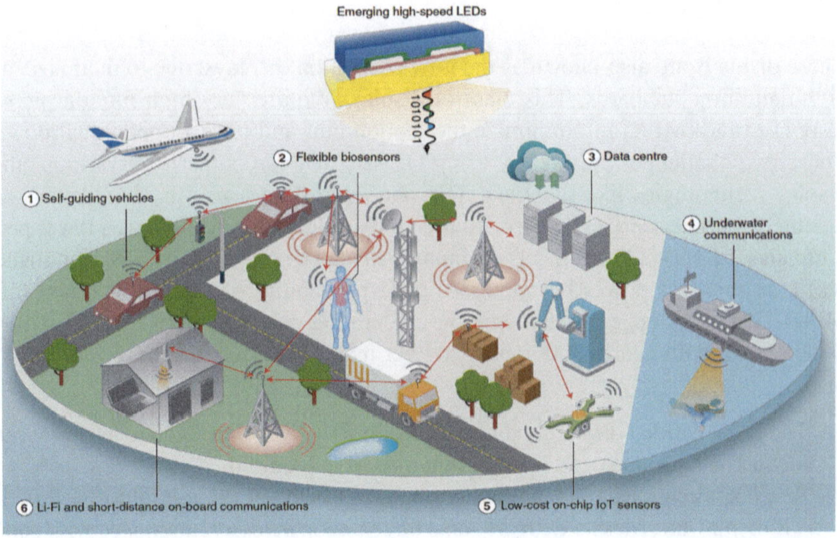

Fig. 5 The communication channel propagation in Light Fidelity (Li-Fi) using QLEDs as source of light. Adapted from [49]

dpi clarity whereas, open sky displays need high luminescence with 8 K resolution. All these prerequisites for concerning display applications are distinctive from one another. In this section, three different types of displays namely Virtual and Augmented reality (VR/AR) displays, Wearable displays, Stand-alone displays are discussed (Fig. 6).

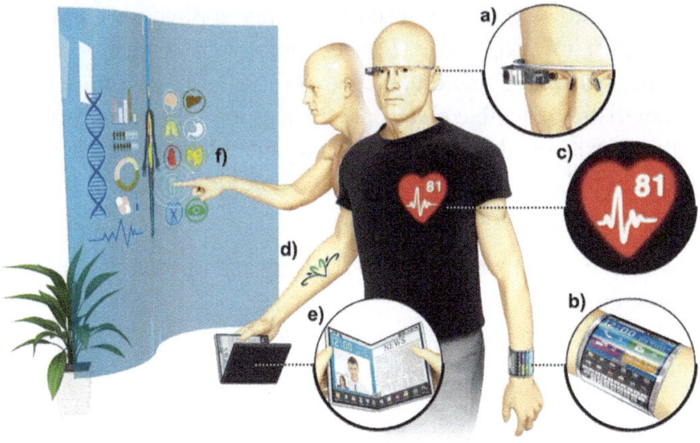

Fig. 6 Future of flexible and wearable displays. **a** Smart glasses with AR/VR. **b** Smart watch with wearable biosensors. **c** fabric display. **d** Ultrathin electronic tattoo. **e** Bendable display. **f** Transparent smart windows. (Adapted from citation [51])

3.3.1 VR Display

These displays are also known as the next generation displays, due to their size and high-definition standards. This class of displays require very high motion picture response time (MPRT) to minimize the motion blur and lowest operational power. These two challenges are met by using QLEDs that can outperform various other display technologies like LCD and TFT. An interesting aspect is VR displays are operated at highly dark and near-vision vicinity of human eye and thus a lower peak luminance of 150–200 cd.m^{-2} is optimum. An added advantage is lower luminance will consume less power but increases error probability of dark spots under high MPRT. This challenge is overcome by moderating chip size between 5–7 μm on 2-inch panel with nearly 3000 ppi (pixel per inch). High luminance is vital for AR displays due to:

a. The information commonly displayed is directly over-portrayed to near environmental backgrounds and thus optimum ACR is required.
b. Space applications requisite high luminance to deliver same luminosity flux the VR display effortlessly delivers same information in lesser luminance with better MPRT.
c. Fast MPRT demands high instant luminance for chip size <7 μm. Organic displays have inversely proportional lifetime to luminance intensity and hence QLEDs have bagged multi-fold advantage to conventional displays.

An example of such lucrative display was reported having ultra-fast MPRT with 5644 ppi and 3500 cd.m^{-2} luminance by *Lu et al.*, in 2019 [52].

3.3.2 Wearable Display

These displays are now commonly used for plenty applications and have gathered ample user interest towards them. The technology backstage is far more versatile than ordinary displays, patternable flexibility is the key to achieve lucrative display characteristics. Flexible electronic devices are fabricated more distinctively than flat or thin electronic devices. Maintaining the robustness of flexible electronic materials requires polymeric assistance that can provide optimum stretchability and retainability. In case of wearable displays:

a. The light source must be arranged as 2D array and thus paving path for emissive displays seen today.
b. The light must possess remarkable off-axis performance, color shifting and ACR coherence are main factors to consider.

The deviation of color rendering in wearable displays was a major challenge in ambient environments that demand high luminance. QLEDs were impregnated with organic ligands to primarily suppress these challenges. Using various InP/ZnSe QD nanomaterials have been reported to dramatically increase the ppi of such wearable displays. Incorporating ITO electrodes over flexible PET surface is an initial step towards developing flexible display [23]. Keeping a track of flexibility fragile ITO was a limitation due to minimum bending radius was later a research subject for flexible displays. eventually Polyimide based electrode material were developed having an average thickness of 18 nm, that acted as semitransparent top electrode for EL. These developments were later used to develop EL-Tattoo like displays by Choi et al in year 2015 [53]. Use of parylene-C films was later approved by Food and Drug Administration (FDA) as good interface for biocompatible skin displays. Wearable displays also required precise waterproofing to expand the applicability. Hence, high IP rating films were introduced as covering or protective layer over electrode layers. More and more intriguing applications of QLED based wearable displays can be seen in the near future.

3.3.3 Standalone Display

As the name suggests these are the displays observed for regular utilizations from household, vehicle, commercials, and industrial purposes. These displays have been part of regular activities to display numerous forms of information which may include high speed motions. In conventional displays the limitation was operation temperatures that caused degradation of information displayed. The advanced technology using LCDs began to fall apart due to crystal ageing and limited viewing angles. Newer versions were introduced with wide viewing angles and a high dynamic range that could cope with the limitations of displaying information. The next generation displays have ultra-high dynamic range and far better resolution up to 8 K readily available for consumers. Eventually these newer displays have found emerging applications as vehicle cluster panels, space flight panels, aircraft displays, and others which now have stand-alone functioning in fast moving scenarios. These utilities are comparatively new for standalone displays and bought new challenges as turbulence handling, rough environment, shock proofing, and lower operational power. QLEDs have delivered substantial support in all the physical conditions due to robust capabilities and far-low operating potential. QLEDs are further being explored to grow in expectations such as HDR performance, field of view, color corrections, and even transparency for head-up display (HUD).

4 Conclusion

The QLEDs in recent years have gained enormous scientific advancements that have promoted multi-industrial applications all around the globe. Quantum dots-based light emitting diodes are highly versatile and application dependent. The electroluminescence of QDLEDs is classified by three different working methods namely, Energy transfer, charge injection, and field induced. Each of these methods have certain advantages one over the other and thus independent designs. The electron and hole transport layer are the two common interfaces in every device, and their compositions vary from organic to inorganic constituents. Numerous studies have been performed to enhance the robustness and diverse functioning of QDLEDs. The improvements are related to electron mobilities, charge separation, and QD-sandwiching procedures. Achieving over 80% PL quantum yield has inspired to develop many applied technological devices for several sectors. The devices are used for sensing physical factors such as temperature, pressure, movements, etc. but not limited to these. Eventually gas sensors were developed that are capable of working at elevated temperatures for multiple gases like Oxygen, Nitrogen, Sulphur, etc. Another remarkable achievement of QLEDs is in display and communication technologies. Modern communication methods such as Ligh Fidelity (Li-Fi), internet of things (IoT), and others are dependent on high frequency QLED-based devices. These enhanced technological applications are expected to soon be expanded in space and aviation industry.

References

1. Schubert, E.F.: Light-Emitting Diodes. Cambridge University Press (2006). https://doi.org/10.1017/CBO9780511790546
2. Vos, J.J.: Colorimetric and photometric properties of a 2° fundamental observer. Color. Res. Appl. **3**, 125–128 (1978)
3. Cho, J., Park, J.H., Kim, J.K., Schubert, E.F.: White light-emitting diodes: history, progress, and future. Laser Photon. Rev. **11**, 1600147 (2017)
4. Method of Measuring and Specifying Colour Rendering Properties of Light Sources, CIE Publication 13.3-1995, Central Bureau of the CIE, Vienna, Austria, 1995,20 pp, $76 (members)/ $114 (nonmembers). Color Res. Appl. **20**, 212–212 (1995)
5. Ohno, Y.: Color rendering and luminous efficacy of white LED spectra. In: Optical Science and Technology, the SPIE 49th Annual Meeting, 2004 (2004). https://doi.org/10.1117/12.565757
6. Wendy Davis, Y.O.: Toward an improved color rendering metric. In: Optics and Photonics 2005, S.D., California, USA (2005). https://doi.org/10.1117/12.615388
7. Ohno, Y.: Spectral design considerations for white LED color rendering. Opt. Eng. **44**, 111302 (2005)
8. Colvin, V.L., Schlamp, M.C., Alivisatos, A.P.: Light-emitting diodes made from cadmium selenide nanocrystals and a semiconducting polymer. Nature (1994). https://doi.org/10.1038/370354a0
9. Mueller, A.H., et al.: Multicolor light-emitting diodes based on semiconductor nanocrystals encapsulated in GaN charge injection layers. Nano Lett. **5**, 1039–1044 (2005)
10. Kim, T.-H., et al.: Full-colour quantum dot displays fabricated by transfer printing. Nat. Photonics **5**, 176–182 (2011)
11. Qian, L., Zheng, Y., Xue, J., Holloway, P.H.: Stable and efficient quantum-dot light-emitting diodes based on solution-processed multilayer structures. Nat. Photonics **5**, 543–548 (2011)
12. Yuan, Q., et al.: A review on the electroluminescence properties of quantum-dot light-emitting diodes. Org. Electron. **90**, 106086 (2021)
13. Kagan, C.R., Murray, C.B., Bawendi, M.G.: Long-range resonance transfer of electronic excitations in close-packed CdSe quantum-dot solids. Phys. Rev. B **54**, 8633–8643 (1996)
14. Xu, L., et al.: Direct observation of resonant energy transfer between quantum dots of two different sizes in a single water droplet. Appl. Phys. Lett. **89**, 033121 (2006)
15. Kagan, C.R., Lifshitz, E., Sargent, E.H., Talapin, D.V.: Building devices from colloidal quantum dots. Science **353**(80) (2016).

16. Yang, Y., et al.: High-efficiency light-emitting devices based on quantum dots with tailored nanostructures. Nat. Photonics **9**, 259–266 (2015)
17. Qian, L., et al.: Electroluminescence from light-emitting polymer/ZnO nanoparticle heterojunctions at sub-bandgap voltages. Nano Today **5**, 384–389 (2010)
18. Mashford, B.S., et al.: High-efficiency quantum-dot light-emitting devices with enhanced charge injection. Nat. Photonics **7**, 407–412 (2013)
19. Anikeeva, P.O., Madigan, C.F., Halpert, J.E., Bawendi, M.G., Bulović, V.: Electronic and excitonic processes in light-emitting devices based on organic materials and colloidal quantum dots. Phys. Rev. B **78**, 085434 (2008)
20. Wager, J.F., Keir, P.D.: ELECTRICAL CHARACTERIZATION OF THIN-FILM ELECTROLUMINESCENT DEVICES. Annu. Rev. Mater. Sci. **27**, 223–248 (1997)
21. Wood, V., et al.: Electroluminescence from Nanoscale Materials via Field-Driven Ionization. Nano Lett. **11**, 2927–2932 (2011)
22. Davidson-Hall, T., Aziz, H.: The role of polyethylenimine in enhancing the efficiency of quantum dot light-emitting devices. Nanoscale **10**, 2623–2631 (2018)
23. Kim, D., et al.: Polyethylenimine ethoxylated-mediated all-solution-processed high-performance flexible inverted quantum dot-light-emitting device. ACS Nano **11**, 1982–1990 (2017)
24. Höfle, S., Schienle, A., Bruns, M., Lemmer, U., Colsmann, A.: Enhanced electron injection into inverted polymer light-emitting diodes by combined solution-processed zinc oxide/polyethylenimine interlayers. Adv. Mater. **26**, 2750–2754 (2014)
25. Ding, K., et al.: Polyethylenimine insulativity-dominant charge-injection balance for highly efficient inverted quantum dot light-emitting diodes. ACS Appl. Mater. Interfaces **9**, 20231–20238 (2017)
26. Kim, H.H., et al.: Inverted quantum dot light emitting diodes using polyethylenimine ethoxylated modified ZnO. Sci. Rep. **5**, 8968 (2015)
27. Shi, Y.-L., et al.: High performance blue quantum dot light-emitting diodes employing polyethylenimine ethoxylated as the interfacial modifier. Nanoscale **9**, 14792–14797 (2017)
28. Son, D.I., Kim, H.H., Hwang, D.K., Kwon, S., Choi, W.K.: Inverted CdSe–ZnS quantum dots light-emitting diode using low-work function organic material polyethylenimine ethoxylated. J. Mater. Chem. C **2**, 510–514 (2014)
29. Yuan, Q. et al.: Efficient $CuInS_2$/ZnS quantum dots light-emitting diodes in deep red region using PEIE modified ZnO electron transport layer. Phys. Status Solidi Rapid Res. Lett. **13** (2019)
30. Yang, X., et al.: High-efficiency all-inorganic full-colour quantum dot light-emitting diodes. Nano Energy **46**, 229–233 (2018)
31. Wang, Y., et al.: Efficient structure for InP/ZnS-based electroluminescence device by embedding the emitters in the electron-dominating interface. J. Phys. Chem. Lett. **11**, 1835–1839 (2020)
32. Green, M.: Solution routes to III–V semiconductor quantum dots. Curr. Opin. Solid State Mater. Sci. **6**, 355–363 (2002)
33. Tamang, S., Lincheneau, C., Hermans, Y., Jeong, S., Reiss, P.: Chemistry of InP nanocrystal syntheses. Chem. Mater. **28**, 2491–2506 (2016)
34. Sharma, N., Wu, H.-F.: The emergence of red fluorescence from two-dimensional nitrogenated-stanene oxide nanosheets. Nanoscale **12**, 10505–10510 (2020)
35. Sharma, N., Chi, C.H., Swaminathan, N., Dabur, D., Wu, H.F.: Introducing Stanene oxyboride nanosheets as white light emitting probe for selectively identifying <5 μm microplastic pollutants. Sens. Actuators B Chem. **348**, 130617 (2021)
36. Zhang, Q., Wang, R., Feng, B., Zhong, X., Ostrikov, K.: Photoluminescence mechanism of carbon dots: triggering high-color-purity red fluorescence emission through edge amino protonation. Nat. Commun. **12**, 6856 (2021)
37. Dasog, M., De los Reyes, G.B., Titova, L.V., Hegmann, F.A., Veinot, J.G.C.: Size vs Surface: tuning the photoluminescence of freestanding silicon nanocrystals across the visible spectrum via surface groups. ACS Nano **8**, 9636–9648 (2014)

38. Rocha, U., et al.: Subtissue thermal sensing based on neodymium-doped LaF 3 nanoparticles. ACS Nano **7**, 1188–1199 (2013)
39. Wang, C., et al.: Surface state-controlled C-dot/C-dot based dual-emission fluorescent nanothermometers for intra-cellular thermometry. Nanoscale **10**, 21809–21817 (2018)
40. Li, W., et al.: Preparation and properties of carbon dot-grafted CaAl 12 O 19: Mn 4+ color-tunable hybrid phosphor. Adv. Opt. Mater. **4**, 427–434 (2016)
41. Wang, J., et al.: Growth of large-size SnS thin crystals driven by oriented attachment and applications to Gas sensors and photodetectors. ACS Appl. Mater. Interfaces **8**, 9545–9551 (2016)
42. Liu, H., et al.: Physically flexible, rapid-response gas sensor based on colloidal quantum dot solids. Adv. Mater. **26**, 2718–2724 (2014)
43. Liu, Y., et al.: Highly sensitive and selective ammonia gas sensors based on PbS quantum dots/ TiO_2 nanotube arrays at room temperature. Sens. Actuators B Chem. **236**, 529–536 (2016)
44. Lei, B., et al.: Luminescent carbon dots assembled SBA-15 and its oxygen sensing properties. Sens. Actuators B Chem. **230**, 101–108 (2016)
45. Wang, L., Zhang, H., Zhou, X., Liu, Y., Lei, B.: Preparation and characterization of a luminescent carbon dots grafted $CaSiO_3$: Eu 3+ phosphor for ratiometric fluorescent oxygen sensing. RSC Adv. **6**, 98554–98562 (2016)
46. Karunatilaka, D., Zafar, F., Kalavally, V., Parthiban, R.: LED based indoor visible light communications: state of the art. IEEE Commun. Surv. Tutorials **17**, 1649–1678 (2015)
47. Dang, S., Amin, O., Shihada, B., Alouini, M.-S.: What should 6G be? Nat. Electron. **3**, 20–29 (2020)
48. Rashidi, A., Monavarian, M., Aragon, A., Rishinaramangalam, A., Feezell, D.: Nonpolar *m*-Plane InGaN/GaN micro-scale light-emitting diode with 1.5 GHz modulation bandwidth. IEEE Electron Device Lett. **39**, 520–523 (2018)
49. Ren, A., et al.: Emerging light-emitting diodes for next-generation data communications. Nat. Electron. **4**, 559–572 (2021)
50. Wang, L. et al.: 1.3 GHz E-O bandwidth GaN-based micro-LED for multi-gigabit visible light communication. Photonics Res. **9**, 792 (2021)
51. Choi, M.K., Yang, J., Hyeon, T., Kim, D.-H.: Flexible quantum dot light-emitting diodes for next-generation displays. npj Flex. Electron. **2**, 10 (2018)
52. Lu, P. et al.: 52-4: highest PPI micro-OLED display sustain for near-eye application. SID Symp. Dig. Tech. Pap. **50**, 725–726 (2019)
53. Choi, M.K., et al.: Wearable red–green–blue quantum dot light-emitting diode array using high-resolution intaglio transfer printing. Nat. Commun. **6**, 7149 (2015)

Quantum Dot Composites for Solar Cell Application

Sampurna Mukherjee, Aaishiki Saha, Ankita Chandra, and Sourav Sarkar

Abstract With the ever-increasing energy demand, solar power as an infinite source of renewable energy without any adverse effects to the environment is a plausible alternative to the conventional energy sources. Solar cell is a photovoltaic device, which turn sunlight, the most fascinating renewable energy source, into electricity, has been extensively studied by a sizable number of researchers. In research labs, several types solar cells have been created, and significant progress has also been made towards commercial manufacturing. Most solar cell technologies, except for Si solar cells, are still far from being used in practical applications because of their generally low power conversion efficiency (PCE), lack of long-term stability, and/or harmful compounds used in the components or during processing. The principles of solar cells were covered in this chapter, from early technologies up to the present day. The identification of novel materials has enabled the development of new solar cell technologies with increasing efficiencies. A fundamental description of several solar cell technologies is provided. The crucial special aspects of the nano crystal QD Solar Cell have been large discussed in this chapter. After discussing the quantum heterostructure, the various designs required for the QD solar cell have been briefly explained. The shortcomings and drawbacks of QD Solar Cells have also been focussed on. After that, the top-down and bottom-up synthesis processes for the QD solar cell have been discussed which includes many methods like Reactive-ion etching or wet chemical etching, Focused Ion Beam Technique (FIB), Electron-beam Lithography, Wet-chemical methods, Sol–gel Process, Microemulsion, Hot-solution Decomposition, Microwaves, Hydrothermal Process, Vapour-phase method etc. Among all the discussed techniques, Hot-solution decomposition method is the cost-effective due to high temperature.

S. Mukherjee · A. Saha · A. Chandra · S. Sarkar (✉)
School of Materials Science and Nanotechnology, Jadavpur University, Kolkata, West Bengal 700032, India
e-mail: souravsarkars@gmail.com

S. Mukherjee · A. Chandra
Thin Film and Nanoscience Laboratory, Department of Physics, Jadavpur University, Kolkata, West Bengal 700032, India

S. Thomas et al. (eds.), *Quantum Dots Based Nanocomposites*, Engineering Materials,
https://doi.org/10.1007/978-3-031-54779-9_20

Keywords QD solar cell · Silicon solar cell · Quantum heterostructures · Semiconductor nanoparticle · Synthesis process · Different designs

1 Introduction of Quantum Dot Solar Cell and Its Composites

The Silicon Solar Cells were first developed in the year of 1953 for the purpose of developing communication in Satellites and Space Crafts. Solar Cells which are presently used in large-scale as Renewable Energy Sources were first developed in 1973 with a standard technology. Mainly there are three Generations of Solar Cells primarily termed as first, second, third generation solar cells [1, 2].

The single crystal Solar Cell comprises the First-Generation Solar Cell. The First-Generation Solar Cells can perform at high temperatures with a high efficiency of near about 20% and they are highly stable to atmospheric changes. The only drawback of them is the high cost preventing the large-scale use of them [3, 4].

After single crystal solar cells, the thin film technologies have been developed which are classified/termed as the Second-Generation Solar Cell. Amorphous Silicon which is also called Hydrogenated Amorphous Silicon mainly comes under the category of Second-Generation Solar Cells. It has an efficiency of around 10% which is lesser than the First-Generation Crystalline Silicon. The only beneficial quality of the Second-Generation Solar Cell as compared to the preceding generation is its less manufacturing cost which can be used in Commercial Solar Panels for better fabrication [5].

Further improvement in technology leads to the discovery of the next generation of solar cells named as The Third-Generation Solar Cell. This generation of solar cell includes Dye Sensitized Solar Cell Technology, Organic Solar Cell Technology, Quantum Dot Technology etc. The Dye Sensitized Solar Cells are much more cost-efficient in terms of manufacturing and they are comparatively less toxic. This kind of Solar Cells displays a good performance even in light conditions and they possess a good incident angle. In addition, they can be used in low intensity and are very light -weighted, flexible semi-transparent and bifacial material. The main disadvantage is its efficiency which is less than 12% [6].

Next, the evolution Quantum Dot Solar Cells takes place. QDs were first discovered in glass crystals in the year 1980 by Russian Physicist Ekimov. However, Quantum Dot Technology had been advanced after 1984 when the Scientist Luis Brus obtained the size and band gap relation in the shape of a Sphere model of the Semiconductor Nanoparticle with the bulk Semiconductor in the form Wave Function. The speed becomes much faster after the synthesis process of Colloidal CdX (X = S, Se, Te). CdX is the most extensively studied QDs due to the excellent Optical and Electrochemical Properties. QDs are minute semiconductor particles with a size varying between few nanometres. The speciality of QD particles is that they have specific optical and electronic properties which are different from the other particles.

By changing the size of QDs we can tune the shape as well as optical and electrical properties of the material. For their excellent absorbing quality, they play a pivotal role in the third generation PV cells like QDSSCs. The quantum dot cells attained the remarkable efficiency of maximum power conversion of 66% approximately. Moreover, Quantum Dots have other properties as well. Among these the most exceptional property of QDs is that with decrease in size of the nanoparticles, we can notice the energy of lowest quantum confinement levels of electron hole pairs increases with increasing band edge energy for optical transition. In a quantum dot solar cell, quantum dots are typically embedded within the active layer of the cell. This integration allows the quantum dots to absorb a broader range of wavelengths of light, including those beyond the visible spectrum. The tunable bandgap of quantum dots enables customization of their light absorption characteristics, making it possible to match the solar spectrum more effectively. One key advantage of quantum dot solar cells is their potential to mitigate energy loss by facilitating the efficient transport of charge carriers. The quantum dots can be engineered to enhance the separation and collection of electrons and holes generated by the absorption of sunlight, leading to improved overall energy conversion efficiency. Researchers are actively exploring various materials and fabrication techniques to optimize quantum dot solar cells for practical applications. This technology holds promise for advancing the field of solar energy by addressing limitations associated with traditional solar cells and opening new possibilities for more efficient and versatile solar power generation [7–19].

Quantum dot composites in solar cells represent a cutting-edge technology that leverages the unique properties of quantum dots to enhance the efficiency and performance of solar energy harvesting. Quantum dots are nanoscale semiconductor particles that exhibit quantum mechanical properties, including size-dependent tunable bandgaps and high absorption coefficients. In the context of solar cells, quantum dots are integrated into composite materials, often in the form of thin films or coatings, to create what is known as quantum dot composites. These composites can be incorporated into various types of solar cells, such as thin-film solar cells or third-generation solar cells, to improve their light-absorbing capabilities and overall energy conversion efficiency. The key advantages of using quantum dot composites in solar cells include their ability to absorb a broader spectrum of light due to the tunable bandgap, enabling better utilization of sunlight. Quantum dots can be engineered to absorb specific wavelengths of light, allowing for customization to match the solar spectrum. Additionally, their small size facilitates the transport of charge carriers, reducing energy loss and improving the overall efficiency of the solar cell. The development of quantum dot composites in solar cell technology holds promise for addressing the challenges associated with traditional solar cells, such as limited absorption range and efficiency. Researchers and engineers are actively exploring ways to optimize quantum dot composites for commercial applications, aiming to create more efficient and cost-effective solar energy solutions for a sustainable future.

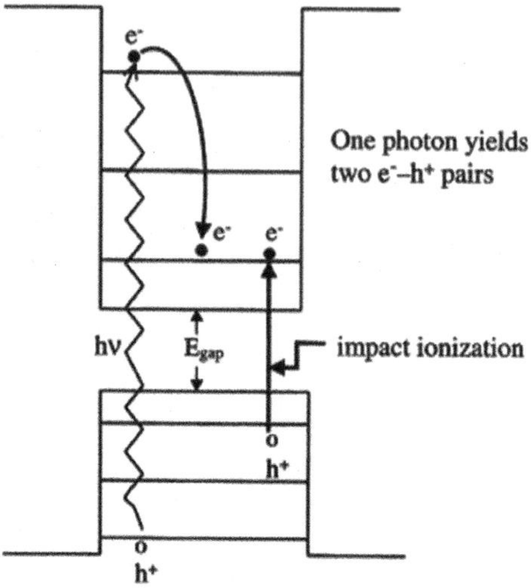

Fig. 1 Impact ionisation (inverse Auger effect)-based increased photovoltaic efficiency in QD solar cells. (Reproduced with permission from Nozik, A.J., 2002. Quantum dot solar cells. Physica E: Low-dimensional Systems and Nanostructures, 14(1–2), pp. 115–120) [20]

2 Quantum Heterostructures

A Quantum heterostructure is defined as the heterostructure in which the size of the nanoparticle has a limitation for the movements of the charger carriers in quantum confinement. It helps to form the separate energy levels in which the carriers can exist. In comparison/contrast to bulk materials, quantum heterostructures do have a better density of states. It has a vital role in the fabrication of small wavelength light emitting diodes, diode lasers and other optoelectronic applications. Quantum Dot Nanoscale Semiconductor Heterostructures are a type of materials that can be well integrated into solar energy conversion devices. This kind of structure can identify and synthesize the given designed materials with their controlled structure and size and show the morphology for different systems individually [21–29] (Fig. 1).

3 Synthesis of Quantum Dots

The two main processes by which we can synthesize Quantum Dots are (i) Top-down approach and (ii) Bottom-up Approach. By applying the former technique bulk materials are gradually reduced to form quantum dots. The latter method mainly includes the wet-chemical and vapor-phase techniques which involve the self-assembly of nanoparticles, aggregation of monomers, wet chemistry, decomposition (Fig. 2).

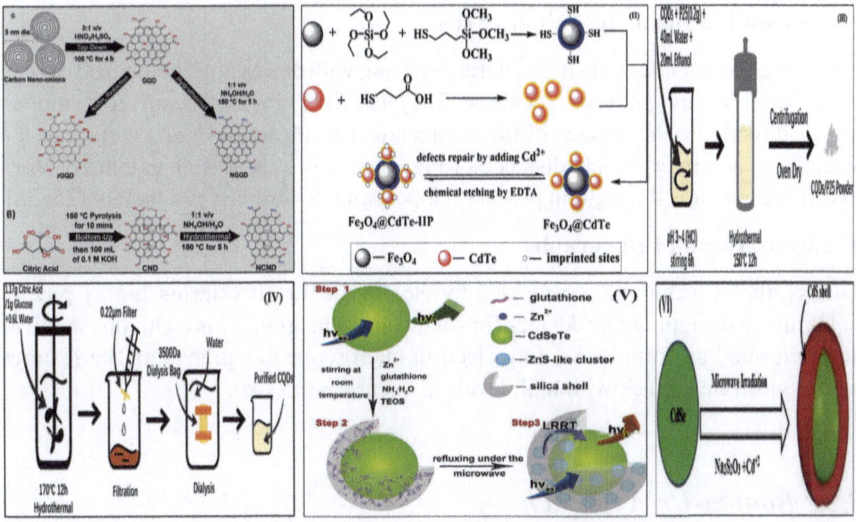

Fig. 2 Different synthesis techniques of quantum dots (Reproduced with the permission from (1) Pillar-Little, T.J., Wanninayake, N., Nease, L., Heidary, D.K., Glazer, E.C. and Kim, D.Y., 2018. Superior photodynamic effect of carbon quantum dots through both type I and type II pathways: Detailed comparison study of top-down-synthesized and bottom-up-synthesized carbon quantum dots. Carbon, 140, pp. 616–623. (2) Luo, X., Guo, B., Wang, L., Deng, F., Qi, R., Luo, S. and Au, C., 2014. Synthesis of magnetic ion-imprinted fluorescent CdTe quantum dots by chemical etching and their visualization application for selective removal of Cd (II) from water. Colloids and Surfaces A: Physicochemical and Engineering Aspects, 462, pp.186–193. (3) Singh, R.K., Kumar, R., Singh, D.P., Savu, R. and Moshkalev, S.A., 2019. Progress in microwave-assisted synthesis of quantum dots (graphene/carbon/semiconducting) for bioapplications: a review. Materials today chemistry, 12, pp.282–314. (4) Shen, T., Wang, Q., Guo, Z., Kuang, J. and Cao, W., 2018. Hydrothermal synthesis of carbon quantum dots using different precursors and their combination with TiO_2 for enhanced photocatalytic activity. Ceramics International, 44(10), pp. 11828–11,834) [21–24]

3.1 Top-Down Approach

The Top-Down Approach helps in producing quantum dots of controllable shapes, sizes with desired packing geometrical capabilities, flexibility, and periodicity for systematic nanostructured design in solar cell applications. A few of the synthesis techniques under this approach have been briefly discussed as follows:

- **Reactive-ion etching or wet chemical etching**

In this process, a radio frequency voltage is applied to create plasma that eventually breaks down the reactive molecules into more reactive fragments inside the etching chamber. After that with the help of masking selective etching is done accordingly. This process is used to produce closed-packed arrays of QD semiconductors [25, 26].

- **Focused Ion Beam Technique (FIB)**

The fabrication of QDs with high lateral precision with dimensions below 100 nm can be achieved by this synthesis technique. They use highly focused beams from molten metals directly on the surface of the semiconductor substrate which determines the size, shape and interparticle distance of the QDs. However, it is an extremely slow, expensive, and low throughput process, i.e., unsuitable for mass production [25, 26].

- **Electron-beam Lithography**

This synthesis technique is followed by etching or lift-off offering highly precise flexibility in designing QD devices for solar cell application. This technique involves an electronic gun from a scanning electron microscope to nanopattern the features on the substrate surface with high levels of accuracy [25, 26].

3.2 Bottom-Up Approach

The bottom-up approach involves the production of most QDs with a high degree of accuracy which exhibit good quantum confinement properties. This quantum confinement effect contributes to the material's unique optical and electronic properties, enhancing the solar cells power conversion efficiency.

A. **Wet-Chemical Methods**

This method involves common precipitation processes with careful control over varying parameters such as concentration and ratios of the species and solvent, temperature, pressure, stabilizers or micelle formation, double layer thickness for proper nucleation sites of the QDs [25, 26].

- **Sol–gel Process**

This process is comprised of three main steps i.e., Hydrolysis, Sol formation (condensation), Gel formation (growth) and is one typical technique that is in use for many years for the formation of QDs. It is a very inexpensive process that can be easily used for scaling up the production but involves broad size distribution subsequently resulting in too many defects [25, 26].

- **Microemulsion**

The microemulsion technique follows synthesizing QDs at room temperature by stirring and rigorous mixing to immiscible liquids (normal emulsion i.e., oil-in-water or reverse emulsion i.e., water-in-oil) until they form emulsions. Utilizing surfactants such as triton-X, aerosol OT (AOT), cetyl trimethyl-ammonium bromide (CTAB), sodium dodecyl sulphate (SDS), and aerosol CTAB, it is possible to disperse nanometre-sized water droplets in n-alkane solutions. Numerous micelles, or very small droplets, are generated in the continuous oil medium because of the surfactants' termination by hydrophilic and hydrophobic groups on opposing ends. The vigorous stirring of these micelles results in the formation of QDs whose growth depends on their size which is again monitored by the molar ratio of water and surfactant [25, 26].

- **Hot-solution Decomposition**

A high-temperature pyrolysis process of organometallic compounds is involved in the hot solution decomposition process that provides sufficient thermal energy to anneal defects, resulting in well-monodispersed QDs. Compared to other processes, the growth of QDs through this method is relatively slower and the size, shape an overall reaction depend on various parameters such as reaction time, temperature, coordinating agents, purity of the coordinating solvents. Though many QDs can be synthesized by this technique, one of the major drawbacks include the toxic effects of organometallic precursors and excessive cost margin due to the use of high temperatures [25, 26].

- **Microwaves**

By Microwave Synthesis, QDs of the dimension 1–5 nm have been successfully produced. The ultrasonic waves or microwaves pass through the precursor mixture through water helping in the dissociation that results in the formation of QDs [25, 26].

- **Hydrothermal Process**

By controlling the temperature, pressure, reaction, and aging time of the reactants inside a stainless-steel autoclave different sorts of QDs of various shapes, sizes, crystallinity can be synthesized [25, 26].

B. **Vapour-Phase Method**

This method involves the growth of QDs by layers of atom-by-atom process, that subsequently appear on the substrate without any nanopatterning. This self-assembly of nanostructure devoid of nanopatterning mainly involves epitaxial growth which is achieved by techniques such as Atomic Layer Epitaxy (ALE), Molecular Beam Epitaxy (MBE), Physical vapor Deposition (PVD), Chemical vapor Deposition (CVD) [B1, B2] (Table 1).

4 Designs for QD Solar Cell

4.1 Optical Design for QD Solar Cell

The quantum confinement effect is the vital concept for QDs. The exciton is confined in zero-dimensional space when the nanoparticles are smaller than the radius of Bohr exciton. Hence the band gap gets increased by decreasing the particle size which in

Table 1 Synthesis of various QDs using different synthesis methods

Synthesis process	Examples	References number
Reactive-ion etching	> The enhanced efficiency of the N-I-P structure amorphous silicon solar cell with nanopillar arrays and CdS QDs layers. The short circuit current is increased from 20.3 ± 0.6 mA/cm^2 to 37.6 ± 0.8 mA/cm^2. The power conversion efficiency is increased from 6.63% to 18.45%. These excellent results are attributed to (i) the antireflection effect of the nano-patterned substrates and the newly rough surface formed by CdS QDs; (ii) CdS QDs absorption and photon down-conversion. The a-Si NPA solar cell with CdS QDs layer is suitable for Si-based solar cells, and other related applications in the future	> [27]
Sol–Gel process	>The sol–gel process approach to spin-coat a thin Si-QDs layer confined in a silica matrix to form composite films. The absorption edge of these films is governed by quantum confinement, meaning that it is a function of the diameter of Si-QDs. The bandgap of these layers of around 1.8 eV is consistent with the use of these films in a tandem photovoltaic cell. Moreover, the bandgap is invariant with Si-QD density in the matrix (in the range 2.0–2.6 10^{18} particles cm^{-3}). The band gap of the nanocomposite layers is not modified by high-temperature treatment for dopant diffusion, which is an essential element for the intended application of solar cells >One step high-temperature synthesis of TiO_2/CdS nanocomposite. The effective heterointerface and high crystallinity, according to scientists, are responsible for the improved visible-light-driven photocatalytic effectiveness. Varying sizes of CdS QDs can control the absorption of TiO_2/CdS nanocomposite. This type of quantum-dot-sensitized solar cell, results in an ideal band gap and conduction band potential	> [28, 29] > [29–31]
Microemulsion	>CdS:Mn/ZnS, Mn doped CdS core, epitaxial ZnS shell by reverse micelle process; Particle Size: 1.5–2.3 nm	> [25, 32–37]
Hot-solution Decomposition	>1996: CdSe/ZnS; single-step synthesis; Core: 350 °C at 1 atm, Ar; Reaction Growth: 310 °C; Shell: 300 °C; Particle size: 2.7–4 nm >1997: CdSe/CdS; two-step process; Core: 300 °C; Shell:100 °C; Particle size: 2.5–4 nm >1996–1999: InAs/InP, InAs/CdSe; two-step process; Core: 200 °C; Shell: dropwise addition at 260 °C; Particle size: 2.5 – 6 nm (InAs); Core/shell: 1.7 nm >2001–2003: CdSe/ZnS, two-step process: Core: reaction and growth: 270°–310°; Shell: slow addition of Zn & S precursor at 180–220 °C	> [25, 38] > [25, 39] > [25, 40–42] > [25, 43, 44]

(continued)

Table 1 (continued)

Synthesis process	Examples	References number
Microwaves	> During synthesis, CdSe seeds were first formed by the reaction of NaHSe and Cd^{2+}, and then alloyed quantum dots (CdSe–CdS) were rapidly produced by releasing sulfide ions from 3-mercaptopropionic acid as sulfide source with microwave irradiation. The alloyed quantum dots synthesized had good optical properties, the quantum yield was up to 25%, and the full width at half maximum of the emission spectrum peak was about 28 nm >Nanocomposites of graphene oxide (GO) or reduced graphene oxide (rGO) with $CuInGaSe_2$ quantum dots (QDs) synthesized by a microwave-assisted method; bulk CIGSe has been successfully applied in thin film solar cells with ca. 20% of efficiency; synthesis of nano-sized CIGSe and other quaternary quantum dots (QDs), aiming at further improving the efficiency of the solar cells and the reduction of costs by using lower amounts of materials; the nanocrystals of 10 nm average size possessed a large absorption tailing and emitted light at 650 nm >Water-dispersed CdSe/CdS/ZnS core/shell/shell QDs with high photoluminescence and a large Stokes shift were synthesized by a facile microwave-assisted route with the highest quantum yield of 25.4%; fabricated CdSe/CdS/ZnS QDs/SiO_2 composite films were applied to Si solar cells as luminescent down-shifting layers and enhances QDs on the photoelectric conversion efficiency of the Si solar cell by spectral down-shifting as compared to the bare glass substrate, and the maximum achieves 16.14% >The B-GQDs are found to have an average lateral width of B6 ± 2 nm and thickness of B0.2–0.3 nm. Nano ZnO is incorporated into these B-GQDs to prepare a B-GQDs-ZnO composite based high-performance photo-anode for Dye-Sensitized Solar Cells (DSSCs). The results showed the highest power conversion efficiency of B3.7% with these new composite-based DSSCs, showing the potential of this composite in real-cell applications	> [25, 45] > [46] > [47] > [48]
Hydrothermal process	>Zero-dimensional graphene quantum dots (GQD) anchored on reduced graphene oxide (rGO) sheets were synthesized by hydrothermal method. CuS and GQD anchored rGO sheet (GRGS/CuS) composite was prepared by simple spin coating and thermal sulfidation techniques and used as counter electrode (CE) material for CdS quantum dot sensitized solar cell (QDSSC). This synergistically helped to reduce the charge transfer resistance of the composite counter electrode with increased current density and to enhance the photovoltaic performance of bare CuS and GRGS CEs. This synergistically helped to reduce the charge transfer resistance of the composite counter electrode with increased current density and to enhance the photovoltaic performance of bare CuS and GRGS CEs, which has a brilliant future for high efficiency solar cells >One-step approach for the synthesis of carbon dots (CDs); CDs can be conveniently utilized by coating the hybrid solution on proper substrates, etc.; based on the luminescent down-shifting behaviour of the composite with CDs, the power conversion efficiency of polymer-fullerene-based bulk heterojunction (BHJ) solar cell is increased by about 12% due to effective light conversion of near ultraviolet and blue-violet portions of sunlight	> [49] > [50]

(continued)

Table 1 (continued)

Synthesis process	Examples	References number
Atomic layer epitaxy, Molecular Beam Epitaxy, physical vapour deposition, chemical vapor deposition	>Kim et al. synthesized ZnSe/ZnS by Stranski-Krastonov mode using CVD technique in ALE mode of operations (mean dot height 1.9 nm) >For indium arsenic/gallium arsenic (InAs/GaAs) systems, In and as atoms are deposited on the GaAs substrate to form an epitaxial layer. A large band gap material, such as InGaAs, is then deposited on top of the InAs epitaxial layer	> [25, 51–60] > [59, 60]

turn affects the band structure and absorption coefficient of the materials. The light path is an important part for optical design in QD Solar Cell, as light gets reflected at the interfaces by reducing the injected reflection of light and increasing the inner reflection of light. The high refractive index is important parameter for controlling the energy level providing Fresnel reflection at a high order at the interfaces, which is the major reason for light loss in QD Solar Cell [35–63].

4.2 *Electrical Design for QD Solar Cell*

The fabrication of superior performance in devices is mainly dependent on the electrical properties of QDs as the ratio of the surface area to volume, interdot distance and transportation of the carrier in QD depends on surface modification and short ligand treatments. The ligands also affect the energy states and fermi level carriers are generated by photon energy via doping. The surface modification helps to trap density and the alignment of the energy in the device. The surrounding ligands provide the QD shifting by producing Dipole Moment. Therefore, we can say surface chemistry modification is the main approach for enhancing the efficiency of solar cells [61].

4.3 *Stability Issue and Design for QD Solar Cell*

The design for the stability is the main role for manufacturing QD Solar Cells. The device of the QD Solar Cell is mainly dependent on the QD materials which comprises of large surface area and organic–inorganic hybrid ligands. QDs synthesized at high temperature possess high crystallinity tend to degrade faster which affects the size dependent characteristics [61, 64, 65].

It can be overserved or noticed the QD solar cell is less efficient than solar cells made up of materials like organics, perovskites and others which are mainly made up of thin films and Si wafers. Hence QDs are insufficient for conversion efficiency so better designs are essential for getting a good result (Fig. 3).

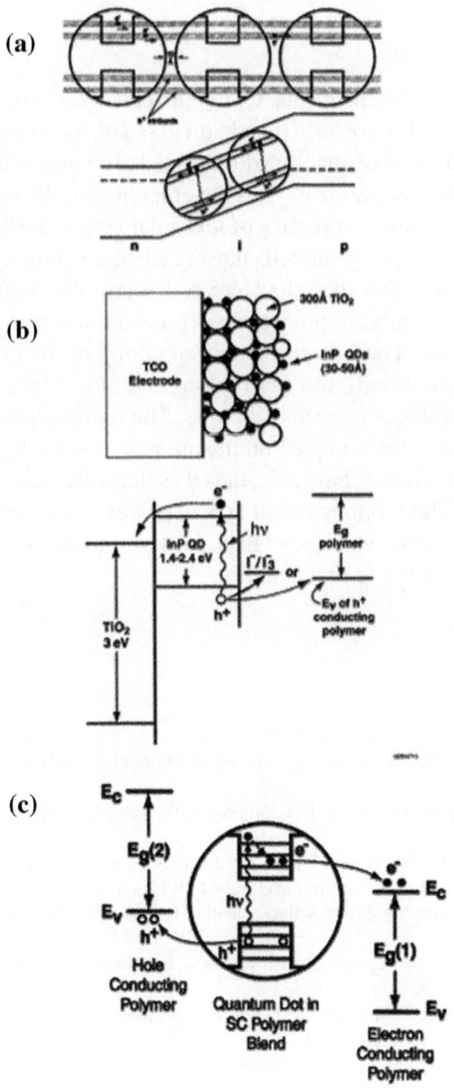

Fig. 3 Different configurations in QD solar cells. **a** a QD array that functions as a photoelectrode for a photoelectrochemical or as the i-region of a p-i-n photovoltaic cell; **b** QDs were used to make a nanocrystalline TiO_2 semiconductor with a wide band gap more sensitive to visible light. This arrangement is comparable to the dye-sensitized solar cell, except that QDs are used in place of the dye; **c** QDs dispersed in a mixture of polymers that conduct electrons and holes. In configurations (**a**), (**b**) and (**c**), Higher photocurrents and better conversion efficiency might result from impact ionisation. In **a**, A higher photo potential could be attained through impact ionisation or hot carrier transport through the QD array's minibands to increase efficiency (Reproduced with the permission from Nozik, A.J., 2002. Quantum dot solar cells. Physica E: Low-dimensional Systems and Nanostructures, 14(1–2), pp. 115–120) [20]

5 Conclusion

In this chapter, advancement in the QD solar cell technology have been briefly discussed. QD Solar cells have been mostly focussed on by contemporary researchers as the fine tuning of the size of quantum dot can lead to change in optical and electronic property. By utilizing this feature of QDs the efficiency of the solar cells can be optimized and in addition other properties of QD solar cells can be modified according to our requirement. The QD solar cells have reached a maximum efficiency of 66% approximately which higher than the solar cells currently used for practical applications. The various synthesis procedures from bottom-up to top-down approach have been focussed on. The size and the morphology of the QDs achieved in each synthesis process have briefly discussed along with the cost efficiency, feasibility, benefits, and disadvantages of each technique. The morphological structure and the size of these dot have a huge impact on the amount of solar light absorption which highly influences the efficacy. However, the QD solar cells are yet to be commercially used due to various shortcomings such as low power consumption, constant power supply, long-term stability which need to be further studied to improve the efficiency of these solar cells and put to practical use.

References

1. Kodati, R.B., Rao, P.N.: A review of solar cell fundamentals and technologies. Adv. Sci. Lett. 26 (2020)
2. Solanki, C.S.: Solar Photovoltaics: Fundamentals, Technologies, and Applications. Phi learning pvt. Ltd. (2015)
3. Kasap, S.O.: Optoelectronics and Photonics. Pearson Education UK (2013)
4. Van Zant, P.: Microchip Fabrication. McGraw-Hill Education (2014)
5. Robertson, N.: Optimizing dyes for dye-sensitized solar cells. Angew. Chem. Int. Ed. **45**(15), 2338–2345 (2006)
6. . Kamat, P.V.: Quantum dot solar cells. The next big thing in photovoltaics. J. Phys. Chem. Lett. **4**(6), 908–918 (2013)
7. Efros, A.L., Brus, L.E.: Nanocrystal quantum dots: from discovery to modern development. ACS Nano **15**(4), 6192–6210 (2021)
8. Ekimov, A.I., Onushchenko, A.A.: Quantum size effect in three-dimensional microscopic semiconductor crystals. ZhETF Pisma Redaktsiiu **34**, 363 (1981)
9. Brus, L.E.: A simple model for the ionization potential, electron affinity, and aqueous redox potentials of small semiconductor crystallites. J. Chem. Phys. **79**(11), 5566–5571 (1983)
10. Rossetti, R., Nakahara, S., Brus, L.E.: Quantum size effects in the redox potentials, resonance Raman spectra, and electronic spectra of CdS crystallites in aqueous solution. J. Chem. Phys. **79**(2), 1086–1088 (1983)
11. Ekimov, A.I., Onushchenko, A.A.: Size quantization of the electron energy spectrum in a microscopic semiconductor crystal. ZhETF Pisma Redaktsiiu **40**, 337 (1984)
12. Brus, L.E.: Electron–electron and electron-hole interactions in small semiconductor crystallites: the size dependence of the lowest excited electronic state. J. Chem. Phys. **80**(9), 4403–4409 (1984)
13. Rossetti, R., Hull, R., Gibson, J.M., Brus, L.E.: Excited electronic states and optical spectra of ZnS and CdS crystallites in the≅ 15 to 50 Å size range: evolution from molecular to bulk semiconducting properties. J. Chem. Phys. **82**(1), 552–559 (1985)

14. Kim, M.R., Ma, D.: Quantum-dot-based solar cells: recent advances, strategies, and challenges. J. Phys. Chem. Lett. **6**(1), 85–99 (2015)
15. Chen, G., Seo, J., Yang, C., Prasad, P.N.: Nanochemistry and nanomaterials for photovoltaics. Chem. Soc. Rev. **42**(21), 8304–8338 (2013)
16. Powell, D.M., Winkler, M.T., Choi, H.J., Simmons, C.B., Needleman, D.B., Buonassisi, T.: Crystalline silicon photovoltaics: a cost analysis framework for determining technology pathways to reach baseload electricity costs. Energy Environ. Sci. **5**(3), 5874–5883 (2012)
17. Green, M.A., Dunlop, E.D., Hohl-Ebinger, J., Yoshita, M., Kopidakis, N., Hao, X.: Solar cell efficiency tables (Version 58). Prog. Photovoltaics Res. Appl. **29**(7), 657–667 (2021)
18. Shah, A., Torres, P., Tscharner, R., Wyrsch, N., Keppner, H.: Photovoltaic technology: the case for thin-film solar cells. Science **285**(5428), 692–698 (1999)
19. Bloss, W.H., Pfisterer, F., Schubert, M., Walter, T.: Thin-film solar cells. Prog. Photovoltaics Res. Appl. **3**(1), 3–24 (1995)
20. Nozik, A.J.: Quantum dot solar cells. Phys. E **14**(1–2), 115–120 (2002)
21. Pillar-Little, T.J., Wanninayake, N., Nease, L., Heidary, D.K., Glazer, E.C., Kim, D.Y.: Superior photodynamic effect of carbon quantum dots through both type I and type II pathways: detailed comparison study of top-down-synthesized and bottom-up-synthesized carbon quantum dots. Carbon **140**, 616–623 (2018)
22. Luo, X., Guo, B., Wang, L., Deng, F., Qi, R., Luo, S., Au, C.: Synthesis of magnetic ion-imprinted fluorescent CdTe quantum dots by chemical etching and their visualization application for selective removal of Cd (II) from water. Colloids Surf. A **462**, 186–193 (2014)
23. Singh, R.K., Kumar, R., Singh, D.P., Savu, R., Moshkalev, S.A.: Progress in microwave-assisted synthesis of quantum dots (graphene/carbon/semiconducting) for bioapplications: a review. Mater. Today Chem. **12**, 282–314 (2019)
24. Shen, T., Wang, Q., Guo, Z., Kuang, J., Cao, W.: Hydrothermal synthesis of carbon quantum dots using different precursors and their combination with TiO_2 for enhanced photocatalytic activity. Ceram. Int. **44**(10), 11828–11834 (2018)
25. Bera, D., Qian, L., Holloway. P.H.: Quantum dots and their multimodal applications: a review. Materials **3**(4), 2260–2345 (2010)
26. Mohamed, W.A., Abd El-Gawad, H., Mekkey, S., Galal, H., Handal, H., Mousa, H., Labib, A.: Quantum dots synthetization and future prospect applications. Nanotechnol. Rev. **10**(1), 1926–1940 (2021)
27. Li, W., Cai, Y., Wang, L., Pan, P., Li, J., Bai, G., Ren, Q.: Fabrication and characteristics of NIP structure amorphous silicon solar cells with CdS quantum dots on nanopillar array. Phys. E **109**, 152–155 (2019)
28. Kintz, H., Paquez, X., Sublemontier, O., Leconte, Y., Herlin-Boime, N., Reynaud, C.: Synthesis and layering of Si quantum dots/SiO_2 composite films for third generation solar cells. Thin Solid Films **593**, 96–101 (2015)
29. Zhao, D., Yang, C.F.: Recent advances in the TiO_2/CdS nanocomposite used for photocatalytic hydrogen production and quantum-dot-sensitized solar cells. Renew. Sustain. Energy Rev. **54**, 1048–1059 (2016)
30. Kouhnavard, M., Ikeda, S., Ludin, N.A., Khairudin, N.A., Ghaffari, B.V., Mat-Teridi, M.A., Ibrahim, M.A., Sepeai, S., Sopian, K.: A review of semiconductor materials as sensitizers for quantum dot-sensitized solar cells. Renew. Sustain. Energy Rev. **37**, 397–407 (2014)
31. Ding, S., Yin, X., Lü, X., Wang, Y., Huang, F., Wan, D.: One-step high-temperature solvothermal synthesis of TiO_2/Sulfide nanocomposite spheres and their solar visible-light applications. ACS Appl. Mater. Interfaces **4**, 306–311 (2011)
32. Yang, H., Holloway, P.H., Cunningham, G., Schanze, K.S.: CdS: Mn nanocrystals passivated by ZnS: synthesis and luminescent properties. J. Chem. Phys. **121**(20), 10233–10240 (2004)
33. Yang, H., Holloway, P.H.: Efficient and photostable ZnS-passivated CdS: Mn luminescent nanocrystals. Adv. Func. Mater. **14**(2), 152–156 (2004)
34. Yang, H., Santra, S., Holloway, P.H.: Syntheses and applications of Mn-doped II–VI semiconductor nanocrystals. J. Nanosci. Nanotechnol. **5**(9), 1364–1375 (2005)

35. Yang, H., Holloway, P.H.: Enhanced photoluminescence from CdS: Mn/ZnS core/shell quantum dots. Appl. Phys. Lett. **82**(12), 1965–1967 (2003)
36. Kortan, A.R., Hull, R., Opila, R.L., Bawendi, M.G., Steigerwald, M.L., Carroll, P.J., Brus, L.E.: Nucleation and growth of CdSe on ZnS quantum crystallite seeds, and vice versa, in inverse micelle media. J. Am. Chem. Soc. **112**(4), 1327–1332 (1990)
37. Mallouk, T.E., Webber, S.E., White, J.M.: Demonstration of a shell-core structure in layered CdSe-ZnSe small particles by X-ray photoelectron and Auger spectroscopies. J. Phys. Chem. **96**, 3812–3811 (1992)
38. Hines, M.A., Guyot-Sionnest, P.: Synthesis and characterization of strongly luminescing ZnS-capped CdSe nanocrystals. J. Phys. Chem. **100**(2), 468–471 (1996)
39. Peng, X., Schlamp, M.C., Kadavanich, A.V., Alivisatos, A.P.: Epitaxial growth of highly luminescent CdSe/CdS core/shell nanocrystals with photostability and electronic accessibility. J. Am. Chem. Soc. **119**(30), 7019–7029 (1997)
40. Peng, X., Wickham, J., Alivisatos, A.P.: Kinetics of II–VI and III–V colloidal semiconductor nanocrystal growth: 'focusing' of size distributions. J. Am. Chem. Soc. **120**(LBNL-41792) (1998)
41. Guzelian, A.A., Banin, U., Kadavanich, A.V., Peng, X., Alivisatos, A.P.: Colloidal chemical synthesis and characterization of InAs nanocrystal quantum dots. Appl. Phys. Lett. **69**(10), 1432–1434 (1996)
42. Cao, Y.W., Banin, U.: Synthesis and characterization of InAs/InP and InAs/CdSe core/shell nanocrystals. Angew. Chem. Int. Ed. **38**(24), 3692–3694 (1999)
43. Talapin, D.V., Rogach, A.L., Kornowski, A., Haase, M., Weller, H.: Highly luminescent monodisperse CdSe and CdSe/ZnS nanocrystals synthesized in a hexadecylamine−trioctylphosphine oxide−Strioctylphospine mixture. Nano Lett. **1**(4), 207–211 (2001)
44. Hikmet, R.A.M., Talapin, D.V., Weller, H.: Study of conduction mechanism and electroluminescence in CdSe/ZnS quantum dot composites. J. Appl. Phys. **93**(6), 3509–3514 (2003)
45. Qian, H., Li, L., Ren, J.: One-step and rapid synthesis of high quality alloyed quantum dots (CdSe–CdS) in aqueous phase by microwave irradiation with controllable temperature. Mater. Res. Bull. **40**(10), 1726–1736 (2005)
46. Mubiayi, K.P., Neto, D.M.G., Morais, A., Nogueira, H.P., de Almeida Santos, T.E., Mazon, T., Moloto, N., Moloto, M.J., Freitas, J.N.: Microwave assisted synthesis of $CuInGaSe_2$ quantum dots and spray deposition of their composites with graphene oxide derivatives. Mater. Chem. Phys. **242**, 122449 (2020)
47. Xuan, T.T., Liu, J.Q., Li, H.L., Sun, H.C., Pan, L.K., Chen, X.H., Sun, Z.: Microwave synthesis of high luminescent aqueous CdSe/CdS/ZnS quantum dots for crystalline silicon solar cells with enhanced photovoltaic performance. RSC Adv. **5**(10), 7673–7678 (2015)
48. Kumar, M.P., Takahashi, C., Kundu, S., Narayanan, T.N., Pattanayak, D.K.: Boron-doped graphene quantum dots: an efficient photoanode for a dye sensitized solar cell. New J. Chem. **43**(36), 14313–14319 (2019)
49. Monika, S., Mahalakshmi, M., Subha, N., Pandian, M.S., Ramasamy, P.: Graphene quantum dots and CuS microflowers anchored rGO composite counter electrode for the enhanced performance of quantum dot sensitized solar cells. Diam. Relat. Mater. **125**, 109033 (2022)
50. Huang, J.J., Zhong, Z.F., Rong, M.Z., Zhou, X., Chen, X.D., Zhang, M.Q.: An easy approach of preparing strongly luminescent carbon dots and their polymer-based composites for enhancing solar cell efficiency. Carbon **70**, 190–198 (2014)
51. Lobo, C., Leon, R.: InGaAs island shapes and adatom migration behavior on (100), (110), (111), and (311) GaAs surfaces. J. Appl. Phys. **83**(8), 4168–4172 (1998)
52. Leonard, D., Krishnamurthy, M., Reaves, C., DenBaars, S.P., Petroff, P.M.: Direct formation of quantum-sized dots from uniform coherent islands of InGaAs on GaAs surfaces. Appl. Phys. Lett. **63**(23), 3203–3205 (1993)
53. Xin, S.H., Wang, P.D., Yin, A., Kim, C., Dobrowolska, M., Merz, J.L., Furdyna, J.K.: Formation of self-assembling CdSe quantum dots on ZnSe by molecular beam epitaxy. Appl. Phys. Lett. **69**(25), 3884–3886 (1996)

54. Kim, Y.G., Joh, Y.S., Song, J.H., Baek, K.S., Chang, S.K., Sim, E.D.: Temperature-dependent photoluminescence of ZnSe/ZnS quantum dots fabricated under the Stranski-Krastanov mode. Appl. Phys. Lett. **83**(13), 2656–2658 (2003)
55. Volmer, M., Weber, A.: Keimbildung in ubersettigten gebilden. Z. Phys. Chem. **119**, 277 (1926)
56. Stranski, I.N., Krastanow, V.L.: Sitzungberichte der akademie der wissenschaften in wien. Akad. Wiss. Lit. Mainz Math-Natur. KI. IIb 1939, 146, 797
57. Eaglesham, D.J., Cerullo, M.: Published 16 April 1990 dislocation-free Stranski-Krastanow growth of Ge on Si(1QQ). Phys. Rev. Lett. 64 (1943)
58. Burda, C., Chen, X., Narayanan, R., El-Sayed, M.A.: Chemistry and properties of nanocrystals of different shapes. Chem. Rev. **105**(4), 1025–1102 (2005)
59. Kang, H.Y.: A review of the emerging nanotechnology industry: materials, fabrications, and applications. Department of Toxic Substances Control Pollution Prevention and Green Technology, California, 64 (2010)
60. Ustinov, V.M., Zhukov, A.E., Maleev, N.A., Kovsh, A.R., Mikhrin, S.S., Volovik, B.V., Musikhin, Y.G., Shernyakov, Y.M., Maximov, M.V., Tsatsul'nikov, A.F., Ledentsov, N.N.: 1.3 μm InAs/GaAs quantum dot lasers and VCSELs grown by molecular beam epitaxy. J. Cryst. Growth **227**, 1155–1161 (2001)
61. Kim, T., Lim, S., Yun, S., Jeong, S., Park, T., Choi, J.: Design strategy of quantum dot thin-film solar cells. Small **16**(45), 2002460 (2020)
62. Lin, L., Li, Z., Feng, J., Zhang, Z.: Indirect to direct band gap transition in ultra-thin silicon films. Phys. Chem. Chem. Phys. **15**(16), 6063–6067 (2013)
63. Kim, T., Jin, X., Song, J.H., Jeong, S., Park, T.: Efficiency limit of colloidal quantum dot solar cells: effect of optical interference on active layer absorption. ACS Energy Lett. **5**(1), 248–251 (2019)
64. Zhang, Y., He, J., Wang, P.N., Chen, J.Y., Lu, Z.J., Lu, D.R., Guo, J., Wang, C.C., Yang, W.L.: Time-dependent photoluminescence blue shift of the quantum dots in living cells: effect of oxidation by singlet oxygen. J. Am. Chem. Soc. **128**(41), 13396–13401 (2006)
65. Cao, Y., Stavrinadis, A., Lasanta, T., So, D., Konstantatos, G.: The role of surface passivation for efficient and photostable PbS quantum dot solar cells. Nat. Energy **1**(4), 1–6 (2016)

Quantum Dots Composites in Supercapacitor Applications

P. Anitha, V. Viswanath, J. Sumalatha, P. Dwarakanadha Reddy, D. Swarna Latha, and D. Vasavi Devi

Abstract This chapter focuses into how quantum dots have been cleverly incorporated into supercapacitor technology. Quantum dots are studied for their potential to improve supercapacitor performance because of their unusual qualities, such as their electrical behaviour that changes with size and the impact of quantum confinement. Quantum dot composites are discussed in this chapter, including their synthesis, characterisation, and electrochemical characteristics. It also explores their potential use in energy storage devices with an eye towards increasing the devices' energy density, charge–discharge dynamics, and durability. This chapter is a great introduction to the intriguing new subject of quantum dot-based supercapacitors, which has great promise for cutting-edge energy storage technologies.

1 Introduction

As traditional fossil fuel reserves appear to be dwindling, concerns have been raised about the future of energy generation and distribution. This unfortunate upheaval has prompted people in recent decades to investigate renewable energy sources like wind, solar, biofuels, etc., which have shown great promise in reducing reliance on traditional fuels like coal and oil. Another possible reason to lean towards nontoxic

P. Anitha · V. Viswanath · P. D. Reddy
Department of Pharmaceutics, Annamacharya College of Pharmacy, New Boyanapalli, Rajampet, Annamayya Dt., A.P 516 126, India

J. Sumalatha
Department of Pharmaceutical Chemistry, P. Rami Reddy Memorial College of Pharmacy, Prakruthi Nagar, Utukur, Kadapa 516003, India

D. S. Latha
Annamacharya College of Pharmacy, New Boyanapalli, Rajampet, Annamayya Dt., A.P 516 126, India

D. Vasavi Devi (✉)
Department of Pharmaceutical Analysis, Annamacharya College of Pharmacy, New Boyanapalli, Rajampet, Annamayya Dt., A.P 516 126, India
e-mail: dvas.reddy@gmail.com

S. Thomas et al. (eds.), *Quantum Dots Based Nanocomposites*, Engineering Materials,
https://doi.org/10.1007/978-3-031-54779-9_21

or very lesser hazardous renewable energy sources is the scarcity of coal and oil-like natural supplies and the damaging influence they have on the environment. Experiments have shown that they pose minimal to no danger to the environment; therefore their widespread use is being examined as a means of making daily energy consumption more exact and secure and of putting an end to the unchecked growth of global warming. In order to make better use of various energy sources, the construction of large energy storage facilities became a possibility. There has been a lot of thought put into this, including research into electrochemical energy storage technologies like batteries, supercapacitors (SCs), and fuel cells, all of which are having a profoundly positive effect on the technological landscape and on people's daily lives. Several studies have proposed widespread use of SCs in energy storage applications as a result of their high power density and high device performance. However, batteries have recently gained a better reputation than SCs in large-scale industrial applications due to their dependable energy density performance. However, batteries have a greater production cost than SCs, and SCs can be made even cheaper by selecting the most cost-effective electrodes. Whereas SCs have a very long cycle life, even batteries whose performance has adequate cycles are still not enough to meet the need of advanced electronics. Batteries also have significant limitations in terms of safety, adaptability, and weight. SCs that are both tiny and flexible, and that use a solid-state, semi-solid, or gel-polymer electrolyte, can readily solve all of these problems [1]. EDLC SCs, pseudo-SCs, faradaic-SCs, and hybrid SCs are just some of the common subtypes of SCs. Carbon-based substances are typically classified as EDLCs [2]. Pseudocapacitors are found wherever metal oxides or chalcogenides behave in this way. Furthermore, the metal oxides/chalcogenides used in batteries store their charges via faradaic processes [3].

In the near future, electrochemical capacitors (also known as supercapacitors) will be the most common type of energy storage device due to their high power density, quick charge–discharge rate, long life, and stability [4]. The poor energy density of electrochemical capacitors, especially electrical double layer capacitors (EDLC), presents a significant barrier to their widespread use. Pseudocapacitive materials, such as metal oxide (RuO_2, MnO_2, Fe_2O_3, V_2O_5, Co_3O_4, NiO_x, etc.) nanoparticles (NPs), can be used to boost energy density [5]. Manganese oxide-based materials have attracted a lot of interest because of their abundance, low cost, and nontoxicity among these metal oxide-based compounds [6]. Dispersing graphene in an aqueous solution remains a difficulty for graphene-MnO_2 nanocomposite production, despite the fact that graphene's scaffolding nanostructures and strong electrical conductivity enable large mass loading of MnO_2 and increase the energy density of supercapacitors [7]. Consequently, graphene oxide (GO) has been used in the synthesis of composites containing MnO_2 NPs, as reported by a number of research [8]. However, GO nanosheets' weak electrical conductivity restricts their capacitance as a nanostructured scaffold. To increase the electric conductivity of nanocomposites for supercapacitor applications, GO must be treated with reductants or exposed to light to create reduced graphene oxide (RGO), which can then be used to form electronic conduction channels [9]. However, the production of MnO_2 NPs and their subsequent deposition on RGO can damage the sp2-hybridized carbon structures of the RGO. Consequently, the use of appropriate reducing agents and the intricate modification of RGO for protection by surfactants or polymers are crucial steps in the composite synthesis process.

Quantum dots (QDs) have gained popularity because of their ultrastable photoluminescence capabilities, tiny size (2–5 nm), and ability to host a variety of functional groups (e.g., hydroxyl, carboxylate, carbonyl) on their surfaces [10–12]. Delocalized electrons are primarily found in the C = C bonds that make up the carbon core of a QD [13–15]. Applications for composites based on quantum dots can be found in many domains, such as biomedical engineering, environmental remediation, photonics, optoelectronics, and electronics [10, 13, 16–23]. Graphene quantum dots and other forms of CDs have recently found use in a variety of energy storage and generation devices, including fuel cells, metal-oxide-based supercapacitors, and lithium-ion and sodium-ion batteries. During the charge–discharge processes in supercapacitors, QD networks in hybrid nanomaterials improve the accessibility of the charged ions, which aids in charge transfer and ionic motion. Supercapacitors and fuel cells can both benefit from the superior capacitance properties and catalytic properties of hybrid nanomaterials synthesised from nitrogen- and sulfur-codoped graphene quantum dots and reduced GO [24].

2 Electrochemical Supercapacitors Based on Carbon Dots

2.1 QDs in Electric Double-Layer Capacitors (EDLCs)

Electric double-layer capacitors (EDLCs), more commonly known as "supercapacitors," are renowned for their high energy density and short charging times. They've found uses everywhere from consumer electronics to alternative energy. Quantum dots (QDs) are one such novel material that has been included into EDLCs in an effort to improve their performance and energy density. Higher energy densities, faster charge–discharge rates, and enhanced overall performance are just some of the benefits that this new field promises. Electric double layers (EDLs) are the fundamental building block of EDLCs, where they develop at the electrode–electrolyte interface. Due to their high surface area and porosity, which give adequate space for ions to accumulate, activated carbon materials have traditionally been utilised as electrodes in EDLCs. Since these materials have a low energy density, scientists have been looking for alternatives, and QDs fit the bill. The unique quantum confinement effects of quantum dots are the result of their nanometer-sized semiconductor particle composition. By carefully manipulating their size and composition, their electrical and optical properties can be finely adjusted. Quantum dots (QDs) added to EDLC electrodes have various benefits. Their increased surface area for ion adsorption is due, in part, to their diminutive size when compared to more conventional materials. This means that the EDLC can store more charges, increasing its energy density. The high electrical conductivity of QDs is also important for the rapid transfer of charges within the electrode. QD-based EDLCs are ideal for high-power applications like electric vehicles and renewable energy systems due to their fast electron-shuttle speed. QD-based EDLCs are an attractive answer to the energy storage problems of the present because of their high charge storage capacity and fast charge–discharge speeds. QDs in EDLCs also stand out due to their electrochemical stability. QDs demonstrate extraordinary endurance, guaranteeing a lengthy cycle life for the supercapacitor, in contrast to some other high-capacity materials which suffer from degradation over time. Having a battery that can withstand repeated charging and discharging for a lengthy period of time is essential for any application. Significant progress has been made in recent years in the production and integration of QDs into EDLC electrodes. Electrode materials based on QDs can have their characteristics modified in a number of ways, including by solution-based deposition, chemical vapour deposition, and electrodeposition. To further improve the performance of EDLCs, researchers have investigated hybrid architectures, which combine QDs with other materials like graphene or carbon nanotubes. Despite these promising developments, EDLCs still face obstacles in fully utilising QDs. For instance, researchers are working to improve QD integration into useful devices and guarantee their stability through extensive electrochemical cycling. In addition, ensuring cost-effectiveness as manufacture of QD-based EDLCs is scaled up for commercial applications are essential.

Sinha et al. reported the development of a hybrid supercapacitor that is sensitive to light. By fusing carbon dots (CDs) to the surface of SWCNT/ZnO nanocomposites, a hybrid electrode material for the supercapacitor was synthesised. By using the supercapacitor while it is lighted (with UV light), researchers have been able to investigate the optical characteristics of CDs and ZnO [25]. At a scan rate of 50 mV/s, the UV light was found to increase the areal capacitance of the manufactured supercapacitor by 41.38%. It has also been investigated how the supercapacitor system operates. Electric double layer capacitance (EDLC) and pseudocapacitance are both shown to contribute to the total capacitance of the hybrid electrode. Dunn method verification of EDLC and pseudocapacitance contributions yields values of 69.35 and 30.65% for EDLC and pseudocapacitance, respectively. As an electrode material for organic electric double-layer capacitors (EDLCs), commercial activated carbon (YP50f) was doped consecutively to increase its electrochemical characteristics [26]. The sequential doping technique involves applying nitrogen doping following an oxygen doping step. We used the sequential doping technique to create a carbon material with a high electrical conductivity and a large specific surface area. Oxygen doping significantly increased the carbon material's specific surface area, whereas nitrogen doping improved its electrical conductivity. Through in situ polymerization of GO/CDs with pyrrole under mild conditions, a ternary composite comprising graphene oxide, carbon dots, and polypyrrole (GO/CDs/PPy) has been synthesised for usage as electrode active material in supercapacitors. By facilitating electron transport in the ternary composite between the GO film and PPy layer, CDs can lower the internal resistance and charge transfer resistance of the electrode [27]. In addition, the dielectric constant of the ternary composite can be improved by using CDs with a high specific surface area at the interface of GO, CDs, and

PPy. For example, at a current density of 0.5 A/g, GO/CDs/PPy displays a specific capacitance of 576 F/g, and a supercapacitor produced with two symmetric GO/CDs/PPy electrodes possesses a high energy density of 30.1 Wh kg1 at a power density of 250 W/kg. Utilising a straightforward interface self-assembly and subsequent activation technique, 2D porous carbon nanosheet (PCN) with high carbon yield was synthesised from graphene quantum dots present in industrial sulfonated pitch waste. The synergistic impact of electrons and ions is fully exploited thanks to its ultrathin thickness of 20 nm, sufficient specific surface area (SSA) of 2429 m2 g1, high packing density of 0.562 g cm3, and great mechanical flexibility. The structural benefits of this PCN make it suitable for usage as a high-performance EDLC electrode. The high volumetric capacitance of 90.1 F/cm at 0.5 A g1 in pure EMIMBF4 electrolyte, along with the exceptional rate capability C20/0.05 of 71% and the good high-voltage endurance lifetime of 88% after 5000 cycles at 1 A/g, make this material very promising. The highest achievable volumetric energy density is 39 Wh/L, which is quite satisfactory. Furthermore, this approach can offer a fresh viewpoint on the design of other 2D carbon materials. To predict thermal electrokinetics in electric double layer capacitors (EDLCs), Ji et al. suggested a new variational, thermodynamically consistent model. Non-isothermal ion transport with temperature inhomogeneity is modelled using modified Nernst-Planck equations, which are based on the least action principle and maximum dissipation principle of non-equilibrium thermodynamics. An equation for the development of temperatures, including contributions from thermal pressure and electrostatic interactions, is derived using the laws of thermodynamics. The ability of the devised model to capture reversible and irreversible heat production is demonstrated by the accurate prediction of temperature oscillation in the charging-discharging processes of EDLCs in numerical simulations. Simulations using cyclic voltammetry allow for a methodical examination of how ionic size and scan rate of surface potential affect ion transport, heat generation, and charge current [28]. The thermal electrokinetics in EDLCs are seen to have delayed dynamics as depicted by hysteresis diagrams, meaning they cannot keep up with the surface potential at high scan rates. As a result, the physics-based prediction of thermal electrokinetics in EDLCs can benefit from the work we've done. Two commercially available activated carbon electrodes (RP20 and MSP20) for an electric double-layer capacitor have their rate capability predicted using electrochemical impedance analysis. To do this, an equivalent circuit is applied to the ac impedance data, which includes the ohmic resistance and pore impedance of the intra-particle. Fitted impedance parameters are used to profile ionic accessibility into intra-particle pores, and these profiles are then converted into usable capacitance graphs as a function of charge–discharge rate, allowing for characterization of the latter. The impedance analysis prediction of the rate capability matches up well with the charge–discharge rate test result [29]. Composite materials were created by Harper et al. by coating CDs homogeneously over lignin-based activated carbons [30]. After applying N-doped CDs and mixing them with lignin-based ACs at an optimised ratio, the CDs were evenly dispersed across the surface of the extremely porous AC matrix, creating a 3D carbon composite with a CD-decorated interior.

Hydrophilic binding sites produced by CDs on the stable AC matrix give many electroactive sites with outstanding cyclability, rather than optimising the inner porous structure. When it comes to attracting ionic species and solvated electrolyte ions in aqueous environments, the superhydrophilic CD sites in the new design of carbon composites have an edge over the undecorated deep and branched micropores. More aqueous electrolytes can move and diffuse into the porous network when CDs are uniformly adhered to the surface of ACs. As a result, the CDs drastically reduce the length of the ionic diffusion path, resulting in more rapid ionic diffusion and less resistance to mass transfer (Fig. 1). This study suggests a straightforward layout for achieving high-performance SCs with a straightforward production technique. We demonstrated the method's efficacy and the 3D carbon composite's viability through the verification of defining measurements; the latter could be used in a wide range of energy-storage devices, from carbon electrodes for EDLCs and ion-storage batteries to catalyst electrodes for fuel cell and photovoltaic cell applications. CD@AC-12 has a specific capacitance of 191.8 F/g, and it grows by 52% when only half as many CDs are added, supporting the idea that the improved capacitance is proportional to the amount of CDs added. These results suggest that modifying ACs with carbon dots is a promising approach to enhancing their performance, though the corresponding study is still in its infancy.

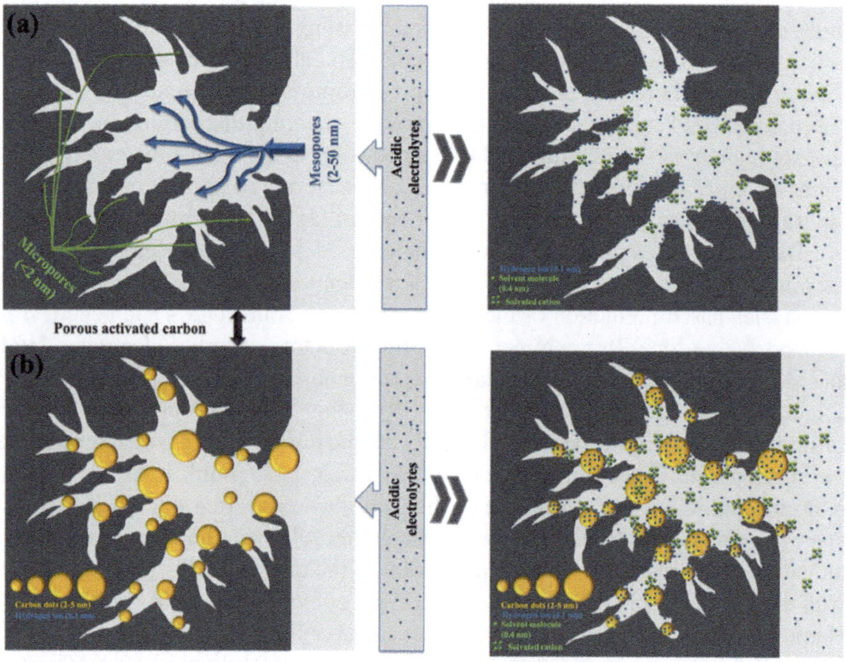

Fig. 1 Schematic illustrations of the charging process in (a) AC and (b) CD@AC-11 electrodes. Reproduced with the permission from Ref. [30] © 2021 American Chemical Society

The EDLC capacitance of GO/CDs/PPy was found to be enhanced [31]. The inclusion of CDs improved the supercapacitive characteristics, as reported by Lima et al. [32].

2.2 *QDs in Asymmetric Capacitors (AC)*

Asymmetric capacitors, also known as supercapacitors and ultracapacitors, are a major technological leap forward in the field of energy storage. Asymmetric capacitors have a different electrode arrangement than ordinary symmetric capacitors, which store energy electrostatically by amassing similar charges on two conducting plates. Typically, one electrode in such a device will have a large surface area, while the other will have a much smaller one. The primary benefit of asymmetric capacitors is that they are able to store more energy than symmetric capacitors. High capacitance and energy density are provided by the low-surface-area electrode, which is typically formed of a metal oxide, while a considerable amount of charge can be accumulated at the high-surface-area electrode, which is typically made of activated carbon or graphene. Because of this imbalance, we may store energy in a way that is more compact and efficient than either conventional capacitors or batteries. Asymmetric capacitors are well-suited for uses like regenerative braking in electric vehicles and peak power demand levelling in renewable energy systems because of their high charge and discharge rates. Asymmetric capacitors are exceptionally long-lasting and economical because they can withstand thousands, if not millions, of charge–discharge cycles before failing. Many industries, from consumer electronics to aerospace, have a need for high-performance energy storage technologies, which might be met by these materials. Asymmetric capacitors have the potential to revolutionise energy storage, ushering in a more eco-friendly and cost-effective era of innovation.

Micro-supercapacitors (MSCs) are a novel type of energy storage device that has garnered a lot of interest due to their many advantages. However, MSCs still fall short due to their limited operating voltage, slow frequency response, and low energy density. As a result, it is still a significant challenge to devise a method to effectively enhance the electrochemical properties of them by innovating upon the design in a number of different ways. For the purpose of fine-tuning the physicochemical properties of 2D materials, surface and structural engineering techniques such as miniaturisation to the quantum dot scale, doping heteroatoms, increasing the number of structural defects, and introducing rich functional groups are used [33]. Both the molybdenum disulfide quantum dots (MoS2-QDs) and nitrogen-doped graphene quantum dots (N-GQDs) produced in this way make for excellent negative and positive electrode materials. The obtained N-GQDs//MoS2-QDs asymmetric MSCs device demonstrates important characteristics, including a high energy density of 0.55 mWh cm3, long-term cycling stability, a high energy density of 1.5 V (far exceeding that of most reported MSCs), an ultrafast frequency response (with a short time constant of 0.087 ms), and ultrafast frequency response (with a short time constant of 0.087 ms). There is a lot of room for improvement in the use of binary metal oxides as electrodes in supercapacitors. $NiCo_2O_4$ nanostructures with a three-dimensional flower shape have been obtained through a cheap and straightforward hydrothermal synthesis and subsequent annealing at various temperatures [34]. Nanostructure structure and electrochemical behaviour as a function of calcination temperature have also been studied. $NiCo_2O_4$ quantum dots (NCO-QDs) with a specific capacitance of 362 F/g at a current density of 0.5 A/g were formed in a highly porous, hollow, and continuous structure at a calcination temperature of 300 °C. These nanostructures may exhibit favourable electrochemical behaviour due to their novel, extremely porous, integrated nanostructures, which serve as an excellent electroactive surface for faradaic redox processes. It has also been demonstrated that an asymmetric-supercapacitor made of NCO-QDs as the positive electrode and reduced graphene oxide (rGO) as the negative electrode can store a significant amount of charge (81 F/g at the scan rate of 5 mV/s) and maintain nearly all of that charge after a thousand cycles. After one thousand charge–discharge cycles, the supercapacitor's energy density and power density were measured at a potential window of 2.5 V. The device's power density was 2.22 kW/ kg at a current density of 1.5 A/ g, and its energy density was 69.5 Wh/kg. Ni_3S_4 quantum dots (QDs) show enormous potential for supercapacitors because of their unique quantum effects, high specific surface area, high water solubility, and superior stability; nevertheless, the current manufacturing technique is time-consuming, hazardous, and laborious. With the help of polyvinylpyrrolidone (PVP) and ethanol, Zheng et al. reported for the first time a simple and eco-friendly synthesis of Ni_3S_4 QDs [35]. XRD analysis of the impact of the synthesis solvent indicated the synergistic process. The specific capacitance at 1 A/g for QDs showed that they were an appropriate electrode material for supercapacitors. Furthermore, a Ni_3S_4 QD//AC asymmetric supercapacitor (ASC) was built, which provided a maximum energy density of 60.4 W hkg. This study introduces novel strategies for preparing QDs and suggests novel approaches to synthesising nickel sulphide. Suzuki et. al. developed tunnel diodes with a few self-assembled InAs quantum dots encased in symmetric and asymmetric AlGaAs barriers to investigate tunnelling current through individual self-assembled quantum dots. Current peaks due to resonant tunnelling through zero-dimensional states in the individual quantum dots have been seen in the current–voltage characteristics of the symmetric barrier diode for both voltage polarities [36]. Doong et al. produced a V_2O_5 nanosheet (VNS) adorned with 0D graphene quantum dots (GQDs), demonstrating long interlayer distances, high stability, and rapid electrochemical kinetics [37]. Flexible, ultrathin, and lightweight next-generation power sources could benefit greatly from the use of free-standing paper-like thin-film electrodes. Paper-like electrodes made from a silver quantum dot (2–5 nm) modified transition metal oxide (including MoO_3 and MnO_2) are created in this work for use in energy storage. Significant enhancements in electrical conductivity, efficient ionic diffusion, and areal capacitances are observed in the hybrid nanostructure electrodes due to the ohmic contact at the interfaces between the silver quantum dots and the MoO_3 nanobelts (or MnO_2 nanowires). The constructed asymmetric supercapacitors with Ag quantum dots/ MoO_3 "paper" as anode and Ag quantum dots/MnO_2 "paper" as cathode and neutral

Na_2SO_4/polyvinyl alcohol hydrogel as electrolyte present significantly enhanced energy and power densities in comparison to those of the supercapacitors without modification of Ag quantum dots on electrodes; present excellent cycling stability at different current densities and good flexibility under cyclic loading conditions [38]. Graphene/carbon nanotube (G/CNT) composites doped with Ni_3S_4 quantum dots (QDs) were synthesised in a single step. Graphene is created by a combination of carbon nanotubes and a Ni catalyst. In addition to preventing Ni_3S_4 QDs from clumping together, G/CNTs can also eliminate the insulating disadvantage brought on by the QDs' interface. Adding G/CNTs to Ni_3S_4 QDs increased the electron interaction, as shown by XPS measurement, and hence facilitated electron transport. In electrochemical tests, G/CNTs were found to boost the rate capability of Ni3S4 QDs from 22 to 82%, and to enhance the specific capacitance of the electrode material by a factor of 4.5 [39]. For use in a high-performance asymmetric supercapacitor device, a three-dimensional (3-D) porous structure based on carbon dot (CD)-supported copper sulphide (CuS)-decorated graphene oxide (GO) hydrogel was developed to increase the specific capacitance, cyclic stability, and energy density of CuS and GO based supercapacitors. Hydrothermal reaction at 180 °C was used to prepare CuS@CD decorated GO hydrogels (CuS@CD-GOH), which were then optimised using a variety of spectroscopic, diffraction, microscopic, and electrochemical techniques. In the 3-D hydrogel structure, CD served as a stabiliser for the CuS nanoparticles and aided in the strong binding of CuS nanoparticles with GO [40]. At a current density of 1 A/g, the CuS@CD-GOH demonstrated a high specific capacitance of 920 F/g. Fabricating an asymmetric supercapacitor with the optimal CuS@CD-GOH as the positive electrode and reduced GO as the negative electrode resulted in the highest energy density (up to 28 W h/kg), longest cycling life (up to 5000 cycles), and highest specific capacitance retention (up to 90%).Increased performance energy storage devices with high areal capacity and outstanding adaptability are in high demand because to the rapid development of portable and wearable electronics. In order to prevent the Ti_3C_2 nanosheets from stacking again, amorphous FeOOH QDs have been created and synthesised into a hybrid film using just electrostatic repulsion. These QDs also contribute significantly to the capacitance of the film itself. The Ti_3C_2/FeOOH QDs hybrid film has excellent cycle stability (94.8% retention over 5000 cycles) in neutral electrolyte and 2.3-times higher areal capacitance (485 mF/cm^2) than the pure Ti_3C_2 film [41]. An asymmetric device, with the hybrid film serving as the negative electrode and MnO_2 on carbon cloth serving as the positive, has been constructed and functions at an output potential difference of 1.6 V. The device maintains 82% of its capacitance after 3000 cycles while providing a maximum energy density of 40 W h/cm^2 and a maximum power density of 8.2 mW/cm^2. Good capacitance and high flexibility are demonstrated by a wet-spun fibre made of $Ti_3C_2T_x$ nanosheets/$Ti_3C_2T_x$ quantum dots/RGO (reduced graphene oxide) (M6M3RG1). Capacitance and pliability of the M6M3RG1 fibre electrode are dose-dependent, and it has a new network topology with a maximum volumetric capacitance of 542 F cm-3. Injecting a homogenous solution of Ti3C2Tx nanosheets and PEDOT:PSS into a bath of 98 wt% H_2SO_4 is how the Ti_3C_2Tx/PEDOT:PSS [poly(3,4-ethylenedioxythiophene):poly(styrenesulfonate)] fibre (M7P3) is made

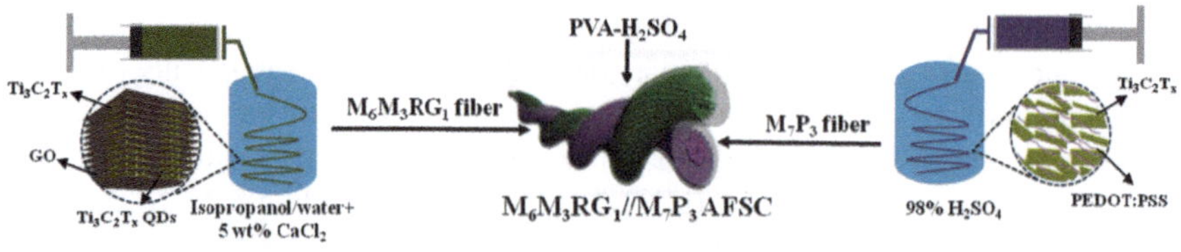

Fig. 2 Preparation schematic diagram of all-solid-state asymmetric fiber supercapacitors. Reproduced with the permission from Ref. [42] © 2020 American Chemical Society

(Fig. 2). A M6M3RG1//M7P3 asymmetric, flexible, solid-state supercapacitor is built in a PVA-H2SO4 gel electrolyte, with the M6M3RG1 fibre serving as the positive electrode and the M7P3 fibre serving as the negative electrode [42].

Next-generation asymmetric supercapacitor devices rely heavily on the advancement of high-performance anode materials. Using a simple two-step technique, a hollow and porous hierarchical anode of bismuth oxide/nitrogen-doped carbon dots (Bi_2O_3/NCDs) was constructed [43]. NCDs are shown to significantly enhance the specific capacitances of Bi_2O_3 porous nanosheets assembled into hollow microspheres, as well as stimulate their assembly. The improved specific capacitance of the Bi_2O_3/NCDs hierarchies is around 1046 F/g at a current density of 1 A/g, which can be attributed to the fascinating features of NCDs and the distinctive structure of Bi_2O_3/NCDs composites. The Bi_2O_3/NCDs hierarchical microspheres were used as the anode and the Ni(OH)2/NCDs nanosheets were used as the cathode in an asymmetric supercapacitor that displays an ultrahigh energy density of about 79.9 Wh/kg at 770.9 W/kg. Xia et al. developed a binder-free supercapacitor cathode electrode by directly growing $MnCO_3$ quantum dots (QDs, 1.2 nm) on $Ni(HCO_3)_2$-$MnCO_3$ hedgehog-like shell@needle composites onto a 3D macro-porous nickel foam [44]. This process was simple and scalable. Maximum specific capacitance of 2641.3 F/g at 3 A/g and 1493.3 F/g at 15 A/g was demonstrated by the $MnCO_3$ QDs/NiH-Mn-CO_3 composite electrode. More so than most reported nickel or manganese oxide/hydroxide-based asymmetric supercapacitors, the $MnCO_3$ QDs/NiH-Mn-CO_3 composites as the positive electrode and graphene as the negative electrode showed an energy density of 58.1 W h/kg at a power density of 900 W/kg and excellent cycling stability with 91.3% retention after 10,000 cycles. Energy storage devices should benefit greatly from using nickel molybdate ($NiMoO_4$), a ternary metal oxide. $NiMoO_4$ electrode has low electrical conductivity, which limits its usefulness [45]. Using a "dipping and drying" technique, silver quantum dots (Ag QDs/$NiMoO_4$) are deposited on the surface of mesoporous $NiMoO_4$ microspheres in yet another study. It's fascinating to see that the AgQDs and $NiMoO_4$ microsphere form an ohmic contact interface (electron anti-barrier layer). The composite electrode exhibits high specific capacitance (3342.7 F/g at 1 mV/s), good rate capability and improved cycling stability thanks to the Ag QDs/$NiMoO_4$ microspheres' high specific surface area, abundant surface active sites, enhanced interfacial conductivity, and porous structure.

2.3 QDs in Hybrid Ion Capacitors (HIC)

In the exciting new field of Hybrid Ion Capacitors (HIC), which combines the best features of batteries and supercapacitors, QDs play a pivotal role? Energy density, charge/discharge speed, and cycle life are all areas where HICs excel above more traditional forms of energy storage. There is hope for improving HIC efficiency by including QDs into design. Quantum dots are semiconductor particles on the nanoscale scale that exhibit unusual electrical and optical characteristics. HICs benefit from increased ion storage capacity and streamlined charge transfer when QDs are incorporated into the electrode materials. This is because QDs have a high surface area and can have their electrical characteristics tuned. Because of this, HICs can store more energy and quickly release it, making them perfect for applications that need power bursts. In addition, QDs can be made to display tailored energy levels, allowing for granular regulation of charge storage methods. By tackling the primary obstacles associated with energy storage, such as short cycle life and sluggish charge/discharge rates, this custom design improves the overall efficiency and durability of HICs. One promising path towards the creation of high-performance energy storage devices is the use of quantum dots in hybrid ion capacitors. By providing solutions that are efficient, long-lasting, and high-energy–density, their integration has the potential to revolutionise many different applications, such as portable devices, electric cars, and grid-level energy storage.

Wu et al. manufacture an innovative SC architecture (ASC) in the shape of an accordion, complete with a hierarchical porous framework [46]. Pseudocapacitive K^+ storage is considerably enhanced by the ASC's highly exposed structure following electro-deposition of extremely electrochemically active nitrogen-doped graphene quantum dots (N-GQDs). Sodium ion hybrid capacitors (SIHCs) are severely limited in their performance due to the kinetics incompatibility between their battery-type anode and capacitive-type cathode. In a first of its kind study, researchers created a SIHCs device by connecting freestanding anode quantum grade vanadium nitride (VN) nanodots (Fig. 3) anchored in one-dimensional N/F co-doped carbon nanofiber cages (VNQDs@PCNFs-N/F) with corresponding activated N/F co-doped carbon nanofiber cages (APCNFs-N/F) as cathode [47]. To overcome this kinetics mismatch, quick sodium storage can be ensured by combining VN quantum dots with N/F co-doped 1D conductive carbon cages, which strongly couples the two materials together. Density functional theory simulations further show that the high concentration of active sites in the VNQDs@PCNFs-N/F configuration is beneficial for enhancing Na + adsorption/reaction activity, which in turn boosts "intrinsic" and "extrinsic" pseudocapacitance and the kinetics of the anode.

The sodium ion hybrid supercapacitor (Na-HSC) is gaining popularity as a form of energy storage because to its high power density and long cycle life. The advancement of this technology relies on the use of anode materials with a high rate of Na^+ embedding. In this study, we used yeast enriched with Nb^{5+} and a straightforward hydrothermal-carbonization method to create multi-chamber microspheres with Nb_2O_5 quantum dots (T-Nb_2O_5@YC). The electrochemical performance of

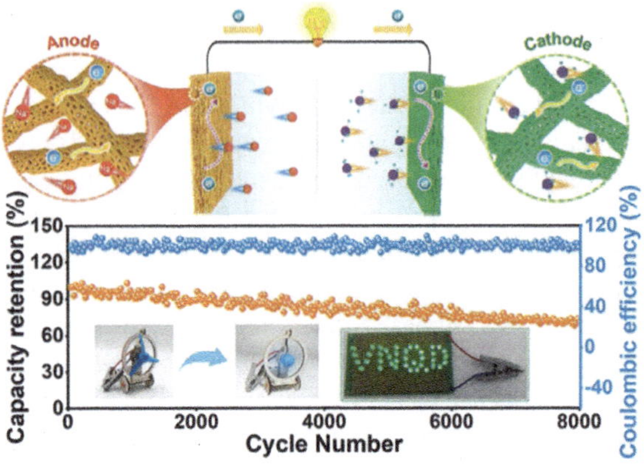

Fig. 3 Schematic illustration of the assembled VNQDs@PCNFs-N/F||APCNFs-N/F device. long-term cycling stability at 1 A g–1 for the VNQDs@PCNFs-N/F||APCNFs-N/F device. Inset: digital pictures of the LEDs and wind trolley powered by our device. Reproduced with the permission from Ref. [47] © 2022 American Chemical Society

Na+ half-cells and Na-HSCs is excellent [48]. Using an easy thermal synthesis process and some old cola soda, a multi-gram quantity of water-soluble, fluorescent carbon nanodots (C-Dots) were synthesised and used as a conductive and ion-donating component in a supercapacitor. The quantum yield of these C-Dots varies from 4.6 to 18.3%. The size, crystal structure, morphology, and surface properties of C-Dots were examined and compared using fluorescence and electrical properties, as were their Raman and electron paramagnetic resonance spectra. Nanorods of polypyrrole (PPy) doped with manganese dioxide (MnO) and adorned with carbon dots (C-Dots) were used as anode material in a supercapacitor experiment. Cathode materials comprised of reduced graphene oxide were doped with a dicationic bis-imidazolium-based ionic liquid to improve charge transfer and surface wetting. Octyl-bis(3-methylimidazolium) diiodide (C8H16BImI), synthesised by N-alkylation process, was used as the electrolyte in the liquid ionic membrane [49]. Research using paramagnetism and impedance spectroscopy has been conducted to trace the hybrid capacitor's performance back to its roots. In particular, the quality of the synthesis, as well as the selection of electrode and electrolyte materials, all contributed to the extraordinarily high capacitance value ($C = 17.3\ F/cm^2$) we observed. Li-ion capacitors (LICs) with battery-type anodes and capacitive cathodes could be the next generation of electrochemical energy storage (EES) devices. This is because they have a high energy density (ED) and a high power density (PD), which is mainly determined by the chemical and physical properties of their anodes and cathodes. We used a one-step solvothermal method to coat $Ti_3C_2T_x$ MXene sheets with SnO_2 quantum dots (QDs, about 4–5 nm) that have oxygen gaps on their surface [50]. The sheets had different amounts of SnO_2 quantum dots. The improved SnO_2 QDs/$Ti_3C_2T_{x\text{-}50}$ composite anode has a high reversible discharge capacity (1000.4 mAh/g@0.2 A/g), excellent cycling stability (the discharge capacity of 1004 mAh/g at the end of 300 cycles at the same current density), and good rate performance. The reaction kinetics is dominated by the fast capacitive process to a large extent, because SnO_2 QDs has fewer Li^+ ions and oxygen vacancies.

3 Conclusions and Summary

It is clear that QDs-based electrode materials have enormous potential in the design of electrode materials in supercapacitors due to their unique quantum size, ease of functionalization on the surface, and abundance of functional groups. In modern supercapacitors, QDs are utilised primarily to alter the electrode materials through the introduction of supplementary chemicals or QDs-derived carbon materials created through the high-temperature treatment of QDs precursors. Keep in mind that the QDs' low electron conductivity and limited BET area mean that they won't provide good electrochemical performance if utilised directly as electrode materials.

QDs have been successfully integrated into supercapacitors, suggesting a promising future for these energy storage devices. Supercapacitors can be made more effective by taking use of QDs' high surface area, customizable electronic characteristics, and efficient charge transfer. Improved energy density, faster charge/discharge rates, and greater long-term stability are the primary benefits of employing QDs in supercapacitors. Several promising future directions for QDs in supercapacitors are emerging. Developing new QD materials and unique electrode designs are two areas where researchers are currently looking for methods to improve QD-based supercapacitors. This has the potential to lead to even higher energy densities and longer cycle life, expanding the applicability of QD-enhanced supercapacitors to uses beyond just renewable energy storage and portable devices, such as electric vehicles. Making these cutting-edge supercapacitors more easily available and environmentally benign through the use of sustainable and inexpensive QD materials is also a continuing priority. Scaling up production processes and improving manufacturing methods are also crucial for bringing QD-based supercapacitors to market. To sum up, Quantum Dots may herald a new era in energy storage technologies and could cause a revolution in the supercapacitor business. We can expect more sustainable and energy-efficient supercapacitors to play a role in the future if research and development efforts are maintained.

References

1. Balaji, T.E., Tanaya Das, H., Maiyalagan, T.: Recent trends in bimetallic oxides and their composites as electrode materials for supercapacitor applications. ChemElectroChem **8**(10), 1723–1746 (2021)
2. Uddin, M.S., et al.: Influence of designed electrode surfaces on double layer capacitance in aqueous electrolyte: Insights from standard models. Appl. Surf. Sci. **449**, 445–453 (2018)
3. Das, H.T., et al.: Performance of solid-state hybrid energy-storage device using reduced graphene-oxide anchored sol-gel derived Ni/NiO nanocomposite. Sci. Rep. **7**(1), 15342 (2017)
4. Miller, J.R., Simon, P.: Electrochemical capacitors for energy management. Science **321**(5889), 651–652 (2008)
5. Wang, G., Zhang, L., Zhang, J.: A review of electrode materials for electrochemical supercapacitors. Chem. Soc. Rev. **41**(2), 797–828 (2012)
6. Wei, W., et al.: Manganese oxide-based materials as electrochemical supercapacitor electrodes. Chem. Soc. Rev. **40**(3), 1697–1721 (2011)

7. Huang, Y., Liang, J., Chen, Y.: An overview of the applications of graphene-based materials in supercapacitors. Small **8**(12), 1805–1834 (2012)
8. Park, K.-W.: Carboxylated graphene oxide–Mn 2 O 3 nanorod composites for their electrochemical characteristics. J. Mater. Chem. A **2**(12), 4292–4298 (2014)
9. Wei, B., et al.: Fabrication of manganese oxide/three-dimensional reduced graphene oxide composites as the supercapacitors by a reverse microemulsion method. Carbon **85**, 249–260 (2015)
10. Das, P., et al.: Immobilization of heteroatom-doped carbon dots onto nonpolar plastics for antifogging, antioxidant, and food monitoring applications. Langmuir **37**(11), 3508–3520 (2021)
11. Ganguly, S., et al.: Advancement in science and technology of carbon dot-polymer hybrid composites: a review. Funct. Compos. Struct. **1**(2), 022001 (2019)
12. Das, P., et al.: Heteroatom doped photoluminescent carbon dots for sensitive detection of acetone in human fluids. Sens. Actuators, B Chem. **266**, 583–593 (2018)
13. Ganguly, S., et al.: Microwave-synthesized polysaccharide-derived carbon dots as therapeutic cargoes and toughening agents for elastomeric gels. ACS Appl. Mater. Interfaces **12**(46), 51940–51951 (2020)
14. Saravanan, A., et al.: Applications of N-doped carbon dots as antimicrobial agents, antibiotic carriers, and selective fluorescent probes for nitro explosives. ACS Appl. Bio Mater. **3**(11), 8023–8031 (2020)
15. Das, P., et al.: Graphene based emergent nanolights: a short review on the synthesis, properties and application. Res. Chem. Intermed.Intermed. **45**, 3823–3853 (2019)
16. Das, P., et al.: Biocompatible carbon dots derived from κ-carrageenan and phenyl boronic acid for dual modality sensing platform of sugar and its anti-diabetic drug release behavior. Int. J. Biol. Macromol.Macromol. **132**, 316–329 (2019)
17. Das, P., et al.: Carbon dots for heavy-metal sensing, pH-sensitive cargo delivery, and antibacterial applications. ACS Appl. Nano Mater. **3**(12), 11777–11790 (2020)
18. Das, P., et al.: MXene/0D nanocomposite architectures: Design, properties and emerging applications. Mater. Today Nano **24**, 100428 (2023)
19. Das, P., et al.: Waste-derived sustainable fluorescent nanocarbon-coated breathable functional fabric for antioxidant and antimicrobial applications. ACS App. Mater. Interfaces (2023)
20. Rajabzadeh, A.R., et al.: Magnetic Quantum Dots for Bioimaging. CRC Press (2023)
21. Das, P., et al.: Carbon-dots-initiated photopolymerization: an in situ synthetic approach for MXene/poly (norepinephrine)/copper hybrid and its application for mitigating water pollution. ACS Appl. Mater. Interfaces **13**(26), 31038–31050 (2021)
22. Das, P., et al.: Tailor made magnetic nanolights: fabrication to cancer theranostics applications. Nanoscale Adv. **3**(24), 6762–6796 (2021)
23. Das, P., et al.: Carbon dot biopolymer-based flexible functional films for antioxidant and food monitoring applications. ACS Appl. Polym. Mater. **4**(12), 9323–9340 (2022)
24. Zhu, Y., et al.: A carbon quantum dot decorated RuO 2 network: outstanding supercapacitances under ultrafast charge and discharge. Energy Environ. Sci. **6**(12), 3665–3675 (2013)
25. Sinha, R., Roy, N., Mandal, T.K.: SWCNT/ZnO nanocomposite decorated with carbon dots for photoresponsive supercapacitor applications. Chem. Eng. J. **431**, 133915 (2022)
26. Yang, I., et al.: Improvement of a commercial activated carbon for organic electric double-layer capacitors using a consecutive doping method. Carbon **160**, 45–53 (2020)
27. Zhang, X., et al.: Design and preparation of a ternary composite of graphene oxide/carbon dots/polypyrrole for supercapacitor application: Importance and unique role of carbon dots. Carbon **115**, 134–146 (2017)
28. Ji, X., et al.: Energetic variational approach for prediction of thermal electrokinetics in charging and discharging processes of electrical double layer capacitors. J. Power. Sources **551**, 232184 (2022)
29. Yoo, H.D., et al.: Impedance analysis of porous carbon electrodes to predict rate capability of electric double-layer capacitors. J. Power. Sources **267**, 411–420 (2014)

30. Xu, D., et al.: Interfacial engineering of 2D/2D MXene heterostructures: face-to-face contact for augmented photodegradation of amoxicillin. Chem. Eng. J. **426**, 131246 (2021)
31. Deng, J.Z., Zhang, Y.: Jingmin Cheng, Mengxin Zhao, Jing Cong, Qi Qi, Ye Xiao, Wei Cong, Ye
32. Lima, R.M.A.P., de Oliveira, H.P.: Carbon dots reinforced polypyrrole/graphene nanoplatelets on flexible eggshell membranes as electrodes of all-solid flexible supercapacitors. J. Energy Storage **28**, 101284 (2020)
33. Liu, W., et al.: Advanced electrode materials comprising of structure-engineered quantum dots for high-performance asymmetric micro-supercapacitors. Adv. Energy Mater. **10**(8), 1903724 (2020)
34. Siwatch, P., Sharma, K., Tripathi, S.: Facile synthesis of NiCo2O4 quantum dots for asymmetric supercapacitor. Electrochim. Acta. Acta **329**, 135084 (2020)
35. Zheng, J., et al.: Synergy of PVP and ethanol to synthesize Ni 3 S 4 quantum dots for high-performance asymmetric supercapacitors. Mater. Chem. Front. **4**(6), 1764–1772 (2020)
36. Suzuki, T., et al.: Tunneling current through self-assembled InAs quantum dots embedded in symmetric and asymmetric AlGaAs barriers. Solid-State Electron. **42**(7–8), 1303–1307 (1998)
37. Ganganboina, A.B., Park, E.Y., Doong, R.-A.: Boosting the energy storage performance of V 2 O 5 nanosheets by intercalating conductive graphene quantum dots. Nanoscale **12**(32), 16944–16955 (2020)
38. Zhang, X., et al.: Silver-quantum-dot-modified MoO3 and MnO2 paper-like freestanding films for flexible solid-state asymmetric supercapacitors. Small **15**(13), 1805235 (2019)
39. Zheng, J., et al.: One-step preparation of Ni3S4 quantum dots composite graphene/carbon nanotube conductive network for asymmetric supercapacitor. J. Alloy. Compd. **859**, 158247 (2021)
40. De, B., et al.: Carbon dot stabilized copper sulphide nanoparticles decorated graphene oxide hydrogel for high performance asymmetric supercapacitor. Carbon **122**, 247–257 (2017)
41. Zhao, K., et al.: Free-standing MXene film modified by amorphous FeOOH quantum dots for high-performance asymmetric supercapacitor. Electrochim. Acta. Acta **308**, 1–8 (2019)
42. Zhou, X., et al.: Ti3C2T x nanosheets/Ti3C2T x quantum dots/RGO (Reduced Graphene Oxide) fibers for an all-solid-state asymmetric supercapacitor with high volume energy density and good flexibility. ACS Appl. Mater. Interfaces **12**(10), 11833–11842 (2020)
43. Ji, Z., et al.: Bismuth oxide/nitrogen-doped carbon dots hollow and porous hierarchitectures for high-performance asymmetric supercapacitors. Adv. Powder Technol. **31**(2), 632–638 (2020)
44. Xia, Q.X., et al.: Facile synthesis of manganese carbonate quantum dots/Ni (HCO 3) 2–MnCO 3 composites as advanced cathode materials for high energy density asymmetric supercapacitors. J. Mater. Chem. A **3**(44), 22102–22117 (2015)
45. Zhang, X., et al.: Mesoporous NiMoO4 microspheres decorated by Ag quantum dots as cathode material for asymmetric supercapacitors: enhanced interfacial conductivity and capacitive storage. Appl. Surf. Sci. **505**, 144513 (2020)
46. Zhang, C., et al.: Nitrogen-doped accordion-like soft carbon anodes with exposed hierarchical pores for advanced potassium-ion hybrid capacitors. Adv. Func. Mater.Func. Mater. **31**(23), 2101470 (2021)
47. Yuan, J., et al.: Pseudocapacitive vanadium nitride quantum dots modified one-dimensional carbon cages enable highly kinetics-compatible sodium ion capacitors. ACS Nano **16**(9), 14807–14818 (2022)
48. Lian, Y., et al.: Nb2O5 quantum dots confined in multi-chamber yeast carbon for sodium ion hybrid capacitors. J. Alloy. Compd. **896**, 163128 (2022)
49. Genc, R., et al.: High-capacitance hybrid supercapacitor based on multi-colored fluorescent carbon-dots. Sci. Rep. **7**(1), 11222 (2017)
50. Wang, Y.-L., et al.: Ti3C2Tx MXene supported SnO2 quantum dots with oxygen vacancies as anode for Li-ion capacitors. Chem. Eng. J. **428**, 131993 (2022)

Quantum Dots Composites for Energy Storage Applications

Abdelrahman M. Ishmael, Ahmed I. Abdel-Salam, Mostafa M. H. Khalil, and Ahmed S. Mansour

Abstract The demand for developing energy conversion and storage systems is increasingly growing due to the escalating environmental problems effects caused by global warming and fossil fuels depletion. The rapid expansion of renewable energy sources such as solar energy, wind energy, biofuels, hydropower, geothermal, and hydrogen energy is expected to be critical to alleviating the energy crisis. Quantum dots (QDs) are nanocrystalline semiconducting materials with a few nanometers in diameter, ranging from 1 to 10 nm. Their electrical and optical characteristics can be manipulated by altering the size up to 10 nm, which is attributed to the quantum confinement effect. However, the size of the quantum dots depends on the semiconducting material because the exciton Bohr radius varies depending on the type of semiconductor. Many literature articles have highlighted the application of quantum dots in supercapacitors (SCs), metal-ion batteries (MIBs), and photocatalytic and electrocatalytic hydrogen evolution. This chapter is dedicated to presenting recent progress regarding the potential application of quantum dots and related composite materials in various energy storage systems.

Keywords Quantum dots · Quantum confinement effect · Li-ion batteries · Li-Metal batteries · Supercapacitors · Hydrogen storage

---

A. M. Ishmael
Nanomaterials Science Program, Faculty of Science, Benha University, Benha 13815, Egypt
e-mail: Abdelrahman199458@fsc.bu.edu.eg

A. I. Abdel-Salam
Nanotechnology Research Centre (NTRC), The British University in Egypt (BUE), El-Shorouk City 11837, Egypt
e-mail: Ahmed.mysara@bue.edu.eg

M. M. H. Khalil
Chemistry Department, Faculty of Science, Ain Shams University, Cairo 11566, Abbassia, Egypt
e-mail: Mostafa_khalil@sci.asu.edu.eg

A. S. Mansour (✉)
National Institute of Laser Enhanced Sciences, Giza, Egypt
e-mail: Ahmed_chemi2012@yahoo.com

Faculty of Nanotechnology for Postgraduates Studies, Giza, Egypt

S. Thomas et al. (eds.), *Quantum Dots Based Nanocomposites*, Engineering Materials,
https://doi.org/10.1007/978-3-031-54779-9_22

Abbreviations

QDs	Quantum dots
SCs	Supercapacitors
MIBs	Metal-ion batteries
0D	Zero-dimensional
SCQDs	Semiconductor QDs
GO	Graphene oxide
CNTs	Carbon nanotubes
CDs-QDs	Cadmium sulfide QDs
CQDs	Carbon quantum dots
LIBs	Lithium-ion batteries
SEI	Solid electrolyte interface
SIBs	Sodium ion batteries
GQDs	Graphene quantum dots
GO	Graphene oxide
SSA	Specific surface area
rGO	Reduced graphene oxide
SWCNTs	Single walled carbon
MNC	Multilayered N-doped nano carbon sheets
BP	Black phosphorous
RP	Red phosphorus
MOFs	Metal organic frameworks
HCP	Hollow carbon polyhedral matrix
PIBs	Potassium-ion batteries
LMBs	Li-metal batteries
LVCF	Lithophilic vertical cactus-like framework
PEO	Polyethylene oxide
LSBs	Lithium-sulfur batteries
CTAB	Cetyltrimethylammonium bromide
PEI	Polyethyleneimine functionalized
CB	Carbon black
OER	Oxygen evolution reaction
ORR	Oxygen reduction reaction
ZIBs	Zinc-ion batteries
ZMO	$ZnMn_2O_4$
EDLC	Electric double-layer capacitors
PCN	Porous carbon nanosheet
ASCs	Asymmetric super capacitors
AC	Activated carbon
ZIF-8	Zeolitic imidazolate framework
FGS	Functionalized graphene nanosheets
HICs	Hybrid super capacitors
MSCs	Micro super capacitors

NF Ni foam

1 Introduction

Decarbonized green energy, including solar, wind, hydro, and geothermal energy, is the solution to the global warming issue, the crisis of energy, and harmful emissions to the environment [1]. In terms of carbon dioxide neutrality, new sources of energy will take over as the main source of electricity, and storing massive amounts of renewable energy will be an immense challenge [2]. According to their intermittent nature, these energy sources should be stored and released when necessary [3]. Electrochemical energy storage represents a promising solution to such a worldwide challenge. Supercapacitors and batteries are suitable examples for these types of storage technologies [4]. Furthermore, encouraging the creation of electrochemical hydrides and photochemical bonding to produce fuel is the foundation of photocatalytic hydrogen evolution, another "clean" method of energy storage [5].

Quantum dots have attracted a lot of research interest recently due to their potential electrochemical energy storage because of their huge specific surface area, controllable size, far-short ion/electron transport channel, non-toxicity, cost-effectiveness, adaptable photoluminescence, and simple functionalization of the surface [6]. Environmental problems are further exacerbated by the usage of fossil fuels and global warming. The field of quantum dots (QDs) is rapidly advancing in terms of energy conversion and storage. Less than 10 nm in diameter, sphere- or quasi-sphere-shaped materials make up the majority of QDs [7]. Synthesis methods of QDs are typically divided into two main categories: "top-down" and "bottom-up" techniques [8]. The top-down approach often begins with a massive amount of materials and progresses to QDs by a variety of techniques, including chemical etching, microwave-assisted irradiation, ultrasonication processing, hydrothermal/solvothermal technique, electrochemical method, and some others [9]. A bottom-up synthesis method can also be used to chemically develop QDs from tiny organic and inorganic precursors [10]. Numerous methods have been used to create QDs, which are zero-dimension materials with relatively large bandgap, extremely small size, and high surface to volume ratio [11]. These materials exhibit unusual physicochemical properties because of their strong quantum confinement and edge effect. During the synthesis of QDs, their structure, size, and heteroatomic functional groups can simply be controlled. As a result, they have more active sites per unit mass, are more physically and chemically tunable, and are more able to hybridize with other nanomaterials [12].

An electrochemical energy storage system's overall efficiency is greatly influenced by the electrode materials used [13]. Heteroatomic functional groups abound on the surface of QDs, providing a wide range of active sites. Considering their superior conductivity of ions, rapid rate of charge–discharge, high capacity, cycle stability, and other characteristics, they can be used in composites for current collectors [14] and active material electrodes [15]. This would significantly improve the quality of electrochemical energy storage systems. Because of their distinct features as electron donors and acceptors as well as their excellent electron transport properties, QDs are an extremely stable and effective photocatalyst for processes involving the generation and storage of hydrogen and the reduction of oxygen [16].

2 Types of QDs for Energy Storage Applications

QDs are zero-dimensional(0D) materials with less than 10 nm in diameter. They are spherical or quasi-spherical morphologies that have attracted a lot of researchers' interest in the context of energy storage and conversion. They can be synthesized using a variety of preparation techniques, such as the chemical method, by assembling the tiny organic and inorganic precursors [17]. These techniques aim to tune physicochemical properties of the materials by adjusting the bandgap that emerges from the strong quantum confinement and controlling the material dimensions to be 0D with ultra-small sizes. In addition to short ion/electron transfer paths, high dispersibility, and variable surface functionalities [18]. The QDs are classified into various types have been using in the energy storage applications as follow.

2.1 Semiconductor QDs (SCQDS)

The remarkable optical characteristics of SCQDs such as their strong photo bleaching resistance and narrow emission band with broad excitation, have recently attracted a lot of attention [19]. Additionally, they have been used in various applications such as biomedical sensors [20], biological cell templates [21], biomolecular imaging [22], bioconjugation [23], solar cells [24], light-emitting diodes [25] and catalysis [26]. Another area where QDs have received little attention is electrochemical energy storage [27]. This is mainly due to the difficulties in producing large amount of stable and monodispersed QDs [28]. To overcome these limitations, specific chemical procedures are used, such as digestive ripening, to create monodisperse QDs from polydisperse particles [29]. Therefore, QD-based materials have the capability to be utilized in electrochemical energy storage applications as a result of their distinctive properties, including their high surface area, electroactive sites, exceptional electrical conductivity, and short ion diffusion length [30]. Additionally, because of their very comparable surface chemistries, their bulk behavior is probably more predictable, which has the benefits of high yield and manufacturability.

The charge storage properties of QDs in terms of capacitance and cycle stability are improved by integrating them with carbonaceous materials like graphene oxide (GO) and carbon nanotubes (CNTs), which results in uniform distribution of the QDs on the surfaces of the GO and CNTs and prevents agglomeration of the QDs, as well as increasing the conductivity of the nanocomposite [31]. Few research studies have been performed on the SCQDs for energy storage applications such as supporting the cadmium sulfide QDs (CdS-QDs) onto the surface of carbon nanotubes to be involved in lithium-sulfur batteries as the sulfur cathode [5], using CdSe/ZnS QDs@CNTs composite for rechargeable photo-electrochemical Li-O_2 batteries [32].

2.2 Metal Oxide QDs

Metal oxide nanomaterials, particularly transition metal oxide QDs, are considered as a prominent popular material for lithium-ion batteries (LIBs) and SCs because of their acceptable cost, natural availability, non-toxicity, and excellent safety. Due to their great density, they can attain high volumetric performance [33]. Furthermore, their specific capacity is usually two to three times higher when compared to their carbon-based counterparts [34]. Nevertheless, they demonstrate reduced overall performance and cycle stability values due to their low electrical conductivity, which restricts their utility in energy storage, so a lot of research studies have been performed to enhance their electrochemical characteristics such as electrical conductivity, cycling stability, and rate performance by decreasing their sizes to the nanoscale and simultaneously integrating with carbonaceous materials [35].

The carbonaceous materials, such as nitrogen-doped graphene, CNT, and CQDs, could prevent the agglomeration of the metal oxide nanoparticles [36], improve the attachment of the metal oxides such as TiO_2, Fe_3O_4, CoO, CuO, and SnO_2 on the carbon surfaces, as well as, increase the electroactive sites [37]. This results in fast charge transfer, good wettability at the solid electrolyte interface (SEI), and high conductivity of electricity. All these outstanding properties give high potential for the metal oxides to be effective electrodes for LIBs or sodium-ion batteries (SIBs) [38].

2.3 Graphene Quantum Dots (GQDs)

GQDs are a relatively recent finding within the allotropic carbonaceous family, which also includes graphite, diamond, nanotubes, fullerene, and graphene. They combine a number of advantageous properties of conventional SCQDs without the drawbacks of elemental shortage or inherent toxicity, as well as requiring labor-intensive, expensive, or ineffective preparation procedures [39]. GQDs are functionalized nanoscale pieces of graphene, typically less than 10 nm in size. Their lateral dimension being greater than their height is the source of their anisotropic morphology.

GQDs can be prepared through two synthetic approaches, each of which relies on either top-down or bottom-up synthetic processes. In brief, top-down synthesis method involves degradation of large carbon source such as carbon fibres, GO, coal, fullerenes, graphite into smaller fragments using concentrated [40]. Because graphene oxide (GO) has many functional groups that contain oxygen atoms which promote chemical cleavage towards nanosized GQDs, GO is often the best starting material. However, unlike graphite, GO is not a naturally occurring substance; rather, it must first be produced by Hummer using a complicated chemical process using a range of materials, including anthracite or coal. The use of graphite as a natural source has been investigated as a potential alternative to GO but has so far had less success due to lower synthetic yields. The common methods for preparing the GQDs are microwave-assisted, hydrothermal or solvothermal, and electrochemical methods. Contrarily, bottom-up approaches based on organic precursors carbonization like citric acid, glucose, and others have poorer yields and difficulty purifying from the unreacted tiny organic components [38].

Despite numerous attempts, conducting studies on the fabrication of GQDs-based various nanomaterials to expand their application in the energy-related domains is still in its early phases. There has been a persistent drive among researchers in this context to direct their attention to the innovative and reasonable fabrication of GQDs-based nanomaterials connected to electrochemical energy storage and electrocatalysis. Consequently, a rapid expansion of new active materials based-GQDs research and its possible applications in the energy sector is anticipated in the near future. As a cutting-edge material for electrodes in electrochemical energy storage devices like batteries and supercapacitors, GQDs have been investigated. For improving the physicochemical properties of the GQDs, they could be integrated with graphene nanosheets to combine the advantages of both materials and give them a significant promise in a variety of catalytic and energy storage applications [41].

2.4 Carbon QDs

Carbon dots attracted great attention as a new derivative member of the carbon family with diameters not larger than 10 nm [42]. They have sp^2 graphitic core structure with oxidized functionalities. Since carbon dots have semiconducting characteristics, they can substitute SCQDs in optoelectronic devices [43]. Carbon dots have great potential for a variety of real-world applications, including photocatalysis, solar cells, bio-imaging, and bio-sensing, because of their exceptional characteristics including huge luminescence, high solubility, biocompatibility, and excellent chemical stability [44]. Additionally, carbon dots have a potential future as next-generation energy storage materials. Since 2013, the number of studies on carbon dots-based electrode materials has increased dramatically.

Early on, a few innovative research teams made some promising advancements by utilizing carbon dots to create electrode active materials for different energy storage systems like MIBs [45], metal-S batteries [46] and supercapacitors [13]. Carbon dots have revealed advantages when used as active electrode materials since they have been improving some characteristics in MIBs, supercapacitors, metal-air batteries such as coulombic efficiency, long cycling life, functional groups, and interfacial wettability at the SEI [26]. Furthermore, integrating the carbon dots in nanocomposites to be involved as electrode active material in SCs results in enhanced electron transport and ionic motion at the SEI, fast faradic reactions, and high specific capacity [41].

3 QDs Based Nanocomposites for Energy Storage Applications

Electrochemical devices to store energy including batteries and supercapacitors serve as the primary solution for addressing the challenge of intermittent energy sources found in nature, so energy can be stored and released later. Recently, QDs and their composites have been widely employed in the field of electrochemical energy storage because of their distinctive characteristics, including a significant specific surface area (SSA), tunable size ranging from 2–10 nm, low cost, non-toxicity, short ion diffusion distance, facile surface functionalization [6]. QDs are synthesized via several methods to become zero-dimensional materials showing unique physicochemical properties based on the strong quantum confinement effect resulting in a relatively large band gap and a considerable surface area-to-volume ratio [35]. The heteroatomic parameters that can be regulated during their synthesis enhance the surface of QDs, allowing for a diverse array of active sites, the ability to tune their physicochemical properties, and the potential for hybridization with other nanomaterials, which greatly provides more flexibility and adaptability during their usage [47]. QDs based composites can be used as a current collector or electrode active materials [48], obtaining superior conductivity of ions, rapid rate of charge–discharge, short ion diffusion length, high capacity, cycle stability, and other characteristics. Another efficient approach to storing energy is through photo-electrocatalytic hydrogen generation that involves forming electrochemical hydrides and bonding them photochemically to create fuel [4].

QDs show special effects during charge–discharge processes such as the low charge transfer resistance, short ion diffusion distance, superior surface characteristics, quantum confinement effect, and QDs structure against volume change [48]. Each of which has received great attention in the field of energy storage. QDs with smaller sizes demonstrate reduced charge transfer resistance and excellent electrical conductivity, which leads to a relatively high capacity of batteries. Wang et al. fabricated ZnS QDs/N-doped reduced graphene oxide (rGO) composite for Li-CO_2 battery which has an extremely low charge transfer resistance when compared to other cathodes without QDs [49]. The ultrasmall size of QDs can efficiently shorten the ion diffusion length, thereby facilitating rapid charge transfer during electrochemical reactions. Cheng et al. reported N-doped carbon decorated with SnO_2 QDs which in turn shortened the ion diffusion distance with superior performance in LIBs and SIBs [50]. The noticeable factor of QDs is the enhanced active surface area and hence, high percentage of edge atoms that improves the supercapacitor's ability to store charges and boosts ion adsorption or desorption activity. Bonu el al. investigated SnO_2 QDs with a substantial surface area-to-volume ratio as a supercapacitor active material.

The constructed supercapacitor possessed good activity, high conductivity with reinforced specific capacity, and no volume variation during the electrochemical process in comparison to other nanoparticles [51]. The quantum confinement phenomenon exhibited by QDs can enhance their electrical conductivity, thereby contributing to the improvement in the performance of energy storage systems. Zhou et al. synthesized Co_3O_4 QDs/graphene composite via a facile microwave irradiation process which was used as anode material for LIBs and possessed clearly improved cycling performance, and high-efficient charge storage kinetics [52]. Liu et al. fabricated hybrids of graphene-Fe_3O_4 QDs as anode materials in SIBs. The hierarchical structure of the obtained G-Fe_3O_4 QDs composite showed superior storage capacity of sodium ions and considerably enhanced its cycling stability [53]. Different types of reported QDs used in different energy storage devices including rechargeable batteries, supercapacitors, and storage of hydrogen will be illustrated in the subsections below in more detail.

3.1 QDs for Rechargeable Batteries

QDs are employed as electrode materials in rechargeable batteries for their beneficial properties towards enhancing capacity, rate capability, conductivity, and cycling stability [54]. Herein, we discuss different QDs based nanocomposites where one or more of the composite components are in the form of QDs for diverse types of batteries.

3.1.1 QDs in Metal Ion Batteries (MIBs)

MIBs storage entails a charge–discharge process commonly referred to as a "rocking chair" mechanism, where the alkali-metal ions (such as Li, Na, and K-ion) move from the cathode to the anode during charging and return during discharging. MIBs have an edge over conventional batteries in terms of extended lifespan, greater energy density, with no memory effect. Nevertheless, MIBs encounter issues related to unstable performance and low power density. Several studies have investigated the utility of QDs-based composites as anode materials in MIBs that can significantly improve the electrochemical capabilities [55].

Lithium-Ion Batteries

Several metal oxide QDs based nanocomposites are used as anode materials for LIBs. ZnO possesses higher electrical conductivity, theoretical capacity of about 978 mA h/g, more abundant than other metal oxide anodes which make it suitable as a LIB anode [56], however, ZnO is rather scarce because of its poor cycle stability. The process by which ZnO stores charge involves a process of conversion with Li^+, before alloying to generate LiZn, consequently resulting in a 228% volume change, which in turn causes the anode to pulverize and a rapid fading of capacity. Thus, ZnO anode needs to be controlled below a critical size with enhanced flexibility to shield the volume change [57]. Sun et al. reported ZnO QDs/Graphene sheets with tunable sizes of 2, 4, and 7 nm. The tension resulting from the change in volume of ZnO QDs is limited by the great mechanical stability of graphene. ZnO QDs are strongly anchored on graphene, which keeps them from clumping together during cycling and creates an extremely stable anode. The optimum electrochemical activity of ZnO QDs is represented by confining their size to about 2 nm with a specific capacity of 560 mA h/g at100 mA/g current [56]. Tan et al. prepared ZnO/rGO composite electrode which achieved outstanding performance in anodes of LIBs [58]. ZnO QDs contained by amorphous carbon have some advantages over crystalline materials, primarily because of the isotropic insertion of Li^+ ions therein, which is more susceptible to structural strain in crystalline materials. Ma et al. implanted ZnO QDs in a matrix of N-doped porous carbon (ZnO@NPC) on a single walled carbon nanotubes (SWCNTs) film. The obtained spongy film of ZnO@NPC/SWCNTs was used as a self-standing and binder-free anode which exhibited a highly stable cycling stability [59]. ZnS QDs were uniformly distributed in multilayered N-doped nano carbon sheets (ZnS QDs@mNC) in a tailored architecture created by Fang et al. This allowed for ZnS QDs expansion in volume during cycling with just a 6.5% total volume expansion [60]. Zhang et al. used an in-situ ion exchange technique to grow ZnO@ZnO QDs/C nanorod clusters on conducting carbon cloth. Because of the synthetic composite nanorod arrays' better structure stability, good electron accumulation efficiency, and reduced Li-ion diffusion length, they demonstrated excellent rate capability and cycle stability [61].

TiO_2 has been investigated for anodes of LIBs. TiO_2 shows poor electronic conductivity that restricts the mobility of Li-ion within TiO_2 crystal. Thus downsizing the TiO_2 to QDs and inserting them in other conductive surfaces have speeded up the reaction kinetics and enhanced the total electrochemical efficacy. Mo et al. synthesized TiO_2 QDs/G composite as an anode active material in LIBs. TiO_2 QDs showed well dispersion on graphene substrates with no aggregation in addition to preventing the stacking of the nanostructure with cycling [62]. Meng et al. reported black phosphorous (BP) QDs/Ti_3C_2 nanosheets as anodes with a dual mechanism of storing energy in LIBs [125]. BP QDs decorated on the Ti_3C_2 nanosheets exhibited of about 910 mA h/g at 100 mA/g, and excellent 100% cycle retention reserved after 2400 cycles. By using a ball-milling technique, Zhang et al. synthesized a unique red phosphorous (RP) QDs/$Ti_3C_2T_x$ composite that demonstrated outstanding electrochemical behavior in LIBs. Because of their intimate attachment to the MXene sheets, the RP QDs improved Li-ion transfer throughout cycling operations [63]. Lee et al. synthesized ZnS quantum dots embedded in N-doped carbon matrix as and employed it in LIBs as shown in Fig. 1 [64]. The ZnS-QD@NC electrode has a reversible specific capacity of 800 $mAhg^{-1}$, a high initial energy round-trip efficiency of 83%, and higher cycling stability and power due to the composite's structural properties and unique electrochemical activity.

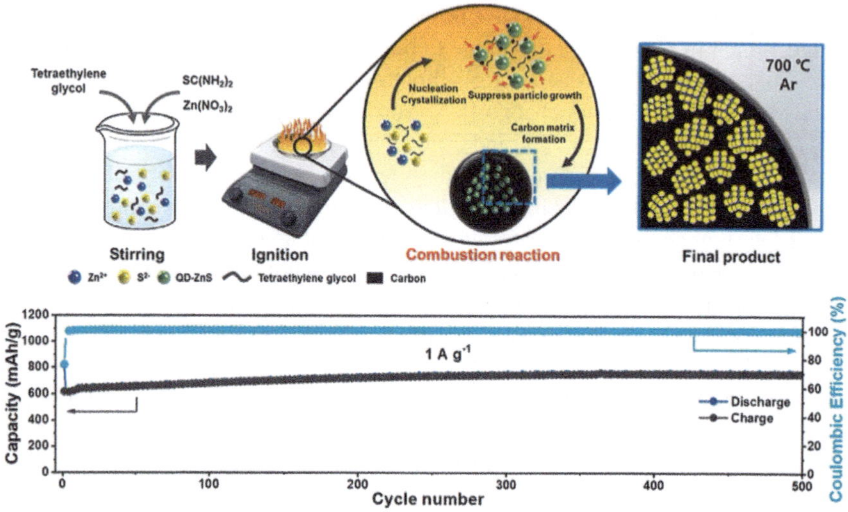

Fig. 1 A schematic illustration of ZnS-QD@N-doped Carbon for Li-ion battery anode, reproduced with permission Copyright (2022) [64]

Various studies concerned with SnO_2 or Sn-dichalcogenides QDs based composites in LIBs to address the challenges of reduced conductivity, high volume expansion > 300% of bulk SnX_2 (X = O, S, Se) during the Li inserting-deinserting process. Carbonaceous substrates are the most suitable conducting hosts of SnX_2 QDs. Zhao et al. demonstrated SnO_2 QDs/rGO by a simple hydrothermal approach where an increased level of oxidation of rGO simplifies the consistent dispersion of QDs [63]. Gao et al. have developed a 3D structure with a flexible network of SnO_2QDs/ negatively charged GO with exceptional flexibility, which could be utilized as a binder-free electrode due to the strong chemical adhesion of SnO_2 QDs to GO surface [65]. Li et al. synthesized a novel composite of SnO_x QDs loaded onto the Zr-metal organic frameworks' (MOFs) surface. During the alloying and dealloying process, SnO_x QDs@Zr-MOFs offers surrounding room to accommodate the volume change of SnOx QDs [66]. Huang et al. stated that $SnSe_2$ QDs/rGO nanocomposite possessed high reversible capacity of 778.5 mA h/g at 50 mA/g, remarkable rate capacity of 324.5 mA h g^{-1} at 5 A g^{-1} current and high cycle retention of 92.2% after 3000 cycle at 5 A g^{-1} [67]. Other metal oxide QDs based carbonaceous materials such as Co_3O_4 QDs, CoO QDs, Fe_3O_4@C QDs, MnO QDs, V_2O_5 QDs are employed as anode materials for LIBs [68].

GQDs, graphitic-type CQDs are most preferred in the surface functionalization of metal oxide cathode active materials. Balogun et al. reported a 3D carbon cloth with VO_2 interlaced nanowires decorated with CQDs. This structure reduced lithium-ion diffusion length and offered 86% as an overall capacity of LIB [69]. Xuewei Zhao et al. have synthesized GQDs@CNTs as a hybrid structural electrodeas illustrated in Fig. 2. The functionalized 3D CNTs sponge with GQDs improved the electrochemical

Fig. 2 Synthesis of GQD@CNTs hybrid sponge reproduced with permission Copyright (2020) from Springer Nature [70]

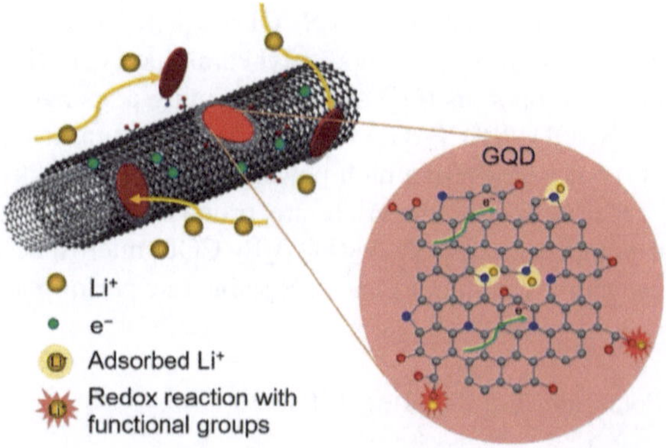

performance [70]. Yingchang et al. have prepared N-doped TiO_2/CQDs that offered 93.6% cycling stability of retention after the 300th cycle [71].

Sodium-Ion Batteries

Rechargeable SIBs are among the most affordable alternatives to LIBs due to the availability of sodium and high theoretical capacity of 1165 mA h g^{-1}. Nevertheless, SIBs feature substantial volume expansion, struggling conductivity, and Na-ion diffusion kinetics. Hence, several studies were performed to overcome the forementioned barriers. Carbonaceous materials functionalized with different QDs are the most favorable alternatives to be used as SIB anodes [58].

Yu et al. synthesized a composite of TiO_2 QDs and rGO to serve as an anode for SIBs. Superior mechanical stability and pathways for the passage of electrons and ions were made possible by the porous structure [72]. Liu et al. synthesized graphene-Fe_3O_4 QDs to serve as an anode active material for SIBs via a hydrothermal/solvothermal technique. Fe_3O_4 QDs were homogenously loaded on a 3D graphene sponge demonstrating high sodium-ion storage capacity of 525 mA h/g at 30 mA/g, good cycling stability of 312 mA h/g after 200 cycles at 50 mA/g, and high-rate performance [52]. Kong et al. reported Fe_3O_4 QDs/MoS_2 nanosheets on graphite flakes with outstanding reversible capacity and good rate performance [73]. Zhang et al. prepared a composite of ZnS QDs/graphene as anode via a one-pot hydrothermal method. This ZnS QDs/G composite demonstrated an adjustable specific capacity of 491 mA h/g at 100 mA/g after the 100th cycle [74]. Wang et al. synthesized SnO_2 QDs/G aerogel nanocomposite which demonstrated better characteristics as anode active materials in SIBs than individual SnO_2 [75].

ZnS Metal sulfide QDs are also of high importance as anode materials for SIBs. Co_9S_8-QDs@NC/Na_3V_2 $(PO_4)_3$ were tested and gave the composites advantages including an exceptional rate capability, a high reversible specific capacity, and outstanding stability [76]. Furthermore, SnS, SnS_2, and ZnS QDs were used as possible anode materials for high-performance SIBs [77]. Chen et al. reported a dual strategy confining Co_9S_8 QDs, where the Co_9S_8 QDs were involved within a mesoporous hollow carbon polyhedral matrix (HCP), and HCP was also embedded in a macroporous rGO structure forming a sponge-like architecture. The obtained Co_9S_8-QDs@HCP/rGO structure caused a significant capacity of 72% at a scan rate of 0.6 mV/s, offering high-power performance electrodes [78].

Carbonaceous materials are promising candidates for SIB anode. Liu et al. developed graded $Na_3V_2(PO_4)_2F_3$ CQD microspheres for SIB cathode with high capacitive capabilities and acceptable rate performance [55].

Potassium-Ion Batteries (PIBs)

PIBs are another promising and attractive energy storage alternative regarding their cost effectiveness and availability. Even though PIBs have a high potassium ionic radius (1.38 Å) that causes them to function inadequately, reducing the active material's size to the QD scale provides a practical way to address the main electrode barrier issues. Gao et al. constructed a CoS QDs@graphene composite via a facile two-step solvothermal/hydrothermal method which offered fast ion/electron transfer capability, good structural stability, and excellent electrochemical performance [79]. Liu et al. fabricated $Co_{0.85}Se$ QDs encapsulated in mesoporous carbon that was used as anode and demonstrated remarkable K-ion storage performance because of its hierarchical structure [80]. Hussain et al. employed Co_3Se_4 QDs decorated with N-doped carbon (QDs@C) as anode for PIBs, which possessed excellent discharge capacity and good cyclic performance [81]. Moreover, constructing bimetallic sulfide QDs for anodes of PIBs can improve its cycle performance. Cao et al. synthesized $Cu_{12}Sb_4S_{13}$ QDs/Ti_3C_2 nanosheets (CAS-Ti_3C_2) as anode for promoting the K-ion storage [82]. Erjin et al. have prepared a 3D structure of CQDs on rGO. PIB with CQDs@rGO gives a remarkable capacity of 310 mA h/g at 0.1 A/g [83].

3.1.2 Quantum Dots in Lithium-Metal Batteries

It is highly desirable to build Li-metal batteries (LMBs) with high energy density due to their large theoretical capacity (3860 mA h/g) [84]. Unlike LIBs that employ intercalated graphite anodes, Pristine Li metal is involved as the anode to increase the energy density of LMBs. The deposited Li ions on the anode's surface form dendrites which can easily penetrate the separator leading to a short circuit. The randomly deposited Li ions causes volume expansion which restricts their practical application in various battery-types including LIBs, LMBs, LSBs, and Li-O_2 batteries as well. Hence, battery safety is of great consideration while studying LMBs [85]. To improve the functionality and safety of Li metal anodes, Liu et al. inserted ZnO QDs/CNTs in a lithiophilic vertical cactus-like framework (LVCF) derived from a Zn/Cu-based coordination polymer as shown in Fig. 3 [84].

Wang et al. prepared a composite of perovskite $CsPbI_3$ QDs with Polyethylene oxide (PEO) that notably improved the current properties and power capabilities by

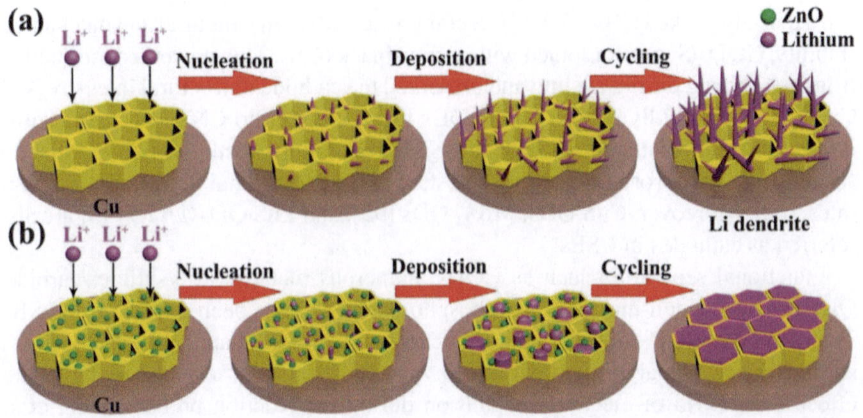

Fig. 3 **a** For hierarchical porous carbon (HPC) with no ZnO, **b** ZnO modified HPC scaffold with permission Copyright (2017) from Elsevier [84]

promoting better conductivity and a higher Li-ion transference number in LMBs. They have demonstrated that the presence of Lewis acid–base interactions plays a crucial role in facilitating the performance of PEO-Perovskite QD solid-state electrolytes. This phenomenon arises from the ability of these interactions to hinder the crystallization process of quantized $CsPbX_3$ and to promote robust electrostatic adsorption of halogen atoms [86].

3.1.3 QDs in Lithium-Sulfur Batteries

Rechargeable Lithium-Sulfur batteries (LSBs) possess a high theoretical capacity of 1675 mA h/g, high energy density 2600 Wh/kg which is 5 times higher than conventional Li-ion counterparts in addition to their reasonable cost [87]. QDs can be engaged as the active material of cathodes and separators in LSBs [88]. Nevertheless, there are still several issues with LSB commercialization, namely the shuttle effect and sluggish reaction kinetics that cause rapid capacity loss during discharge.

Several studies concerned the modification of sulfur cathodes of LSBs using QDs based nanocomposites. Gao et al. followed a facile strategy to synthesize TiO_2 QDs on thin MXene ($Ti_3C_2T_x$) nanosheets via hydrothermal method using cetyltrimethylammonium bromide (CTAB). The resulting TiO_2 QDs/MXene composite exhibits strong conductivity to guarantee quick ion diffusion and prevents the shuttle effect of polysulfide [89]. Hu et al. developed polyethyleneimine functionalized carbon quantum dots (PEI/CQDs) to enhance the performance of Li–S batteries that have high sulfur loading. Enhancement of Li-ion flow characteristics of polysulfides in the cathode with PEI/CQD modification [90]. Park et al. synthesized oxygenated GQDs from carbon fiber using a modified Hummers' technique, demonstrating that they possess good sulphidic characteristics. To create a homogeneous GQD/S coating and a densely packed CB@GQDS@S composite with high structural and mechanical stability, GQDs/S was combined with carbon black (CB). This prevented the cathode from pulverizing during cycling and preserved the cathode's structural integrity [91]. Cai et al. successfully designed CdS QDs incorporated into CNTs to enable sulfur infiltration as the cathode material in LSBs. This unique configuration exhibited a strong ability to adsorb polysulfide and showcased exceptional performance at high rates. [62]. Moreover, CuS QDs, MoS_2 QDs [92], and ZnS QDs@rGO [93] are also referred as cathodes in LSBs.

Functional separators such as CQDs, numerous metal oxides/sulfides/carbides QDs, and even non-metallic like phosphorus QDs, have been involved in LSBs. Typically, these QDs are adorned on a carbonaceous substrate and employed as a sulfur host or to enable the separator to work [94]. In order to mitigate the shuttle effect and get rid of electrode expansion during the reaction process, Ding et al. created a separator of TiO_2 QDs/MWCNTs; that change could boost the stability of LSBs [95].

3.1.4 Quantum Dots in Metal-Air Batteries

Metal-air batteries have the fascinating capability to utilize oxygen from the air at the cathode, converting it into electricity and generating water as a byproduct. This unique process allows for zero pollution. It is worth noting that metal-air batteries possess a remarkably high theoretical energy density, surpassing 1000 Wh/kg [96]. The components of metal-air batteries are metal anode, electrolyte, and air cathode. Typically, iron, magnesium, zinc, aluminum, and lithium make up the anode. [97]. Li-air batteries' extremely high energy density has drawn increased attention in recent years. Higher-performance Li-O_2 batteries require a catalyst that can effectively catalyze both the oxygen evolution reaction (OER) and the oxygen reduction reaction (ORR). Sun et al. have synthesized rich-edge MoS_2 QDs with S vacancy (Vs) employing a unique top-down approach. In a Li-O_2 battery, the innovative Vs-MoS_2 QDs showed impressive ORR/OER activities according to either experimental or theoretical research [98]. To address the issue of over potential, Veeramani et al. synthesized CdSe/ZnS core/shell QDs with CNTs frameworks for a high-performance Li-O_2 battery. The suggested structure provided routes for the transfer of electrons and holes, O_2 gas, and Li ions, respectively. Mn_3O_4 QDs placed on N-doped MWCNTs were developed by Huang and his coauthors as an inexpensive and effective ORR catalyst. After being discharged for 105 h, the composite catalyst showed a specific capacity of 745 mAh/g at 10 mA/cm without voltage loss. It was utilized in a Zn-air battery [99]. Combining $NiFe_2O_4$ QDs with CNTs, which demonstrated a large specific surface area and outstanding conductivity, Xu et al. created a high activity bifunctional ORR/OER electrocatalysts that were employed as the cathode in Zn-air and Mg-air batteries. [100]. CeO_2 QDs on N-doped carbon were prepared by Liu et al., and the Al-air batteries fitted with the resulting catalyst in cathodes demonstrated superior discharge performance [95].

3.1.5 Quantum Dots in Zinc-Ion Batteries

Because of their excellent safety, affordability, and sustainability, rechargeable zinc-ion batteries (ZIBs) have garnered a lot of interest [101]. $ZnMn_2O_4$ (ZMO) QDs in carbon frameworks were reported by Deng et al. as cathode material. Zn/ZMO QD@C batteries' resulting structure significantly improved ZIBs' electrochemical performance [102]. According to Zhang et al., dendritic-free ZIBs are made possible by GQDs because of their improved Zn^{2+} binding. After the 600th cycle at 1 A g^{-1}, the built ZIBs with the vanadium cathode show exceptional performance with a capacity of 164.3 mAh g^{-1} [103].

3.2 QDs in Supercapacitors

SCs are electrochemical energy storage systems that have become of great concern in recent years regarding their outstanding properties including exceptional power density, higher energy density than traditional capacitors, longer cycle life than secondary batteries, good security, and fast charge and discharge abilities [104]. SCs consist of three main components, namely, electrodes, electrolyte, and separator. The most essential component influencing the electrochemical parameters and overall performance of SCs is the electrode material [105]. The practical deployment of SC devices has been struggling due to their poor energy density. According to the energy density (E) equation, SCs' energy density may be improved by increasing either the voltage window (V) or the specific capacitance (C). Furthermore, the kind and mechanism of electrode materials have a considerable influence on the electrochemical characteristics of SCs. Electric double-layer capacitors (EDLC), pseudocapacitors, and hybrid capacitors are the three major types of SCs [106].

3.2.1 QDs in Symmetric Supercapacitors

Electric double-layered capacitors (EDLCs) store charges in a non-faradaic process where the charges are split at the interface between electrode and electrolyte when the electrode is inserted in an electrochemical cell according to the Helmholtz model. The quick reversible insertion/desertion of electrolyte ions onto the electrode surface distinguishes these devices, leading to a high charge/discharge rate and great cyclability. Electrode materials used in EDLCs are mainly carbonaceous material with large surface area which in turn is responsible for enlarging the specific capacitance of EDLC. Nonetheless, because of the low degree of specificity of industrial activated carbon, the energy density of EDLCs is limited, limiting its future practical utility [107]. In EDLCs, the application of QDs and their composites is to embed QDs in carbon electrode materials to boost the specific surface area of carbon materials, resulting in better SC performance.

Quing et al. employed CQDs as nanofillers on graphene fiber in micro-supercapacitors. The interactions inside the composite demonstrated a tremendous boost in mechanical strength and electrical stability via hydrogen bonds between CQDs and the sheets of graphene. The CQDs/graphene fiber assembled device exhibited a high specific capacitance of 607 mF/cm^2, with cycling stability after 10,000 cycles. The inclusion of CD facilitated the rise of active sites and increased capacitance by 22.1%, opening the door for additional ionic channels for conduction [108].

Chang et al. created a 2D porous carbon nanosheet (PCN) made of graphene quantum dots using a facile interface themselves and a subsequent activation technique from a by-product of the commercial sulfonated pitch. It has an extremely thin thickness of 20 nm, a sufficient SSA of 2429 m^2 g^{-1}, a high packing density of 0.562 g cm^{-3}, a high electron-conductivity of 89 S m^{-1}, and a high mechanical elasticity. This PCN may be employed as a powerful EDLC electrode due to its structural benefits [109].

Harper et al. created composite materials by homogeneously coating CDs onto lignin-based activated carbons and decorating the inner surface with CDs binding sites. The obtained CDs have a consistent size distribution range of 2 nm to 5 nm, allowing the crystal lattice to be observed. After the CDs modify the surface of the activated carbons, multiple CDs domains with crystalline structures are evenly distributed on the AC matrix. The composed electrode offers a high capacitance of 301.7 F/g at 0.15 A/g, which is twice larger than that of the AC electrode (125.8F g^{-1}). The specific capacitance of this electrode is 191.8F g^{-1}, with 52% growth achieved with half the number of CDs added, indicating that the increased capacitance is proportional to the quantity of CDs added. These findings suggest that carbon dot modification is an effective technique to increase the performance of ACs, however current research is in its early stages [110].

The surface faradic redox reaction with electrode materials such as metal oxides and chalcogenides is mostly used by pseudocapacitors. Metal oxides QDs and chalcogenides QDs have recently been demonstrated to be viable electrode materials for pseudocapacitor applications. Cong et al. created nonstoichiometric tungsten oxide QDs, $WO_{3\text{-}x}$ ($x = 0.06$), with an average size of 1.6 nm for rapid pseudocapacitor applications. The conversion of the octylamine ligand to pyridine resulted in outstanding conductivity and hydrophilic properties, as well as extraordinary electrochemical activity. In QD electrodes, reversible effective ion transport may be seen during the charging and discharge process. The electrode may profit from a short diffusion path and easy mass/charge transfer due to the QD size and pyridine ligand. Furthermore, 2D layered metal chalcogenides QDs are promising options for capacitive sensors [111].

Yin et al. prepared WS_2 QDs using a simple method and tested their pseudocapacitance features. Surface treatment with 1,2-ethanedithiol improved the characteristics. As a result, a high specific capacitance (457 F g^{-1}) was produced, as well as exceptional cyclic stability over 8000 cycles and Coulombic efficiency (81%) [112].

Elsaid et al. used a simple hydrothermal technique to prepare $FeCo_2S_4$ rod-like structures on nickel foam. The $FeCo_2S_4$ nanorods electrochemically demonstrated a high specific capacitance of 4035 F g^{-1} at 1 A g^{-1} as a current density in 6 M KOH solution, with outstanding cycling stability of 80.4% capacitance retention and 100% coulombic efficiency after 5000 cycles. The symmetric device has the highest specific energy (37.72 W h kg^{-1}) and specific power (603.6 W kg^{-1}) at a current density of 0.5 A/g, as well as extraordinary cycling stability of 80.3% capacitance retention after 5000 cycles and high coulombic efficiency of 94% [113].

3.2.2 QDs in Asymmetric Supercapacitors

Synthesizing anode materials with high specific capacitance remains a big difficulty for building asymmetric supercapacitors (ASCs) with high energy density. Among the many approaches, producing ultrafine QDs distributed in a carbon substrate and creating ASCs can be a useful way to address the current problems. Zhihui et al. created a C-$NiCo_2O_4$/C@Ni-Co-S electrode by connecting the core of 0D/2D carbon-modified $NiCo_2O_4$ and the exterior sheath of Ni-Co-S nanosheets. The 0D carbon quantum dot doping efficiently avoids agglomeration of $NiCo_2O_4$ nanoparticles while improving electric conductivity. The use of 2D carbon nanosheets as an intermediate layer can increase the durability of the electrode. Ni-Co-S nanosheets can enhance the electrode's area-specific capacitance and hence the energy density of the device. The formed electrode offers a remarkable specific capacitance of 2396 F g^{-1} (7.9 F cm^{-2}) at 3 A/g and good cycle stability (92% after 10,000 cycles). A semi solid asymmetric supercapacitor (ASC) employing this composite as a positive electrode against activated carbon (AC) as the negative electrode offers a high energy density of 95 Wh kg^{-1} at 0.9 kW kg^{-1} and an remarkable cycle performance [114].

Zhang et al. created a precursor using a zeolitic imidazolate framework (ZIF-8) and carbon nanotubes (CNTs). As electrodes for ASC, both ZnO QDs/C/CNTs and N-doped carbon/CNTs were used. The device achieved a working potential of 1.7 V while providing high energy density (23.6 Wh kg^{-1}) and power density (16.9 kW kg^{-1}) [115].

Fe_2O_3 is also a good anode material for ASCs. However, the Fe2O3-based anodes continue to have a poor specific capacitance. Using a simple method, Xia et al. created a hybrid material containing Fe_2O_3 QDs evenly placed onto functionalized graphene nanosheets (FGS). In 1 M Na2SO4, the hybrid material displayed a significant specific capacitance of 347 F g^{-1} between -1 and 0 V vs Ag/AgCl. A high-performance ASC operating at 2 V was built using Fe_2O_3 QDs/FGS as the anode and MnO_2/FGS as the cathode. At 100 W kg^{-1}, the supercapacitor had a high energy density of 50.7 Wh kg^{-1}, as well as strong cycle stability and power capabilities [116].

Jinfeng et al. created a composite of Ni_3S_4 (QDs)@(G/CNTs). The addition of G/CNTs to Ni_3S_4 QDs has increased electron interaction, promoting electron transport in the material. G/CNTs enhanced the specific capacitance of Ni_3S_4 QDs by 4.5 times and boosted their rate capability from 22 to 82%. The energy density of an asymmetric supercapacitor built with Ni_3S_4 QDs@G/CNTs and activated carbon electrodes was 50 Wh kg^{-1} [117].

Gamma radiation was employed by Abdel Maksoud et al. to aid in the manufacture of Ag adorned nanoplates spinel-type oxide NiMn2O4 as a potential electrode material. The Ag@ $NiMn_2O_4$ electrode has a capacitance of 622 Fg^{-1} at 1 Ag^{-1} current density while preserving outstanding cycle stability (capacitance retention of about 88.8% and coulombic efficiency of 97.6% at 10 Ag^{-1} after 5000 cycles). An asymmetric supercapacitor (Ag@ $NiMn_2O_4$ //rGO) with a maximum specific density of 30.44 Wh kg^{-1} and a specific power of 10.3006 kW kg^{-1} within a broad potential window of 1.6 V has been built. Furthermore, the capacitance of the Ag@ $NiMn_2O_4$// rGO hybrid supercapacitor device is around 94.8% retained and exhibits almost 100% coulombic efficiency [118].

Some recent research has employed quantum dots directly as electrodes, which have substantial benefits as electrode materials for micro-supercapacitors (MSCs) [109]. Molybdenum disulfide quantum dots (MoS_2-QDs) and nitrogen-doped GQDs (N-GQDs) were utilized as cathode and anode materials, respectively, by Liu et al., and the resulting (MSCs) showed outstanding electrochemical performance [119].

3.2.3 QDs in Hybrid Supercapacitors

Both faradaic and non-faradaic processes are exploited for benefit in hybrid supercapacitors. Manufacturing hybrid SCs with the purpose of merging the features of high energy in batteries and high power in SCs is particularly encouraging. Hybrid supercapacitors generally employ a capacitor-type positive electrode and a battery-type negative electrode [120]. Diab et al. used a simple solvent evaporation technique followed by a hydrothermal reaction for supercapacitors to create a hybrid structured GQDs/CuO. CuO aggregates are created by employing a Cu-MOF complex as the structural and compositional predecessor, and GQDs as an external support. The combined GQDs/CuO electrode provides a high specific capacitance of 729 F g^{-1} at a current density of 1 A/g and good-rate capability, as well as increased cyclic performance (82.2% retention after 3000 cycles) as a supercapacitor cathode material. The as-fabricated ASC has an energy density of 32.2 W h kg^{-1} at a power density of 748.9 W/kg and an enhanced cyclability of over 8000 cycles, thanks to the enriched electrochemical active sites and intimated connection between Cu species and doped GQDs [121].

Elsaid et al. prepared a new nano coral-reef $ZnFeNiCo_2S_4$ quaternary sulfide on a Ni Foam (NF) substrate in two simple hydrothermal phases for enhanced battery-type SCs. On the Ni foam, the $ZnFeNiCo_2S_4$/NF electrode was effectively produced as a binder-free electrode. At 6 M KOH as the electrolyte, the electrochemical analysis of the $ZnFeNiCo_2S_4$/NF revealed a high specific capacity of 1778 C g^{-1} at a current density of 3 A g^{-1} with outstanding cycle stability of 86.1% capacity retention and 100% coulombic efficiency. The specific capacitance of the battery-type symmetric supercapacitor was 162.3 C g^{-1} at a current density of 2 A g^{-1} with a specific energy of 33.8 Wh kg^{-1} and a specific power of 2519.25 W kg^{-1} [122].

Xiang et al. developed a ternary composite of (GO/CQDs/Polypyrrole) that may be employed as an electrode-active material in a supercapacitor. At a current density of 0.5 A g^{-1}, GO/CDs/PPy has a specific capacitance of 576 F g^{-1}, and a supercapacitor made of two symmetric GO/CDs/PPy electrodes has a high energy density of 30.1 Wh kg^{-1} at a power density of 250 W kg^{-1}. More notably, after 5000 cycles, the as-fabricated supercapacitor maintains remarkable cycle stability [123].

3.3 *QDs for Hydrogen Storage*

Photocatalytic hydrogen evolution is an environmentally friendly technique of storing energy. QDs are significant because of the quantum confinement effect, which allows them to create numerous excitons, enhancing charge separation and transmission, and boosting photocatalytic activity [124]. There are three types of hydrogen storage technologies: physical, chemical, and hybrid. Now, compressed hydrogen storage is the most established and widely accessible technique. Graphene QDs have a high specific surface area and can be readily doped with transition and alkali metals, which increases hydrogen storage capacity significantly. Few research attempts to dope Pb [125], Cu [126], and other metal QDs to increase hydrogen storage performance. No QDs composite materials for hydrogen storage have been developed.

4 Challenges and Future Motivation

QDs may play an important role in the design of microsupercapacitors or micro-batteries in the future. The extra functional groups on the surface of QDs can be exploited to manipulate their shape. The improved chemical stability and photostability of various types of QDs can be investigated further to find a role in future photochargeable supercapacitors or batteries. Because of their synthesis techniques, the large-scale synthesis of QDs remains a source of worry. This might stymie the commercialization of QDs. Researchers are looking at natural precursors as a source for environmentally benign, low-cost, and non-toxic QDs. This might pave the way for a wide range of studies employing cleaner technology.

5 Conclusion

In summary, this chapter provides an overview of quantum dot composites' substantial capabilities in the development of high-performance energy storage devices. With the advancement of improved technologies and characterization methodologies, interesting new research fields based on QDs may emerge.

References

1. Zhao, N., You, F.: Can renewable generation, energy storage and energy efficient technologies enable carbon neutral energy transition? Appl. Energy **279**, 115889 (2020)
2. Denholm, P., et al.: The potential for battery energy storage to provide peaking capacity in the United States. Renew. Energy **151**, 1269–1277 (2020)
3. Yang, Y., et al.: Battery energy storage system size determination in renewable energy systems: a review. Renew. Sustain. Energy Rev. **91**, 109–125 (2018)
4. Adamopoulos, P.M., et al.: Photoelectrocatalytic hydrogen production using a TiO2/WO3 bilayer photocatalyst in the presence of ethanol as a fuel. Catalysts **9**, 976 (2019)
5. Samsudin, M.F.R., Sufian, S.: Hybrid 2D/3D g-C3N4/BiVO4 photocatalyst decorated with RGO for boosted photoelectrocatalytic hydrogen production from natural lake water and photocatalytic degradation of antibiotics. J. Mol. Liq. **314**, 113530 (2020)
6. Wu, P., et al.: The research development of quantum dots in electrochemical energy storage. Small **14**, 1801479 (2018)
7. Wang, J., et al.: 0D/2D interface engineering of carbon quantum dots modified Bi2WO6 ultrathin nanosheets with enhanced photoactivity for full spectrum light utilization and mechanism insight. Appl. Catal. B **222**, 115–123 (2018)
8. Yang, J., et al.: A top–down strategy towards monodisperse colloidal lead sulphide quantum dots. Nat. Commun. **4**, 1695 (2013)
9. Hang, D.-R., et al.: Facile bottom-up preparation of WS 2-based water-soluble quantum dots as luminescent probes for hydrogen peroxide and glucose. Nanoscale Res. Lett. **14**, 1–15 (2019)
10. Xu, N., et al.: Zero-dimensional mxene-based optical devices for ultrafast and ultranarrow photonics applications. Adv. Sci. **7**, 2002209 (2020)
11. Liu, Q., et al.: Graphene quantum dots for energy storage and conversion: from fabrication to applications. Mater. Chem. Front. **4**, 421–436 (2020)
12. Li, B., et al.: Nitrogen-doped activated carbon for a high energy hybrid supercapacitor. Energy Environ. Sci. **9**, 102–106 (2016)
13. Zhao, Z., Xie, Y.: Enhanced electrochemical performance of carbon quantum dots-polyaniline hybrid. J. Power. Sources **337**, 54–64 (2017)
14. Li, J., et al.: Three-dimensional nitrogen and phosphorus co-doped carbon quantum dots/ reduced graphene oxide composite aerogels with a hierarchical porous structure as superior electrode materials for supercapacitors. J. Mater. Chem. A **7**, 26311–26325 (2019)
15. Dai, L.: Carbon-based catalysts for metal-free electrocatalysis. Curr. Opin. Electrochem. **4**, 18–25 (2017)
16. Xu, Q., et al.: Recent progress of quantum dots for energy storage applications. Carbon Neutrality **1**(1), 13 (2022)
17. Yang, J., et al.: A top–down strategy towards monodisperse colloidal lead sulphide quantum dots. Nat. Commun. **4**(1), 1695 (2013)
18. Abdel-Salam, A.I., et al.: Investigating the impact of growth time of CdSe quantum dots on the structure and optical properties of its nanocomposites with SiO2 for improvement of optical devices. J. Alloy. Compd. **925**, 166729 (2022)
19. Haes, A.J., et al.: A localized surface plasmon resonance biosensor: First steps toward an assay for Alzheimer's disease. Nano Lett. **4**(6), 1029–1034 (2004)
20. Kotov, N.A., et al.: Inverted colloidal crystals as three-dimensional cell scaffolds. Langmuir **20**(19), 7887–7892 (2004)
21. Klostranec, J.M., Chan, W.C.: Quantum dots in biological and biomedical research: recent progress and present challenges. Adv. Mater. **18**(15), 1953–1964 (2006)
22. Ding, S.Y., et al.: Quantum dot molecules assembled with genetically engineered proteins. Nano Lett. **3**(11), 1581–1585 (2003)
23. Abdel-Salam, A.I., et al.: Anisotropic CuInSe2 nanocrystals: synthesis, optical properties and their effect on photoelectric response of dye-sensitized solar cell. Revista mexicana de física **66**(1), 14–22 (2020)

24. Choi, M.K., et al.: Flexible quantum dot light-emitting diodes for next-generation displays. npj Flex. Electron. **2**(1), 10 (2018)
25. Pal, A., et al.: Quantum dots in visible-light photoredox catalysis: reductive dehalogenations and C-H arylation reactions using aryl bromides. Chem. Mater. **29**(12), 5225–5231 (2017)
26. Han, C., et al.: V2O5 quantum dots/graphene hybrid nanocomposite with stable cyclability for advanced lithium batteries. Nano Energy **2**(5), 916–922 (2013)
27. Zhang, Q., et al.: CuO nanostructures: synthesis, characterization, growth mechanisms, fundamental properties, and applications. Prog. Mater. Sci. **60**, 208–337 (2014)
28. Shimpi, J.R., Sidhaye, D.S., Prasad, B.L.V.: Digestive ripening: a fine chemical machining process on the nanoscale. Langmuir **33**(38), 9491–9507 (2017)
29. Talluri, B., et al.: Nanocomposites of digestively ripened copper oxide quantum dots and graphene oxide as a binder free battery-like supercapacitor electrode material. Electrochim. Acta **321**, 134709 (2019)
30. Jinlong, L., Meng, Y., Tongxiang, L.: Enhanced performance of NiMoO4 nanoparticles and quantum dots and reduced nanohole graphene oxide hybrid for supercapacitor applications. Appl. Surf. Sci. **419**, 624–630 (2017)
31. Veeramani, V., et al.: CdSe/ZnS QD@ CNT nanocomposite photocathode for improvement on charge overpotential in photoelectrochemical Li-O2 batteries. Chem. Eng. J. **349**, 235–240 (2018)
32. Zhu, J., et al.: General strategy to synthesize highly dense metal oxide quantum dots-anchored nitrogen-rich graphene compact monoliths to enable fast and high-stability volumetric lithium/sodium storage. ACS Appl. Energy Mater. **2**(5), 3500–3512 (2019)
33. Zhou, X., et al.: High rate magnesium–sulfur battery with improved cyclability based on metal–organic framework derivative carbon host. Adv. Mater. **30**(7), 1704166 (2018)
34. Li, Y., et al.: Li-ion and Na-ion transportation and storage properties in various sized TiO 2 spheres with hierarchical pores and high tap density. J. Mater. Chem. A **5**(9), 4359–4367 (2017)
35. Guo, X., et al.: Nanostructured graphene-based materials for flexible energy storage. Energy Storage Mater. **9**, 150–169 (2017)
36. Chen, M., et al.: Nitrogen-doped mesoporous carbon-encapsulation urchin-like Fe3O4 as anode materials for high performance Li-ions batteries. Electrochim. Acta **195**, 94–105 (2016)
37. Li, Y., et al.: Towards fast and ultralong-life Li-ion battery anodes: embedding ultradispersed TiO2 quantum dots into three-dimensional porous graphene-like networks. Electrochim. Acta **246**, 1183–1192 (2017)
38. Zahir, N., et al.: Recent advances on graphene quantum dots for electrochemical energy storage devices. Energy Environ. Mater. **5**(1), 201–214 (2022)
39. Shen, S., et al.: Graphene quantum dots with high yield and high quality synthesized from low cost precursor of aphanitic graphite. Nanomaterials **10**(2), 375 (2020)
40. Liu, W., et al.: Graphene quantum dots-based advanced electrode materials: design, synthesis and their applications in electrochemical energy storage and electrocatalysis. Adv. Energy Mater. **10**(29), 2001275 (2020)
41. Wei, J.-S., et al.: A new generation of energy storage electrode materials constructed from carbon dots. Mater. Chem. Front. **4**(3), 729–749 (2020)
42. Devadas, B., Imae, T.: Effect of carbon dots on conducting polymers for energy storage applications. ACS Sustain. Chem. Eng. **6**(1), 127–134 (2018)
43. Li, X., et al.: Carbon and graphene quantum dots for optoelectronic and energy devices: a review. Adv. Func. Mater. **25**(31), 4929–4947 (2015)
44. Chao, D., et al.: Graphene quantum dots coated VO2 arrays for highly durable electrodes for Li and Na ion batteries. Nano Lett. **15**(1), 565–573 (2015)
45. Chabu, J.M., et al.: Simple approach for the preparation of nitrogen and sulfur codoped carbon dots/reduced graphene oxide as host for high-rate lithiumsulfur batteries. Mater. Chem. Phys. **229**, 226–231 (2019)
46. Zhai, Y., et al.: Carbon dots as new building blocks for electrochemical energy storage and electrocatalysis. Adv. Energy Mater. **12**(6), 2103426 (2022)

47. Yu, Y., Ma, T., Huang, H.: Semiconducting quantum dots for energy conversion and storage. Adv. Funct. Mater. 2213770 (2023)
48. Wang, H., Xie, K., You, Y., Hou, Q., Zhang, K., Li, N., Yu, W., Loh, K.P., Shen, C., Wei, B.: Realizing interfacial electronic interaction within ZnS quantum dots/N-rGO heterostructures for efficient Li–CO2 batteries. Adv. Energy Mater. **9**(34), 1901806 (2019)
49. Cheng, Y., Wang, S., Zhou, L., Chang, L., Liu, W., Yin, D., Yi, Z., Wang, L.: SnO2 quantum dots: rational design to achieve highly reversible conversion reaction and stable capacities for lithium and sodium storage. Small **16**(26), 2000681 (2020)
50. Bonu, V., Gupta, B., Chandra, S., Das, A., Dhara, S., Tyagi, A.K.: Electrochemical supercapacitor performance of SnO2 quantum dots. Electrochim. Acta **203**, 230–237 (2016)
51. Zhou, X., Shi, J., Liu, Y., Su, Q., Zhang, J., Du, G.: Microwave irradiation synthesis of Co3O4 quantum dots/graphene composite as anode materials for Li-ion battery. Electrochim. Acta **143**, 175–179 (2014)
52. Liu, H., Jia, M., Zhu, Q., Cao, B., Chen, R., Wang, Y., Wu, F., Xu, B.: 3D–0D graphene-Fe3O4 quantum dot hybrids as high-performance anode materials for sodium-ion batteries. ACS Appl. Mater. Interfaces **8**(40), 26878–26885 (2016)
53. Wei, Z., Wang, D., Yang, X., Wang, C., Chen, G., Du, F.: From crystalline to amorphous: an effective avenue to engineer high-performance electrode materials for sodium-ion batteries. Adv. Mater. Interfaces **5**(19), 1800639 (2018)
54. Meng, R., Huang, J., Feng, Y., Zu, L., Peng, C., Zheng, L., Chen, Z., Liu, G., Chen, B., Yang, J.: Black phosphorus quantum dot/Ti3C2 MXene nanosheet composites for efficient electrochemical lithium/sodium-ion storage. Adv. Energy Mater. **8**(26), 1801514 (2018)
55. Sun, X., Zhou, C., Xie, M., Sun, H., Hu, T., Lu, F., Scott, S.M., George, S.M., Lian, J.: Synthesis of ZnO quantum dot/graphene nanocomposites by atomic layer deposition with high lithium storage capacity. J. Mater. Chem. A **2**(20), 7319–7326 (2014)
56. Li, Y., Zhang, M., Huang, Q., Zhou, P., Xu, P., Guo, Z., Dai, K.: Enhanced electrochemical properties of ZnO encapsulated in carbon nanofibers as anode material for lithium-ion batteries. Ionics **26**, 4351–4361 (2020)
57. Tan, Q., Kong, X., Guan, X., Wang, C., Xu, B.: Crystallization of zinc oxide quantum dots on graphene sheets as an anode material for lithium ion batteries. CrystEngComm **22**(2), 320–329 (2020)
58. Ma, X., Li, Z., Chen, D., Li, Z., Yan, L., Li, S., Liang, C., Ling, M., Peng, X.: Nitrogen-doped porous carbon sponge-confined ZnO quantum dots for metal collector-free lithium ion battery. J. Electroanal. Chem. **848**, 113275 (2019)
59. Fang, D., Chen, S., Wang, X., Bando, Y., Golberg, D., Zhang, S.: ZnS quantum dots@ multilayered carbon: geological-plate-movement-inspired design for high-energy Li-ion batteries. J. Mater. Chem. A **6**(18), 8358–8365 (2018)
60. Zhang, G., Hou, S., Zhang, H., Zeng, W., Yan, F., Li, C.C., Duan, H.: High-performance and ultra-stable lithium-ion batteries based on MOF-derived ZnO@ ZnO quantum Dots/C core–shell nanorod arrays on a carbon cloth anode. Adv. Mater. **27**(14), 2400–2405 (2015)
61. Mo, R., Lei, Z., Sun, K., Rooney, D.: Facile synthesis of anatase TiO2 quantum-dot/graphene-nanosheet composites with enhanced electrochemical performance for lithium-ion batteries. Adv. Mater. **26**(13), 2084–2088 (2014)
62. Zhang, S., Li, X.Y., Yang, W., Tian, H., Han, Z., Ying, H., Wang, G., Han, W.Q.: Novel synthesis of red phosphorus nanodot/Ti3C2T x MXenes from low-cost Ti3SiC2 MAX phases for superior lithium-and sodium-ion batteries. ACS Appl. Mater. Interfaces **11**(45), 42086–42093 (2019)
63. Lee, S., Kim, S., Gim, J., Alfaruqi, M.H., Kim, S., Mathew, V., Sambandam, B., Hwang, J., Kim, J.: Ultra-small ZnS quantum dots embedded in N-doped carbon matrix for high-performance Li-ion battery anode. Compos. B Eng. **231**, 109548 (2022)
64. Gao, L., Wu, G., Ma, J., Jiang, T., Chang, B., Huang, Y., Han, S.: SnO2 quantum dots@ graphene framework as a high-performance flexible anode electrode for lithium-ion batteries. ACS Appl. Mater. Interfaces **12**(11), 12982–12989 (2020)

65. Li, W., Li, Z., Yang, F., Fang, X., Tang, B.: Synthesis and electrochemical performance of SnO x quantum dots@ UiO-66 hybrid for lithium ion battery applications. ACS Appl. Mater. Interfaces **9**(40), 35030–35039 (2017)
66. Jin, R., Meng, Y., Li, G.: Multiwalled carbon nanotubes@ C@ SnO2 quantum dots and SnO2 quantum dots@ C as high rate anode materials for lithium-ion batteries. Appl. Surf. Sci. **423**, 476–483 (2017)
67. Zhang, L., Xia, G., Huang, Y., Wei, C., Yu, Y., Sun, D., Yu, X.: MnO quantum dots embedded in carbon nanotubes as excellent anode for lithium-ion batteries. Energy Storage Mater. **10**, 160–167 (2018)
68. Balogun, M.S., Luo, Y., Lyu, F., Wang, F., Yang, H., Li, H., Liang, C., Huang, M., Huang, Y., Tong, Y.: Carbon quantum dot surface-engineered VO2 interwoven nanowires: a flexible cathode material for lithium and sodium ion batteries. ACS Appl. Mater. Interfaces **8**(15), 9733–9744 (2016)
69. Zhao, X., Wu, Y., Wang, Y., Wu, H., Yang, Y., Wang, Z., Dai, L., Shang, Y., Cao, A.: High-performance Li-ion batteries based on graphene quantum dot wrapped carbon nanotube hybrid anodes. Nano Res. **13**, 1044–1052 (2020)
70. Yang, Y., Ji, X., Jing, M., Hou, H., Zhu, Y., Fang, L., Yang, X., Chen, Q., Banks, C.E.: Carbon dots supported upon N-doped TiO 2 nanorods applied into sodium and lithium ion batteries. J. Mater. Chem. A **3**(10), 5648–5655 (2015)
71. Yu, J., Huang, H., Bian, F., Liang, C., Xia, Y., Zhang, J., Gan, Y., Zhang, W.: Supercritical CO2-fluid-assisted synthesis of TiO2 quantum dots/reduced graphene oxide composites for outstanding sodium storage capability. ACS Appl. Energy Mater. **1**(12), 7213–7219 (2018)
72. Kong, D., Cheng, C., Wang, Y., Huang, Z., Liu, B., Von Lim, Y., Ge, Q., Yang, H.Y.: Fe 3 O 4 quantum dot decorated MoS 2 nanosheet arrays on graphite paper as free-standing sodium-ion battery anodes. J. Mater. Chem. A **5**(19), 9122–9131 (2017)
73. Zhang, R., Wang, Y., Jia, M., Xu, J., Pan, E.: One-pot hydrothermal synthesis of ZnS quantum dots/graphene hybrids as a dual anode for sodium ion and lithium ion batteries. Appl. Surf. Sci. **437**, 375–383 (2018)
74. Wang, Y., Jin, Y., Zhao, C., Duan, Y., He, X., Jia, M.: SnO2 quantum dots/graphene aerogel composite as high-performance anode material for sodium ion batteries. Mater. Lett. **191**, 169–172 (2017)
75. Hu, X., Jia, J., Wang, G., Chen, J., Zhan, H., Wen, Z.: Reliable and general route to inverse opal structured nanohybrids of carbon-confined transition metal sulfides quantum dots for high-performance sodium storage. Adv. Energy Mater. **8**(25), 1801452 (2018)
76. Zhang, R., Xu, J., Jia, M., Pan, E., Zhou, C., Jia, M.: Ultrafine ZnS quantum dots decorated reduced graphene oxide composites derived from ZIF-8/graphene oxide hybrids as anode for sodium-ion batteries. J. Alloy. Compd. **781**, 450–459 (2019)
77. Gao, H., Zhou, T., Zheng, Y., Zhang, Q., Liu, Y., Chen, J., Liu, H., Guo, Z.: CoS quantum dot nanoclusters for high-energy potassium-ion batteries. Adv. Func. Mater. **27**(43), 1702634 (2017)
78. Liu, Z., Han, K., Li, P., Wang, W., He, D., Tan, Q., Wang, L., Li, Y., Qin, M., Qu, X.: Tuning metallic Co 0.85 Se quantum dots/carbon hollow polyhedrons with tertiary hierarchical structure for high-performance potassium ion batteries. Nano-Micro Lett. **11**, 1–14 (2019)
79. Hussain, N., Li, M., Tian, B., Wang, H.: Co3Se4 quantum dots as an ultrastable host material for potassium-ion intercalation. Adv. Mater. **33**(26), 2102164 (2021)
80. Cao, Y., Zhang, Y., Chen, H., Qin, S., Zhang, L., Guo, S., Yang, H.: Cu12Sb4S13 quantum dots/few-layered Ti3C2 nanosheets with enhanced K+ diffusion dynamics for efficient potassium ion storage. Adv. Func. Mater. **32**(6), 2108574 (2022)
81. Zhang, E., Jia, X., Wang, B., Wang, J., Yu, X., Lu, B.: Carbon dots@ rGO paper as freestanding and flexible potassium-ion batteries anode. Adv. Sci. **7**(15), 2000470 (2020)
82. Jin, C., Sheng, O., Luo, J., Yuan, H., Fang, C., Zhang, W., Huang, H., Gan, Y., Xia, Y., Liang, C., Tao, X.: 3D lithium metal embedded within lithiophilic porous matrix for stable lithium metal batteries. Nano Energy **37**, 177–186 (2017)

83. Liu, T., Chen, S., Sun, W., Lv, L.P., Du, F.H., Liu, H., Wang, Y.: Lithiophilic vertical cactus-like framework derived from Cu/Zn-based coordination polymer through in situ chemical etching for stable lithium metal batteries. Adv. Func. Mater. **31**(14), 2008514 (2021)
84. Wang, Y., Li, W., Xu, Z., Xie, Y., Wang, Y., Zhao, H., Huang, J., Yang, W., Zhang, H.: Perovskite quantum dots for lewis acid-base interactions and interface engineering in lithium-metal batteries. ACS Appl. Energy Mater. **4**(10), 11470–11479 (2021)
85. Zhang, Q., Sun, C., Fan, L., Zhang, N., Sun, K.: Iron fluoride vertical nanosheets array modified with graphene quantum dots as long-life cathode for lithium ion batteries. Chem. Eng. J. **371**, 245–251 (2019)
86. Ding, H., Zhang, Q., Liu, Z., Wang, J., Ma, R., Fan, L., Wang, T., Zhao, J., Ge, J., Lu, X., Lu, B.: TiO2 quantum dots decorated multi-walled carbon nanotubes as the multifunctional separator for highly stable lithium sulfur batteries. Electrochim. Acta **284**, 314–320 (2018)
87. Gao, X.T., Xie, Y., Zhu, X.D., Sun, K.N., Xie, X.M., Liu, Y.T., Yu, J.Y., Ding, B.: Ultrathin MXene nanosheets decorated with TiO2 quantum dots as an efficient sulfur host toward fast and stable Li–S batteries. Small **14**(41), 1802443 (2018)
88. Hu, Y., Chen, W., Lei, T., Zhou, B., Jiao, Y., Yan, Y., Du, X., Huang, J., Wu, C., Wang, X., Xiong, J.: Carbon quantum dots–modified interfacial interactions and ion conductivity for enhanced high current density performance in lithium–sulfur batteries. Adv. Energy Mater. **9**(7), 1802955 (2019)
89. Park, J., Moon, J., Kim, C., Kang, J.H., Lim, E., Park, J., Lee, K.J., Yu, S.H., Seo, J.H., Lee, J., Sung, Y.E.: Graphene quantum dots: structural integrity and oxygen functional groups for high sulfur/sulfide utilization in lithium sulfur batteries. NPG Asia Mater. **8**(5), e272–e272 (2016)
90. Wei, H., Ding, Y., Li, H., Zhang, Q., Hu, N., Wei, L., Yang, Z.: MoS2 quantum dots decorated reduced graphene oxide as a sulfur host for advanced lithium-sulfur batteries. Electrochim. Acta **327**, 134994 (2019)
91. Shen, C., Zhang, K., You, Y., Wang, H., Ning, R., Qi, Y., Li, N., Ding, C., Xie, K., Wei, B.: Inducing rapid polysulfide transformation through enhanced interfacial electronic interaction for lithium–sulfur batteries. Nanoscale **12**(26), 13980–13986 (2020)
92. Wang, J.G., Xie, K., Wei, B.: Advanced engineering of nanostructured carbons for lithium–sulfur batteries. Nano Energy **15**, 413–444 (2015)
93. Liu, Q., Han, X., Park, H., Kim, J., Xiong, P., Yuan, H., Yeon, J.S., Kang, Y., Park, H.S.: Layered double hydroxide quantum dots for use in a bifunctional separator of lithium–sulfur batteries. ACS Appl. Mater. Interfaces **13**(15), 17978–17987 (2021)
94. Xu, F., Wang, J., Zhang, Y., Wang, W., Guan, T., Wang, N., Li, K.: Structure-engineered bifunctional oxygen electrocatalysts with Ni3S2 quantum dot embedded S/N-doped carbon nanosheets for rechargeable Zn-air batteries. Chem. Eng. J. **432**, 134256 (2022)
95. Liu, Y., Zhang, Y., Chen, Z., Li, Z., Pan, Q., Li, Z., Du, Y., Li, W., Li, J.: CeO2 quantum dots embedded in 3D hierarchical porous foliaceous N-doped carbon as an efficient oxygen reduction electrocatalyst for metal-air battery. J. Alloy. Compd. **905**, 164063 (2022)
96. Sun, Z., He, J., Yuan, M., Lin, L., Zhang, Z., Kang, Z., Liao, Q., Li, H., Sun, G., Yang, X., Zhang, Y.: Li+-clipping for edge S-vacancy MoS2 quantum dots as an efficient bifunctional electrocatalyst enabling discharge growth of amorphous Li2O2 film. Nano Energy **65**, 103996 (2019)
97. Huang, Z., Qin, X., Gu, X., Li, G., Mu, Y., Wang, N., Guo, Z., Shi, Z., Shao, M.: Mn3O4 quantum dots supported on nitrogen-doped partially exfoliated multiwall carbon nanotubes as oxygen reduction electrocatalysts for high-performance Zn–Air batteries. ACS Appl. Mater. Interfaces **10**(28), 23900–23909 (2018)
98. Xu, N., Zhang, Y., Zhang, T., Liu, Y., Qiao, J.: Efficient quantum dots anchored nanocomposite for highly active ORR/OER electrocatalyst of advanced metal-air batteries. Nano Energy **57**, 176–185 (2019)
99. Chen, Z., Yang, Q., Mo, F., Li, N., Liang, G., Li, X., Huang, Z., Wang, D., Huang, W., Fan, J., Zhi, C.: Aqueous zinc–tellurium batteries with ultraflat discharge plateau and high volumetric capacity. Adv. Mater. **32**(42), 2001469 (2020)

100. Deng, S., Tie, Z., Yue, F., Cao, H., Yao, M., Niu, Z.: Rational design of ZnMn2O4 quantum dots in a carbon framework for durable aqueous zinc-ion batteries. Angew. Chem. Int. Ed. **61**(12), e202115877 (2022)
101. Zhang, H., Guo, R., Li, S., Liu, C., Li, H., Zou, G., Hu, J., Hou, H., Ji, X.: Graphene quantum dots enable dendrite-free zinc ion battery. Nano Energy **92**, 106752 (2022)
102. Das, H.T., Barai, P., Dutta, S., Das, N., Das, P., Roy, M., Alauddin, M., Barai, H.R.: Polymer composites with quantum dots as potential electrode materials for supercapacitors application: a review. Polymers **14**(5), 1053 (2022)
103. Noori, A., El-Kady, M.F., Rahmanifar, M.S., Kaner, R.B., Mousavi, M.F.: Towards establishing standard performance metrics for batteries, supercapacitors and beyond. Chem. Soc. Rev. **48**(5), 1272–1341 (2019)
104. Xiao, J., Momen, R., Liu, C.: Application of carbon quantum dots in supercapacitors: a mini review. Electrochem. Commun. **132**, 107143 (2021)
105. Volfkovich, Y.M.: Electrochemical supercapacitors (a review). Russ. J. Electrochem. **57**, 311–347 (2021)
106. Li, Q., Cheng, H., Wu, X., Wang, C.F., Wu, G., Chen, S.: Enriched carbon dots/graphene microfibers towards high-performance micro-supercapacitors. J. Mater. Chem. A **6**(29), 14112–14119 (2018)
107. Chang, P., Yang, F., Xie, Q., Li, T., Dong, J.: 2D porous carbon nanosheet from sulfonated pitch-based graphene quantum dots for high volumetric performance EDLCs. J. Power. Sources **479**, 228825 (2020)
108. Yu, L., Hsieh, C.T., Keffer, D.J., Chen, H., Goenaga, G.A., Dai, S., Zawodzinski, T.A., Harper, D.P.: Hierarchical lignin-based carbon matrix and carbon dot composite electrodes for high-performance supercapacitors. ACS Omega **6**(11), 7851–7861 (2021)
109. Cong, S., Tian, Y., Li, Q., Zhao, Z., Geng, F.: Single-crystalline tungsten oxide quantum dots for fast pseudocapacitor and electrochromic applications. Adv. Mater. **26**(25), 4260–4267 (2014)
110. Yin, W., He, D., Bai, X., William, W.Y.: Synthesis of tungsten disulfide quantum dots for high-performance supercapacitor electrodes. J. Alloy. Compd. **786**, 764–769 (2019)
111. Elsaid, M.A., Hassan, A.A., Mohamed, S.G., Sayed, A.Z., Ashmawy, A.M., Waheed, A.F.: Synthesis and electrochemical performance of porous FeCo2S4 nanorods as an electrode material for supercapacitor. J. Energy Storage **44**, 103330 (2021)
112. Xu, Z., Li, X., Sun, S., Wei, Z., Li, H., Yin, S.: Multidimensional carbon-modified NiCo2O4/Ni–Co–S nanocomposite electrode material for high-energy asymmetric supercapacitors. Energy Fuels **35**(11), 9692–9704 (2021)
113. Zhang, Y., Lin, B., Wang, J., Tian, J., Sun, Y., Zhang, X., Yang, H.: All-solid-state asymmetric supercapacitors based on ZnO quantum dots/carbon/CNT and porous N-doped carbon/CNT electrodes derived from a single ZIF-8/CNT template. J. Mater. Chem. A **4**(26), 10282–10293 (2016)
114. Lu, X.F., Huang, Z.X., Tong, Y.X., Li, G.R.: Asymmetric supercapacitors with high energy density based on helical hierarchical porous Na x MnO 2 and MoO 2. Chem. Sci. **7**(1), 510–517 (2016)
115. Zheng, J., Lian, X., Wu, M., Zheng, F., Gao, Y., Niu, H.: One-step preparation of Ni3S4 quantum dots composite graphene/carbon nanotube conductive network for asymmetric supercapacitor. J. Alloy. Compd. **859**, 158247 (2021)
116. Maksoud, M.A., Elsaid, M.A., Abd Elkodous, M.: Gamma radiation induced synthesis of Ag decorated NiMn2O4 nanoplates with enhanced electrochemical performance for asymmetric supercapacitor. J. Energy Storage **56**, 105938 (2022)
117. Liu, W., Zhang, M., Li, M., Li, B., Zhang, W., Li, G., Xiao, M., Zhu, J., Yu, A., Chen, Z.: Advanced electrode materials comprising of structure-engineered quantum dots for high-performance asymmetric micro-supercapacitors. Adv. Energy Mater. **10**(8), 1903724 (2020)
118. Deng, X., Zou, K., Cai, P., Wang, B., Hou, H., Zou, G., Ji, X.: Advanced battery-type anode materials for high-performance sodium-ion capacitors. Small Methods **4**(10), 2000401 (2020)

119. Khalafallah, D., Miao, J., Zhi, M., Hong, Z.: Structuring graphene quantum dots anchored CuO for high-performance hybrid supercapacitors. J. Taiwan Inst. Chem. Eng. **122**, 168–175 (2021)
120. Elsaid, M.A., Hassan, A.A., Sayed, A.Z., Ashmawy, A.M., Waheed, A.F., Mohamed, S.G.: Fabrication of novel coral reef-like nanostructured ZnFeNiCo2S4 on Ni foam as an electrode material for battery-type supercapacitors. Electrochim. Acta **434**, 141320 (2022)
121. Zhang, X., Wang, J., Liu, J., Wu, J., Chen, H., Bi, H.: Design and preparation of a ternary composite of graphene oxide/carbon dots/polypyrrole for supercapacitor application: Importance and unique role of carbon dots. Carbon **115**, 134–146 (2017)
122. Wu, H.L., Li, X.B., Tung, C.H., Wu, L.Z.: Semiconductor quantum dots: an emerging candidate for CO2 photoreduction. Adv. Mater. **31**(36), 1900709 (2019)
123. Sharma, V., Kagdada, H.L., Wang, J., Jha, P.K.: Hydrogen adsorption on pristine and platinum decorated graphene quantum dot: A first principle study. Int. J. Hydrogen Energy **45**(44), 23977–23987 (2020)
124. Malček, M., Bučinský, L.: On the hydrogen storage performance of Cu-doped and Cu-decorated graphene quantum dots: a computational study. Theoret. Chem. Acc. **139**(11), 167 (2020)
125. Xiang, C., Li, A., Yang, S., Lan, Z., Xie, W., Tang, Y., Xu, H., Wang, Z., Gu, H.: Enhanced hydrogen storage performance of graphene nanoflakes doped with Cr atoms: a DFT study. RSC Adv. **9**(44), 25690–25696 (2019)
126. Manthiram, A., Fu, Y., Su, Y.S.: Challenges and prospects of lithium–sulfur batteries. Acc. Chem. Res. **46**(5), 1125–1134 (2013)

Life Cycle Assessment of Quantum Dots and Its Composites

M. Jeevitha, Jeetendra Kumar Gupta, Alapati Sahithi, Anil Kumar Tallam, Venkata Lakshamana Sagar Dantinapalli, Selvaraja Elumalai, Jeslin Devadhas, and Yallamalli Indira Muzib

Abstract In this chapter, we take a close look at the environmental friendliness of quantum dots (QDs) and their composites via the lens of a life cycle assessment (LCA). The importance of QDs and their widespread practicality are first elucidated. The environmental effects of extracting raw materials, producing, transporting, using, and disposing of QDs are all considered in the LCA analysis. This chapter focuses on the trade-offs between performance and sustainability, and it explores the environmental effects of various QD composite materials. QDs' use in low-power uses like lighting and screens is also highlighted. Carbon emissions, resource use, and trash output are only some of the environmental measures used to calculate the

M. Jeevitha
Department of Pharmaceutics, Shri Venkateswara College of Pharmacy, 13A, Pondy, Main Road, Ariyur, Puducherry 605102, India

J. K. Gupta
Institute of Pharmaceutical Research, GLA University, Uttar Pradesh, 17 Km Stone, NH-2, Mathura-Delhi Road Mathura, Chaumuhan 281406, India

A. Sahithi
Department of Pharmaceutical Analysis, School of Pharmacy, Nalla Narasimha Reddy Educational Society's Group of Institutions, Hyderabad 500088, Telangana, India

A. K. Tallam
Department of Pharmacy, Shri Venkateshwara University, Uttar Pradesh, Rajabpur, NH-24, Venkateshwara Nagar, Gajraula 244236, India

V. L. S. Dantinapalli
Analytical Scientist, 435 Devon Park Dr Suite 400, Wayne, PA 19087, USA

S. Elumalai
Senior Director Quality Control, Ambiopharm Inc., Dittman ct, Beech Island, SC 29842, USA

J. Devadhas
Department of Pharmaceutics, Faculty of Pharmacy, Sree Balaji Medical College & Hospital Campus, Bharath Institute of Higher Education & Research, Chromepet, Chennai 600044, Tamil Nadu, India

Y. I. Muzib (✉)
Institute of Pharmaceutical Technology, Sri Padmavati Mahila Visvavidyalayam, Tirupati 517502, Andhra Pradesh, India
e-mail: yindira1415@gmail.com

S. Thomas et al. (eds.), *Quantum Dots Based Nanocomposites*, Engineering Materials,
https://doi.org/10.1007/978-3-031-54779-9_23

extent of these effects. This chapter provides a thorough analysis of the environmental impacts of quantum dots (QDs) in an effort to educate researchers, industry professionals, and policymakers so that they can make educated decisions about the research, application, and management of QDs and their composites.

1 Introduction

The environmental impact of composites is measured across their full life cycle using a method called life cycle assessment (LCA). It's an all-encompassing method that considers the composite material's lifecycle from raw material extraction to final disposal [1]. The fundamental objective of a life cycle assessment (LCA) is to provide an all-encompassing view of the environmental effects connected with composites by quantifying and assessing the energy consumption, emissions, resource utilisation, and waste generation at each stage. Composite materials' life cycle kicks off with raw material extraction and processing. Carbon fibres, glass fibres, and different resins are common components of composites [2]. The mining, refining, and processing of these basic materials all have environmental implications that may be assessed using an LCA at this point. It takes into account things like energy use, water use, and pollution caused by these procedures. Composite material fabrication and manufacturing is the next step. Manufacturing processes in this sector range from pultrusion and filament winding to resin transfer moulding. The LCA analyses the resources used, pollution released, and trash produced by these activities [3]. Energy is expended to shape and bond materials and emissions are released into the atmosphere [4]. If you want to know how alternative manufacturing methods could affect sustainability, you need to know the environmental impact of the production phase [5]. Distribution and transport are also crucial components of a life cycle assessment. At this point, we factor in the emissions and energy used to move raw materials to production facilities and completed composite products to consumers [6–8]. The environmental impact of composites can be greatly affected by factors like as the mode of transportation utilised and the geographic distance between suppliers, manufacturers, and consumers. After composites have been put into practise, their effect on the natural world is still being studied. Composite parts are commonly used in the aerospace sector, for instance [9]. The environmental benefits of employing composites instead of traditional materials can be determined by evaluating aspects such as fuel economy, energy savings, and emissions reduction during the use phase. The life cycle of composites also includes the crucial step of maintenance and repairs [10]. Consequences to the environment, such as the energy and materials needed for repairs or replacements, should be taken into account when dealing with composites. The long-term durability of composites can be seriously compromised by these actions [11, 12]. Composites' end-of-life management is the last phase of their LCA. In this stage, we examine the processes used to dispose of or recycle composite materials once they have served their purpose. It considers the ecological effects of various waste management options such landfilling, incineration, and recycling. To lessen the impact that composites have on the environment, proper end-of-life management is crucial.

2 Background of LCA of Materials

Rising global consumption of plastics over the past few years has resulted in a rise in plastic trash. About half of all plastics are put to use in packaging and agricultural films are two examples of single-use usage [13]. Pipes, cable coatings, and structural materials only account for 20–25% of all plastic usage. The rest goes into applications with a shorter to medium-term lifespan in the consumer market, such as electronics, furniture, and auto parts [14]. The short useful lives of plastic products contribute significantly to the difficulties inherent in plastics disposal. Plastic packaging is just one example of an area where this period can be significantly less than a month.

Because of their low density and widespread use in hollow objects (making their apparent density even lower), plastics are a glaring problem in our landfills and oceans [15]. However, while plastics may account for 20–30% of the volume weight fraction of MSW, they account for just 7–9% of the overall MSW mass [16]. However, plastic trash can be found in much larger concentrations in certain streams, such as those from the manufacturing and service sectors. The fact that plastics are typically not biodegradable means they will stay in the environment for a very long time and contribute to the problem [17]. The public's awareness of the plastics problem has grown as a result of their pervasiveness. In response, numerous LCAs have been conducted to analyse the effects of plastic items over their entire lifespan because of public interest in the topic. Recently, economic evaluations have been added to the mix to round out the research.

3 LCA of Polymer Composites

Cradle-to-factory gate and cradle-to-grave are the two most popular systems utilised in LCA research. All processes from raw material and fuel extraction to final product delivery at the factory gate are factored into a cradle-to-factory-gate LCA analysis [18]. Material manufacturers frequently provide Cradle to Factory Gate studies. Every stage, from initial creation to final disposal, is accounted for in the cradle-to-grave system. The full life cycle can be accounted for in cradle-to-grave studies. Since waste management varies from nation to country and not all waste treatment alternatives may be considered, the results of cradle-to-grave evaluations for the same product can vary greatly [19]. Cradle-to-factory-gate assessments can provide initial insight into environmental implications if comparisons across different waste disposal strategies are unavailable. Multiple formats exist for reporting LCA findings. Results from the "middle" of the range are typically given in studies. At this stage, information from the life cycle inventory is transformed into environmental effect categories, such as the contribution to global warming or acidification. The final environmental score for a product is determined by adding up the results from several impact categories using appropriate weighting factors. Single-score analyses are another name for analyses with a focus on the final score. The weighting factors for the different effect categories are determined based on subjective assessment. When one product is demonstrably superior to another in all impact categories (or is comparable to the other option in all categories), it may be possible to draw conclusions from a comparative LCA research based on results from the middle of the range. If, on the other hand, some types of environmental consequences are more severe than others, then prioritisation decisions must be made among the many types of impacts.

A natural target of the movement to get rid of everything that isn't "green" appears to be plastics [20]. The huge amount of energy they contain and their pervasiveness as litter are typically highlighted when discussing their impact on the environment. Recent quantitative investigations, however, rarely back up this prejudice [21]. In fact, plastics typically show very favourable life cycle (LC) profiles when compared to other materials in studies evaluating the environmental and economic effect of alternative materials. In terms of Global Warming Potential (GWP) and Total Energy Use (TEU) [22]. The current significance of greenhouse gases enhancement led to the selection of these classes of environmental impact. Additional areas of environmental impact are reported by some research. These include the possibility for ozone depletion, photochemical oxidation, acidification, and eutrophication. The findings reveal that, contrary to popular belief, ordinary polymers typically cause fewer adverse effects on the environment in terms of GWP and TEU than do other materials. Reuse, which prevents the use of nonrenewable resources, has also shown to have a positive effect on both measures of environmental sustainability.

Many engineering projects hinge on the material choice made, since it affects the longevity, cost, and manufacturability of the end result. In addition, manufacturers are under increased regulatory pressure from government agencies to reduce the environmental impact of their operations and the goods they produce. Since recycling potential and/or end-of-life disposal options differ from material to material, material selection can be particularly important for green design. Identifying multiple mechanical, electrical, chemical, thermal qualities, environmental impact variables, and life cycle costs of candidate materials is the first step in modern integrated design processes (IDP) for systematically selecting the optimal material for a specific application (Fig. 1). For an IDP to be successful, multidisciplinary design teams must collaborate from the start of a project to provide solutions with many uses [23]. However, when several criteria from several fields must be met in a material selection problem, complications arise due to potential conflicts between the criteria and/or the relative relevance of the various criteria. Additionally, correct indices within each area require specialist expertise to define [24]. For a wing spar's leaf spring/beam, for instance, it may be preferable that it not only be light but also strong enough to withstand a specified bending force without buckling.

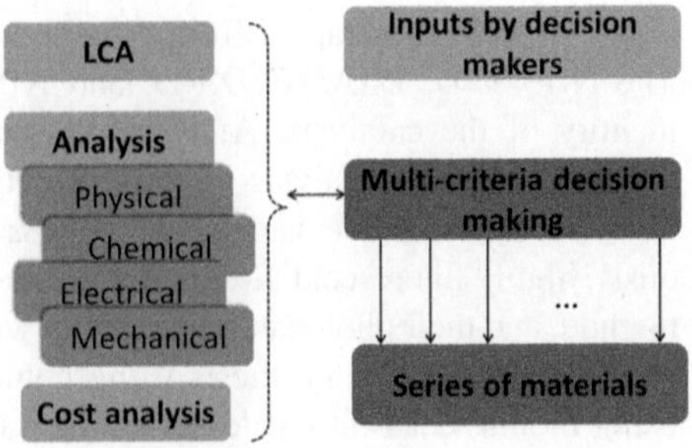

Fig. 1 A diagram showing how to choose materials in an integrated design process

4 LCA of Quantum Dots and Its Composites

Specifically designed and manufactured materials having dimensions between 1 and 100 nm are known as engineered nanomaterials (ENMs) [25]. In addition, ENMs show unique characteristics in comparison to bulk materials of the same composition [26]. The environmental implications of ENM production and use are unknown, despite the fact that their commercial production volumes have increased [27]. Therefore, in order to guarantee their sustainable production and utilisation, it is crucial to have a thorough awareness of the impacts made by ENMs throughout their life cycle. Since LCA attempts to quantify the environmental consequences of a system over the course of its life cycle [28], it is the most suited instrument for tackling this challenge [29]. However, the environmental evaluation of a significant and unique ENM, carbon dots (CDs), has not been the subject of any prior research. Carbon nanospheres are a new type of nanoparticle with a round shape and a size between 1 and 10 nm [30]. Nonetheless, making them in a way that doesn't harm the environment hasn't received much attention. Consequently, it is important and essential to provide information on the potential environmental implications of representative synthesis processes for the creation of CDs [31]. Here, we apply a cradle-to-gate life cycle assessment (LCA) to compare and comprehend the environmental effects of carbon dots (CDs) produced using six distinct bottom-up synthetic techniques. The most common methods now used to synthesise CDs are hydrothermal synthesis and microwave-assisted synthesis, both of which use citric acid as a starting material (and sometimes urea as well). According to the findings, power consumption is the most significant environmental factor in hydrothermal synthesis, while citric acid is the main source of pollution in microwave-assisted synthesis. Rescaling results using the CDs' fluorescence quantum yield allowed for a performance-based comparison as well. This method substantially altered the previous preferred order across all classes. The most environmentally friendly method, according to a recent study, was the microwave-assisted synthesis of citric acid-derived CDs; however, today both urea- and citric acid-derived CDs can be synthesised (by hydrothermal or microwave-assisted treatment) [32]. Different acid catalysts based on transition metal phosphates, such as vanadium and niobium, have been evaluated for their performance in the hydrothermal synthesis of carbon dots (CDs). Commercial xylose and liquor of xylose extracted from olive pits were used as the carbohydrate sources. The NH3-TPD, DTA/TG, XRD, and XPS methods were used to determine the identity of the catalysts. At a temperature of 180 degrees Celsius, the reaction was run for 4 h. Analysis of the properties and characteristics of CDs nanoparticles revealed their existence, regardless of the carbohydrate source. Simultaneously, highly fluorescent N-doped CDs were synthesised via the same hydrothermal method, and their photocatalytic activity was studied. To compare the environmental consequences of the synthesis using commercial xylose to those of the synthesis using biomass, a Life Cycle Assessment (LCA) was performed on both synthesis methods [33]. Organic light-emitting diodes (OLEDs) and inorganic nanoparticle-based quantum LEDs (QLEDs) are two rival forms of light-emitting technology used in displays. For these purposes, innovative nanomaterials and organics can be found in the next substance classes of perovskites and Q-OLED displays. However, due to their complexity, intrinsically diverse structures, and rapid growth in the literature, assessing the safety and viability of these emissive compounds in a timely and comprehensive manner is challenging. To compare these potential alternatives to incombent cadmium-containing quantum dots, we propose adopting an alternatives assessment centred on danger, cost, and performance. Chemical substitution is a growing trend, and this type of assessment is employed by both industry and government. It makes advantage of existing data while highlighting crucial information gaps that must be considered. The low quantities required in their application for display make OLEDs cost-effective, although performance evaluations do not single out a superior option. According to the risk assessment, there is no better option because each unique nanomaterial or organic compound has its own set of drawbacks [34]. The study provides a framework for future researchers to analyse their own unique drugs, and the results highlight the need for a low-hazard high-performance replacement substance. New domains of application can be explored with the advent of emerging photovoltaic systems (EPVs) such organic solar cells, dye-sensitized solar cells, perovskite solar cells, and quantum dots solar cells. Life cycle analysis is essential for determining the environmental impact of EPV technology developments and sophisticated materials. Materials and production methods responsible for the bulk of a product's environmental effect can be isolated with the help of life cycle evaluations (LCAs). Recent life cycle assessments (LCAs) have shown that EPV production may result in reduced energy consumption and faster energy payback time compared to conventional PV technologies, although these results are sometimes hard to compare due to differences in methodology and system boundaries. However, the examined LCAs also identify some "environmental hotspots" regarding the materials, energy, and chemicals used. The existing use of vital raw materials, precious metals, and hazardous as well as energy-intensive products means there is still opportunity for optimisation in terms of environmental sustainability and the circular economy [35]. Carbohydrates, lipids, phenolic compounds, and proteins are only few of the many organic substances found in spent coffee grounds (SCGs). As a result, we looked at them as a feasible option for obtaining carbon dots (CDs) via a nanotechnology method. In this study, CDs made by SCGs were compared to those made from more conventional precursors like citric acid and urea. The SCG-based CDs were made by carbonising solid samples in a single pot without using any solvents, yielding particles on the nanoscale (2.1–3.9 nm). These nanoparticles displayed the signature blue fluorescence of carbon dots as well as modest quantum yields (2.9–5.8%) and excitation-dependent emission. The SCG-based CDs showed promise as fluorescent probes for Fe3+ in water that are also environmentally relevant. Furthermore, life cycle assessment studies confirmed that making CDs out of SCG samples is a more environmentally friendly way than making CDs out of classic reported precursors, when looking at it from both a weight-based and a function-based perspective [36]. Among the many potential uses for quantum dot nanoparticles (NPs) is in photon upconversion devices, which boost the solar panels' ability to convert light into power. In this study, we present ready-to-use LCA unit process data for four NPs (cadmium selenide, cadmium sulphide, lead selenide, and lead sulphide) that are well-suited for photon upconversion applications. The information is presented for two potential futures: an optimistic and a pessimistic one. The effectiveness of the NPs in mitigating climate change is evaluated using an impact assessment, which reveals that solvent-related operations, like steam production for recycling and hazardous waste treatment, are major contributors to this issue. To demonstrate the relevance of the findings, an upconversion-layer solar module is evaluated prospectively to determine whether adding more solar modules or retrofitting existing ones with upconversion devices is preferable from a climate perspective [37]. The evaluation reveals that, depending on the circumstance, solar modules would need to increase their efficiency by 0.05 to 2 percentage points per gramme of applied NPs for the upconversion layer to be desirable.

5 Conclusions

The environmental implications of well-established processes and products can be effectively evaluated with the help of life cycle assessment (LCA). However, incorporating it into decision-making for the long-term success of cutting-edge technology presents a formidable obstacle. It is challenging to do LCA assessments at the outset of product design for new nano-enabled products (NEPs) due to the high levels of uncertainty and lack of data along the whole value chain. Data scarcity and quality difficulties are common for LCA practitioners working on developing technologies due to the proprietary nature of industrial data, necessitating some assumptions based on prior scientific literature and industry reports. These assumptions are standard practise in LCA, and while they are not false, they are likely to add some degree of uncertainty to the findings. To facilitate the growth of collaborative research across the life cycle and the generation of fresh datasets, the dLCA framework has been established. By feeding the experimental data back into the LCA model, uncertainty in various LCA stages can be reduced with each iteration, while also pointing experimentalists in the direction of questions they should be asking in the future. Although this research focuses on quantum dot (QD) applications, the suggested dLCA framework and the interdisciplinary cooperation it encourages can be applied to the estimation of environmental impacts across a wide range of developing technologies.

Future developments in assessment methodologies for Life Cycle Assessment (LCA) of quantum dots (QDs) and their composites will take into account shifting synthesis methods, expanding applications, shifting regulations on nanomaterials, and a heightened emphasis on circular economy principles and recycling practises. In addition to promoting better data gathering, transparency, and accessibility, future LCA studies will cover a wider range of uses across a variety of sectors. The results of life cycle assessments (LCAs) will also be used in real-time decision making to promote ethically sound product and policy creation. Researchers hope that by effectively disseminating their findings, LCA studies will help raise public awareness of the environmental impact of QDs and ultimately lead to policy changes that promote sustainability and environmental responsibility in the QD industry.

References

1. Bachmann, J., Hidalgo, C., Bricout, S.: Environmental analysis of innovative sustainable composites with potential use in aviation sector—A life cycle assessment review. Sci. China Technol. Sci. **60**, 1301–1317 (2017)
2. Karuppannan Gopalraj, S., Kärki, T.: A review on the recycling of waste carbon fibre/glass fibre-reinforced composites: Fibre recovery, properties and life-cycle analysis. SN Appl. Sci. **2**(3), 433 (2020)
3. Tapper, R.J., et al.: An evaluation of life cycle assessment and its application to the closed-loop recycling of carbon fibre reinforced polymers. Compos. B Eng. **184**, 107665 (2020)
4. Das, P., et al.: Dual doped biocompatible multicolor luminescent carbon dots for bio labeling, UV-active marker and fluorescent polymer composite. Luminescence **33**(6), 1136–1145 (2018)
5. Ead, A.S., et al.: Life cycle analysis for green composites: a review of literature including considerations for local and global agricultural use. J. Eng. Fibers Fabr.Fibers Fabr. **16**, 15589250211026940 (2021)
6. Ardente, F., et al.: Energy performances and life cycle assessment of an Italian wind farm. Renew. Sustain. Energy Rev. **12**(1), 200–217 (2008)
7. Das, P., et al.: One-step synthesis of fluorescent carbon dots for bio-labeling assay. In: Macromolecular Symposia. Wiley Online Library (2018)
8. Das, P., et al.: Green approach to photoluminescent carbon dots for imaging of gram-negative bacteria Escherichia coli. Nanotechnology **28**(19), 195501 (2017)
9. Das, S.: Life cycle assessment of carbon fiber-reinforced polymer composites. Int. J. Life Cycle Assess. **16**, 268–282 (2011)
10. Blanco, I., Ingrao, C., Siracusa, V.: Life-cycle assessment in the polymeric sector: a comprehensive review of application experiences on the Italian scale. Polymers **12**(6), 1212 (2020)
11. Maruthapandi, M., et al.: Biocompatible N-doped carbon dots for the eradication of methicillin-resistant S. aureus (MRSA) and sensitive analysis for europium (III). Nano-Struct. Nano-Objects **26**, 100724 (2021)
12. Saravanan, A., et al.: Green synthesis of multifunctional carbon dots with antibacterial activities. Nanomaterials **11**(2), 369 (2021)
13. Dissanayake, N.P., Summerscales, J.: Life cycle assessment for natural fibre composites. Green Compos. Nat. Resour. **8**, 157–186 (2013)

14. Bosner, A., Poršinsky, T., Stankić, I.: Forestry and Life Cycle Assessment. InTech Rijeka (2012)
15. Achilias, D., et al.: Chemical recycling of plastic wastes made from polyethylene (LDPE and HDPE) and polypropylene (PP). J. Hazard. Mater. **149**(3), 536–542 (2007)
16. Scott, D., et al.: Fast pyrolysis of plastic wastes. Energy Fuels **4**(4), 407–411 (1990)
17. Das, S.C., La Rosa, A.D., Grammatikos, S.A.: Life cycle assessment of plant fibers and their composites. In: Plant Fibers, their Composites, and Applications, pp. 457–484. Elsevier (2022)
18. Cao, C.: Sustainability and life assessment of high strength natural fibre composites in construction. In: Advanced High Strength Natural Fibre Composites in Construction, pp. 529–544. Elsevier (2017)
19. Karjalainen, T., et al.: Energy, carbon and other material flows in the life cycle assessment of forestry and forest products. Joensuu: European Forest Institute Discussion Paper, vol. 10, pp. 1–67 (2001).
20. Papanek, V.: The Green Imperative: Ecology and Ethics in Design and Architecture. Thames & Hudson (2022)
21. Liboiron, M., Lepawsky, J.: Discard Studies: Wasting, Systems, and Power. MIT Press (2022)
22. Bernardo, C., Simões, C.L., Pinto, L.M.C.: Environmental and economic life cycle assessment of polymers and polymer matrix composites: a review. Ciência & Tecnologia dos Materiais **28**(1), 55–59 (2016)
23. Milani, A., et al.: Multiple criteria decision making with life cycle assessment for material selection of composites. Express Polym. Lett. **5**(12) (2011)
24. Akin, Ö.: Necessity of Cognitive Modeling in BIM's Future. Building Information Modeling: BIM in Current and Future Practice, pp. 17–27 (2015)
25. Auffan, M., et al.: Towards a definition of inorganic nanoparticles from an environmental, health and safety perspective. Nat. Nanotechnol.Nanotechnol. **4**(10), 634–641 (2009)
26. Feijoo, S., et al.: Comparative life cycle assessment of different synthesis routes of magnetic nanoparticles. J. Clean. Prod. **143**, 528–538 (2017)
27. Salieri, B., et al.: Life cycle assessment of manufactured nanomaterials: where are we? NanoImpact **10**, 108–120 (2018)
28. Upadhyayula, V.K., et al.: Life cycle assessment as a tool to enhance the environmental performance of carbon nanotube products: a review. J. Clean. Prod. **26**, 37–47 (2012)
29. Ramos, A., Afonso Teixeira, C., Rouboa, A.: Environmental analysis of waste-to-energy—A Portuguese case study. Energies **11**(3), 548 (2018)
30. Das, P., et al.: Tailor made magnetic nanolights: fabrication to cancer theranostics applications. Nanoscale Adv. **3**(24), 6762–6796 (2021)
31. Das, P., et al.: Graphene based emergent nanolights: a short review on the synthesis, properties and application. Res. Chem. Intermed.Intermed. **45**, 3823–3853 (2019)
32. Sendao, R., et al.: Comparative life cycle assessment of bottom-up synthesis routes for carbon dots derived from citric acid and urea. J. Clean. Prod. **254**, 120080 (2020)
33. Rodríguez-Carballo, G., et al.: Nitrogen doped carbon dots as a photocatalyst based on biomass. A life cycle assessment. J. Clean. Prod. **423**, 138728 (2023)
34. Bechu, A., et al.: Are substitutes to Cd-based quantum dots in displays more sustainable, effective, and cost competitive? An alternatives assessment approach. ACS Sustain. Chem. Eng. **10**(7), 2294–2307 (2022)
35. Gressler, S., et al.: Advanced materials for emerging photovoltaic systems – Environmental hotspots in the production and end-of-life phase of organic, dye-sensitized, perovskite, and quantum dots solar cells. Sustain. Mater. Technol. **34**, e00501 (2022)
36. Crista, D.M., et al.: Turning spent coffee grounds into sustainable precursors for the fabrication of carbon dots. Nanomaterials **10**(6), 1209 (2020)
37. Wickerts, S., et al.: Prospective life-cycle modeling of quantum dot nanoparticles for use in photon upconversion devices. ACS Sustain. Chem. Eng. **9**(14), 5187–5195 (2021)

Theory, Modeling and Simulation of Quantum Dots and the Multiphase Systems

Siddiqui Fatima Sarah and Sunita Bhosle

Abstract To understand the complex behaviour of quantum dots in multiphase systems, this chapter of a book delves into the theoretical foundations, advanced modelling approaches, and sophisticated simulations used to accomplish so. Nano-sized structures known as quantum dots show great promise for a variety of uses due to their exceptional electrical and optical characteristics. To fully appreciate and make use of these attributes, a thorough theoretical foundation is required. The confinement effects and energy levels that determine the behaviour of quantum dots are first explored in this chapter, which delves into the fundamental quantum mechanical principles regulating quantum dots. Next, it delves into cutting-edge modelling techniques, such Density Functional Theory, for making precise predictions about quantum dot properties. The effect of quantum dots' interactions with various materials on optical, electrical, and thermal properties is discussed, and extended to multiphase systems. This chapter utilises state-of-the-art simulations to provide light on the design and optimisation of devices based on quantum dots, with a particular emphasis on fine-tuning their features for use in next-generation photonics, electronics, and renewable energy technologies.

Keywords Quantum dots · Theoretical foundation · Modelling · Photonics · Electronics

1 Introduction

Quantum dots (QDs) are nanometer-sized systems that exhibit quantum confinement in all three spatial dimensions due to their small size [1–5]. Typically smaller than 20 nm in size, colloidal QDs are fluorescent semiconducting nanocrystals having a radius comparable to the Bohr exciton radius of the material [6]. Colloid chemistry has advanced greatly in recent decades, allowing for the manufacture of high-quality

S. F. Sarah (✉) · S. Bhosle
Balbhim Arts, Science and Commerce College, Balbhim Nagar Path, Bir, Beed, Maharashtra 431122, India
e-mail: fatimasarah25@gmail.com

S. Thomas et al. (eds.), *Quantum Dots Based Nanocomposites*, Engineering Materials,
https://doi.org/10.1007/978-3-031-54779-9_24

semiconductor nanocrystal QDs. Compared to organic dyes and polymers, quantum dots have improved chemical and photo-stability as well as a high photoluminescence quantum yield and a broad absorption spectrum that overlaps with the solar spectrum [7, 8]. In highly sensitive non-isotopic detection methods, colloidal QDs have been frequently used as biological tags. It is a viable contender for use in light-emitting diodes, nonlinear optics, laser media, and photovoltaics due to its adjustable emission, strong photo-oxidation stability, and electron conductivity [9].

The energy and spacing of discrete electron levels are strongly dependent on crystal size in nanocrystals, making them accessible to scientific observation and practical use. Since quantum dots have electron states on a higher scale than the lattice constant of a crystal, they are sometimes referred to as mesoscopic atoms or artificial atoms. A quantum dot (QD) can be any size up to and including a macromolecule or a single impurity atom in a crystal. Changes in transport and optical properties are profound when electron energy are quantized in nanometer-sized crystals. The dependency of the fluorescence wavelength on the size and material composition of the nanocrystals is illustrated in Fig. 1. Using materials with varying band gaps (CdSe for blue, InP for green, and InAs for red) leads to significant changes in emission wavelength. Within each colour, the size of the QDs determines the degree of fluorescence emission.

Discrete energies occur when the size of the system is on par with the Bohr exciton radius at the nanoscale, which results in a high effective band gap and a blue shift in the optical transition. Enclosing the QD within an organic surfactant is a common method for achieving quantum confinement in QDs. Therefore, the typical QD has an inorganic core surrounded by organic surfactant ligands. Type II–VI, IV–VI, and III–V semiconductors, including (CdS) ZnS, (CdSe) ZnS, (CdSe) CdS, and (InAs) CdSe, are usually used to construct core–shell quantum dot (QD) systems [10]. The perovskite solar cell has emerged as a significant member of the core–shell QDs family in recent years [11]. The most promising optical gain medium for the advancement of solar cells was thought to be the organic–inorganic halide perovskite material ($CH_3NH_3PbX_3$, X = Cl, Br, I), due to its huge optical absorption coefficient and high carrier mobility [12].

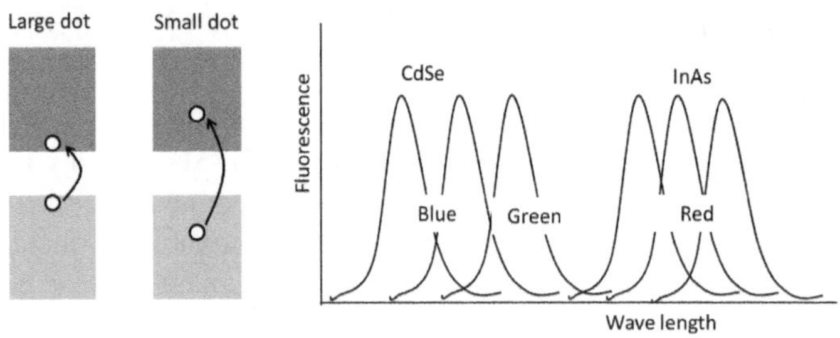

Fig. 1 The band gaps of the materials cause significant variations in the fluorescence wavelength. The varying diameters of the NCQDs define the wavelength for each of the three primary colours

The study of QDs primarily makes use of two methods: experiment and computational simulation. The optical, electrical, and thermal properties of QDs, as well as their sensitivity to changes in size, shape, and surface passivation, have been the subject of extensive experimental and theoretical study. The eventual design and production of colloidal QDs relies heavily on experimentation, however this approach presents significant obstacles due to its high cost and lengthy development cycle. In order to keep the optical, electrical, and thermal properties of typical QDs intact despite their susceptibility to environmental changes, special attention must be taken in their design. Extreme difficulty is posed by the fact that experimental examinations of QDs' surface configurations may be costly and time-consuming. Core–shell QDs frequently use chemical functionalization, although it is challenging to realise surface decorations with specified functional groups. QDs' electrical states, on the other hand, are crucial to comprehending their characteristics. Common experimental methods, however, do not yield usable electronic images at the atomic scale. Because of the nanoscale nature of QDs, computational simulation has emerged as a crucial tool in our quest to learn more about them. Numerical simulations from basic principles and classical molecular dynamics (MD) have been employed extensively in difficult issues for a long time. Modern solutions to the issues encountered in experimental research can be found in computational approaches. They can consider the bulk surface and defects on an equal footing and add electronic correlation effects using the first principles method, which yields an accurate atomistic description of the electronic structure.33 Using the density functional theory (DFT) technique, researchers have proposed a new device based on graphene QD for DNA sequencing, which should serve as useful guidance for future experiments [13].

2 Quantum Dot Simulations and Common Computational Methods

Quantum dots (QDs) have had their electrical and optical properties, which are regulated by the density of states (DOS), studied extensively through first principles computational technique [14]. The ab initio method, or first principles method, is a technique for numerically solving the Schrodinger problem. In order to numerically solve the Schrodinger equation, or more accurately the Kohn–Sham equation, the DFT method is commonly employed [15]. Traditional self-consistent first principles local-density approximation (LDA) calculation is limited to calculations involving a few dozen atoms at most, and is written as [16];

$$E_{xc}^{LDA} = \int \rho(r)\varepsilon_{xc}(\rho)dr,$$

where ρ is the per-particle exchange correlation energy in a homogeneous electron plasma of density.

Calculating the near-band-gap single electron states relevant to the optical and transport properties of thousand-atom colloidal QDs has been proposed as a much faster alternative to LDA that scales linearly with system size, mitigating the extensive computational effect brought by traditional self-consistent first principles method [17].

It has been noted that the energy gap may be underestimated by LDA and generalised gradient approximation (GGA) methods when calculating principal energy [18].

So, the modified Becke Johnson (BJ) and the Engel-Vosko generalised gradient approximation (EVGGA) formalism48 were put forward. It was stated that the exchange correlation energy and its charge derivative could not be accurately reproduced using LDA or GGA. The exchange energy can be reproduced with less precision using EVGGA and BJ, but they are still capable of replicating the exchange potential [19].

It is possible to determine the QD potential by;

$$P_{QD} = \frac{b}{a} E_g \times 10^{-3} \lambda,$$

Parameter for semiconductors, where E_g is the energy gap and b/a is a constant. The indirect energy gap can be computed using DFT by employing the full potential-linearized augmented plane wave approach. Within the non-overlapping spheres, the wave function and charge density are extended using spherical harmonic functions, while outside, the wave function and charge density are expanded using a basis set of plane waves. The exchange correlation potential within a GGA framework can be used for total energy estimates.

It provides the maximum thermal conductivity of a material and can accurately represent ballistic thermal transfer. Thermal transport can be thought of as ballistic and characterised by the Landauer formula when the system size of a nanomaterial is smaller than its phonon mean free path [20].

$$I = \int_0^\infty \frac{d\omega}{2\pi} \mathrm{h}\omega \tau[\omega](f_L - f_R),$$

Here $\tau[\omega]$ is the transmission coefficient.

3 Optical Properties by Computational Simulations

Colloidal semiconducting QDs are promising emitters for luminescent solar concentrators with potential uses in high-efficiency and cost-effective photovoltaics due to their broad absorption spectra spanning the ultraviolet to the near infrared. QDs can have their absorption and emission spectra tuned to minimise energy loss due to reabsorption. Because QD's quantum confinement is size, shape, and chemical composition dependant; its optical properties can be precisely tuned through design. Despite intensive work on the core–shell QD systems' fabrication [21], The development of well-behaved colloidal QDs may be hampered by the lack of in-depth research into the optical properties of these core–shell QDs, which has thus far hampered progress in these areas. In order to improve photocatalytic capabilities, it is essential to absorb solar energy and undergo photo-induced charge separation and transfer. Photon losses at subgap energies and the decay of electron–hole pairs limit the highest efficiency of conventional solar cells to the Shockley-Queisser (SQ) limit (33%). Using intermediate band (IB) as a subgap photons absorber is a viable approach to breaking the SQ limit [22]. The optical performance of QD structures was intensively worked on using computational methods. In order to apply the IB idea to solar cells, colloidal QDs have recently been proposed as a platform. The IB states may be clearly distinguished from the margins of the valence and conduction bands in arrays of surface-reconstructed CdSe dots, as shown by Vörös et al. using a first-principles approach. Cobaltocene-doped CdSe QDs have their electron affinity and ionisation potential measured by their total energy difference [23]. When the electron affinity for the cobaltocene is greater than the ionisation potential of the CdSe, electron transfer occurs. Cubic binding energy of 0.16 eV/QD for CdSe QDs indicates stability in the absence of ligands at ambient temperature.

In addition to the aforementioned core–shell QD systems, there is a plethora of carbon nanodots based QDs that vary greatly in size, edge topology, and functional groups. After simulating 2D graphene QDs with high-resolution transmission electron microscopy and performing tight-binding computations of their optical spectra using a time-dependent density functional, Ghosh et al. suggested multiple atomistic models of these materials. Light absorption at the visible range has been decoded by locating the dominant electron and hole orbitals [24]. Armchair edges contribute heavily to the DOS and result in a clear band gap in the graphene QDs model at the nanoscale length scale. It is also determined that the introduction of a controlled number of hydroxyl (OH) or carbonyl (COOH) groups into graphene quantum dots is an efficient way to fine-tune their excitation energies in the visible spectrum.

4 Electronic Properties by Computational Simulations

QDs have a highly variable electrical structure that is mostly determined by their size and surface conditions. Traditional QDs, consisting of a semiconductor or metal core or shell, may find usage in electronics. Although polymeric materials are simple to manipulate, they also have a low dielectric field. Ceramic materials, on the other hand, have a greater dielectric constant but are more difficult to synthesise. Engineers can more easily access the polymers' inside through a core–shell structure that combines it with a ceramic. The high capacitance of these materials makes them useful in a variety of electrical contexts.

By coating the core particle in a shell, we can boost its colloidal stability and shield it from photodegradation. Many inorganic components, including silicon and silica, are employed in core–shell QDs. Using a first-principles approach, Reeves et al. investigated the effect of surface termination on phonon-mediated relaxation of an excited electron in a silicon-based QD structure [25]. Surface passivation for silicon QD with fluorine atoms was found to significantly slow down hot electron relaxation, it was stated. The relaxation period is mostly determined by the rate at which the heated electron is shuttled between energetically neighbouring states, and specific electronic states are able to hold the electron for longer than others.

The behaviour of the electrons during relaxation in QDs can be characterised by [26]:

$$\langle E(t)\rangle = \sum_i \varepsilon_i(t)p_i(t),$$

where E(t) is the ensemble-averaged energy, ε_i is the electronic state *i* and $p_i(t)$ is the state probability.

The impact of silicon QD size on hot electron relaxation was studied by Wong et al.86. Large QD have a lesser surface area to volume ratio than small QD, hence the effect of fluorine passivation is mitigated [27]. Decoherence and the impact of quantum dot size on this result connected with fluorine passivation were also explored. The relaxation time constant decreases dramatically, and the characteristic hot electron shuttling behaviour disappears, as the size of the silicon quantum dot increases. However, decoherence can halt the hot electron relaxation by a factor of two or more. Our research shows that the quantum Zeno effect, in addition to the surface-specific vibronic effect for tiny quantum dots, can cause gradual hot electron relaxation.

Core–shell QDs can be split in half based on their electrical states. If the minimum of the conduction band and the maximum of the valence band are located in the same material, we say that the core–shell structure is of type-I. They are considered type-II if they are notable changes in their electrical structures. The core–shell CdSe/CdS and CdSe/CdTe QDs electronic states were studied by Li et al. using a first-principles method. CdSe/CdS and CdSe/CdTe QDs are proven to be type-I and type-II heterostructures, respectively, based on their inherent band-offset alignments [28]. CdSe/CdTe QDs have an extremely distinct hole state from CdS/CdS QDs, with the hole wave function being localised within the shell in the former and the core in the latter.

Size, shape, and edge morphology all play important roles in determining the energy gap in 2D monolayer based QDs such silicene, graphene, and hexagonal boron nitride (h-BN), in addition to the core–shell structure [29]. The influence of edge passivation with OH and CN groups and H, F, O, and S elements on the electrical characteristics of hexagonal and triangular shaped silicene QDs with armchair and zigzag borders is studied using first principles DFT approach [30]. Edge passivation by various atomic and molecular families is studied for its effect on the stability and electrical characteristics of hexagonal and triangular silicene quantum dots. Depending on the pieces or groups that are attached, the structures undergo significant shape changes. Sulfur-passivated zigzag triangular flakes and flakes with OH groups linked to their edge atoms exhibit the most striking morphological changes. A significant total dipole moment is manifested in a spherical form. Positive binding energies have been calculated for all the clusters that have been investigated, proving their stability. In zigzag hexagonal-H (zigzag triangular-H) and zigzag hexagonal-S (zigzag triangular-OH), a flexible structural transformation from insulator (conductor) to conductor (insulator) is obtained. Flakes with an even number of Si atoms will be antiferromagnetic, while flakes with an odd number of Si atoms can be either ferromagnetic or antiferromagnetic, depending on the attached element or group, due to the parity of the total number of Si atoms. As a result, silicene flakes can have their characteristics modified for specific uses by selecting the appropriate functional groups or components to attach to them. Clusters passivated with S or OH are insensitive to edge defects and have tunable electronic properties, making them promising in semiconductor device applications; in contrast, hydrogenated or fluorinated flakes are highly interactive with the surrounding environment and can be used for sensor applications.

5 Thermal Properties by Computational Simulations

Especially in nanotechnology, the thermal characteristics of new materials are critical to their design and performance. Composites made from integrating quantum dots into a matrix material are an exciting new field of study with numerous potential uses. By controlling their thermal behaviour, these composites can be optimised for a wide variety of uses. To investigate and comprehend these materials' thermal properties at the nanoscale, computational simulations are an indispensable tool. The importance of computational simulations in investigating the thermal characteristics of quantum dot-based composites is explored in this article. The ability of a substance to conduct heat is described by its thermal conductivity, a fundamental thermal attribute. The thermal conductivity of composites based on quantum dots is crucial because it has a direct impact on the performance of these materials in a wide range of contexts. Small size and quantum confinement factors mean that quantum dots can have dramatically different thermal conductivity from bulk materials. Computational simulations, and in particular Molecular Dynamics (MD) simulations, are used by scientists to learn about and make predictions about the thermal conductivity of these composites. In MD simulations, atom mobility and energy transmission within the composite are modelled and monitored. The thermal conductivity can be determined by measuring the atomic vibrations and energy transfer. These simulations can shed light on the impact of variables like quantum dot size and spacing, as well as dot-matrix interactions, on thermal transport inside the composite.

The quantity of heat energy needed to increase the temperature of a material by a given amount is known as its "specific heat capacity." Knowing the specific heat capacity of composites based on quantum dots is crucial for uses in temperature regulation and management. The electrical structure of quantum dots and their interactions with the surrounding matrix can be predicted with the help of computational simulations, in particular Density Functional Theory (DFT). To determine the composite's specific heat capacity, this data is necessary. In addition, the lattice contribution to specific heat can be understood by classical MD simulations. Since the specific heat capacity of quantum dots is highly application-dependent, computational simulations are an invaluable tool for optimising this feature for particular uses. The term "thermal expansion" is used to refer to the enlargement or contraction of a substance as a result of changes in temperature. In order to design materials for specific applications, knowledge of thermal expansion is essential for quantum dot-based composites. The phonon frequencies and vibrational modes of both the quantum dots and the surrounding matrix can be predicted using computational simulations, in particular those using lattice dynamics and Density Functional Theory. The thermal expansion coefficient can be calculated by observing the motion of atoms within the composite when it is heated and cooled. Computational simulations allow us to probe and manipulate the thermal expansion of quantum dots by changing their size, shape, and interaction with the matrix material.

In addition to these standard thermal characteristics, composites based on quantum dots may display unusual behaviours as a result of quantum mechanical processes. The electrical and phonon structure of quantum dots, as well as their interactions with the surrounding matrix, can be understood with the help of computational tools like Density Functional Theory. Understanding the effects of quantum confinement and quantum dot size on heat transit and storage inside the composite requires an accurate assessment of its thermal properties, which may be achieved with the help of the data presented here. It is also important to carefully analyse the simulation boundary conditions you choose, such as periodic boundary requirements. To get reliable findings, one must carefully evaluate the simulation cell size and the impact of finite size effects on thermal parameters. In order to characterise temperature-dependent behaviour in quantum dot-based composites, it is necessary to run simulations over a wide temperature range. In addition, the thermal characteristics of the composite might be drastically altered by the presence of flaws or impurities. Researchers can evaluate the effects of these flaws on heat transport, thermal expansion, and specific heat capacity by extending computational models to include them. Knowing that these flaws can be present in practical applications is crucial for understanding how real-world materials behave. The examination of simulation data has also benefited from the use of machine learning methods. Conventional analytic approaches may miss trends, patterns, and correlations that can be uncovered by applying machine learning algorithms to simulation findings. These methods can be used to foretell how different quantum dot composites would behave at high temperatures. This will help in the design of novel materials with controllable thermal properties. Computational results must be compared with experimental observations to verify the reliability of simulations and improve the quality of the underlying models. Researchers can verify their models and increase their faith in their accuracy by comparing them to experimental data, which provides the real-world context for the computational predictions.

Numerous promising applications could be realised with a better understanding of the ability to manipulate the thermal characteristics of composites based on quantum dots. For optimal performance in thermoelectric applications, for instance, it is essential to optimise heat conductivity while keeping good electrical conductivity. To keep devices from overheating and failing, effective heat management is crucial in nanoelectronics. By manipulating their thermal behaviour, composites can improve the efficacy of photothermal therapy in killing cancer cells. To better understand the physics at play and to direct the development of such materials with specific thermal characteristics, computational simulations play a crucial role.

In conclusion, the investigation of the thermal characteristics of composites based on quantum dots requires the use of computational models. Quantum mechanical effects, thermal conductivity, and specific heat capacity can all be better understood with the help of these models. Quantum dot size, flaws, and temperature dependence are all important considerations that contribute to a fuller picture of the nanoscale behaviour of these materials. Electronics and medicine are only two of the many fields that stand to benefit from advances in our capacity to anticipate and manipulate the thermal properties of composites based on quantum dots. Computational simulations are a crucial tool in the development of innovative materials, especially when combined with machine learning approaches and experimental validation.

The invention of thermoelectric materials with a high figure of merit is one of the most difficult tasks in modern material research. It has been suggested that a QDSL, which is made by clustering quantum dots into patterned composites, could be a good choice for thermoelectric applications due to its lower lattice thermal conductivities. Reduced lattice thermal conductivity due to acoustic phonon scatterings at the interface and the quantum confinement of carriers contribute to the enhancement of ZT in QDSL. By manipulating the cluster configurations, one can change the Fermi level positions and hence the electronic characteristics of QDSL. Thermoelectric figure-of-merit enhancement in quantum dot superlattices was proposed by Balandin et al. For the p-type regulated superlattice of Ge dots on Si, a proof-of-concept calculation has been performed. It is demonstrated that the thermoelectric figure-of-merit can be improved by tuning carrier transport in such structures, under circumstances for miniband formations, resulting in high carrier mobility, a high Seebeck coefficient, and a high thermoelectric performance. Maximising the gain requires setting the settings of the quantum dot superlattice so that the majority of the current flows in the wide, well-separated minibands [31].

6 Summary and Conclusions

This chapter provides a thorough examination of the theoretical underpinnings, sophisticated modelling approaches, and simulations required to comprehend quantum dot behaviour in multiphase systems. The unique electrical and optical features that quantum dots, as nanoscale semiconductor structures, offer have tremendous potential in a wide range of applications. The confinement effects and energy level structures that characterise quantum dots' behaviour are introduced as the chapter's first topic of discussion. Next, it explores state-of-the-art modelling techniques, such as Density Functional Theory, that pave the road for reliable prediction of quantum dot properties. The effects of quantum dots on optical, electrical, and thermal characteristics are discussed, and their application to multiphase systems is investigated. This chapter uses state-of-the-art simulations to show how to develop and optimise systems based on quantum dots, with a focus on customising their properties for use in photonics, electronics, and renewable energy.

The importance of theory, modelling, and simulation in furthering our knowledge of quantum dots and their behaviour in multiphase systems is highlighted in this chapter. Quantum dots are interesting prospects for a wide range of applications due to their unique features, but a strong theoretical foundation is required to properly exploit their potential. Researchers can precisely foretell quantum dot features using Density Functional Theory and other state-of-the-art modelling approaches, paving the way for the custom construction of materials with desired electrical and optical properties. The interactions of quantum dots with other materials have a major effect on the characteristics of multiphase systems in which they are incorporated. This chapter highlights the significance of these interactions and the use of simulations in achieving optimal performance in quantum dot-based devices for use in next-generation photonics, electronics, and renewable energy applications. Finally, the book chapter offers helpful advice for scientists and engineers who want to use quantum dots in various applications, thereby boosting the fields of nanoscience and materials engineering.

QD modelling and simulation is set to become a rapidly developing and game-changing industry in the near future. Accurate and comprehensive simulations are in high demand as the number of potential uses for quantum dots in nanoelectronics, photonics, and quantum technologies increases. Researchers will be able to optimise QD features for specific uses if they can model them more accurately using cutting-edge quantum mechanical techniques. QDs' behaviour in different contexts can be better understood with the use of multiscale models that combine quantum mechanics and classical techniques. In addition, understanding the role that ligands and solvents play in the interactions between QDs and their environments is essential for developing QDs with desirable characteristics. Advanced materials and devices will be easier to create with the help of simulations as the use of QD assemblies, heterostructures, and complicated systems grows in popularity. Unlocking the full potential of quantum dots in developing technologies will require a multidimensional strategy, including the integration of machine learning techniques and an emphasis on examining environmental and toxicological concerns.

References

1. Das, P., et al.: MXene/0D nanocomposite architectures: design, properties and emerging applications. Mater. Today Nano **24**, 100428 (2023)
2. Rajabzadeh, A.R., et al.: Magnetic Quantum Dots for Bioimaging. CRC Press (2023)
3. Das, P., et al.: Waste-derived sustainable fluorescent nanocarbon-coated breathable functional fabric for antioxidant and antimicrobial applications. ACS Appl. Mater. Interfaces (2023)
4. Das, P., et al.: Naturally derived carbon dots in situ confined self-healing and breathable hydrogel monolith for anomalous diffusion-driven phytomedicine release. ACS Appl. Bio Mater. **5**(12), 5617–5633 (2022)
5. Das, P., et al.: Carbon dot biopolymer-based flexible functional films for antioxidant and food monitoring applications. ACS Appl. Polym. Mater. **4**(12), 9323–9340 (2022)
6. Alivisatos, A.P.: Semiconductor clusters, nanocrystals, and quantum dots. Science **271**(5251), 933–937 (1996)
7. Regulacio, M.D., Han, M.-Y.: Composition-tunable alloyed semiconductor nanocrystals. Acc. Chem. Res. **43**(5), 621–630 (2010)
8. Zhao, H., Chaker, M., Ma, D.: Bimodal photoluminescence during the growth of PbS quantum dots. J. Phys. Chem. C **113**(16), 6497–6504 (2009)
9. Huang, S.-C., et al.: Investigation of luminescence enhancement and decay of QD-LEDs: interface reactions between QDs and atmospheres. ACS Appl. Mater. Interfaces **11**(2), 2516–2525 (2018)
10. Zeng, Z., et al., Linear and nonlinear optical properties of ZnO/ZnS and ZnS/ZnO core shell quantum dots: Effects of shell thickness, impurity, and dielectric environment. J. Appl. Phys. **114**(2), (2013)
11. Chen, H., et al.: Tunable photoluminescence of CsPbBr 3 perovskite quantum dots for their physical research. Appl. Surf. Sci. **465**, 656–664 (2019)
12. Zhao, F., et al.: Highly flexible organometal halide perovskite quantum dot based light-emitting diodes on a silver nanowire–polymer composite electrode. J. Mater. Chem. C **5**(3), 531–538 (2017)
13. Rastgoo, M., Tabatabaei, S.-M., Fathipour, M.: A first-principles study on DNA sequencing using graphene quantum dot. Eur. Phys. J. B. **91**, 1–6 (2018)
14. Lee, D., DuBois, J.L., Kanai, Y.: Importance of excitonic effect in charge separation at quantum-dot/organic interface: First-principles many-body calculations. Nano Lett. **14**(12), 6884–6888 (2014)
15. Kohn, W., Sham, L.J.: Self-consistent equations including exchange and correlation effects. Phys. Rev. **140**(4A), A1133 (1965)
16. Babaei, S., Niad, M.: Chemical reactivity descriptors as a tool of prediction in the synthesis of sandwich type polyoxometalate organic–inorganic hybrid compounds. Polyhedron **188**, 114710 (2020)
17. Wang, L.-W., Li, J.: First-principles thousand-atom quantum dot calculations. Phys. Rev. B **69**(15), 153302 (2004)
18. Dufek, P., Blaha, P., Schwarz, K.: Applications of Engel and Vosko's generalized gradient approximation in solids. Phys. Rev. B **50**(11), 7279 (1994)
19. Becke, A.D., Johnson, E.R.: A simple effective potential for exchange. J. Chem. Phys. **124**(22) (2006)
20. Landauer, R.: Spatial variation of currents and fields due to localized scatterers in metallic conduction (and comment). J. Math. Phys. **37**(10), 5259–5268 (1996)
21. Sagarzazu, G., et al.: Auger recombination dynamics in hybrid silica-coated CdTe nanocrystals. Phys. Chem. Chem. Phys. **13**(8), 3227–3230 (2011)
22. Luque, A., Martí, A.: Increasing the efficiency of ideal solar cells by photon induced transitions at intermediate levels. Phys. Rev. Lett. **78**(26), 5014 (1997)
23. Voros, M., Galli, G., Zimanyi, G.T.: Colloidal nanoparticles for intermediate band solar cells. ACS Nano **9**(7), 6882–6890 (2015)

24. Ghosh, S., et al.: Graphene quantum dots with visible light absorption of the carbon core: insights from single-particle spectroscopy and first principles based theory. 2D Materials **3**(4), 041008 (2016)
25. Reeves, K.G., et al.: Role of surface termination on hot electron relaxation in silicon quantum dots: a first-principles dynamics simulation study. Nano Lett. **15**(10), 6429–6433 (2015)
26. Li, L., Kanai, Y.: Excited electron dynamics at semiconductor-molecule type-II heterojunction interface: first-principles dynamics simulation. J. Phys. Chem. Lett. **7**(8), 1495–1500 (2016)
27. Wong, J.C., Li, L., Kanai, Y.: Size Dependence and role of decoherence in hot electron relaxation within fluorinated silicon quantum dots: a first-principles study. J. Phys. Chem. C **122**(51), 29526–29536 (2018)
28. Li, J., Wang, L.-W.: First principle study of core/shell structure quantum dots. Appl. Phys. Lett. **84**(18), 3648–3650 (2004)
29. Zhang, J., et al.: Dynamic response of graphene to thermal impulse. Phys. Rev. B **84**(23), 235416 (2011)
30. Abdelsalam, H., et al.: Stability and electronic properties of edge functionalized silicene quantum dots: a first principles study. Phys. E **108**, 339–346 (2019)
31. Balandin, A.A., Lazarenkova, O.L.: Mechanism for thermoelectric figure-of-merit enhancement in regimented quantum dot superlattices. Appl. Phys. Lett. **82**(3), 415–417 (2003)

Future Outlook and Present Market

Qurratulayen Siddiqui and Shaikh Yasmeen Aabeed

Abstract Composites containing quantum dots (QDs) have recently grown in importance, driving innovation in a wide range of sectors. This abstract explores the present and future market dynamics of composites based on quantum dots and the uses they have. With the increasing need for high-performance materials, QDs offer an unmatched opportunity to improve the electrical, optical, and catalytic characteristics of composites. Industries are investigating new uses in fields including electronics, healthcare, and energy, which have led to a boom in R&D efforts and the current state of the industry. When creating composites with specific functions in mind, quantum dots are an absolute must due to their size-tunable characteristics and quantum confinement effects. Composites based on quantum dots have the potential to shape the next generation of high-tech materials and technologies, which emphasizes the revolutionary effect of QDs on the field of materials research.

1 Background

Nanostructured materials are intriguing due to their ability to connect the macroscopic and molecular scales, opening up novel possibilities for applications, particularly in the fields of electronics, optoelectronics, and biology [1]. Nanostructures can be classified into three categories: (1) two-dimensional, such as thin films or quantum wells, (2) one-dimensional, such as quantum wires, and (3) zero-dimensional, also known as dots. In recent years, there has been significant interest in studying the optoelectronic characteristics of nanostructured semiconductors, namely quantum dots (Qdots). This is because many fundamental features of these materials are influenced by their size, particularly when it falls inside the nanometer range. A Qdot is considered to be zero-dimensional compared to the bulk material. Due to the limited amount of electrons, the density of states (DOS) for nonaggregated zero-dimensional objects exhibits discontinuous quantized energies [2]. (While the box is considered to be zero-dimensional in terms of its bulk, it is treated as a box in the field of quantum

Q. Siddiqui (✉) · S. Y. Aabeed
Department of Zoology, Dr. Rafiq Zakaria College for Women, Aurangabad, India
e-mail: ayenisiddiqui28@gmail.com

S. Thomas et al. (eds.), *Quantum Dots Based Nanocomposites*, Engineering Materials,
https://doi.org/10.1007/978-3-031-54779-9_25

mechanics. The size of the box is significant and will be examined further.) Occasionally, the existence of a single electronic charge in the Qdots causes the avoidance of an additional charge and results in a stair-shaped I-V curve and DOS. The size of each step in the staircase is directly proportional to the inverse of the radius of the Qdots. The transition between the bulk, Qdot, and atomic characteristics of a material is determined by the composition and crystal structure of the compound or elemental solid. A wide array of fundamental qualities can be achieved by altering the size while maintaining a constant composition, and some of these properties are discussed. Quantum dots can be classified into two main categories: elemental systems and compound systems. This review focuses on compound semiconductor-based nanostructured materials and their diverse applications, which leverage their optoelectronic and optical capabilities.

QDs exhibit exceptional potential in diverse applications owing to their distinct optical and electrical characteristics. Nanocomposites including QDs, which entail the integration of QDs into composite materials, have garnered considerable interest due to their potential applications in various disciplines. In this chapter we will provide a speculative outlook for the potential development of nanocomposites, incorporating QDs, focusing on their design, fabrication procedure, and growing applications [3]. The distinctive optical features of QDs have rendered them unparalleled possibilities for imaging applications both in living organisms (in vivo) and in laboratory settings (in vitro). The optical qualities of the QDs arise from their inorganic core, which is accountable for the fundamental semiconducting capabilities and optical properties. Before capping a QD with ligands, it is common to passivate the active core surface with an additional inorganic shell [4]. The optical properties of QDs are enhanced by the larger band gap of the shell in comparison to the core material, effectively preventing the migration of electrons and holes into the shell [5]. QDs exhibit distinctive optical characteristics as a result of the confinement of electron–hole pairs (excitons) within the limits of the nanocrystal grains. The unique photo-physical characteristics of QDs, such as wide absorption spectra and precise emission spectra that can be adjusted by size and composition, along with their ability to emit light, high fluorescent efficiency, strong absorption capacity, resistance to fading under light exposure, large difference between absorption and emission wavelengths, and reduced intermittent fluorescence, have established them as suitable options for a wide range of biomedical uses [6].

2 Tailored Characteristics

Scientists are projected to prioritize the development of QDs with customized characteristics, including sizes, shape, and composition, to attain precise functionality [7–14]. Future designs of nanocomposites may use various quantum dot varieties or other nanomaterials to generate multifunctional nanocomposites that exhibit improved performance.

3 Fabrication Procedures

The focus should be on finding efficient and economical fabrication methods for QDs and their nanocomposites to enable mass production[15]. Nanofabrication techniques, such as molecular self-assembly and template-assisted procedures, can allow accurate control of the arrangement of QDs within nanocomposite structures [16–18].

4 Emerging Applications

QDs possess exceptional optical characteristics, making them suitable for food industry and medicinal purposes such as improved imaging and tailored therapy[19–22]. Future advancements could encompass the development of more effective methods for administering drugs and the improvement of contrast agents.

Quantum dot-based displays have already been introduced in the market, providing enhanced colour accuracy and brightness compared to conventional displays [23]. Continuing research has the potential to result in display technologies that are even more efficient in terms of energy consumption and produce more vivid visuals.

QDs have the potential to be used in very effective energy harvesting devices, such as solar cells. Research may concentrate on maximising the efficacy of energy conversion and investigating innovative methods for harvesting solar energy[24].

5 Environmental Sensing

Sensor Technologies: Nanocomposites based on QDs have the potential to be utilised in the development of advanced sensor technologies for the purpose of monitoring the environment, identifying contaminants, and assuring the safety of atmospheric and water resources [25–27].

Smart Materials: The incorporation of QDs into smart materials may make it possible to create materials that are responsive and able to adjust to the shifting conditions of their surrounding environment[28].

6 Optical Computing

QDs provide distinctive optical properties that can be utilised to enhance optical computing, leading to improved speed and efficiency in computational solutions.

7 Regulatory and Safety Considerations

To address any environmental and health risks that are linked with the utilization of these nanocomposites, there may be an increased emphasis placed on the establishment of safety standards and regulations as quantum dot applications grow more widespread.

The future of QDs-based nanocomposites depends on ongoing scientific advancements, technical advances, and the effective transition of laboratory-scale progress into real-world applications[29, 30]. The potential of QDs in diverse fields can only be realized by continuous multidisciplinary collaboration among materials scientists, physicists, biologists, and engineers.

QDs have become widely adopted in the display industry [31]. Quantum dot displays, commonly referred to as QLED displays, are being integrated into televisions, monitors, and other consumer electronic products. These displays have benefits such as higher colour accuracy, expanded colour range, and increased brightness in comparison to conventional LED displays [32]. QDs are now being investigated for their potential use in a variety of biomedical applications, including diagnostics and imaging. As a result of their one-of-a-kind optical features, QDs are extremely useful for fluorescence imaging in biological systems. Investigations are also being conducted into applications in the fields of drug delivery and theranostics, which refers to combination diagnosis and therapy.

Ongoing research and development efforts are centred on enhancing the performance and stability of QDs, investigating new techniques of synthesis, and broadening the scope of applications for these QDs [33]. There is a lot of competition in the market for QDs, and many different companies and research organizations are actively contributing to the growth of quantum dot technology. The companies considered to be significant participants in this industry include well-established technological companies and newer startups that focus on nanotechnology and materials science [34]. Anticipated advancements and partnerships are expected to propel the use of QDs in many industries as the market for this technology progresses. The future trajectory of the QDs industry will be significantly influenced by the development of production methods that are both scalable and cost-effective, as well as the resolution of environmental and safety concerns.

QDs based nanocomposites have been actively investigated in a wide variety of practical research domains ever since their discovery. These fields include energy to electromagnetic interference (EMI) and catalysis to sensing, tribology, and biomedicine. However, despite their widespread application, they continue to have bottlenecks and drawbacks that need to be resolved and overcome in order to realize their full potential in the applications that were described earlier [35]. Every one of these applications has the potential to benefit from the advantages of the synergistic effect that is brought about by the rational design of functional zero dimensional materials, as well as surface-modified and hybrids based on mixed-dimensional materials properties [36].

8 QDs in the Healthcare Market

The global market for quantum dots (QDs) is now valued at USD 4 billion in 2021 and is projected to reach USD 8.6 billion during the next five years [37]. While display devices and LED applications have been the main focus of the business, there has been a growing interest in the biomedical uses of quantum dots (QDs).The user's text is "Citation83". Within this particular section, our aim is to emphasise the prominent market participants and their notable advancements in utilising quantum dots for biological purposes [38]. NANOCOTM holds a significant position in the healthcare market for quantum dots (QDs), with a projected market size of USD 1 billion in 2022 as per the company's prediction. HEATWAVETM is one of NANOCO's commercially available products that utilises quantum dots (QDs) technology for medical biosensor applications. This instrument can be utilised across a wide electromagnetic spectrum (400–1650 nm) to accurately measure several proteins in human blood, such as haemoglobin (at 575 nm), bilirubin (at 455 nm), and glucose (at 1650 nm), without the need for intrusive procedures. VIVODOTS® is a commercial product that utilises quantum dots (QDs) for the purpose of mapping tumorous tissues during surgery and preventing the removal of healthy tissues unnecessarily. Abdellatif led a Phase I clinical experiment where carboxylic acid-functionalized CdS/ZnS core–shell type quantum dots (QDs-COOH) were linked to veldoreotide, a specific ligand for somatostatin receptors [39]. The quantum dots (QDs) were included into a topical cream to explore their potential for bioimaging and as a treatment for breast and skin malignancies. Stability concerns encompass the significant susceptibility to deterioration and clumping that may occur during storage. Industrial problems occur due to the intricate scaling processes and the environmental risks posed by the heavy metal constituents. Issues encountered in vivo include the significant reduction of the administered dose through renal clearance due to the extremely small particle size, lack of specificity due to non-specific interactions with tissues and cellular membranes, and toxicity within cells caused by the production of reactive oxygen species (ROS) or damage to DNA.

From a pharmacological perspective, quantum dots (QDs) are extremely small colloidal particles with a large surface area and metallic properties. This makes them prone to aggregation, deterioration, hygroscopicity, and chemical redox changes. Even little alterations in the physico-chemical characteristics of quantum dots (QDs) can have a significant effect on their optical qualities [40]. Considerable endeavours have been made to enhance the physical and chemical stability of quantum dots (QDs). Zhang et al. introduced a chemical technique to stabilise PbS quantum dots (QDs) that are prone to degradation due to the oxidation of chalcogen atoms on their surface. A surface reaction using trioctylphosphine was employed to passivate surface S atoms with lead mono-carboxylate. The proposed technique successfully shielded quantum dots (QDs) from oxygen and enhanced their quantum characteristics, regardless of the particle size or the initial surface ligands [41]. CdSe/ZnCdS core/shell quantum dots (QDs) and InP/ZnSeS core/shell QDs were shielded against high temperatures and oxidation by applying a double polymeric shell made of thiol-terminated poly(methyl methacrylate-b-glycidyl methacrylate) (P(MMA-b-GMA)-SH) block copolymer ligands. The external layer consisted of clear PMMA, which did not affect the optical characteristics of QDs. Meanwhile, the inside shell, which was cross-linked, provided protection against oxidation [42]. Another issue that emerges due to the low solubility in water is the use of organometallic synthesis, which is a very efficient and often employed process for producing quantum dots (QDs). Proposed solutions include replacing hydrophobic ligands on the surface with hydrophilic thiol-containing compounds, applying hydrophilic shells as a coating, and utilising aqueous-based production [3].

9 Summary

In the past decade, numerous synthesis techniques and tactics have been developed to improve the properties of quantum dots (QDs) for specific applications. This class of materials has lately demonstrated significant promise in various biomedical applications, including cancer therapeutics, tissue engineering, and regenerative medicine. Moreover, quantum dots (QDs) have lately been discovered to be valuable instruments in particular scientific domains, including the simultaneous administration of drugs to two targets in real-time and the visualisation of adipose regions in obese individuals by molecular imaging. QDs have a very wide surface area and a unique chemical structure, which allows them to successfully interact with various biomolecules. These interactions can be crucial occurrences that impact the ultimate destiny of both specialised and undifferentiated cells in the human body. The utilisation of quantum dots (QDs) in cancer care has garnered significant attention due to their notable contribution to the advancement of chemotherapy and protein-based medicines. Furthermore, quantum dots (QDs) possess optical properties specifically designed for bioimaging and biosensing, making them potentially transformative in the field of medicine. In recent years, the primary obstacle has been the severe and prolonged toxicity of quantum dots (QDs) in biological organisms. The primary organs impacted by the accumulation and resulting toxicity of QDs are the lungs, kidneys, and liver. However, the brain has been found to be more resilient to the toxic effects of QDs. This discovery encourages further experimentation to assess the potential of different types of QDs in the treatment and imaging of neuroregenerative diseases.

Through the use of chemical, physical, and biological methodologies, together with revolutionary strategies such as surface functionalization approaches, it is anticipated that quantum dots (QDs) may soon find practical applications in therapeutic settings. The primary objective of these approaches should be to address the persistent toxicity of quantum dots (QDs) in order to develop a pharmaceutical QD product suitable for clinical and industrial use. Additional difficulties may arise from the need to achieve consistent results in the production of quantum dots (QDs) and the establishment of rigorous quality control measures to minimise discrepancies between different batches. Furthermore, it is crucial to thoroughly evaluate and assess the administration routes and protocols for introducing the manufactured quantum dots (QDs) into the body to ensure compliance with the necessary clinical standards. Recent advancements in our understanding of molecular interactions within the body have revealed the promise of using quantum dots (QDs) for personalised medicine. However, the issue of QDs binding to molecular compartments (such as proteins) in cells and tissues without specificity still poses a challenge. Researchers worldwide are urged to focus greater attention on this matter. An further crucial concern about the widespread use of quantum dots (QDs) pertains to the inconsistencies in the structure and composition of QDs, whether they are in their bare form or functionalized. These variabilities pose challenges in reaching definitive conclusions about the potential of QDs in theranostic applications. Further study is necessary to establish dependable techniques. While the issues surrounding the optimal use of QDs in clinical applications have not been completely resolved, recent advancements have significantly propelled these materials towards their therapeutic potential.

References

1. Ghaffarkhah, A., Hosseini, E., Kamkar, M., Sehat, A.A., Dordanihaghighi, S., Allahbakhsh, A., van der Kuur, C., Arjmand, M.: Synthesis, applications, and prospects of graphene quantum dots: a comprehensive review. Small **18**(2), 2102683 (2022)
2. Raut, N.S., Umekar, M.J., Dhoble, S.J.: Quantum dots: novel approach for biological imaging. Quantum Dots 477–500 (2023)
3. Kargozar, S., Hoseini, S.J., Milan, P.B., Hooshmand, S., Kim, H.W., Mozafari, M.: Quantum dots: a review from concept to clinic. Biotechnol. J. **15**(12), 2000117 (2020)
4. Reimann, S.M., Manninen, M.: Electronic structure of quantum dots. Rev. Mod. Phys. **74**(4), 1283 (2002)
5. Reshma, V., Mohanan, P.: Quantum dots: applications and safety consequences. J. Lumin.Lumin. **205**, 287–298 (2019)
6. Shukla, S.: Recent developments in biomedical applications of quantum dots. Adv Mater Rev **1**(1), 2–12 (2014)
7. Rajabzadeh, A.R., Srinivasan, S., Das, P., Ganguly, S.: Magnetic Quantum Dots for Bioimaging. CRC Press (2023)
8. Das, P., Ganguly, S., Saravanan, A., Margel, S., Gedanken, A., Srinivasan, S., Rajabzadeh, A.R.: Naturally derived carbon dots in situ confined self-healing and breathable hydrogel monolith for anomalous diffusion-driven phytomedicine release. ACS Appl. Bio Mater. **5**(12), 5617–5633 (2022)
9. Das, P., Ganguly, S., Ahmed, S.R., Sherazee, M., Margel, S., Gedanken, A., Srinivasan, S., Rajabzadeh, A.R.: Carbon dot biopolymer-based flexible functional films for antioxidant and food monitoring applications. ACS Appl. Polym. Mater. **4**(12), 9323–9340 (2022)
10. Das, P., Ganguly, S., Margel, S., Gedanken, A.: Tailor made magnetic nanolights: fabrication to cancer theranostics applications. Nanoscale Adv. **3**(24), 6762–6796 (2021)
11. Maruthapandi, M., Das, P., Saravanan, A., Natan, M., Banin, E., Kannan, S., Michaeli, S., Luong, J.H., Gedanken, A.: Biocompatible N-doped carbon dots for the eradication of methicillin-resistant S. aureus (MRSA) and sensitive analysis for europium (III). Nano Struct. Nano Objects **26**, 100724 (2021)
12. Das, P., Ganguly, S., Saha, A., Noked, M., Margel, S., Gedanken, A.: Carbon-dots-initiated photopolymerization: an in situ synthetic approach for MXene/poly (norepinephrine)/copper hybrid and its application for mitigating water pollution. ACS Appl. Mater. Interfaces **13**(26), 31038–31050 (2021)

13. Das, P., Ganguly, S., Banerjee, S., Das, N.C.: Graphene based emergent nanolights: A short review on the synthesis, properties and application. Res. Chem. Intermed.Intermed. **45**, 3823–3853 (2019)
14. Ganguly, S., Das, P., Banerjee, S., Das, N.C.: Advancement in science and technology of carbon dot-polymer hybrid composites: a review. Funct. Compos. Struct. **1**(2), 022001 (2019)
15. Ganguly, S., Das, P., Itzhaki, E., Hadad, E., Gedanken, A., Margel, S.: Microwave-synthesized polysaccharide-derived carbon dots as therapeutic cargoes and toughening agents for elastomeric gels. ACS Appl. Mater. Interfaces **12**(46), 51940–51951 (2020)
16. Bacon, M., Bradley, S.J., Nann, T.: Graphene quantum dots. Part. Part. Syst. Charact.Charact. **31**(4), 415–428 (2014)
17. Bimberg, D., Pohl, U.W.: Quantum dots: promises and accomplishments. Mater. Today **14**(9), 388–397 (2011)
18. Molaei, M.J.: Principles, mechanisms, and application of carbon quantum dots in sensors: a review. Anal. Methods **12**(10), 1266–1287 (2020)
19. Das, P., Ganguly, S., Margel, S., Gedanken, A.: Immobilization of heteroatom-doped carbon dots onto nonpolar plastics for antifogging, antioxidant, and food monitoring applications. Langmuir **37**(11), 3508–3520 (2021)
20. Saravanan, A., Maruthapandi, M., Das, P., Luong, J.H., Gedanken, A.: Green synthesis of multifunctional carbon dots with antibacterial activities. Nanomaterials **11**(2), 369 (2021)
21. Maruthapandi, M., Saravanan, A., Das, P., Natan, M., Jacobi, G., Banin, E., Luong, J.H., Gedanken, A.: Antimicrobial activities of Zn-doped CuO microparticles decorated on polydopamine against sensitive and antibiotic-resistant bacteria. ACS Appl. Polym. Mater. **2**(12), 5878–5888 (2020)
22. Arul, P., Huang, S.-T., Mani, V., Huang, C.-H.: Graphene quantum dots-based nanocomposite for electrocatalytic application of L-cysteine in whole blood and live cells. Electrochim. Acta. Acta **428**, 140937 (2022)
23. Yu, M., Saeed, M.H., Zhang, S., Wei, H., Gao, Y., Zou, C., Zhang, L., Yang, H.: Luminescence enhancement, encapsulation, and patterning of quantum dots toward display applications. Adv. Funct. Mater. **32**(13), 2109472 (2022)
24. Das, P., Ganguly, S., Maity, P.P., Srivastava, H.K., Bose, M., Dhara, S., Bandyopadhyay, S., Das, A.K., Banerjee, S., Das, N.C.: Converting waste Allium sativum peel to nitrogen and sulphur co-doped photoluminescence carbon dots for solar conversion, cell labeling, and photobleaching diligences: a path from discarded waste to value-added products. J. Photochem. Photobiol. B.Photochem. Photobiol., B **197**, 111545 (2019)
25. Das, P., Maruthapandi, M., Saravanan, A., Natan, M., Jacobi, G., Banin, E., Gedanken, A.: Carbon dots for heavy-metal sensing, pH-sensitive cargo delivery, and antibacterial applications. ACS Appl. Nano Mater. **3**(12), 11777–11790 (2020)
26. Saravanan, A., Maruthapandi, M., Das, P., Ganguly, S., Margel, S., Luong, J.H., Gedanken, A.: Applications of N-doped carbon dots as antimicrobial agents, antibiotic carriers, and selective fluorescent probes for nitro explosives. ACS Appl. Bio Mater. **3**(11), 8023–8031 (2020)
27. Ganguly, S., Das, P., Das, S., Ghorai, U., Bose, M., Ghosh, S., Mondal, M., Das, A.K., Banerjee, S., Das, N.C.: Microwave assisted green synthesis of Zwitterionic photolumenescent N-doped carbon dots: an efficient 'on-off'chemosensor for tracer Cr (+ 6) considering the inner filter effect and nano drug-delivery vector. Colloids Surf. A **579**, 123604 (2019)
28. Das, P., Ganguly, S., Agarwal, T., Maity, P., Ghosh, S., Choudhary, S., Gangopadhyay, S., Maiti, T.K., Dhara, S., Banerjee, S.: Heteroatom doped blue luminescent carbon dots as a nano-probe for targeted cell labeling and anticancer drug delivery vehicle. Mater. Chem. Phys. **237**, 121860 (2019)
29. Puttaswamy, R., Pai, R.K., Ghosh, D.: Recent progress in quantum dots based nanocomposite electrodes for rechargeable monovalent metal-ion and lithium metal batteries. J. Mater. Chem. A **10**(2), 508–553 (2022)
30. Bera, J., Betal, A., Sharma, A., Shankar, U., Rath, A.K., Sahu, S.: CdSe quantum dot-based nanocomposites for ultralow-power memristors. ACS Appl. Nano Mater. **5**(6), 8502–8510 (2022)

31. Wu, X.-G., Ji, H., Yan, X., Zhong, H.: Industry outlook of perovskite quantum dots for display applications. Nat. Nanotechnol.Nanotechnol. **17**(8), 813–816 (2022)
32. Mishra, S.P., Panda, R., Tripathy, S.K., Naik, K.K.: Enhancement of light extraction from the LED using PDMS-ZnO quantum dots layer. J. Mater. Sci. Mater. Electron. **34**(5), 435 (2023)
33. Ruffino, A., Yang, T.-Y., Michniewicz, J., Peng, Y., Charbon, E., Gonzalez-Zalba, M.F.: A cryo-CMOS chip that integrates silicon quantum dots and multiplexed dispersive readout electronics. Nat. Electron. **5**(1), 53–59 (2022)
34. Mo, Y., Feng, X., Zhang, L., Han, R., Bao, R., Pan, C.: Tuning the light emission of a Si micropillar quantum dot light-emitting device array with the strain coupling effect. NPG Asia Mater. **14**(1), 83 (2022)
35. Park, J.Y., Kwak, Y., Lim, H.-R., Park, S.-W., Lim, M.S., Cho, H.-B., Myung, N.V., Choa, Y.-H.: Tuning the sensing responses towards room-temperature hypersensitive methanol gas sensor using exfoliated graphene-enhanced ZnO quantum dot nanostructures. J. Hazard. Mater. **438**, 129412 (2022)
36. Karim, S., Ahmad, N., Hussain, D., Mok, Y.S., Siddiqui, G.U.: Active removal of anionic azo dyes (MO, CR, EBT) from aqueous solution by potential adsorptive capacity of zinc oxide quantum dots. J. Chem. Technol. Biotechnol.Biotechnol. **97**(8), 2087–2097 (2022)
37. Triana, M.A., Hsiang, E.-L., Zhang, C., Dong, Y., Wu, S.-T.: Luminescent nanomaterials for energy-efficient display and healthcare. ACS Energy Lett. **7**(3), 1001–1020 (2022)
38. Pawar, R.S., Upadhaya, P.G., Patravale, V.B.: Quantum dots: novel realm in biomedical and pharmaceutical industry. In: Handbook of Nanomaterials for Industrial Applications, pp. 621–637. Elsevier (2018)
39. Training, S., Khorasani, M.M.: Biomedical Applications of Quantum Dots
40. Moon, H., Lee, C., Lee, W., Kim, J., Chae, H.: Stability of quantum dots, quantum dot films, and quantum dot light-emitting diodes for display applications. Adv. Mater. **31**(34), 1804294 (**2019**).
41. Zhang, X., Chen, Y., Lian, L., Zhang, Z., Liu, Y., Song, L., Geng, C., Zhang, J., Xu, S.: Stability enhancement of PbS quantum dots by site-selective surface passivation for near-infrared LED application. Nano Res. **14**, 628–634 (2021)
42. Ko, J., Jeong, B.G., Chang, J.H., Joung, J.F., Yoon, S.-Y., Lee, D.C., Park, S., Huh, J., Yang, H., Bae, W.K.: Chemically resistant and thermally stable quantum dots prepared by shell encapsulation with cross-linkable block copolymer ligands. NPG Asia Mater. 12(1), 19 (2020)

GPSR Compliance

The European Union's (EU) General Product Safety Regulation (GPSR) is a set of rules that requires consumer products to be safe and our obligations to ensure this.

If you have any concerns about our products, you can contact us on ProductSafety@springernature.com

In case Publisher is established outside the EU, the EU authorized representative is:

Springer Nature Customer Service Center GmbH
Europaplatz 3
69115 Heidelberg, Germany

Batch number: 10761599

Printed by Printforce, the Netherlands